Study Guide

for
Kotz, Treichel, and Weaver's

Chemistry & Chemical Reactivity

Sixth Edition

John R. Townsend
West Chester University of Pennsylvania

THOMSON

BROOKS/COLE

Australia • Brazil • Canada • Mexico • Singapore • Spain • United Kingdom • United States

Printed in the United States of America
1 2 3 4 5 6 7 09 08 07 06 05

Printer: Globus Printing

0-534-99851-8

Thomson Higher Education
10 Davis Drive
Belmont, CA 94002-3098
USA

For more information about our products,
contact us at:
Thomson Learning Academic Resource Center
1-800-423-0563

For permission to use material from this text or
product, submit a request online at
http://www.thomsonrights.com.
Any additional questions about permissions can be
submitted by email to **thomsonrights@thomson.com.**

CONTENTS

to

My wonderful wife, Laurie

Preface

Welcome to your study of general chemistry! I am sure that you wish to succeed in studying chemistry. My goal in writing this brand new study guide is to help you do this.

Why should you want to use this study guide, particularly when you have an excellent textbook like *Chemistry and Chemical Reactivity*? First of all, this study guide is not a replacement for your textbook; it is meant to be used together with your textbook. You should want to use this study guide because it will help focus your attention on the things that you need to know how to do.

Each chapter in the study guide has the same basic structure. First, there is a chapter overview followed by a list of the key terms and their definitions. The majority of the text for each chapter deals with the Chapter Goals that are listed at the end of each chapter in your textbook. These Chapter Goals are the items that the authors of your textbook believe you should know and be able to do after studying that chapter. The commentaries in this study guide are designed to call attention to the important points and to supplement what has been stated in the textbook. In addition, this study guide contains numerous examples with very detailed solutions provided for you. It is suggested that you not only read these examples but that you actively solve them on your own as well. Finally, you are often instructed to work a similar problem in your textbook. It is important that you work these problems. Until you can solve a problem on your own, you do not really know the material.

Even though my name appears on the cover of this book, it is not just the work of one person. First of all, I wish to thank the authors of your textbook (Jack Kotz, Paul Treichel, and Gabriela Weaver) for writing their excellent textbook (from which I have, of course, borrowed heavily for this study guide), for their confidence in me and for suggesting that I should undertake this task. I am also indebted to Paul W. W. Hunter, the author of the study guides for the previous editions, for establishing the reputation of the study guides that accompany this textbook; I hope that this brand new study guide lives up to the standard he set. I also wish to thank all those at Thomson Brooks/Cole involved with this project. Thanks to David Harris and Peter McGahey for their oversight of the project. Special thanks go out to the two editors with whom I worked, Annie Mac and Ellen Bitter. Annie was the editor for the beginning of the project, and Ellen took over midway through. The patience that they have had with someone new to the world of authoring has been immense. The support they have shown has been great. Without their support this book would not be what it is today. I would also like to thank Barbara Mowery who proofread the study guide. Her work and suggestions have been invaluable. The study guide is much freer of errors because of her. Finally, I wish to thank my family, friends, and particularly my wife Laurie for all of their help and patience throughout this project. I could not have done this without them.

Even with all of this help, there may be some errors that remain in this study guide. The responsibility for these is mine alone. If you find one, please feel free to let me know. My e-mail address is jtownsend@wcupa.edu.

Thank you for purchasing this book. I wish you the greatest possible success in your study of chemistry. An amazing way of thinking about the universe will hopefully be opened up for you.

John Townsend

CHAPTER 1: Matter and Measurement

Chapter Overview

In this first chapter, you will begin your study of chemistry, the study of matter and the changes matter undergoes. We start by talking about different ways to classify a sample of matter. The first way we discuss is according to its state of matter, whether the material is a solid, liquid, or gas. You will then learn the basics of one of the major theories in chemistry: the kinetic-molecular theory with which we can explain some of the properties of the different states of matter. You will also be introduced to the way that chemists tend to carry out experiments on the macroscopic scale but then construct explanations based on the submicroscopic world of atoms, molecules, and ions and then represent these explanations on paper using symbols. Following this, you will learn about further ways to classify matter. Matter can be divided into pure substances and mixtures. There are two types of pure substances: elements and compounds. An element is composed of only one type of atom, whereas compounds are composed of two or more elements combined in a set way. We will then turn our attention to physical and chemical properties and physical and chemical changes. In a physical change, all that changes are physical properties; no new substances are produced. In a chemical change, new substances are produced. After dealing with these qualitative ways of looking at matter and its changes, we will introduce quantitative measurements and how to solve mathematical problems in chemistry. The system of measurements used in chemistry is the modern incarnation of the metric system called SI. You will need to memorize the various prefixes commonly used in the metric system as well as some of the common units that we shall use throughout the course. In order to carry out unit conversions, you will learn a very important problem solving technique called dimensional analysis in which the units will be treated like factors in an equation. If the same unit appears both on the top and on the bottom of a fraction, then it may be canceled. You will gain familiarity with carrying out temperature conversions, unit conversions involving the metric prefixes, and density problems. Finally, we will take a look at accuracy and precision and quantitative measures of these: percent error for accuracy and standard deviation for precision.

Key Terms

In this chapter, you will need to learn and be able to use the following terms:

Absolute zero: the lowest temperature believed to be possible. It is defined to be zero kelvin and has a value on the Celsius scale of $-273.15°C$.

Accuracy: the agreement of a measurement with the accepted value for the quantity.

Atom: the smallest particle of an element that retains the characteristic chemical properties of that element.

Boiling point: the temperature at which a liquid boils.

Chemical bond: an interaction between two atoms that holds them together in a molecule, ionic crystal, or covalent network.

Chemical change: a change that produces one or more new substances.

Chemical formula: a listing of the chemical symbols for the elements present in a compound and the relative number of atoms of each element present (written as subscripts).

Chemical property: a characteristic of a material that describes whether and sometimes how readily one material will react with another material.

Chemical symbol: an abbreviation used to represent an element.

Density: indicates how much matter is packed into a given space; mathematically it is calculated by dividing the mass of a material by its volume ($D = m/V$).

Dimensional analysis: a technique that pays attention to the units in a calculation. The units are treated like factors that can be cancelled if they appear both in the numerator and denominator of a calculation.

Ductility: the ability of a material to be drawn into a wire.

Electric conductivity: a property that indicates whether a material will conduct electricity and the extent to which it will do so.

Element: a pure substance in which all of the atoms are of the same type.

Extensive property: a property of a substance that depends on the amount of the substance present.

Gas: the physical state of matter that has neither a fixed volume nor a fixed shape – both the volume and shape of a gas is determined by its container. The volume of a gas varies with temperature and pressure.

Heterogeneous: not uniform throughout.

Homogeneous: uniform throughout.

Intensive property: a property of a substance that does not depend on the amount of the substance present.

Ion: an electrically charged atom or group of atoms.

Kinetic energy: the energy of motion.

Kinetic molecular theory: a theory of matter used to explain the properties of solids, liquids, and gases. This theory supposes that all matter consists of extremely small particles (atoms, molecules, or ions), which are in constant motion.

Liquid: the physical state of matter with a volume that changes little as temperature and pressure change but that has no definite shape of its own – a liquid takes on the shape of the portion of the container that it fills.

Macroscopic: large-scale; the macroscopic view in chemistry is the world in which we live and make observations.

Malleability: the ability of a material to be deformed (bent).

Matter: anything that has mass and takes up space.

Melting point: the temperature at which a solid melts.

Mixture: a material formed by physically mixing together two or more elements, two or more compounds, or a combination of elements or compounds. A mixture can be made using any relative amounts of the components. The components do not chemically bond to each other in the mixture.

Molecule: the smallest particle normally found in nature for those elements whose atoms chemically bond together to form small discrete particles; also, the smallest discrete particle of a nonionic compound.

Periodic table: a table that lists the elements in order of increasing atomic number and places elements with similar properties in the same vertical column. Each box on the periodic table usually includes the symbol for an element as well as other useful information about the element.

Physical change: a change in physical properties; a physical change does not produce new substances.

Physical property: a characteristic of a material that can be observed without changing the composition of a substance.

Precision: indicates how well several determinations of the same quantity agree.

Pure substance: a material that has its own set of unique properties and that cannot be separated into two or more different species by any physical technique.

Pyrometer: a device used to measure temperature; it measures the wavelength of light emitted from an object and relates that to the object's temperature.

Qualitative observation: an observation that does not involve measurements and numbers.

Quantitative observation: an observation that involves numerical information.

SI: the Système International d'Unités, the modern incarnation of the metric system used in scientific work.

Significant figures: the digits in a measured quantity that are known exactly plus one digit that has some uncertainty; in the answer to a calculation, one should report only digits that are significant.

Solid: the physical state of matter that has a rigid shape and a fixed volume that changes little as temperature and pressure change.

Solubility: the amount of a material that will dissolve in a given amount of another material.

Solution: a homogenous mixture in which the components are evenly distributed in each other down to an atomic or molecular scale.

State: the physical form a material is in: solid, liquid, gas.

Submicroscopic (particulate): the level of matter dealing with atoms, molecules, and ions.

Temperature: a property that determines the direction of heat transfer; heat transfers from something at a higher temperature to something at a lower temperature.

Viscosity: a measure of how easily a liquid flows; the more viscous a liquid is, the less readily it flows.

Chapter Goals

By the end of this chapter you should be able to:

- **Classify matter.**

  a) Recognize the different states of matter (solids, liquids, and gases) and give their characteristics (Section 1.1).

  A solid has a fixed shape and a fixed volume. A liquid has a fixed volume, but it takes on the shape of the portion of the container that it fills. A gas takes on both the shape and volume of its container.

  Most materials are solids if the temperature is reduced sufficiently. If the temperature is increased, most solids will go through transformations first into a liquid and then into a gas.

  For a given mass of a solid, there is usually a small increase in volume on melting to form a liquid (water is an exception – its volume actually decreases on melting). For a given mass of a liquid, there is a large increase in volume that occurs when the liquid changes from a liquid to a gas.

b) Appreciate the difference between pure substances and mixtures and the difference between homogeneous and heterogeneous mixtures (Section 1.1).

Matter is anything that has mass and takes up space. It is the "stuff" of the universe. We can classify matter into different categories. Matter that is uniform throughout is referred to as being homogeneous whereas matter that is not uniform throughout is referred to as being heterogeneous. In a heterogeneous sample, different parts of the material have different compositions. In the noodle soup example in your textbook, the noodle soup is a heterogeneous material because some parts of the soup are chunks of noodles and some parts are the liquid soup. In addition, the noodles are not evenly distributed throughout the soup (there may be more noodles at the bottom of the container than at the top). There are clearly regions in the soup where the distribution of the materials is not uniform.

A sample of matter may also be classified as being either a pure substance or a mixture. A pure substance is a material that has its own set of unique properties and that cannot be separated into two or more different species by any physical technique. A pure substance may also be classified further as being either an element or a compound (we'll deal with these in a little bit). A pure substance is an example of homogeneous matter.

A mixture is a combination of two or more pure substances in which the pure substances simply mix together but do not chemically react to form a new chemical substance. The individual components of a mixture retain their original properties. We can make a mixture of the elements carbon (one form is graphite – a black material used in pencil "leads") and sulfur (a yellow solid). The mixture will have black specks and yellow specks. The carbon stays carbon, and the sulfur stays sulfur. No chemical change has occurred. In addition, we can make a mixture of carbon and sulfur in which we use twice as much sulfur as carbon or make a mixture in which we use twice as much carbon as sulfur or use any ratio we choose to make a mixture of carbon and sulfur. A mixture can be made using any proportion of its components. Mixtures can be separated into their original pure substances by the use of physical means because the components have not chemically bonded to each other.

Mixtures can be either homogeneous or heterogeneous. A homogeneous mixture in which the components are evenly dispersed throughout each other even down to an atomic or molecular scale is called a solution.

c) Recognize the importance of representing matter at the macroscopic level and at the particulate level (Section 1.1).

Chemists carry out experiments at the macroscopic level, they think about chemistry at the particulate level, and they represent what they have observed using symbols.

As scientists, we perform experiments and make observations. These observations are made at the macroscopic level, the large-scale level on which we live. For example, we dissolve sugar in water. We see that we start out with a white solid (sugar) and a clear colorless liquid (water). As the sugar dissolves, the white solid appears to disappear. At the end of the process, we have a clear colorless liquid that tastes sweet.

Chemists, in particular, attempt to predict and visualize what is going on at the submicroscopic or particulate level. This is the level of atoms, molecules, and ions. We could picture sugar molecules packed together in a solid crystal and water molecules in the liquid state before we mix them. We could then imagine mixing the sugar and water. We would imagine water molecules setting up forces of attraction with a sugar molecule that eventually are enough to pull the sugar molecule away from the sugar crystal and out into the bulk water. By repeating this process over and over, all of the sugar molecules could then go out into the water. Eventually, we end up with the sugar molecules evenly distributed among the water molecules.

We might represent what has gone on by symbols. We might write the following to describe the process we have just been describing.

$$C_{12}H_{22}O_{11} \ (s) \rightarrow C_{12}H_{22}O_{11} \ (aq)$$

This equation indicates that we start out with solid table sugar ($C_{12}H_{22}O_{11}$) and end up with a solution of sugar in water (the aq stands for aqueous which means "in water"). To a chemist, this equation means the same thing as the macroscopic observations that were noted and conjures up the picture we described for the particulate view.

As you progress through this course, you will be learning about the submicroscopic view and how it relates to the macroscopic world in which we live.

• Apply the kinetic-molecular theory to the properties of matter.

Understand the basic ideas of the kinetic-molecular theory (Section 1.1).

A theory is a unifying principle that explains a body of facts and the laws based on them. The kinetic-molecular theory is a useful theory for explaining the properties of solids, liquids, and gases. It supposes that all matter consists of extremely small particles (atoms, molecules, or ions), which are in constant motion. Another key feature of the kinetic-molecular theory is that the kinetic energy of the particles increases as the temperature is increased.

In a solid, the particles are pictured as being packed closely together. They are locked into their positions vibrating about their average positions but not able to slide past one another. When the temperature is raised to the melting point of the solid, the particles have sufficient energy to push out of their positions in the solid and to move around. The forces of attraction between the particles are enough to keep the particles close to each other. In the resulting liquid, the particles can slide past one another but stay in contact other particles of the liquid. As the temperature is raised even further, the particles have sufficient kinetic energy to overcome the forces of attraction between the particles. The particles break away and move into the gaseous state. In a gas, the particles are far apart and are moving extremely rapidly.

• Recognize elements, atoms, compounds, and molecules.

a) Identify the name or symbol for an element, given its symbol or name (Section 1.2).

The elements are listed on the periodic table. Each element has an abbreviation called its chemical symbol. The first letter of an element's chemical symbol is always capitalized. Any other letters are lowercase. Throughout this course, we will be using chemical symbols and expecting you to know the names of the elements and vice versa. You should begin memorizing the names and symbols for some of the elements. The following table lists 50 elements and their symbols, just under half of the elements. As a start, begin by learning these names and symbols. Your instructor may modify this list to include more elements or eliminate others; do what your instructor indicates, but if he/she does not give an indication, this list will get you started. Perhaps the best way to learn these is by using flash cards. On one side, write the name of the element; on the other side, write the symbol. Practice going through the flash cards both ways. Look at the name and ask yourself the symbol. Look at the symbol and ask yourself the name. Do this until you have mastered these.

Element	Symbol		Element	Symbol
hydrogen	H		iron	Fe
helium	He		cobalt	Co
lithium	Li		nickel	Ni
beryllium	Be		copper	Cu
boron	B		zinc	Zn
carbon	C		arsenic	As
nitrogen	N		bromine	Br
oxygen	O		krypton	Kr
fluorine	F		rubidium	Rb
neon	Ne		strontium	Sr
sodium	Na		silver	Ag
magnesium	Mg		cadmium	Cd
aluminum	Al		tin	Sn
silicon	Si		iodine	I
phosphorus	P		xenon	Xe
sulfur	S		cesium	Cs
chlorine	Cl		barium	Ba
argon	Ar		tungsten	W
potassium	K		platinum	Pt
calcium	Ca		gold	Au
scandium	Sc		mercury	Hg
titanium	Ti		lead	Pb
vanadium	V		bismuth	Bi
chromium	Cr		radon	Rn
manganese	Mn		uranium	U

Here are some hints so that you can avoid some of the more common mistakes:

The chemical symbol for carbon is C, and that for calcium is Ca.

The chemical symbol for nitrogen is N, and that for nickel is Ni.

The chemical symbol for fluorine is F, not Fl.

The chemical symbol for magnesium is Mg, and that for manganese is Mn.

The chemical symbol for copper is Cu, and that for cobalt is Co.

The name of the element with chemical symbol F is spelled fluorine not flourine.

The name of the element with chemical symbol Cl is spelled chlorine (don't forget the h).

The name of the element with chemical symbol P is spelled phosphorus, not phosphorous (the ending is just –us, not –ous).

Example 1-1: What is the name of the element with chemical symbol Cr?

This just has to be memorized. The name of this element is chromium.

Example 1-2: What is the chemical symbol for the element mercury?

Once, again, this just has to be memorized. The symbol is Hg.

Try Study Questions 1 and 3 in Chapter 1 of your textbook now!

b) Use the terms atom, element, molecule, and compound correctly (Sections 1.2 and 1.3).

There are two types of pure substances: elements and compounds. There is only one type of atom in an element. An atom is the smallest particle of an element that retains the characteristic chemical properties of that element. For example, in a sample of the element sodium, all of the atoms are one type of atom: sodium atoms. In a sample of oxygen gas, all of the atoms are oxygen atoms. The elements are listed on the periodic table.

Some elements are found in nature as isolated atoms. For example, in helium, the atoms are all helium atoms and occur as individual atoms traveling around by themselves. In some other elements, the atoms of the element hook together in some way. One way that they can do this is to form molecules of the element. A molecule is a small discrete particle consisting of more than one atom. For example, the element chlorine normally occurs in nature as pairs of chlorine atoms bonded together. In a sample of chlorine, therefore, we have a whole bunch of pairs of chlorine atoms traveling around. Chlorine is still an element, however, because all of the atoms present are the same type of atom, chlorine atoms.

A compound is composed of two or more different elements held together by chemical bonds. An example of a compound is water. You may know that the formula for water is H_2O. There are thus two elements present: hydrogen and oxygen. A compound will have a particular composition. The formula for water tells us that there are two hydrogen atoms for each oxygen atom. If we had a material with a different proportion of hydrogen to oxygen, it would not be water. Finally, the hydrogen and oxygen are chemically bonded to each other. They are linked together to form a new entity, a water molecule, which is different from just a physical mixture of the two elements. The only way to get the elements hydrogen and oxygen back from water is to carry out a chemical reaction. We thus say that compounds are chemically separable into the elements from which they were formed. The properties of a compound are very different from those of the elements that combined to form the compound. For example, the properties of water are different from those of hydrogen and oxygen gas. Some compounds are composed of molecules, and others are composed of ions.

Students sometimes get confused about the differences between mixtures and compounds. Some of their key differences are summarized in the following table:

Material	Composed of	Ratio of Components	Bonding
Compound	Two or more elements	Fixed ratio of elements	Atoms chemically bonded.
Mixture	Two or more elements, two or more compounds, or a mixture of elements and compounds	Materials can be present in any proportion	Materials physically mixed; the components are not chemically bonded to each other.

Example 1-3: Classify each of the following as an element, a compound, or a mixture.

       a. Nitrogen, N_2

Nitrogen, chemical symbol N, is listed on the periodic table. It is an element. A sample of nitrogen is composed of only nitrogen atoms. Even though we are indicating that they come in pairs, they are still all the same type of atom. It is an element.

       b. Nitrogen monoxide, NO

Nitrogen monoxide is composed of two elements: nitrogen and oxygen. These are combined in molecules consisting of one atom of nitrogen per atom of oxygen. This is a compound.

c. You
There are many different compounds present in your body. Your body is also not uniform throughout; some parts are different from other parts, and some compounds are present in greater or lesser amount in different parts. You are a mixture and, more specifically, a heterogeneous mixture.

Try Study Question 5 in Chapter 1 of your textbook now!

• Identify physical and chemical properties and changes.

a) List commonly used physical properties of matter (Section 1.4).

Physical properties are characteristics of a material that can be observed without changing the composition of a substance. Some examples of physical properties are color, state of matter, melting point, boiling point, density, solubility, electric conductivity, malleability, ductility, and viscosity. You should review the definitions of any of these physical properties that you do not know. In contrast to physical properties, chemical properties indicate whether and sometimes how readily a material will react with another material.

b) Identify several physical properties of common substances (Section 1.4).

Example 1-4: Identify some physical properties of copper metal.

The key to listing physical properties is to list properties that do not involve the material chemically reacting with another. Of course, in order to do this, we would need to know something about copper and this task would be easiest if we had a sample of copper that we could examine. Some of the things we might list would be that copper at room temperature is a solid, has a reddish-orange color, has a density of 8.94 g/cm^3, has a high electrical conductivity, and is both very malleable and ductile. Copper has a melting point of 1083°C, and a boiling point of 2567°C.

Example 1-5: Identify whether the following properties are physical properties or chemical properties.

a. Molasses is a very viscous liquid.
Viscosity is a measure of how easily a liquids flows. The more viscous a liquid is, the less readily it flows. This property does not involve reacting the material with another substance so this is a physical property.

b. Wood is flammable.
In the process of burning, the wood reacts with oxygen. This property thus describes reacting the material with another substance (oxygen). This is a chemical property.

Try Study Question 7 in Chapter 1 of your textbook now!

c) Use density to connect the volume and mass of a substance (Sections 1.4, 1.6, and 1.8).

The density of a material is an indication of how much matter there is in a given volume of the material, how tightly packed the matter is. It may be calculated by using the equation:

$$\text{Density} \ = \ \frac{\text{mass}}{\text{volume}}$$

$$D \ = \ \frac{m}{V}$$

Example 1-6: What is the volume of a sample of copper that has a mass of 28.7 g? The density of copper is 8.94 g/cm^3.

In this problem, we are given the mass of copper and asked to solve for the volume. The mass is 28.7 g. We can view this as

$$\frac{28.7 \text{ g Cu}}{1}$$

The grams starts out on top in the given. Density connects mass and volume. We know that the density is 8.94 g/cm^3. Thus, 8.94 g of copper will have a volume of 1 cm^3. From this, we can write two expressions:

$$\frac{8.94 \text{ g Cu}}{1 \text{ cm}^3 \text{ Cu}} \text{ and } \frac{1 \text{ cm}^3 \text{ Cu}}{8.94 \text{ g Cu}}$$

These correspond to us saying that there are 8.94 g of copper for every cubic centimeter of volume or to saying there is 1 cubic centimeter for every 8.94 g of copper. They are both true statements. Using the dimensional analysis method for solving problems discussed below, we can use these expressions to solve some problems involving density. In order to cancel grams, we need to have grams on the bottom of the factor we shall use. This means that we shall use the factor

$$\frac{1 \text{ cm}^3 \text{ Cu}}{8.94 \text{ g Cu}}$$

The solution becomes

$$\frac{28.7 \text{ g Cu}}{1} \ \text{x} \ \frac{1 \text{ cm}^3 \text{ Cu}}{8.94 \text{ g Cu}} \ = \ 3.21 \text{ cm}^3 \text{ Cu}$$

Example 1-7: What is the mass of a block of copper that has a volume of 34.2 cm^3?

In this case, we are given the volume and asked to solve for the mass. The volume is 34.2 cm^3. We can view this as

$$\frac{34.2 \text{ cm}^3 \text{ Cu}}{1}$$

The connection between mass and volume is still density. Now, we have cm^3 on top. In order to cancel cm^3, we need to have cm^3 in the bottom of the factor that we use. We will express the density as

$$\frac{8.94 \text{ g Cu}}{1 \text{ cm}^3 \text{ Cu}}$$

The solution is

$$34.2 \text{ cm}^3 \text{ Cu x } \frac{8.94 \text{ g Cu}}{1 \text{ cm}^3 \text{ Cu}} \ = \ 306 \text{ g Cu}$$

Try Study Question 13 in Chapter 1 of your textbook now!

d) Explain the difference between chemical and physical changes (Sections 1.4 and 1.5).

In a physical change, only physical properties are changed. No new substances are produced. For example, imagine that you rip a piece of paper. You started with paper. You ended up with paper. This was a physical change. A chemical change produces new substances. For example, imagine that you burn some wood. The wood gradually disappears leaving a black residue, and smoke goes off into the air. You started with wood. What you end up with is definitely not wood. This was a chemical change.

<u>Example 1-8:</u> Determine whether the following changes are chemical or physical changes.

      a. Boiling water.

In this process, liquid water is changed into gaseous water. The chemical composition, H_2O, has not changed. We started with H_2O; we ended up with H_2O. This is a physical change.

      b. Combining zinc metal and hydrochloric acid.

Zinc metal is a shiny metal. Hydrochloric acid is a corrosive solution of the gas hydrogen chloride in water. When the two are mixed, the solution begins to bubble, and the zinc metal disappears. Further examination determines that the gas produced is hydrogen gas and that, along with chloride ions in the solution, there are now also zinc ions present. We began with zinc metal (Zn) and hydrochloric acid (HCl). We ended up with zinc chloride ($ZnCl_2$) in solution and hydrogen gas (H_2). We have changed which atoms are bonded to each other in the substances, forming different substances than were present at the start of the reaction. This is a chemical change.

Try Study Question 8 in Chapter 1 of your textbook now!

The answers to this question are not in the back of the textbook. The correct answers are as follows:

8a. Chemical Change. Even though this might at first appear to be a physical change because it involves the change of the physical property color, it is nonetheless a chemical change. The reason that the color is different is that we have carried out a chemical reaction of the bleach with the dye to convert the dye into a new compound, which is not colored. This is a chemical change.

b. Physical Change. We begin with H_2O in the gaseous state and end with H_2O in the liquid state. We have not changed the chemical composition. This is a physical change.

c. Chemical Change. Carbon dioxide and sugar are different substances. This is a chemical change.

d. Physical Change. We start with solid butter. We end with liquid butter. No change in chemical composition has occurred. This is a physical change.

e) Understand the difference between extensive and intensive properties and give examples of them (Section 1.4).

An extensive property of a substance depends on the amount of the material present. An intensive property does not depend on the amount of the material present. In determining whether a property is extensive or intensive, simply ask yourself whether the property would remain the same if the amount of the material were changed. The mass of an object clearly depends on how much of the substance is present; it is an extensive property. The volume likewise clearly depends on how much of the substance is present; it is an extensive property. However, the ratio of the two (density) does not depend on the amount of the substance present; it is an intensive property. The density of copper is the same whether we have a small pellet of copper or we have a large sheet of it. Some other examples of intensive properties are the melting point and boiling point of a substance.

- ## Use metric units correctly.

a) Convert between temperatures on the Celsius and Kelvin scales (Section 1.6).

This type of conversion is one of the easier conversions we shall do because the size of a Celsius degree and a kelvin is the same. The conversion simply involves either adding or subtracting 273.15 to the value given. The only trick comes in remembering whether to add or subtract 273.15. Keep in mind that 0°C = 273.15 K because absolute zero (0 K) is –273.15°C. Thus the value of a temperature on the Kelvin scale will always be greater than the value on the Celsius scale. To convert from degrees Celsius to kelvins, therefore, you will add 273.15. To convert from kelvins to degrees Celsius, you will subtract 273.15.

Example 1-9: Perform the following conversions.

a. Human body temperature is 37.0°C. What is this temperature in kelvins?
We are converting from °C to K. The temperature in K will always have a higher value than in °C, so we will add 273.15.

$$T(K) = \frac{1 \text{ K}}{1 °C} \left(37.0°C + 273.15°C\right) = 310.2 \text{ K}$$

b. Convert 190 K to °C.
We are converting from K to °C. The temperature in K will always have a higher value than in °C so we will subtract 273.15.

$$T(°C) = \frac{1 °C}{1 \text{ K}} \left(190 \text{ K} - 273.15 \text{ K}\right) = -83°C$$

Try Study Question 19 in Chapter 1 of your textbook now!

b) Recognize and know how to use the prefixes that modify the sizes of metric units (Section 1.6).

The key prefixes you need to know are listed in Table 1.3 in your textbook. You should memorize them and their meanings. Always try to keep in mind their relative sizes and which are bigger than the base units and which are smaller. The meaning of kilo- is 10^3. It is bigger than the base unit. Thus, 1 kg = 10^3 g, and there are 1000 g in one kilogram. The meaning of milli is 10^{-3}. It is smaller than the base unit. Thus, 1 mg = 10^{-3} g, and there are 1000 mg in one gram. It would never make sense to say that there are 1000 g in 1 mg (which students have been known to do on tests) because you know that a gram is bigger than a milligram.

- ## Understand and use the mathematics of chemistry.

a) Use dimensional analysis to carry out unit conversions. Perform other mathematical operations (Section 1.8).

The ability to use dimensional analysis is a very valuable skill that you will find useful not only in this course but in virtually all of your science courses. You are probably familiar with the idea that if we were to carry out the following calculation

$$\frac{3}{5} \times \frac{7}{3}$$

we could simplify the problem because the threes would cancel, leaving us with 7/5. Thus, if we have the same number on the top in one fraction and on the bottom in another, then that number will cancel. In dimensional analysis, we will treat units in the same way. If we have the same unit on the top and on the bottom in a fraction, then it will cancel.

The basic strategy we will use to convert between units is to multiply the measurement we wish to change by a conversion factor that expresses the relationship between the unit we wish to change and the unit that we desire. We will decide which unit goes on the top and which unit goes on the bottom in our conversion factor by placing the portion of the conversion factor involving the unit we wish to get rid of in such a way that it will cancel that unit and leave the unit we desire.

Example 1-10: Convert 12,400 g to kg.

We are given 12,400 g. We wish to get rid of grams and end up with kg. Do we know the relationship between g and kg? Yes. We know that 1000 g = 1 kg. From this, we can write two possible conversion factors:

$$\frac{1000 \text{ g}}{1 \text{ kg}} \quad \text{and} \quad \frac{1 \text{ kg}}{1000 \text{ g}}$$

Which should we use? Begin by writing your given

12,400 g

This could also be written as

$$\frac{12,400 \text{ g}}{1}$$

The grams are on the top in the given. We need to get rid of grams so we will use the conversion factor that has grams on the bottom. Grams will then cancel, leaving behind kg.

$$\frac{12,400 \text{ g}}{1} \times \frac{1 \text{ kg}}{1000 \text{ g}} = 12.4 \text{ kg}$$

Multiplying all of the things on the top and dividing by all of the things on the bottom, we come out with our answer of 12.4 kg.

Example 1-11: Convert 670. m/min to m/s.

We are given 670. m/min. We wish to keep m, get rid of min, and end up with s on the bottom. The relationship between the unit we wish to get rid of, min, and the unit we want, s, is 60 s = 1 min. The two conversion factors we could construct are

$$\frac{60 \text{ s}}{1 \text{ min}} \quad \text{and} \quad \frac{1 \text{ min}}{60 \text{ s}}$$

Write the given and set up the conversion factor so the units we want to cancel will cancel:

$$\frac{670. \text{ m}}{\text{min}} \times \frac{1 \text{ min}}{60 \text{ s}} = 11.2 \text{ m/s}$$

Note that in this case, we used the conversion factor with minutes on the top because we wanted to cancel the minutes that were on the bottom in our given information.

Sometimes, we will need to use multiple conversion factors to accomplish our goal.

Example 1-12: Convert 2.4 km to mm.

In this case, we are given 2.4 km and we wish to end up with mm. We probably have not memorized the relationship between km and mm. We do know the relationship between km and m (1000 m = 1 km) as well as the relationship between m and mm (1000 mm = 1 m).

$$\frac{2.4 \text{ km}}{1} \text{ x } \frac{1000 \text{ m}}{1 \text{ km}} \text{ x } \frac{1000 \text{ mm}}{1 \text{ m}} = 2.4 \text{ x } 10^6 \text{ mm}$$

In this case, we placed the 1 km on the bottom of the first conversion factor so that it would cancel the km in the given information. This gives our answer in m. For the second conversion factor, we need to cancel m, leaving mm so we place the 1 m on the bottom of the second conversion factor and place the 1000 mm on the top. After doing the arithmetic indicated, we end up with our answer of $2.4 \text{x} 10^6$ mm.

Try Study Question 21 in Chapter 1 of your textbook now!

Yet another twist to this type of problem arises when we need a conversion factor that involves a unit raised to some power, such as m^2 or cm^3.

Example 1-13: Convert 75 cm^3 to mm^3.

We need to go from cm^3 to mm^3. You probably do not know the conversion factor between these units, but you might know that 10 mm = 1 cm. We can cube this whole expression to obtain the relationship between mm^3 and cm^3.

$$(10 \text{ mm})^3 = (1 \text{ cm})^3$$
$$1000 \text{ mm}^3 = 1 \text{ cm}^3$$

Notice that both the numbers and the units get cubed. We can now solve our problem in the usual way:

$$\frac{75 \text{ cm}^3}{1} \text{ x } \frac{1000 \text{ mm}^3}{1 \text{ cm}^3} = 7.5 \text{ x } 10^4 \text{ mm}^3$$

Sometimes, instead of figuring out the conversion factor and setting up the conversion as separate steps, you can simply do it all at once

$$\frac{75 \text{ cm}^3}{1} \text{ x } \left(\frac{10 \text{ mm}}{1 \text{ cm}}\right)^3 = 7.5 \text{ x } 10^4 \text{ mm}^3$$

but if you do this, you must remember that in the conversion factor both the numbers and units on both the top and bottom get cubed.

Try Study Question 23 in Chapter 1 of your textbook now!

Dimensional analysis is one of the most important problem solving skills to master. You should try to do as many problems in the textbook as possible that require this technique until you have it well mastered. If you do not master this technique now, it will haunt you for the rest of the course!

b) Know the difference between precision and accuracy and how to calculate percent error (Section 1.7).

Precision and accuracy are not the same thing. Accuracy is the agreement of a measurement with the accepted value for the quantity. In order to comment upon the accuracy of a measurement, therefore, the accepted value must be known. Sometimes, we express the accuracy of a measurement by using a quantitative measure called percent error:

$$\text{Percent Error} = \frac{\text{Experimentally Determined Value} - \text{Accepted Value}}{\text{Accepted Value}} \text{ x } 100$$

Example 1-14: A determination of the density of copper is performed. The experimenter determines the density to be 8.82 g/mL. The accepted value for the density of copper is 8.94 g/mL. What was the percent error of this determination?

$$\text{Percent Error} = \frac{\text{Experimentally Determined Value} - \text{Accepted Value}}{\text{Accepted Value}} \times 100$$

$$= \frac{8.82 \text{ g/mL} - 8.94 \text{ g/mL}}{8.94 \text{ g/mL}} \times 100$$

$$= -1.3\%$$

Try Study Question 29 in Chapter 1 of your textbook now!

Precision indicates how well several determinations of the same quantity agree. It indicates how close the different determinations are to each other. A measure of precision is the standard deviation. If one goes above the average by one standard deviation and goes below the average by one standard deviation, then approximately 68% of the measurements are included in that range. If the standard deviation is small, then the data are much more closely packed together than if the standard deviation is large.

The equation for standard deviation is

$$\text{Standard Deviation} = \sqrt{\frac{\sum \left(x - \bar{x}\right)^2}{n}}$$

Let's break this formula down into parts. To calculate the standard deviation, first calculate all of the deviations, the difference between a measurement (x) and the average ($\bar{x}$). For example, if the average of a series of measurements was 5 and one of those measurements was 3, then the deviation for that measurement is $3 - 5 = -2$. Then square all of the deviations. For the case just stated, this square would be $(-2)^2 = 4$. Then add together all of these squared deviations and divide by the number of measurements. Finally, take the square root of this number. Let's work an example:

Example 1-15: What is the standard deviation for the following series of measurements carried out to determine the density of copper?

Trial	Density (g/ml)
1	8.82
2	8.92
3	8.96
4	8.97
5	8.91

First, calculate the average:

$$\text{average} = \frac{8.82 \text{ g/mL} + 8.92 \text{ g/mL} + 8.96 \text{ g/mL} + 8.97 \text{ g/mL} + 8.91 \text{ g/mL}}{5}$$

$$= 8.92 \text{ g/mL}$$

Next, set up a table to calculate the squares of the deviations.

Trial	Density, g/mL	Deviation $(x - \bar{x})$	Square of Deviations $(x - \bar{x})^2$
1	8.82	−0.10	0.0100
2	8.92	0.00	0.0000
3	8.96	0.04	0.0016
4	8.97	0.05	0.0025
5	8.91	−0.01	0.0001

The sum of the values in the last column is 0.0142.

We can then carry out the rest of the calculation

$$\text{Standard Deviation} = \sqrt{\frac{\sum(x - \bar{x})^2}{n}} = \sqrt{\frac{0.0142}{5}} = 0.05$$

The units for the standard deviation are the same as the units of the measurements. The standard deviation in this example is 0.05 g/mL. This indicates that about 68% of the data is contained in the interval from 8.87 to 8.97 g/mL.

Many calculators and spreadsheet applications contain functions to easily calculate the standard deviation for a set of inputted data.

Warning: In much (most) scientific work where we are dealing with a fairly small data set, a slightly different form of the equation for standard deviation is used. In this other form, instead of dividing by the number of measurements, we divide by one less than this number. In our example, this would have resulted in dividing by 4 instead of 5 in the last step. We would have obtained a standard deviation of 0.06 in that case. The reason for doing this is that often we do not have a large enough data set for all of the assumptions to be true that must be true to use the equation given in the text. The larger value obtained using n-1 in the denominator for the standard deviation is a better reflection of the distribution of the data.

c) Understand the use of significant figures (Section 1.8).

When you read a number in a problem, how do you determine how many significant figures it contains? First of all, if it is an exact number, you don't have to worry about it. Exact numbers will not affect the number of significant figures you will be able to report after a calculation. For example, there are exactly 12 inches in 1 foot. We could add as many zeros after the decimal place, and the expression would still be true. Using this figure in a calculation will not affect how many significant figures we can report in our answer.

If, however, the number is a measured value, we must take into account significant figures. Nonzero digits are significant. All of the rest of the rules deal with zeros:

1) Zeros between two other significant digits are significant. For example, the zero in 103 m is significant. After all, if the 1 is significant and the 3 is significant, then the zero must also be significant.

2) Zeros to the right of a nonzero number and also to the right of a decimal place are significant. For example, in the number 1.30 m, the zero is significant. The only reason for writing such a zero is to indicate that the number contains significant digits out to that decimal place, the hundredths place in this case.

3) Zeros that are just placeholders are NOT significant. There are two types of zeros that fall into this category. The first are zeros in a decimal number that occur before the first nonzero digit. For example, in 0.0013 m, only the 1 and the 3 are significant; the zeros are not. The other zeros that fall under this rule are zeros that are trailing zeros that must be there to indicate the magnitude of the number. In the number 13,000 m the zeros fall into this category. Such zeros may or may not be significant; it depends on whether they were measured or not. They must be written even if the value was not measured in those places simply to distinguish that the number is thirteen thousand and not thirteen or one hundred thirty. The rule used in the textbook and in this study guide is that we will assume they are not significant unless we place a decimal point to the right of the last zero. Thus, we would say that 13,000 m has only two significant figures but that 13,000. m has five. To avoid confusion on your test papers, we suggest that you use scientific notation to make it clear which digits are significant in such a number. For example 1.300×10^4 m clearly indicates four significant figures whereas 1.3×10^4 m clearly indicates two and 1.3000×10^4 m indicates five.

Example 1-16: How many significant figures are there in the following measurements?

   a. 146 seconds
All of the digits are nonzero so all are significant. There are three significant figures.

   b. 0.1030 m
We shall proceed from left to right. The zero before the decimal point is not significant; it is just a placeholder. The 1 is a nonzero digit so it is significant, the next zero is between two nonzero digits so it is significant, the 3 is a nonzero digit so it is significant, and the final zero is to the right of a nonzero digit and to the right of the decimal place so it is significant. There are four significant figures.

   c. 1.0042 years
Proceeding from left to right, the 1 is significant, the next two zeros are between two nonzero digits so they are significant, the 4 and the 2 are nonzero digits so they are significant. There are five significant figures.

   d. 0.0040130
The first three zeros are just placeholders so they are not significant. The 4 is significant. The next zero is between two nonzero digits so it is significant. The 1 and the 3 are significant. The final zero is to the right of the decimal and to the right of a nonzero digit so it is significant. There are five significant figures.

   e. 250 mL
The 2 and the 5 are significant. The zero may or may not be significant. There is no decimal point marked so we will assume it is not significant. There are two significant figures.

The other more important issue with significant figures is to determine how many significant figures should be reported as the result of a calculation. There are two separate rules: one for addition and subtraction and one for multiplication and division. It is worth repeating this: there are two separate rules. Students sometimes get confused by trying to use only one rule for all cases. You must use one rule if you are performing addition or subtraction and another rule if you are performing multiplication or division.

Addition and Subtraction Rule: When adding or subtracting numbers, the number of decimal places in the answer is the same as the number of decimal places in the number that goes out the least amount.

Example 1-17: Add the following numbers: 0.47 m + 12.3 m + 0.842 m and report the answer to the correct number of significant figures.

The calculator reads 13.612. What should we report? The first number we added went out to the hundredths place, the second to the tenths, and the third to the thousandths. The one that went out to the fewest decimal places was 12.3, which only went out to the tenths place. Our answer should be rounded off to the tenths place. The answer is 13.6 m.

Notice that we do not do anything with counting the number of significant figures in each number for an addition and/or subtraction problem.

Multiplication and Division: In multiplication or division, we count the number of significant figures in each number. The number of significant figures in the answer will be the same as that present in the number that had the least number of significant figures.

Example 1-18: Carry out the following calculation: 0.47 m x 12.3 m x 0.842 m and report the answer to the correct number of significant figures.

The calculator reads 4.867602. What should we report? The first number had two significant figures. The second had three significant figures. The third had three significant figures. The smallest number of significant figures was two. Our answer should therefore be reported to only two significant figures. The answer is 4.9 m^3.

Other Notes

1. Qualitative observations do not involve measurements and numbers. Quantitative observations involve numerical information.

2. Converting between the Celsius and Fahrenheit scales is a little trickier than converting between the Celsius and Kelvin scales. On the Celsius scale, the freezing point of water is 0°C, and the boiling point of water is 100°C. There are 100°C between these two points. On the Fahrenheit scale, the freezing point of water is 32°F, and the boiling point of water is 212°F. There are 180° between these two points. For this same interval of temperature, there are 100°C and 180°F. The other thing we need to keep in mind is that the Fahrenheit scale has 32°F as the freezing point of water, not 0°F, so there will be an additional correction of 32°F that we need to take into account.

To convert from Fahrenheit to Celsius, we will first need to subtract this 32°F and then take into account the relationship between a Celsius and Fahrenheit degree.

$$T(°C) = (T(°F) - 32°F)\frac{100°C}{180°F} = (T(°F) - 32°F)\frac{5°C}{9°F}$$

To go the other way, use the relationship between Celsius and Fahrenheit degrees and then add 32°F.

$$T(°F) = \left(T(°C)\frac{180°F}{100°C}\right) + 32°F = \left(T(°C)\frac{9°F}{5°C}\right) + 32°F$$

3. Here are two useful volume equivalences to know:
   1 cm^3 = 1 mL
   1 dm^3 = 1 L

CHAPTER 2: Atoms and Elements

Chapter Overview

In Chapter 2, you will learn much more about atoms, elements, and the periodic table. We will first take a look at atomic structure and some of the experiments that were used to devise our modern model of the atom. Atoms are composed of three major subatomic particles: protons, neutrons, and electrons. Protons have a positive charge, have a mass of approximately 1 atomic mass unit, and are located in the nucleus of the atom. Neutrons are neutral in charge, have a mass of approximately 1 atomic mass unit, and are located in the nucleus of the atom. Electrons are negative in charge, have a much smaller mass than protons and neutrons, and travel around outside the nucleus of the atom. The atomic number is the number of protons in an atom. The mass number is the sum of protons and neutrons in the nucleus of an atom. Isotopes are atoms of the same element with different numbers of neutrons. We will also go over what the atomic weights shown on the periodic table represent and how to calculate them given the masses and percent abundances of the isotopes of an element. We will then move from the submicroscopic world of atoms into the macroscopic world of the laboratory. The chemical counting unit on the macroscopic level is the mole. One mole of an element has a mass equal to the numerical value of the atomic weight of the element, but the unit is grams instead of atomic mass units. Thus one mole of hydrogen gas has a mass of 1.008 g while the average atomic weight of a hydrogen atom is 1.008 atomic mass units. Just as there are two atoms of hydrogen per one atom of oxygen in a water molecule, there are two moles of hydrogen per one mole of oxygen in one mole of water. One mole contains Avogadro's number of the particles (atoms, molecules, etc.) making up that material. You will learn how to convert between mass and amount in moles as well as how to convert between amount in moles and atoms. You will use dimensional analysis to do these calculations. Finally, we will take a closer look at the periodic table, defining various regions on the table. Elements with similar properties are in the same vertical column on the table. Vertical columns are called groups or families, whereas horizontal rows are called periods. You will learn the names of some of the groups on the periodic table. Distinctions will also be made between main group elements, transition elements, lanthanides, and actinides as well as between metals, nonmetals, and metalloids. Some of the properties of various groups and elements within those groups will also be discussed.

Key Terms

In this chapter, you will need to learn and be able to use the following terms:

Actinide: an element with an atomic number anywhere from 90 to 103; the actinides are usually shown (along with the lanthanides) at the bottom of the periodic table separated from the main body of elements.

Alkali metal family: group 1A on the periodic table (excluding hydrogen).

Alkaline earth metal family: group 2A on the periodic table.

Allotropes: different forms of the same element that exist in the same physical state under the same conditions of temperature and pressure; they have different molecular formulas or crystal structures.

Alpha rays: particles produced when certain nuclei decay; they consist of a helium nucleus; they have a relative mass of 2 and have a +2 charge.

Atomic mass unit: one atomic mass unit is one twelfth the mass of a carbon-12 atom.

Atomic number (Z): the number of protons in an atom.

Atomic weight: a weighted average of all of the isotopes of a particular element. The atomic weight is often shown on the periodic table; it is the number that contains the decimal places.

Avogadro's number: the number of particles in a mole of that substance, $6.02214155 \times 10^{23}$ particles (to nine significant figures).

Beta rays: particles produced when certain nuclei decay; they consist of electrons. A beta particle has a charge of −1 and a very small mass.

Canal rays: positively charged particles that travel toward the negative cathode in a cathode ray tube. These particles are positive ions.

Cathode ray tube: a glass tube from which most of the air has been removed and which contains two metal electrodes; when a sufficiently high voltage is applied to the electrodes, cathode rays flow from the cathode to the anode.

Cathode rays: a stream of electrons that flows from the negative cathode of a cathode ray tube to the positive anode.

Chalcogens: sulfur, selenium, and tellurium.

Electron: one of the major subatomic particles. Electrons have a mass much smaller than that of protons and neutrons, have a relative charge of −1, and travel around outside the nucleus of the atom.

Essential element: an element required for life; a deficiency or absence of an essential element may result in death, developmental abnormalities, or other ailments.

Family: a vertical column on the periodic table, sometimes called a group.

Gamma rays: a type of radiation given off when certain nuclei decay; they consist of electromagnetic radiation. Gamma rays have neither mass nor charge.

Group: a vertical column on the periodic table, sometimes called a family.

Halogen family: group 7A on the periodic table.

Isotopes: atoms of the same element that have different mass numbers and thus different numbers of neutrons in their nuclei.

Lanthanide: an element with an atomic number anywhere from 58 to 71; the lanthanides are usually shown (along with the actinides) at the bottom of the periodic table separated from the main body of elements.

Law of chemical periodicity: the properties of the elements are periodic functions of atomic number.

Main group elements: elements found in the taller columns on the periodic table (the A groups on the periodic table most often used in the United States).

Mass defect: the difference between the sum of the masses of the protons and neutrons that go into forming a nucleus and the actual mass of the nucleus, which will always be less. This difference is due to matter being converted into energy when the subatomic particles come together to form the nucleus.

Mass number (A): the sum of the number of protons and number of neutrons in an atom.

Mass spectrometry: a technique that can be used to determine the mass of an isotope.

Metal: an element to the left of the line separating the metals and nonmetals on the periodic table; they tend to be solids at room temperature, conduct electricity, and are ductile and malleable.

Metalloid: an element with some properties of metals and some properties of nonmetals; they are found along the zigzag line on the periodic table. The metalloids are B, Si, Ge, As, Sb, and Te.

Molar mass: the mass in grams of one mole of a material.

Mole: the chemical counting unit on our macroscopic level. One mole of a substance contains as many elementary entities (atoms, molecules, or other particles) as there are in exactly 12 g of the carbon-12 isotope. One mole contains Avogadro's number of elementary entities.

Neutron: one of the major subatomic particles. Neutrons have a relative mass of 1. They are electrically neutral and are located in the nucleus of an atom.

Noble gas family: group 8A on the periodic table.

Nonmetal: an element to the right of the line separating the metals and nonmetals on the periodic table. Some are solids, one is a liquid, and some are gases at room temperature. Almost all nonmetals do not conduct electricity.

Nucleus: the minuscule center of an atom containing the protons and neutrons.

Period: a horizontal row on the periodic table.

Periodicity: the property of repeating at regular intervals.

Proton: one of the major subatomic particles. Protons have a relative mass of 1 and a relative charge of +1. They are located in the nucleus of an atom.

Radioactivity: particles or radiation given off by atoms of certain elements.

Subatomic particle: a particle that makes up an atom. The key subatomic particles to know about for most of chemistry are protons, neutrons, and electrons.

Transition elements: elements found in the shorter columns in the main body of the periodic table (the B groups on the periodic table most often used in the United States).

Chapter Goals

By the end of this chapter you should be able to:

- ## Describe atomic structure and define atomic number and mass number.

   ### a) Explain the historical development of the atomic theory and identify some of the scientists who made important contributions (Section 2.1).

   Toward the beginning of the 1800s, John Dalton proposed that all matter is composed of small indivisible, indestructible particles called atoms. We now believe that there are subatomic particles. In other words, we now believe that atoms are not indivisible.

   In 1896, Henri Becquerel discovered that a uranium ore could darken a photographic plate. This was due to something being given off by the uranium. Marie Curie called this phenomenon radioactivity and proposed that radioactivity was caused by atoms disintegrating. Three major types of radiation were detected in early experiments: alpha, beta, and gamma. (Alpha radiation consists of helium nuclei; alpha particles have a relative mass of 2 and a charge of +2. Beta radiation consists of electrons having an almost negligible mass and a charge of –1. Gamma rays consist of electromagnetic radiation and have neither mass nor a charge associated with them.) The phenomenon of radioactivity thus provided evidence that atoms are not indestructible and that they are not indivisible.

Further evidence for electrons came from work with cathode ray tubes. When a high voltage is applied to a cathode ray tube, cathode rays are produced. They are called cathode rays because they travel from the cathode to the anode. Because the anode is positive and the cathode is negative, it was determined that cathode rays must be negative in charge. The negatively charged particles making up cathode rays were called electrons. Working with cathode ray tubes, J. J. Thomson experimentally proved the existence of electrons and was able to determine the charge to mass ratio of an electron. Robert Milliken, interpreting his oil drop experiment, was able to determine the charge on a single electron. Combining Thomson's and Milliken's work, the mass of an electron could thus be determined.

Because atoms are neutral, it became clear that there must also be positively charged particles in an atom. Ernest Rutherford determined that the hydrogen atom has only one unit charge. He also found that when nitrogen was bombarded with alpha particles, positively charged particles were given off. These particles had the same charge to mass ratio as hydrogen. These were protons.

James Chadwick obtained experimental proof for the existence of a neutral subatomic particle, known as the neutron.

In addition to determining what subatomic particles are present in an atom, scientists also wished to determine how they are arranged in the atom. Dalton had proposed that atoms were small hard spheres. Once electrons were known, J. J. Thomson proposed that his electrons were present in the spheres floating around in a sort of positive goo. This model is sometimes called the plum pudding model of the atom. Ernest Rutherford and his colleagues tested this model by firing alpha particles at a piece of gold foil. If the plum pudding model were correct, then these alpha particles should not have been affected much by the spread out positive charge in an atom. It was expected that the alpha particles should just pass through the foil. Most of the particles did this. But some were deflected and others even bounced back. How could the positive charge in the atom do this? A new model was needed. It was proposed that the positive charge was not spread out but instead was concentrated in a very small center that we call the nucleus. The electrons were proposed to be present outside the nucleus and spread out, not concentrated in one place. Let's now consider the different results Rutherford obtained. Most of the alpha particles passed through the gold foil. This happens because in most of the atom, there is no positive charge to repel the alpha particles. The negative charge of the electrons is spread out in the space around the nucleus so it is not concentrated enough to affect the alpha particles greatly either. Most of the alpha particles thus pass through the foil unaffected. When the few alpha particles that happen to pass close to a nucleus do so, the concentrated positive charge of the nucleus repels the positively charged alpha particles and causes their paths to change, deflecting them. For those very few alpha particles that approach a nucleus head-on, there is enough repulsion to cause the alpha particles to turn around and head back in the direction from which they came. The observations that Rutherford obtained could thus be explained using this new nuclear model for the atom.

b) Describe electrons, protons, and neutrons, and the general structure of the atom (Section 2.1).

The basic picture of the atom is that protons and neutrons are located in the nucleus of the atom. Outside the nucleus, the electrons are moving around.

Some of the important characteristics of protons, neutrons, and electrons are summarized in the following table:

Subatomic Particle	Location in Atom	Relative Mass	Relative Charge
proton	nucleus	1	+1
neutron	nucleus	1	0
electron	moving around outside the nucleus	Almost negligible (about 1/2000 the mass of a proton)	−1

In a neutral atom, the number of electrons must equal the number of protons.

The atomic number of an element, sometimes abbreviated by the letter Z, is the number of protons in the nucleus of an atom of that element. The atomic number for each element is listed on most periodic tables. It is the whole number value in each box. The atomic number of lithium is 3. There are three protons in the nucleus of a lithium atom. If we have an atom with an atomic number of 3, then it must be an atom of the element lithium.

Example 2-1: What is the atomic number of carbon?

To find the atomic number, we look on the periodic table. There are two numbers listed in the block for carbon: 6 and 12.011. The atomic number is the one without the decimals. The atomic number is 6.

Example 2-2: A particular atom has 23 protons. This is an atom of which element?

We look on the periodic table for the element with atomic number 23. This element has the chemical symbol V. The element is vanadium.

Try Study Question 7 in Chapter 2 of your textbook now!

For a neutral atom, the number of protons must equal the number of electrons, so the atomic number will also give the number of electrons in a neutral atom of an element.

c) Understand the relative mass scale and the atomic mass unit (Section 2.2).

One atomic mass unit, 1 u, is defined to be one-twelfth the mass of an atom that contains six protons and six electrons. It is thus one-twelfth of what we will call a carbon-12 atom. On this scale, a proton has a mass of a little over but very close to 1 u. Likewise, a neutron has a mass of a little over but very close to 1 u. An atom that contains 3 protons and 4 neutrons will therefore have a mass close to 7 u. The atomic mass unit is simply a convenient scale to use with atoms. It is not usually important what the relationship is between atomic mass units and the actual mass of an atom in grams, but the conversion factor is that 1 u is equal to 1.661×10^{-24} g.

• Understand the nature of isotopes and calculate atomic weight from isotope abundances and isotopic masses.

a) Define isotope and give the mass number and number of neutrons for a specific isotope (Sections 2.2 and 2.3).

The mass number of an atom, sometimes abbreviated by the letter A, is the sum of the protons and neutrons in the atom.

A = mass number = number of protons + number of neutrons

Example 2-3: An atom contains 6 protons and 7 neutrons. What is its mass number?

The mass number is the sum of the protons and neutrons in an atom. The mass number for this atom is

A = number of protons + number of neutrons = 6 + 7 = 13.

Try Study Question 9 in Chapter 2 of your textbook now!

Isotopes are atoms of the same element that have different masses. To be atoms of the same element, the atoms must contain the same number of protons. In order to have different masses, they must therefore contain different numbers of neutrons.

Sometimes, we use the following types of symbols to represent an isotope of an element:

$$\text{Mass Number} \rightarrow A$$
$$\text{Atomic Number} \rightarrow Z} \, X \leftarrow \text{Element Symbol} \qquad \text{or simply } {}^{A}X$$

For example, the isotope of oxygen (atomic number 8) with 10 neutrons and thus a mass number of 18 can be designated as ${}^{18}_{8}O$ or as ${}^{18}O$. It is referred to as oxygen-18.

Example 2-4: How many neutrons are present in the isotope of lithium having a mass number of 7, ${}^{7}Li$?

The mass number is the sum of the protons and neutrons. We can determine the number of protons in the atom by using the periodic table. Lithium has an atomic number of 3; there are 3 protons.

A = Number of Protons + Number of Neutrons
Number of Neutrons = A − Number of Protons
Number of Neutrons = 7 − 3
Number of Neutrons = 4

Try Study Question 13 in Chapter 2 of your textbook now!

b) Do calculations that relate the atomic weight (atomic mass) of an element and isotopic abundances and masses (Section 2.4).

The atomic weight for an element is the average weight of all of the isotopes of the element. It is a special type of average, however. It is what is called a weighted average. Let's see what we mean by this. There are two major isotopes of the element chlorine, ${}^{35}Cl$, which has a mass close to 35, and ${}^{37}Cl$, which has a mass close to 37. Your gut reaction might be that the average atomic weight should be around 36 because that is the average of 35 and 37. BUT this would only be true if there were an equal number of ${}^{35}Cl$ atoms and ${}^{37}Cl$ atoms. This is not the case. About 3/4 of the atoms of chlorine are ${}^{35}Cl$ and only about 1/4 are ${}^{37}Cl$. The ${}^{35}Cl$ should therefore count more; the average mass should be closer to 35 than to 37. We must take into account both the masses and the abundances.

Let us go through this problem in more detail. The equation that is used to determine the average atomic weight is

$$\text{atomic weight} = \left(\frac{\% \text{ abundance isotope 1}}{100} \right)(\text{mass of isotope 1}) +$$
$$\left(\frac{\% \text{ abundance isotope 2}}{100} \right)(\text{mass of isotope 2}) + \dots$$

Example 2-5: Calculate the atomic weight of chlorine given the following data:

^{35}Cl mass = 34.96885 u; percent abundance = 75.77%
^{37}Cl mass = 36.96590 u; percent abundance = 24.23%

We must take into account both the masses and the abundances.

$$\text{atomic weight} = \left(\frac{\text{\% abundance isotope 1}}{100}\right)(\text{mass of isotope 1}) +$$

$$\left(\frac{\text{\% abundance isotope 2}}{100}\right)(\text{mass of isotope 2})$$

$$\text{atomic weight} = \left(\frac{75.77}{100}\right)(34.96885 \text{ u}) + \left(\frac{24.23}{100}\right)(36.96590 \text{ u})$$

$$= 35.45 \text{ u}$$

Try Study Question 21 in Chapter 2 of your textbook now!

A more challenging problem algebraically involves solving for the abundances of two isotopes given the overall atomic weight (from the periodic table) and the masses of the individual isotopes. In order to solve this type of problem, we will use the same equation but also use the fact that the percent abundances must add up to 100%.

Example 2-6: Boron has two naturally occurring isotopes, ^{10}B and ^{11}B, with masses of 10.0129 u and 11.0093 u, respectively. Calculate the percent abundances of these isotopes.

$$\text{atomic weight} = \left(\frac{\text{\% abundance isotope 1}}{100}\right)(\text{mass of isotope 1}) +$$

$$\left(\frac{\text{\% abundance isotope 2}}{100}\right)(\text{mass of isotope 2})$$

We know from the periodic table that the atomic mass of boron is 10.811 u.

$$10.811 = \left(\frac{\text{\% abundance } ^{10}\text{B}}{100}\right)(10.0129) + \left(\frac{\text{\% abundance } ^{11}\text{B}}{100}\right)(11.0093)$$

We also know that the percent abundances must add up to equal 100%. If we let x = the % abundance of ^{10}B, then the % abundance of ^{11}B will be 100 − x.

$$10.811 = \left(\frac{x}{100}\right)(10.0129) + \left(\frac{100 - x}{100}\right)(11.0093)$$

We can now distribute in the masses.

$$10.811 = 0.100129x + 11.0093 - 0.110093x$$

Combining common terms, we obtain

$$-0.198 = -0.009964x$$

Solving for x,

$$x = 19.9$$

The percent abundance of ^{10}B is thus 19.9% and that of ^{11}B is 100 − 19.9 = 80.1%.

Try Study Question 21 in Chapter 2 of your textbook now!

• Explain the concept of the mole and use molar mass in calculations.

a) Understand that the molar mass of an element is the mass in grams of Avogadro's number of atoms of that element (Section 2.5).

We are going to be using the concept of the mole for the rest of the course. It is worth taking some time to make sure that you understand this important concept.

In one molecule of water (H_2O), there are 2 atoms of hydrogen for each atom of oxygen. This is a relationship that exists on the submicroscopic level of atoms, ions, and molecules. We want to find a way to use this type of relationship in our macroscopic world. The key is that no matter how many water molecules there are, this 2:1 ratio of hydrogen to oxygen will always be maintained.

In two molecules of water, there are 2(2) = 4 atoms of hydrogen for 2(1) = 2 atoms of oxygen.

In ten molecules of water, there are 10(2) = 20 atoms of hydrogen for 10(1) = 10 atoms of oxygen.

In 100 molecules of water, there are 100(2) = 200 atoms of hydrogen for 100(1) = 100 atoms of oxygen.

In 1,000,000 molecules of water, there are 1,000,000(2) = 2,000,000 atoms of hydrogen for 1,000,000(1) = 1,000,000 atoms of oxygen.

In 6.022×10^{23} molecules of water, there are $6.022 \times 10^{23}(2)$ atoms of hydrogen for $6.022 \times 10^{23}(1)$ atoms of oxygen.

Let us now define this quantity of something, 6.022×10^{23}, to be equal to one mole of the material. Therefore, in 1 mole of water, there are two moles of hydrogen atoms and 1 mole of oxygen atoms. Using the mole relationship, we are back to the exact same ratio as we had with atoms and molecules. Thus, 1 molecule of water contains 2 atoms of hydrogen and 1 atom of oxygen, and 1 mole of water contains 2 moles of hydrogen atoms and 1 mole of oxygen atoms.

This number, 6.022×10^{23}, is called Avogadro's number. One mole of something contains Avogadro's number of that thing. It is a counting number, like a dozen. 12 doughnuts = 1 dozen doughnuts. 12 soda bottles = 1 dozen soda bottles. 6.022×10^{23} hydrogen atoms is 1 mole of hydrogen atoms. 1 mole of sodium contains 6.022×10^{23} sodium atoms. 1 mole of water contains 6.022×10^{23} water molecules.

Why choose such a strange value to define as the mole? After all, we could have chosen any collection of atoms to be 1 mole, and we would have gotten the same result, that there are two moles of hydrogen atoms per one mole of oxygen atoms in one mole of water. Why choose 6.022×10^{23}? The first reason is simply that by the time we get to this number of atoms or molecules, we are dealing with a number of atoms that corresponds to an amount we can deal with on our macroscopic level. For the second reason, take note that if I have 6.022×10^{23} atoms of sodium, 1 mole of sodium atoms, then the mass of that sample is 22.99 g. This is the same value as the atomic weight of sodium on the periodic table but where the unit is grams, instead of atomic mass units which we used for one atom. Similarly, the mass of 6.022×10^{23} atoms of chlorine, 1 mole of chlorine atoms, is 35.45 g. Once again, this is the atomic weight of chlorine with units of grams. The average atomic weight of one atom of boron is 10.811 atomic mass units. The mass of a mole of boron is 10.811 g.

One mole of an element contains 6.022 x 10^{23} atoms. (More specifically, one mole contains as many elementary entities (atoms, molecules, or other particles) as there are atoms in exactly 12 g of the carbon-12 isotope. This corresponds to containing Avogadro's number of elementary entities – 6.02214155 x 10^{23} to nine significant figures.) The mass of this collection of atoms will have a mass equal to the atomic weight of the element shown on the periodic table but where the mass unit is grams instead of atomic mass unit. The mass of one mole of something is simply called the molar mass of that thing. The units for molar mass are grams per mole (g/mole). The molar mass of sodium is thus 22.99 g/mole.

b) Know how to use the molar mass of an element and Avogadro's number in calculations (Section 2.5).

You will find that problems involving moles are another case where dimensional analysis will be very useful. To convert between mass and amount in moles, we use the molar mass.

$$\text{mass in grams} \quad \longleftarrow \quad \begin{matrix}\text{MOLAR}\\\text{MASS}\end{matrix} \quad \longrightarrow \quad \text{amount in moles}$$

Example 2-7: How many moles of boron are there in 14.3 g of boron?

In this problem we are given a mass and are asked to calculate the number of moles that it represents. The connection between mass and moles is the molar mass. We can look up the molar mass of boron on the periodic table. It is 10.811 g/mole. From this molar mass, we can construct two conversion factors:

$$\frac{10.811 \text{ g B}}{1 \text{ mole B}} \quad \text{and} \quad \frac{1 \text{ mole B}}{10.811 \text{ g B}}$$

The given is the mass, 14.3 g B. We can write this as

$$\frac{14.3 \text{ g B}}{1}$$

The grams are on the top in our given. In order to get grams to cancel, we will need to have grams on the bottom in our conversion factor, so we will use the conversion factor involving molar mass that we wrote on the right above.

$$\frac{14.3 \text{ g B}}{1} \times \frac{1 \text{ mole B}}{10.811 \text{ g B}} = 1.32 \text{ moles B}$$

Note that the grams cancel, leaving moles on top, which is what we want. Our answer is therefore that this mass of boron contains 1.32 moles of boron.

We can go in the other direction very easily as well.

Example 2-8: What mass of copper would need to be weighed out in order to have 3.50 moles of copper?

In this case, we are given the amount in moles and asked to solve for the mass. The molar mass is still the connecting link between mass and moles. The molar mass of copper is 63.55 g/mole. The two possible conversion factors are

$$\frac{63.55 \text{ g Cu}}{1 \text{ mole Cu}} \quad \text{and} \quad \frac{1 \text{ mole Cu}}{63.55 \text{ g Cu}}$$

The given in this problem is the amount in moles, 3.50 moles Cu.

$$\frac{3.50 \text{ moles Cu}}{1}$$

We want to cancel the moles and end up with grams on top, so we select the conversion factor with moles on the bottom and grams on top.

$$\frac{3.50 \text{ moles Cu}}{1} \times \frac{63.55 \text{ g Cu}}{1 \text{ mole Cu}} = 222 \text{ g Cu}$$

Try Study Questions 27 and 29 in Chapter 2 of your textbook now!

The connection between mass and moles is the molar mass. The connection between number of particles and moles is Avogadro's number.

The conversion factors that can be set up for atoms are the following:

$$\frac{6.022 \times 10^{23} \text{ atoms}}{1 \text{ mole}} \quad \text{and} \quad \frac{1 \text{ mole}}{6.022 \times 10^{23} \text{ atoms}}$$

<u>Example 2-9</u>: How many sodium atoms are there in 3.40 moles of Na?

We are given moles and asked for number of atoms. The connecting link between atoms and moles is Avogadro's number. Our given is

$$\frac{3.40 \text{ moles Na}}{1}$$

We will use the conversion factor involving Avogadro's number in which the term involving moles is on the bottom (to cancel the moles on top in our given) and in which the term involving atoms is on the top.

$$\frac{3.40 \text{ moles Na}}{1} \times \frac{6.022 \times 10^{23} \text{ atoms Na}}{1 \text{ mole Na}} = 2.05 \times 10^{24} \text{ atoms Na}$$

Sometimes, a problem will combine the two types of problems we have just gone over (mass-moles and atoms-moles) into one larger problem.

<u>Example 2-10</u>: How many atoms of phosphorus are there in 16.3 g of phosphorus?

We are given mass and asked for number of atoms. We do not have a conversion factor that connects these two quantities. But we do have a way to go from mass to moles (molar mass) and a way to go from moles to atoms (Avogadro's number). In this particular problem, we are given the mass in grams:

$$\frac{16.3 \text{ g P}}{1}$$

To go from grams to moles, we will use the molar mass of phosphorus, 30.97 g/mole, obtained from the periodic table. Because grams in our given was on the top, we will want grams on the bottom and moles on top in our conversion factor.

$$\frac{16.3 \text{ g P}}{1} \times \frac{1 \text{ mole P}}{30.97 \text{ g P}}$$

If we were to do our arithmetic now, our answer would be in moles of phosphorus. We are asked for atoms of phosphorus. The connection between moles and atoms is Avogadro's number. In our conversion factor, we will want moles on the bottom and atoms on top.

$$\frac{16.3 \text{ g P}}{1} \text{ x } \frac{1 \text{ mole P}}{30.97 \text{ g P}} \text{ x } \frac{6.022 \text{ x } 10^{23} \text{ atoms P}}{1 \text{ mole P}} = 3.17 \text{ x } 10^{23} \text{ atoms P}$$

The key to solving this problem was to recognize that we did not have a single conversion factor to go from grams to atoms and to recognize that we could piece together our two conversion factors to go first from grams to moles and then from moles to atoms.

Try Study Question 33 in Chapter 2 of your textbook now!

- ## Know the terminology of the periodic table.

   a. Identify the periodic table locations of groups, periods, metals, metalloids, nonmetals, alkali metals, alkaline earth metals, halogens, noble gases, and the transition elements (Section 2.6 and 2.7).

   The vertical (up and down) columns on the periodic table are called groups or families. The horizontal (across) rows on the periodic table are called periods.

   Elements with similar chemical and physical properties are in the same vertical column on the periodic table. Some of these families have special names:
   >   Group 1A: Alkali Metal Family (except hydrogen)
   >   Group 2A: Alkaline Earth Metal Family
   >   Group 7A: Halogen Family
   >   Group 8A: Noble Gas Family

   Example 2-11: Name the member of the halogen family that occurs in the fourth period on the periodic table.

   The periods on the periodic table are numbered from the top of the periodic table to the bottom. Hydrogen and helium comprise the first period. The elements from lithium to neon comprise the second period, and so forth. The fourth period consists of the elements from potassium through krypton. The halogen family is the vertical column labeled 7A. The element in this column that is in the fourth period is bromine.

   The main group elements are the elements found in the tall columns on the periodic table. In the numbering system used on the periodic tables most often used in the United States, these groups have a number followed by an A (example: Group 7A). The transition elements (sometimes called the transition metals) are the elements in the shorter columns in the main body of the periodic table. In the numbering system used on the periodic tables most often used in the United States, these groups have a number followed by a B.

   The lanthanides and actinides are the elements in the two rows at the very bottom of the periodic table. These elements are really interspersed in the middle of the sixth and seventh periods. The lanthanides are elements 58-71 and occur right after the element lanthanum. The actinides are elements 90-103 and occur right after the element actinium.

Example 2-12: Classify each of the elements in the fourth period as being main group or transition elements.

Potassium and calcium are in the tall columns labeled 1A and 2A. These are main group elements. The elements from scandium through zinc are in the shorter B columns; these are transition elements. The elements from gallium through krypton are in the tall A columns and are main group elements.

There is a zigzag diagonal line on the periodic table. The elements to the left of this line are called metals. Elements to the right of this line are called nonmetals. The easiest way to keep the direction straight is to think of a metal you know such as iron, Fe, and look for it on the periodic table. It is to the left of the zigzag line so metals must be to the left of the line. Some of the elements that are found along the zigzag line (boron, silicon, germanium, arsenic, antimony, and tellurium) have some properties of metals and others of nonmetals. These elements are called metalloids or semi-metals.

Properties of Metals: Most metals are solids at room temperature (mercury is a liquid), conduct electricity, are ductile, are malleable, and can form alloys.

Properties of Nonmetals: Nonmetals are much more varied than metals. Some are solids, one is a liquid (bromine), and some are gases at room temperature. All except carbon (in the form of graphite) do not conduct electricity.

Try Study Questions 35, 39, and 53 in Chapter 2 of your textbook now!

b. Recognize similarities and differences in properties of some of the common elements of a group.

Group 1A: Alkali Metals

All are solids at room temperature.

All are very reactive. Because of this, they are not found in their free state in nature, only in compounds. The free elements even react with water, producing hydrogen gas and an alkaline solution.

Group 2A: Alkaline Earth Metals

All are solids at room temperature.

They are reactive. Just like the alkali metals, they are found only in compounds in nature, not as free elements. All except beryllium will react with water to produce hydrogen gas and alkaline solutions.

Most of their oxides react with water to form alkaline solutions.

Group 3A

Aluminum is the most abundant metal in the earth's crust.

Group 4A

The elements in this group show more variation in properties than most other groups on the periodic table. One is a nonmetal (C), two are metalloids (Si and Ge), and two are metals (Sn and Pb).

Carbon has three allotropes: graphite, diamond, and fullerenes. In graphite, each carbon atom is bonded to three others forming flat sheets of carbon. The sheets are attracted to each other by weak forces and can easily slide past one another. In diamond, each carbon atom is bonded to four others in a three-dimensional network. Fullerenes consist of molecules of carbon in which the carbon atoms are joined in five- and six-membered rings to form large cages. The most common fullerene is C_{60}, which has the same shape as a soccer ball.

Oxides of silicon are the basis of many minerals, such as clay, quartz, and many gemstones.

Group 5A

Nitrogen is the most abundant gas in the atmosphere, comprising about three quarters of it. Elemental nitrogen occurs as a diatomic molecule.

The most important allotropes of phosphorus are white phosphorus and red phosphorus. White phosphorus ignites spontaneously in air. White phosphorus consists of molecules containing four atoms of phosphorus (P_4). Red phosphorus is used in the striking strips on matchbooks.

Group 6A

Oxygen is found in two allotropic forms: O_2 and O_3. About 20% of the earth's atmosphere is O_2. It is the gas that we use when we breathe. O_3 is called ozone.

Sulfur is normally found as a yellow solid around room temperature and at normal pressure. It has many allotropes. The most common is based on rings of eight sulfur atoms joined in a structure that resembles a crown.

Group 7A: Halogens

These elements occur as diatomic molecules. Fluorine is a pale yellow gas, chlorine is a yellow-green gas, bromine is a red-orange liquid, and iodine is a grey-purple solid. Both bromine (orange) and iodine (purple) vapors are clearly visible over the liquid or solid.

These elements are among the most reactive of the elements.

Group 8A: Noble Gases

All of these elements are gases. These elements are not very reactive. It was not until 1962 that chemists succeeded in making any compounds containing the noble gases.

Other Notes

1. Isotopes of Hydrogen (neutral atoms)

Isotope	Protons	Neutrons	Mass Number (= protons + neutrons)	Electrons (= # of protons)
protium	1	0	1	1
deuterium	1	1	2	1
tritium	1	2	3	1

2. Mass spectrometry is a technique that can be used to determine the mass of an isotope. In one mass spectrometry technique, the atoms in the sample are converted to charged particles called ions. In the mass spectrometer the ions then move into an analyzing chamber where a magnetic field causes the path of the ions to bend. By varying the magnetic field, the ions' paths can be bent to aim them at a device that detects when an ion hits it. The actual magnetic field needed to aim a particular ion so that it will hit the detector depends upon the mass to charge ratio of the ion. It is possible to determine the mass of the ions by using the largest mass to charge ratios, for which the charge will be 1+. The mass will thus be the same as these mass to charge ratios.

3. Dmitri Mendeleev is given much of the credit for the development of the periodic table. Mendeleev arranged the elements then known in order of increasing atomic weight and placed elements with similar properties into the same group. He left blanks for some elements then not known. In the modern periodic table, the elements are arranged not in order of increasing atomic weight but instead in order of increasing atomic number.

4. The elements that occur in nature predominantly as diatomic molecules are hydrogen (H_2), nitrogen (N_2), oxygen (O_2), fluorine (F_2), chlorine (Cl_2), bromine (Br_2), and iodine (I_2). To remember these, start on the periodic table with hydrogen, jump to nitrogen, and form an upside-down L by going over to group 7A and then going down this group.

5. Essential Elements

Four elements (carbon, hydrogen, nitrogen, and oxygen) make up approximately 99% of the total atoms in the human body.

The next 0.9% of the atoms in the human body are atoms of sodium, calcium, phosphorus, sulfur, chlorine, potassium, and magnesium.

There are about 17 other elements that are required by most biological organisms. These are required in only trace amounts, but are very important, nonetheless.

CHAPTER 3: Molecules and Compounds

Chapter Overview

In this chapter, you will learn more about molecules and compounds. There are different ways that we can represent the formula for a molecular compound. You will also learn about ions and how they are formed. One of the most important parts of this chapter is to teach you how to name and write the formulas for ionic compounds and binary molecular compounds. We will then move from the submicroscopic view of molecules and crystal lattices of ions to the macroscopic level where we work in the laboratory and live. We will extend the concept of the mole to compounds. You will learn how to determine the molar mass of a compound and how to interconvert between mass and moles. You will also learn how to calculate the percent composition of a compound and how to determine the empirical and molecular formulas of compounds. Finally, we will discuss hydrated compounds and how to determine the number of waters of hydration in a formula unit of an ionic hydrate.

Key Terms

In this chapter, you will need to learn and be able to use the following terms:

Anhydrous: without water; the anhydrous form of a hydrated compound is the form in which the waters of hydration have been driven off (usually by heating).

Anion: a negative ion; anions are formed by atoms gaining electrons.

Ball-and-stick model: a molecular model in which the balls represent atoms and the sticks represent the bonds holding them together.

Binary compound: a compound composed of two elements.

Cation: a positive ion; cations are formed by atoms losing electrons.

Condensed formula: a molecular formula in which certain atoms are grouped together to better indicate the structure of a compound (example: CH_3CH_2OH).

Crystal lattice: a solid three-dimensional network of particles in which the particles are present in a repeating pattern.

Electrostatic forces: forces of attraction or repulsion that occur when charged substances are brought together.

Empirical formula: the formula for a compound that gives the smallest whole number ratio of atoms (or moles) of each element in the compound.

Hydrated compound: a compound in which molecules of water are associated with the ions of the compound.

Ion: an atom or group of atoms that has a positive or negative charge.

Ionic compound: compounds composed of ions, not molecules.

Isomers: compounds having the same molecular formula but different structures.

Mass spectrometry: an experimental technique that can be used to determine the molar mass of some compounds.

Molar mass of a compound: the mass of one mole of a compound, sometimes called the molecular weight. For an ionic compound, it is sometimes called the formula mass or formula weight.

Molecular formula: a formula that tells the elements and the number of atoms of each element present in a molecule (example: C_2H_6O).

Molecule: the smallest particle of a nonionic compound and of those elements whose atoms chemically bond together to form small discrete particles.

Monatomic ion: an ion consisting of only one atom.

Percent composition: the mass percent of an element in a compound.

Polyatomic ion: an ion made up of two or more atoms.

Space-filling model: a molecular model in which the balls hook up to each other and the sizes of the balls attempt to represent the relative sizes of the atoms.

Structural formula: a formula that shows which atoms are connected to which; lines are used to represent chemical bonds.

Chapter Goals

By the end of this chapter you should be able to:

- **Interpret, predict, and write formulas for ionic and molecular compounds.**

     a) Recognize and interpret molecular formulas, condensed formulas, and structural formulas (Section 3.1).

     Review the definitions of molecular formula, condensed formula, and structural formula.

     Example 3-1: The structural formula for the amino acid alanine is

     What is the molecular formula for this compound?

     The molecular formula simply lists the elements present and the number of atoms of each in the molecule. For many compounds, the elements are listed in the order C, H, and then other elements in alphabetical order. This compound contains 3 carbon atoms, 7 hydrogen atoms, 1 nitrogen atom, and 2 oxygen atoms. The formula is thus $C_3H_7NO_2$. (The condensed formula groups together certain atoms that are connected to the same atom. The condensed formula for this compound is NH_2CHCH_3COOH.)

Try Study Question 3 in Chapter 3 of your textbook now!

b) Recognize that metal atoms commonly lose one or more electrons to form positive ions, called cations, and nonmetal atoms often gain electrons to form negative ions, called anions (see Figure 3.7).

In forming ions from neutral atoms, electrons are lost or gained. Keep in mind that electrons are negative in charge. A neutral atom contains an equal number of protons (positive) and electrons (negative). If an atom loses an electron (or electrons), then the atom will have more positive charge than negative charge. The resulting ion is thus positive and is called a cation. Metals tend to form cations. On the other hand, if an atom gains electrons, then it will have more negative charge than positive charge. The resulting ion in that case is negative in charge. Nonmetals tend to form anions.

c) Recognize that the charge on a metal cation in Groups 1A, 2A, and 3A is equal to the group number in which the element is found in the periodic table (M^{n+}, n= Group Number) (Section 3.3). Charges on transition metal cations are often 2+ or 3+, but other charges are observed.

You should know the charges of all of the ions shown in Figure 3.7.

In learning the charges of the Main Group cations, let the periodic table help you.
   Group 1A => 1+ charge (Examples: Li^+, Na^+, K^+) Note that for a cation with a 1+ charge, we do not write the 1 in the symbol.
   Group 2A => 2+ charge (Examples: Mg^{2+}, Ca^{2+}, Sr^{2+})
   Group 3A => 3+ charge (Example: Al^{3+})

You must memorize the charges of the transition metal cations. Many of these have more than one possible charge, and many have charges of 2+ or 3+, but some do not. You must also memorize the charges of some of the metals at the bottom of the periodic table to the right of the transition metals.

The name of a monatomic cation formed from an element that can form only one possible cation is simply the name of the element followed by the word ion. For example, Na^+ is the sodium ion. If a metal can form more than one possible positive ion, then we must have a way to designate which ion is actually present when we name the ion. To do this, we write the name of the element and then place in parentheses the charge of the ion in Roman numerals. For example: Fe^{3+} is named the iron(III) ion.

Example 3-2: Write the chemical symbol(s) for the commonly observed monatomic ions of the following elements.

   a. calcium
Calcium has the chemical symbol Ca. It is located in Group 2A. Elements in Group 2A have a charge of 2+. The correct symbol is thus Ca^{2+}.

   b. copper
The chemical symbol for copper is Cu. It is a transition metal. The charges of the transition metals must be memorized. In this case, copper has two commonly formed ions: 1+ and 2+. The symbols are thus Cu^+ and Cu^{2+}.

Try Study Question 5 in Chapter 3 of your textbook now!

d) Recognize that the negative charge on a single-atom or monatomic anion, X^{n-}, is given by $n = 8 -$ Group Number (Section 3.3).

The nonmetals in Groups 5A, 6A, and 7A tend to form anions. Once again, the periodic table will help you determine their charges: Value of Charge = 8 – Group Number.

The name of a monatomic anion is formed by taking the root of the name of the element and adding the suffix –ide. For example, Cl^- is derived from chlorine. The root for chlorine is chlor-. We then add –ide to get chloride.

> <u>Example 3-3:</u> Write the chemical symbol for the selenide ion.
>
> The selenide ion is derived from the element selenium, Se, in Group 6A. The value of the charge will be $8 - 6 = 2$. The charge is thus 2–, and the symbol is Se^{2-}.

e) Write formulas for ionic compounds by combining ions in the proper ratio to give no overall charge (Section 3.4).

In writing formulas, list the positive ion first. A compound is neutral in charge. There must thus be an equal amount of positive charge and negative charge. The number of each type of ion present in the compound is indicated by subscripts. If more than one of a particular polyatomic ion is needed, place the formula for the polyatomic ion in parentheses and outside the parentheses write the subscript indicating how many of this polyatomic ion is present in the formula of the compound For example, in the compound copper(II) nitrate, there are two nitrate ions; the formula is $Cu(NO_3)_2$.

> <u>Example 3-4:</u> Write the correct chemical formula for each of the following compounds.
>
>   a. calcium chloride
> Calcium is located in Group 2A. Its charge is 2+. Chloride is a Group 7A anion. $8 - 1 = 1$ so the charge of chloride is 1–. We have Ca^{2+} and Cl^-. The charges involved are 2+ and 1–. We need to get equal positive and negative charges. The least common multiple of 2 and 1 is 2. Thus, if we have a total positive charge of 2+ and a total negative charge of 2–, then we will have equal positive and negative charges. We need one Ca^{2+} ion to get a 2+ charge. We need two Cl^- ions to get a 2– charge. The formula is thus $CaCl_2$.
>
>   b. aluminum oxide
> We have the following two ions: Al^{3+} and O^{2-}. The least common multiple of 3 and 2 is 6. Thus we want a total positive charge of 6+ and a total negative charge of 6–. We need two Al^{3+} ions and three O^{2-} ions. The formula is Al_2O_3.
>
>   c. ammonium sulfide
> We have the following two ions: NH_4^+ (see next section) and S^{2-}. The least common multiple of 1 and 2 is 2. We want a total positive charge of +2 and a total negative charge of –2. We need two NH_4^+ ions and one S^{2-} ions. The formula is $(NH_4)_2S$.

Try Study Question 21 in Chapter 3 of your textbook now!

- **Name compounds.**

   a) Give the names or formulas of polyatomic ions, knowing their formulas or names, respectively (Table 3.1 and Section 3.3).

You MUST learn the formulas and charges of the polyatomic ions in Table 3.1. If you do not, you will miss many points on tests throughout the course. Make up flash cards. On one side write the name of a polyatomic ion (example: sulfate). On the other, write the formula including the charge (example: SO_4^{2-}). Run through these flash cards many times until you have mastered this information.

A useful hint. If there is a series of polyatomic ions containing a particular element and oxygen, memorize the formula for the one ending in –ate. All of these ions will have the same charge. The formulas for the rest can be derived according to the following table. Add one oxygen for each move up in the table; remove one oxygen for each move down.

General Form	Example	Formula
per...ate	perchlorate	ClO_4^-
...ate	chlorate	ClO_3^-
...ite	chlorite	ClO_2^-
hypo...ite	hypochlorite	ClO^-

Example 3-5: What is the formula for the nitrite ion?

You should memorize that nitrate is NO_3^-. Nitrite will have the same charge, 1–. To get to the ...ite spot in the table, you move down one step. Thus nitrite has one fewer oxygen than does nitrate. Its formula is NO_2^-.

Try Study Question 7 in Chapter 3 of your textbook now!

b) Name ionic compounds and simple binary compounds of the nonmetals (Sections 3.3 and 3.4).

A binary compound is a compound consisting of only two elements. In naming binary compounds, there are three major systems you will need to know. This will allow you to properly name many, but not all, compounds. Which system to use depends upon what the first element is.

i. If the first element is a metal that can only have one possible charge (Group 1A, Group 2A, aluminum, and zinc) then the first word of the name is simply the name of the first element, The second word of the name is the root of the name of the second element's name followed by the suffix –ide. The general form for these compounds is thus **name rootide**. Note that we do not use prefixes or Roman numerals for these compounds.

ii. If the first element is a metal whose ions can have more than one possible charge (transition metals or metals on the right side of the periodic table beyond the transition metals), then we must state which ion is actually present in the compound. We do this by placing the <u>charge</u> of the ion in Roman numerals in parentheses immediately after the element's name. The nonmetal portion is named exactly as in the previous case. The general form for these compounds is thus **name(Roman Numeral) rootide**. Note that we still do not use prefixes.

iii. If the first element is a nonmetal, then we use prefixes to indicate the number of atoms of each element present. The general form for these compounds is thus **prefixname prefixrootide**. The prefixes are given in following table:

Number of Atoms	Prefix	Number of Atoms	Prefix
1	mono	6	hexa
2	di	7	hepta
3	tri	8	octa
4	tetra	9	nona
5	penta	10	deca

A couple of caveats: 1) If there is only one atom of the first element, then the prefix mono- is omitted (PCl_5 = phosphorus pentachloride). 2) If the name of the element begins with a vowel, the final o of the prefix mono- is sometimes omitted (N_2O = dinitrogen monoxide).

Example 3-6: Name the following compounds.

    a. NaCl

Sodium is a group 1A metal. We just list its name. The root for chlorine is chlor-. The name is sodium chloride.

    b. Fe_2O_3

Iron is a transition metal. Its common charges are 2+ and 3+. We must indicate which is present in this compound. The charge of oxygen is 2–. The total negative charge is thus 6–. The positive charge must be 6+. There are two iron ions to divide this up. Each iron ion must be 3+. The root for oxygen is ox-. The name is thus iron(III) oxide.

    c. N_2O_5

Nitrogen is a nonmetal. We will thus use the prefix system. There are two nitrogens and five oxygens. The name is thus dinitrogen pentaoxide.

In naming ionic compounds that contain a polyatomic ion, follow rule i or ii above but simply substitute the name of the polyatomic ion for either the first or second word in the name depending on whether the ion is positive or negative.

Example 3-7: Name the following compounds.

    a. K_2SO_4

Potassium is in Group 1A. It has only one common charge. We do not use prefixes or Roman numerals. The name of this compound is potassium sulfate.

    b. $Cu(NO_3)_2$

Copper is a transition metal. Its common charges are 1+ and 2+. The nitrate ion has a charge of 1–. There are two nitrates giving a total of 2–. The total positive charge must be 2+. There is only one copper ion so it must have a charge of 2+. The name of the compound is copper(II) nitrate.

Try Study Questions 19 and 27 in Chapter 3 of your textbook now!

- **Understand some properties of ionic compounds.**

  Understand the importance of Coulomb's law (Equation 3.1) which describes the electrostatic forces of attraction and repulsion of ions. Coulomb's law states that the force of attraction between oppositely charged species increases with electric charge and with decreasing distance between species (Section 3.3).

Opposite charges attract. Like charges repel. The force of attraction between oppositely charged ions is given by the following form of Coulomb's law:

$$\text{Force of Attraction} = k \frac{\left(n^+ e\right)\left(n^- e\right)}{d^2}$$

The key implications of this equation for you to know are that the force of attraction between ions gets larger 1) as the ion charges increase and 2) as the distance between the ions gets smaller. A more highly charged smaller ion will thus have a greater attraction to a particular oppositely charge ion than will a less highly charged larger ion.

Ionic compounds do not exist as molecules but instead are found in an extended three-dimensional network of millions of ions in a crystal lattice.

Some properties of ionic compounds are that they tend to have high melting temperatures, be hard solids, and be able to be cleaved cleanly along a sharp boundary.

• Calculate and use molar mass.

a) Understand that the molar mass of a compound (often called the molecular weight) is the mass in grams of Avogadro's number of molecules (or formula units) of a compound (Section 3.5). For ionic compounds, which do not consist of individual molecules, the sum of atomic masses is often called the formula mass (or formula weight).

The molar mass of a compound is simply the mass of one mole of the compound.

b) Calculate the molar mass of a compound from its formula and a table of atomic weights (Section 3.5).

One mole of K_2SO_4 contains two moles of K, one mole of S, and four moles of O. We can therefore simply add up the masses of two moles of K, one mole of S, and four moles of O to obtain the molar mass of K_2SO_4. The masses for one mole of each element are obtained from the periodic table.

Example 3-8: Calculate the molar mass of K_2SO_4.

$$2K = 2 \times 39.0983 = \quad 78.1966$$
$$1S = 1 \times 32.066 = \quad 32.066$$
$$4O = 4 \times 15.9994 = \quad \underline{63.9976}$$
$$174.260 \text{ g/mole } K_2SO_4$$

Example 3-9: Calculate the molar mass of $BaCl_2 \cdot 2H_2O$.

$$1Ba = 1 \times 137.327 = \quad 137.327$$
$$2Cl = 2 \times 35.4527 = \quad 70.9054$$
$$4H = 4 \times 1.0079 = \quad 4.0316$$
$$2O = 2 \times 15.9994 = \quad \underline{31.9988}$$
$$244.263 \text{ g/mole } BaCl_2 \cdot 2H_2O$$

Try Study Questions 31 and 33 in Chapter 3 of your textbook now!

c) Calculate the number of moles of a compound that is represented by a given mass, and vice versa (Section 3.5).

To convert between mass in grams and amount of a compound in moles, we use the molar mass of the compound.

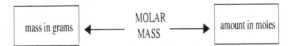

The units for molar mass are grams/mole. We do the calculation exactly the same way as we did when we just had elements. Once again, let your units be your guide. We want to set up the conversion factor so that the units we want to get rid of will cancel, leaving behind the units we desire.

Example 3-10: How many moles of K_2SO_4 are there in 12.4 g of K_2SO_4?

We are given the mass of K_2SO_4 and asked for the number of moles. The molar mass is the connecting link between mass and amount in moles. In Example 3-8, we calculated the molar mass of K_2SO_4, 174.260 g/mole. If we had not done this, then we would need to calculate it in order to solve this problem. Once we have the molar mass, the solution is set up as follows:

$$12.4 \text{ g K}_2\text{SO}_4 \text{ x } \frac{1 \text{ mole K}_2\text{SO}_4}{174.260 \text{ g K}_2\text{SO}_4} = 0.0712 \text{ moles K}_2\text{SO}_4$$

Notice that the conversion factor was set up so that the grams would cancel, leaving moles.

Example 3-11: What is the mass of 1.28 moles of $BaCl_2 \cdot 2H_2O$?

This type of problem is the reverse of the last example. We are now given moles and asked for mass in grams. Once again, the connecting link between mass and moles is the molar mass. In Example 3-9, we calculated the molar mass of $BaCl_2 \cdot 2H_2O$ to be 244.263 g/mole. We can use this to obtain our answer:

$$1.28 \text{ moles BaCl}_2 \cdot 2\text{H}_2\text{O } \text{ x } \frac{244.263 \text{ g BaCl}_2 \cdot 2\text{H}_2\text{O}}{1 \text{ mole BaCl}_2 \cdot 2\text{H}_2\text{O}} = 313 \text{ g BaCl}_2 \cdot 2\text{H}_2\text{O}$$

In this case, we want the moles to cancel so the expression involving moles is placed in the denominator of the conversion factor and that involving grams is placed in the numerator.

Just as there are 6.022×10^{23} atoms in one mole of atoms, there are 6.022×10^{23} molecules in one mole of a molecular compound and 6.022×10^{23} formula units in one mole of an ionic compound.

Example 3-12: How many molecules of water are there in 355 g of water?

We do not have a single conversion factor to go between mass and molecules. But we do know how to go from mass to moles (molar mass) and from moles to molecules (Avogadro's number). First we will calculate the molar mass of H_2O:

$$2H = 2 \times 1.0079 = \quad 2.0158$$
$$1O = 1 \times 15.9994 = \underline{15.9994}$$
$$18.0152 \text{ g/mole}$$

Now we can use this to go from grams to moles:

$$355 \text{ g H}_2\text{O} \ \times \ \frac{1 \text{ mole H}_2\text{O}}{18.0152 \text{ g H}_2\text{O}} \ = \ 19.7 \text{ moles H}_2\text{O}$$

We can then use Avogadro's number to go from moles to molecules.

$$19.7 \text{ moles H}_2\text{O} \ \times \ \frac{6.022 \times 10^{23} \text{ molecules H}_2\text{O}}{1 \text{ mole H}_2\text{O}} \ = \ 1.19 \times 10^{25} \text{ molecules H}_2\text{O}$$

Try Study Questions 35 and 39 in Chapter 3 of your textbook now!

- **Calculate percent composition for a compound and derive formulas from experimental data.**

    a) Express the composition of a compound in terms of percent composition (Section 3.6).

    The percent composition is a mass percent comparing the mass due to one element in a sample of the compound to the total mass of the compound. Just like other percents, it is obtained numerically as the part (one element) divided by the whole (the compound), which is then multiplied by 100.

    More specifically, to calculate the % composition of an element in a compound:

    1. Calculate the molar mass of the compound.

    2. Divide the portion of the molar mass due to the element in question by the total molar mass and multiply by 100.

Example 3-13: What is the percent composition of hydrogen in methane, CH_4?

First we calculate the molar mass of methane.
$$1C = 1 \times 12.011 = 12.011$$
$$4H = 4 \times 1.0079 = \underline{4.0316}$$
$$16.043 \text{ g/mole CH}_4$$

We are asked for the percent composition of hydrogen in this compound. We will take the portion due to hydrogen, 4.0316, divide it by the whole, 16.043 and then multiply by 100.
$$\% \text{ composition of hydrogen} = \frac{4.0316 \text{ g}}{16.043 \text{ g}} \times 100 = 25.130\%$$

Try Study Questions 41 in Chapter 3 of your textbook now!

b) Use percent composition or other experimental data to determine the empirical formula of a compound (Section 3.6).

Standard Empirical Formula Problems
An empirical formula tells us how many atoms or ions of one element are present relative to how many atoms or ions of another element. For example, H_2O tells us that there are two hydrogen atoms to every one oxygen atom. On our macroscopic scale, the empirical formula also tells us how many moles of one element there are to another. H_2O thus also tells us that there are two moles of hydrogen to every one mole of oxygen. It is this latter definition that we shall find more useful in the calculations that follow. To determine the empirical formula of a compound from percent composition data, we follow the following steps:

1. Use the percent composition data, but express them in grams. Thus 87.42% would become 87.42 g.

2. Convert the mass of each element to moles.

3. Find the mole ratio of elements. To do this, divide each of the amounts you calculated in step 2 (the number of moles) by the one that is smallest. In this step, we are assuming a subscript of 1 for the element with the smallest number of moles.

Example 3-14: A particular compound was found to be composed of 30.4% nitrogen and 69.6% oxygen by mass. What is the empirical formula of this compound?

The compound is 30.4% N and 69.6% O. A 100 g sample thus contains 30.4 g N and 69.6 g O. We shall now convert these masses to moles.

$$30.4 \text{ g N} \times \frac{1 \text{ mole N}}{14.01 \text{ g N}} = 2.17 \text{ moles N}$$

$$69.6 \text{ g O} \times \frac{1 \text{ mole O}}{16.00 \text{ g O}} = 4.35 \text{ moles O}$$

The smaller number of moles is 2.17. We divide the number of moles of O by this number.

$$\text{Mole Ratio} = \frac{4.35 \text{ moles O}}{2.17 \text{ moles N}} = 2.00 \frac{\text{moles O}}{\text{mole N}}$$

This tells us that in the compound there are 2 moles of O to every one mole of N. The empirical formula is thus NO_2.

Molecular Formula Calculations

In order to determine a molecular formula, more information needs to be given: the molar mass of the compound. To determine the molecular formula, carry out the following steps:

1. Calculate the molar mass of the empirical formula.

2. Divide the given molar mass of the actual compound by the molar mass of the empirical formula to obtain a factor, X.

3. Multiply each subscript in the empirical formula by X.

Example 3-15: The molar mass of the compound in Example 3-14 is 92.0 g/mole. What is the molecular formula of this compound?

The molar mass of the empirical formula determined in Example 3-14, NO_2, is

$$1N = 1 \times 14.01 = 14.01$$
$$2O = 2 \times 16.00 = \underline{32.00}$$
$$46.01 \text{ g/mole}$$

$$\frac{\text{Molar Mass of Molecular Formula}}{\text{Molar Mass of Empirical Formula}} = \frac{92.0}{46.01} = 2.00$$

Each subscript in the empirical formula must be multiplied by this factor. The molecular formula is thus N_2O_4.

Special Cases

Sometimes, in an empirical formula problem, instead of obtaining a whole number when the mole ratio calculation step is finished, we obtain an integer followed by a decimal that is not within error of a whole number. This simply means that our assumption that the subscript for the element with the smallest number of moles will be a one is not correct; it will be larger than one.

Example 3-16: A compound containing aluminum and oxygen has a percent composition of 52.9% Al and 47.1% O. What is the empirical formula of this compound?

$$52.9 \text{ g Al x } \frac{1 \text{ mole Al}}{26.98 \text{ g Al}} = 1.96 \text{ moles Al}$$

$$47.1 \text{ g O x } \frac{1 \text{ mole O}}{16.00 \text{ g O}} = 2.94 \text{ moles O}$$

$$\text{Mole Ratio} = \frac{2.94 \text{ moles O}}{1.96 \text{ moles Al}} = 1.50 \frac{\text{moles O}}{\text{mole Al}}$$

The way to deal with this is to convert the number to a fraction. 1.50 is equal to 1 1/2, which is equal to 3/2. The mole ratio is thus 3 moles of O to 2 moles of Al. The empirical formula is thus Al_2O_3.

Some more examples of mole ratios like this are 1.25 = 5/4, 1.33 = 4/3, 1.67 = 5/3, 1.75 = 7/4, etc.

A couple of other twists to the empirical formula problem:

      1. Sometimes an actual mass analysis is given instead of % composition data. This is actually easier since the data are already in some unit of mass.

      2. Sometimes, the authors of the problem will provide the % composition data or the mass data for all of the elements except one. The final element's percent composition can be obtained using the fact that all of the % compositions must add up to 100 or that the total mass of all of the elements must add up to the total mass of the sample.

      3. Sometimes the data given will need to be massaged in other ways. The key is make sure that you figure out a way to determine how many moles of each element are present in the sample. Once you know this, you can go ahead with a normal empirical formula calculation.

Try Study Questions 53 in Chapter 3 of your textbook now!

c) Understand how mass spectrometry can be used to find a molar mass (Section 3.6).

Mass spectrometry is one technique that can be used to determine the molar mass of some compounds. In one mass spectrometry technique, the sample is bombarded with electrons that can knock other electrons from molecules of the sample. This forms positively charged ions which themselves can fragment into other smaller ions. In the mass spectrometer the ions then move into an analyzing chamber. A magnetic field causes the path of the ions to bend. By varying the magnetic field, the ions' paths can be bent to aim them at a device that detects when an ion hits it. The actual magnetic field needed to aim a particular ion so that it will hit the detector depends upon the mass to charge ratio of the ion. The mass of the ion can thus be determined.

d) Use experimental data to find the **number of water molecules in a hydrated compound** (Section 3.7).

This type of problem is similar in some ways to the empirical formula problems. The goal here is to find the mole ratio of moles of water per mole of anhydrous compound.

Example 3-17: You want to know how many waters of hydration there are per mole of $BaCl_2$ in hydrated barium chloride. You weigh out 1.270 g of the hydrated $BaCl_2$ and heat it in a crucible to drive off the waters of hydration. After all of the water has been driven off, the mass of the anhydrous barium chloride is 1.083 g. What is the formula for hydrated barium chloride?

We know the mass of $BaCl_2$ in the sample, 1.083 g. The hydrated form contains this same mass of $BaCl_2$ but also water. We can determine the mass of water present in the hydrated sample by subtraction:

Mass of hydrated barium chloride	1.270 g $BaCl_2 \cdot xH_2O$
− Mass of anhydrous barium chloride	− 1.083 g $BaCl_2$
Mass of water	0.187 g H_2O

We can then determine the number of moles of anhydrous compound present and the number of moles of water present. Their ratio will give the number of waters of hydration.

Molar Mass of $BaCl_2$
$1Ba = 1 \times 137.33 = 137.33$
$2Cl = 2 \times 35.4527 = \underline{70.905}$
208.24 g/mole $BaCl_2$

Number of Moles of $BaCl_2$

$$1.083 \text{ g } BaCl_2 \times \frac{1 \text{ mole } BaCl_2}{208.24 \text{ g } BaCl_2} = 5.201 \times 10^{-3} \text{ moles } BaCl_2$$

Molar Mass of H_2O
$2H = 2 \times 1.0079 = 2.0158$
$1O = 1 \times 15.999 = \underline{15.999}$
18.015 g/mole H_2O

Number of Moles of H_2O

$$0.187 \text{ g } H_2O \times \frac{1 \text{ mole } H_2O}{18.015 \text{ g } H_2O} = 0.01038 \text{ moles } H_2O$$

Mole Ratio

$$\frac{0.01038 \text{ moles } H_2O}{5.201 \times 10^{-3} \text{ moles } BaCl_2} = 1.996 \frac{\text{moles } H_2O}{\text{moles } BaCl_2}$$

The formula is $BaCl_2 \cdot 2H_2O$.

Try Study Questions 57 in Chapter 3 of your textbook now!

Other Notes

1. The properties of compounds are very different from the properties of the elements that react together to form the compound.

2. A main group metal atom tends to lose enough electrons so that it will end up with the same number of electrons as the noble gas that comes before it on the periodic table. A nonmetal atom tends to gain enough electrons to end up with the same number of electrons as the noble gas that comes after it on the periodic table.

3. Most metal-containing compounds are ionic. If there is no metal in the formula of a compound, then it is probably a molecular compound. Compounds that contain polyatomic ions are ionic (with the exception of their compounds with hydrogen).

4. At room temperature, ionic compounds are generally solids, whereas molecular compounds can be gases, liquids, or solids.

5. In naming an ionic hydrate, the ionic part is named as a normal ionic compound. To indicate the number of waters of hydration present per formula unit, we use the word hydrate and use a prefix to indicate the number of water molecules of water present per formula unit. The prefixes used are the same ones we learned for naming molecular compounds. $CuSO_4 \cdot 5H_2O$ is thus called copper(II) sulfate pentahydrate.

CHAPTER 4: Chemical Equations and Stoichiometry

Chapter Overview

This chapter is a little different from some of the previous chapters in that not nearly as many new terms are introduced, but there are, nonetheless, many new skills to master. This is a very mathematical chapter, and it is essential that you learn how to balance chemical equations and how to perform stoichiometry calculations. Both of these skills will be important for you to have mastered as you proceed in the course. The chapter begins by discussing chemical equations. The underlying reason we balance chemical equations is the law of conservation of matter. Matter is neither created nor destroyed in chemical reactions; therefore, if an atom appears on the reactant side in a chemical reaction, it must also be present somewhere on the product side. In a chemical reaction, we change around which atoms are connected to which others, but we do not create or destroy atoms. We next turn our attention to a major use of balanced chemical equations: stoichiometry. A chemical equation is like a recipe. It tells us the correct proportion of reactants that will react to form certain amounts of products. If we know how much of one material is present in a chemical reaction, we can use the balanced equation to predict how much of a reactant is needed or how much of a product could be produced. Chemical equations are written in terms of moles of reactants and products. If we are given data in units other than moles, we need to convert those units into moles in order to use the chemical equation to relate the amount of one material to that of another. You will learn how to work stoichiometry problems involving data provided in units of mass and where you are asked about the mass of another reactant or product. You will also learn how to solve problems involving a limiting reactant. Theoretical yield, actual yield, and percent yield will also be discussed. Finally, you will learn some more complex applications of stoichiometry to problems involving mixtures and to problems to determine the empirical formula of a compound based on combustion analysis data.

Key Terms

In this chapter, you will need to learn and be able to use the following terms:

Actual yield: the quantity of product that is actually obtained in the laboratory when a particular chemical reaction is carried out.

Aqueous: in water solution.

Balanced: a chemical equation is said to be balanced when an equal number of atoms of each element is present on both sides of the equation.

Chemical equation: a symbolic representation of a chemical reaction. The reactants are shown on the left side, the products are shown on the right side, and an arrow connects the reactants and products. A chemical equation should be balanced: the same number of atoms of each element should be present on both sides of the equation.

Combustion: a reaction of a material with molecular oxygen to form products in which all of the elements are combined with oxygen.

Hydrocarbon: a compound composed of hydrogen and carbon.

Law of conservation of matter: matter is neither created nor destroyed in a chemical reaction.

Limiting reactant: in a chemical reaction, the reactant present in limited supply that determines the amount of product produced.

Percent yield: a comparison of the actual yield to the theoretical yield; it is calculated by dividing the actual yield by the theoretical yield and multiplying by 100.

Product: an ending material in a chemical reaction.

Reactant: a starting material for a chemical reaction.

Stoichiometric coefficient: a number printed in normal-size writing (not as a subscript or superscript) that comes before a chemical formula in a chemical equation. The coefficients in a chemical equation tell us how many formula units (or moles) of a particular substance are needed for the chemical reaction.

Stoichiometry: the relationship between the quantities of chemical reactants and products in a chemical reaction.

Theoretical yield: the maximum quantity of product that can be obtained from a chemical reaction.

Chapter Goals

By the end of this chapter you should be able to:

- **Balance equations for simple chemical reactions.**

    a) **Understand the information conveyed by a balanced chemical equation (Section 4.1).**

    Chemical reactions are the heart of chemistry. In a chemical reaction, one set of materials is converted into another set of materials. We represent chemical reactions using chemical equations. If chemical symbols are like the letters in our chemical language and if chemical formulas are the words, then chemical reactions are the sentences of this fascinating language. Chemical reactions express an action.

    A balanced chemical equation shows the formulas of the materials we start out with (the reactants) on the left and the formulas of the materials we end up with (the products) on the right. Often, we indicate the physical state of the reactants and products by writing an abbreviation in parentheses following the formula. The abbreviations used are "*s*" for solid, "*l*" for liquid, "*g*" for gas, and "*aq*" for aqueous (in water solution). In between the reactants and products, we draw an arrow. The arrow is read "yield." For example, the following equation indicates that aqueous silver nitrate and aqueous sodium chloride (the reactants) yield aqueous sodium nitrate and solid silver chloride (the products).

    $$AgNO_3 \ (aq) + NaCl \ (aq) \rightarrow NaNO_3 \ (aq) + AgCl \ (s)$$

    In addition to providing qualitative information about what reacts with what, a chemical equation provides quantitative information as well. A chemical equation must be balanced. This means that there must be the same number of atoms of each element on both the reactants side and on the products side. In a chemical reaction, we rearrange which atoms are connected to which other atoms, but we do not change the number of atoms of each element present. This is because of the law of conservation of matter. This law states that matter can neither be created nor destroyed. In other words, atoms are neither created nor destroyed in chemical reactions. Consider the reaction of gaseous hydrogen and gaseous oxygen to form liquid water. The balanced chemical equation is

    $$2H_2 \ (g) + O_2 \ (g) \rightarrow 2H_2O \ (l)$$

This chemical equation tells us that two hydrogen molecules in the gaseous state react with one oxygen molecule in the gaseous state to yield two water molecules in the liquid state. (As you will learn in this chapter, this equation also means that two moles of hydrogen gas react with one mole of oxygen gas to form two moles of liquid water.) This equation is balanced. There are two H_2 molecules on the left. This gives us a total of four H atoms. In the two water molecules on the right, there are also four H atoms. The hydrogens are balanced. There are two oxygen atoms present in the oxygen molecule on the left side. Each water molecule contains one oxygen atom so two water molecules contain two oxygen atoms. The oxygens are balanced. Since all of the elements are balanced, the equation is balanced.

b) Balance simple chemical equations (Section 4.2).

To write a balanced chemical equation:

1. Write the correct formulas for each of the reactants and products. Without the correct formulas, any equation you come up with is meaningless.

2. Once the correct formulas are written, do not change any subscripts. Changing subscripts changes the identities of the materials. Equations are balanced using coefficients. For now, we will simply play around with the coefficients until we come across a set that works. Usually, it is best first to balance elements that are in only one formula on the left and one formula on the right before moving on to elements that appear in more than one species on the same side of the equation.

Example 4-1: Balance the following equation: Na *(s)* + Cl_2 *(g)* → $NaCl$ *(s)*

Starting with the sodium, it looks like we are all right. There is one sodium on the left, and there is one sodium on the right. When we get to the chlorine, however, we see that the equation is not balanced. There are two chlorines on the left but only one on the right. We need to have two chlorines on the right so we will put a 2 in front of the NaCl.

Na *(s)* + Cl_2 *(g)* → $2NaCl$ *(s)*

Note that we must multiply the entire formula for sodium chloride by two; we cannot say $Na2Cl$. We also cannot say $NaCl_2$ because $NaCl_2$ is not the formula for sodium chloride. Now that the chlorines are balanced, we must recheck the sodiums. There is one sodium on the left but two sodiums on the right so balancing the chlorines messed up our balancing of sodium. In order to get the same number on each side of the equation, we need to have two sodiums on the left.

$2Na$ *(s)* + Cl_2 *(g)* → $2NaCl$ *(s)*

You should always double check that everything is balanced. There are two sodium atoms on the left and two sodium atoms on the right. There are two chlorine atoms on the left and two chlorine atoms on the right. The equation is balanced.

Example 4-2: Balance the following equation: S_8 *(s)* + Br_2 *(g)* → SBr_4 *(g)*

We will start with the sulfur. There are eight sulfurs on the left side and only one sulfur on the right side. We need eight on the right side so we will place a coefficient of 8 in front of the formula for sulfur tetrabromide.

$$S_8 \; (s) + Br_2 \; (g) \rightarrow 8SBr_4 \; (g)$$

There are two bromines on the left side and 32 bromines on the right side. We need 32 bromines on the left side. On the left, the bromines come in packets of two so we will need 16 of these to get 32 total bromines on the left.

$$S_8 \; (s) + 16Br_2 \; (g) \rightarrow 8SBr_4 \; (g)$$

There are now eight sulfurs on the left and eight sulfurs on the right. There are 32 bromines on the left and 32 bromines on the right. The equation is balanced.

Example 4-3: Write a balanced chemical equation for the following reaction: aqueous solutions of sodium sulfate and barium chloride react to form aqueous sodium chloride and solid barium sulfate.

The first step is to write the equation showing the correct formulas for the reactants and products. The charge of sodium ions is 1+, the formula for the polyatomic ion sulfate is SO_4^{2-}, the charge of barium ions is 2+, and the charge of chloride ions is 1–. This allows you to write the following preliminary equation (unbalanced):

$$Na_2SO_4 \; (aq) + BaCl_2 \; (aq) \rightarrow NaCl \; (aq) + BaSO_4 \; (s)$$

To write these formulas we made sure that there was an equal number of positive and negative charges for each compound. We do this using subscripts to indicate if more than one of a particular ion is needed. Once we have these correct formulas, however, we will not change the subscripts anymore. To balance the equation we make sure that there is the same number of atoms of each element on both sides of the equation. We use coefficients to balance. We have two sodium ions on the left but only one on the right. We will put a coefficient of 2 in front of the NaCl on the right.

$$Na_2SO_4 \; (aq) + BaCl_2 \; (aq) \rightarrow 2NaCl \; (aq) + BaSO_4 \; (s)$$

In this reaction, the sulfate ion appears to have stayed together as a unit so instead of balancing the sulfur and oxygen separately, we can balance for sulfate as a whole. There is one sulfate on the left and one on the right. The sulfates are balanced. There is one barium on the left and one on the right. Barium is balanced. There are two chlorines on the left and two chlorines on the right. Chlorine is balanced. The entire equation is balanced.

Example 4-4: Write a balanced chemical equation for the following reaction: iron reacts with oxygen to form iron(III) oxide.

The reactants are Fe and O_2. In the product, the iron has a charge of 3+ and the oxide has a charge of 2–; the formula is Fe_2O_3.

$$Fe \; (s) + O_2 \; (g) \rightarrow Fe_2O_3 \; (s)$$

We have one iron on the left and two irons on the right. We shall place a 2 in front of the Fe.

$$2Fe \; (s) + O_2 \; (g) \rightarrow Fe_2O_3 \; (s)$$

We have two oxygens on the left and three on the right. How can we get equal amounts on each side? The least common multiple of 2 and 3 is 6, so we will use three O_2 on the left and two Fe_2O_3 on the right to get six oxygens on each side.

$$2Fe \text{ (s)} + 3O_2 \text{ (g)} \rightarrow 2Fe_2O_3 \text{ (s)}$$

This has messed up our previous balancing of iron. Now there are two on the left and 4 on the right. This is easy to fix. We will change the 2 on the left to a 4.

$$4Fe \text{ (s)} + 3O_2 \text{ (g)} \rightarrow 2Fe_2O_3 \text{ (s)}$$

The equation is balanced.

There is a special type of reaction gone over in this chapter: combustion of a hydrocarbon. A combustion reaction is the reaction of a material with molecular oxygen to form products in which all of the elements are combined with oxygen. A hydrocarbon is a compound composed of the elements hydrogen and carbon. If the combustion of the hydrocarbon takes place in plenty of oxygen, then the products will be carbon dioxide and water.

Example 4-5: Write the balanced chemical equation for the complete combustion of heptane, C_7H_{16}.

The reactants will be C_7H_{16} (a liquid) and oxygen. The products will be carbon dioxide and water (probably a gas under the conditions at which heptane burns).

$$C_7H_{16} \text{ (l)} + O_2 \text{ (g)} \rightarrow CO_2 \text{ (g)} + H_2O \text{ (g)}$$

There are seven carbon atoms on the left but only one on the right. We shall place a 7 in front of the CO_2.

$$C_7H_{16} \text{ (l)} + O_2 \text{ (g)} \rightarrow 7CO_2 \text{(g)} + H_2O \text{ (g)}$$

There are sixteen hydrogens on the left but only two on the right. We will need 8 water molecules on the right to get sixteen hydrogen atoms on the right.

$$C_7H_{16} \text{ (l)} + O_2 \text{ (g)} \rightarrow 7CO_2 \text{(g)} + 8H_2O \text{ (g)}$$

This leaves the oxygens to balance. There are two on the left and a total of 22 on the right (14 from the CO_2 and 8 from the H_2O). In order to get 22 on the left, we will place an 11 in front of the O_2. The equation is now balanced.

$$C_7H_{16} \text{ (l)} + 11O_2 \text{ (g)} \rightarrow 7CO_2 \text{(g)} + 8H_2O \text{ (g)}$$

Example 4-6: Write the balanced chemical equation for the combustion of octane, C_8H_{18}.

The reactants will be C_8H_{18} (l) and O_2 (g). The products will be CO_2 and H_2O.

$$C_8H_{18} \text{ (l)} + O_2 \text{ (g)} \rightarrow CO_2 \text{ (g)} + H_2O \text{ (g)}$$

There are 8 carbons on the left and only one on the right. We will place an 8 in front of the CO_2.

C_8H_{18} *(l)* + O_2 *(g)* → $8CO_2(g)$ + H_2O *(g)*

There are 18 hydrogens on the left and two on the right. We will place a 9 in front of the H_2O.

C_8H_{18} *(l)* + O_2 *(g)* → $8CO_2(g)$ + $9H_2O$ *(g)*

There are two oxygens on the left and a total of 25 on the right. The problem is that we have an even number on the left and an odd number on the right. We cannot multiply the 2 on the left by any whole number and come out with the odd number 25. What we want to do is to multiply the O_2 by 25/2.

C_8H_{18} *(l)* + 25/2 O_2 *(g)* → $8CO_2(g)$ + $9H_2O$ *(g)*

While fractional coefficients will be useful in some chemical equations, we do not usually want to report a balanced equation involving fractional coefficients. To get rid of the fraction, we shall multiply each of the coefficients in the equation by 2.

$2C_8H_{18}$ *(l)* + $25O_2$ *(g)* → $16CO_2(g)$ + $18H_2O$ *(g)*

Let's double-check the balancing. There are 16 carbons on the left and 16 carbons on the right. There are 36 hydrogens on the left and 36 hydrogens on the right. There are 50 oxygens on the left and 32 + 18 = 50 oxygens on the right. The equation is balanced.

Try Study Questions 1, 3, and 5 in Chapter 4 of your textbook now!

- **Perform stoichiometry calculations using balanced chemical equations.**

   a) Understand the principle of the conservation of matter, which forms the basis of chemical stoichiometry (Section 4.3).

   Matter is conserved in chemical reactions. This means that matter is neither created nor destroyed in chemical reactions. In chemical reactions, atoms are conserved. If an atom of a material is present at the beginning of a reaction then it must also be present at the end of the reaction. This is the reason why we balance chemical equations. We may switch around which atoms are connected to which others, but we will not change the number and kinds of atoms. Another consequence of the law of conservation of matter is that mass is also conserved. The total mass of all of the reactants in a chemical reaction must equal the total mass of all of the products in the chemical reaction. With a balanced chemical equation, given the mass of one substance involved in the reaction, we can predict how much of a reactant will be needed or how much of a product will be formed.

   b) Calculate the mass of one reactant or product from the mass of another reactant or product by using the balanced chemical equation (Section 4.3).

   Being able to perform stoichiometry problems is one of the most important goals of this chapter so we shall spend quite a bit of time going over this skill.

   A balanced chemical reaction is like a recipe. Consider the following recipe for a sandwich.

   1 slice turkey + 1 slice cheese + 3 slices tomato + 2 slices bread → 1 sandwich

If we start out with 5 slices of turkey and plenty of the other ingredients, how many sandwiches can we make? That's pretty easy. We have 5 slices of turkey, and we get 1 sandwich for each slice of turkey so we could get 5 sandwiches. Let's do another problem like this. How many sandwiches could we make if we started with 16 slices of bread and plenty of turkey, cheese, and tomato? We have 16 slices of bread and we get 1 sandwich for every 2 slices of bread. You can probably do this problem in your head. The answer is 8 sandwiches. We could show what you did mathematically in your head by the following equation using dimensional analysis:

$$16 \text{ slices bread} \times \frac{1 \text{ sandwich}}{2 \text{ slices bread}} = 8 \text{ sandwiches}$$

We wanted to cancel the slices of bread so we put the slices of bread on the bottom in our conversion factor. We were asked for sandwiches so we put sandwiches on top. Similarly, how many slices of tomato we would need if we had 16 slices of bread? This one is tougher to do in your head, but follows the same pattern using dimensional analysis:

$$16 \text{ slices bread} \times \frac{3 \text{ slices tomato}}{2 \text{ slices bread}} = 24 \text{ slices tomato}$$

The reason that all of these problems were not too difficult was that my recipe was given in terms of counting units, slices and sandwiches. A balanced chemical reaction is a recipe. The counting unit in chemistry on the macroscopic scale is the mole. We can do similar calculations using chemical equations and moles. Consider the chemical reaction:

$$2Na \text{ (s)} + Cl_2 \text{ (g)} \rightarrow 2NaCl \text{ (s)}$$

The chemical equation is written in terms of moles. According to this equation, 2 moles of sodium will react with 1 mole of chlorine gas to form two moles of sodium chloride. If we start with 4 moles of sodium metal and plenty of chlorine gas, how many moles of sodium chloride could we form? Everything is in terms of our counting unit. We can just use the mole ratio from the balanced equation in a dimensional analysis problem like we did above (although you can probably do this one in your head since our recipe tells us that 2 moles of sodium lead to 2 moles of sodium chloride):

$$4 \text{ moles Na} \times \frac{2 \text{ moles NaCl}}{2 \text{ moles Na}} = 4 \text{ moles NaCl}$$

How many moles of chlorine gas do we need to react with 4 moles of Na in this reaction?

$$4 \text{ moles Na} \times \frac{1 \text{ moles Cl}_2}{2 \text{ moles Na}} = 2 \text{ moles Cl}_2$$

The connecting link between moles of one material and moles of another material is the mole ratio of the stoichiometric coefficients in the balanced equation.

As you will see, in most stoichiometry problems, this key step will be the same. All that will differ will be the way that we get to moles of A from some given information and the way that we move away from moles of B to some other unit.

Let's return to our sandwich example for a tougher type of problem. What if we asked you how many sandwiches we could get from 520 g of bread? What would we do? Our recipe is not in terms of grams. It is in terms of slices. In order to use our recipe, we will need to figure out how many slices of bread there are in 520 g of bread. We could figure this out if we knew the mass for one slice of bread. For this particular loaf of bread, the slice mass is 20. g/slice. Using this, we can figure out how many slices of bread there are:

$$520 \text{ g bread} \times \frac{1 \text{ slice bread}}{20. \text{ g bread}} = 26 \text{ slices bread}$$

Now that we know how many slices of bread there are, we can figure out how many sandwiches we can make:

$$26 \text{ slices bread} \times \frac{1 \text{ sandwich}}{2 \text{ slices bread}} = 13 \text{ sandwiches}$$

The key to solving this problem was that we took the given information (the mass of bread) and converted it into our counting unit (slices). Once we had our units in terms of our counting unit, we could use our recipe to relate one item in the recipe to another.

In stoichiometry, it is very similar. In the lab, there is no device that measures out in units of moles. We can measure the mass of something, for example. What we need to do is to convert from the units we are given (grams for example) to our counting unit (moles). The only way we know so far to get to moles is the molar mass (we will learn others as we proceed through the course). Once we have expressed our given information in our counting unit, then we can use our recipe (the balanced equation) to relate one substance in the equation to any other; we will simply use the mole ratio that the balanced equation indicates. If we are asked for an answer in some unit other than moles, then we will need to convert from moles of the asked for substance to whatever unit we are asked for. Schematically, we can show the problem as follows:

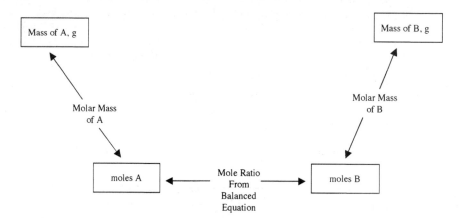

In working a stoichiometry problem, it is strongly suggested that you set up a table to help organize your thinking like the one below:

	aA	+	bB	→	cC	+	dD
measured							
conversion factor							
moles							

In this table, measured is the quantity that can be measured in the lab. For now, it will be the mass of some reactant. The "conversion factor" is the way that you will convert from the measured units to moles. For now, this conversion factor will be the molar mass. Let's work some examples.

Example 4-7: Potassium chlorate decomposes to potassium chloride and oxygen gas. What mass of oxygen gas could be obtained from the decomposition of 3.00 g of potassium chlorate?

To solve a stoichiometry problem, we must start with the balanced chemical equation. If we do not know the proper recipe, we cannot relate one material to another. The balanced chemical equation for this reaction is

$$2KClO_3 \ (s) \rightarrow 2KCl \ (s) + 3O_2 \ (g)$$

The next step is to set up our table with the information that we know. In this case we know that we start out with 3.00 g of potassium chlorate. We are asked for the mass of oxygen that could be formed. We do not care about the KCl in this case.

	$2KClO_3 \ (s)$	$\rightarrow$	$2KCl \ (s)$	+	$3O_2 \ (g)$
measured	3.00 g				x g
conversion factor					
moles					

To go from grams of potassium chlorate to moles, we will need the molar mass of potassium chlorate, and eventually to go from moles of oxygen to grams of oxygen, we will need the molar mass of oxygen. We can get the molar mass of $KClO_3$ (122.55 g/moles) by adding together the molar mass of K, the molar mass of Cl, and three times the molar mass of O. To get the molar mass of O_2 (32.00 g/mole), we use two times the molar mass of O. (Of course, you would not rewrite the entire table as we have done below but would simply fill in the appropriate information.)

	$2KClO_3 \ (s)$	$\rightarrow$	$2KCl \ (s)$	+	$3O_2 \ (g)$
measured	3.00 g				x g
conversion factor	122.55 g/mole				32.00 g/mole
moles					

We can now calculate the moles of potassium chlorate by using the mass of $KClO_3$ and its molar mass:

$$3.00 \text{ g } KClO_3 \times \frac{1 \text{ mole } KClO_3}{122.55 \text{ g } KClO_3} = 0.0245 \text{ moles } KClO_3$$

	$2KClO_3 \ (s)$	$\rightarrow$	$2KCl \ (s)$	+	$3O_2 \ (g)$
measured	3.00 g				x g
conversion factor	122.55 g/mole				32.00 g/mole
moles	0.0245 moles				

We can calculate the number of moles of oxygen produced by using the mole ratio from our balanced equation:

$$0.0245 \text{ moles } KClO_3 \times \frac{3 \text{ moles } O_2}{2 \text{ moles } KClO_3} = 0.0367 \text{ moles } O_2$$

	2KClO$_3$ *(s)*	→	2KCl *(s)*	+	3O$_2$ *(g)*
measured	3.00 g				x g
conversion factor	122.55 g/mole				32.00 g/mole
moles	0.0245 moles				0.0367 moles

Finally, we can use the molar mass of oxygen to calculate the mass of oxygen formed.

$$0.0367 \text{ moles O}_2 \times \frac{32.00 \text{ g O}_2}{1 \text{ mole O}_2} = 1.18 \text{ g O}_2$$

	2KClO$_3$ *(s)*	→	2KCl *(s)*	+	3O$_2$ *(g)*
measured	3.00 g				1.18 g
conversion factor	122.55 g/mole				32.00 g/mole
moles	0.0245 moles				0.0367 moles

You can see that the table helps us organize our information, but that the steps are exactly those we talked about before. Go back to the diagram with boxes and arrows that traces out the path for a mass-mass stoichiometry problem (printed before this example) and trace through the steps that we followed in this example. We started with a balanced chemical equation. We then took our given mass of KClO$_3$ and converted it to moles of KClO$_3$ using the molar mass of KClO$_3$ as the conversion factor. We then went from moles of KClO$_3$ to moles of O$_2$ by using the mole ratio from our balanced equation. Finally we went from moles of O$_2$ to the mass of O$_2$ by using the molar mass of O$_2$.

Example 4-8: Solid sodium reacts with chlorine gas to form solid sodium chloride. What is the minimum mass of chlorine required to react with 2.50 g of sodium in this reaction?

This is another stoichiometry problem. We are given the mass of one reactant and asked for the mass of another reactant. The first step in any stoichiometry problem is to write a balanced chemical equation. We have already looked at this reaction; its balanced equation is
$$2\text{Na } (s) + \text{Cl}_2 (g) \rightarrow 2\text{NaCl } (s)$$

We are given 2.50 g of sodium and are looking for the mass of chlorine. We will set these up in our table.

	2Na *(s)*	+	Cl$_2$ *(g)*	→	2NaCl *(s)*
measured	2.50 g		x g		
conversion factor					
moles					

We need to get to moles because our balanced chemical equation is in terms of moles. To do this, we will need the molar mass of sodium. Eventually, we will need the molar mass of chlorine gas. These are filled in to the table below:

	2Na *(s)*	+	Cl$_2$ *(g)*	→	2NaCl *(s)*
measured	2.50 g		x g		
conversion factor	22.99 g/mole		70.90 g/mole		
moles					

We can now convert from grams of sodium to moles of sodium using the molar mass of sodium:

$$2.50 \text{ g Na} \times \frac{1 \text{ mole Na}}{22.99 \text{ g Na}} = 0.109 \text{ moles Na}$$

	2Na *(s)*	+	Cl$_2$ *(g)*	→	2NaCl *(s)*
measured	2.50 g		x g		
conversion factor	22.99 g/mole		70.90 g/mole		
moles	0.109 moles				

We then use the mole ratio from the balanced equation to go from moles of sodium to moles of chlorine:

$$0.109 \text{ moles Na} \times \frac{1 \text{ mole Cl}_2}{2 \text{ moles Na}} = 0.0544 \text{ moles Cl}_2$$

	2Na *(s)*	+	Cl$_2$ *(g)*	→	2NaCl *(s)*
measured	2.50 g		x g		
conversion factor	22.99 g/mole		70.90 g/mole		
moles	0.109 moles		0.0544 moles		

Finally, we can go from moles of Cl$_2$ to grams of Cl$_2$ by using the molar mass of chlorine.

$$0.0544 \text{ moles Cl}_2 \times \frac{70.90 \text{ g Cl}_2}{1 \text{ mole Cl}_2} = 3.85 \text{ g Cl}_2$$

	2Na *(s)*	+	Cl$_2$ *(g)*	→	2NaCl *(s)*
measured	2.50 g		3.85 g		
conversion factor	22.99 g/mole		70.90 g/mole		
moles	0.109 moles		0.0544 moles		

Try Study Questions 9 and 11 in Chapter 4 of your textbook now!

c) Use amounts tables to organize stoichiometric information.

An amounts table is another tool that can help you see the changes that occur in a chemical reaction. An amounts table is different from the tables shown so far in this study guide for solving stoichiometry problems. At the top of an amounts table, write the chemical equation. Under this, there are three rows: initial amount, change, and amount after reaction. The values entered into the table are always moles.

<u>Example 4-9</u>: Construct an amount table corresponding to the situation in Example 4-7.

The general form of the amounts table is the following:

	2KClO$_3$ *(s)*	→	2KCl *(s)*	+	3O$_2$ *(g)*
Initial amount					
Change					
Amount after reaction					

In Example 4-7, we calculated that the initial amount of $KClO_3$ present was 0.0245 moles. At the beginning of the reaction, there was no potassium chloride or oxygen present.

	2KClO$_3$ *(s)*	→	2KCl *(s)*	+	3O$_2$ *(g)*
Initial amount	0.0245 moles		0 moles		0 moles
Change					
Amount after reaction					

All of the $KClO_3$ reacts, so the change that occurs for $KClO_3$ is that its amount goes down by 0.0245 moles. We use a minus sign to indicate that the amount is going down.

	2KClO$_3$ *(s)*	→	2KCl *(s)*	+	3O$_2$ *(g)*
Initial amount	0.0245 moles		0 moles		0 moles
Change	−0.0245 moles				
Amount after reaction					

Based on this change, we can calculate the change that occurs for the KCl and the O_2 by using the appropriate stoichiometric factors. The result for KCl is

$$0.0245 \text{ moles } KClO_3 \times \frac{2 \text{ moles KCl}}{2 \text{ moles KClO}_3} = 0.0245 \text{ moles KCl}$$

We already did the calculation for O_2 in Example 4.7. The answer was 0.0367 moles. Both the KCl and the O_2 are formed in this reaction, so their changes are labeled with a + sign.

	2KClO$_3$ *(s)*	→	2KCl *(s)*	+	3O$_2$ *(g)*
Initial amount	0.0245 moles		0 moles		0 moles
Change	−0.0245 moles		+0.0245 moles		+0.0367 moles
Amount after reaction					

To calculate the amounts after reaction, we do the arithmetic indicated in each column. For example, for the $KClO_3$, 0.0245 moles − 0.0245 moles = 0 moles. The final table is

	2KClO$_3$ *(s)*	→	2KCl *(s)*	+	3O$_2$ *(g)*
Initial amount	0.0245 moles		0 moles		0 moles
Change	−0.0245 moles		+0.0245 moles		+0.0367 moles
Amount after reaction	0 moles		0.0245 moles		0.0367 moles

Try Study Question 15 in Chapter 4 of your textbook now!

- **Understand the impact of a limiting reactant on a chemical reaction.**

  ### a) Determine which of two reactants is the limiting reactant (Section 4.4).

  Let's do another sandwich example. Given the following recipe,

  1 slice turkey + 1 slice cheese + 2 slices bread → 1 sandwich

how many sandwiches could we make if we started out with 6 slices of turkey, 8 slices of cheese, and 10 slices of bread? With 6 slices of turkey, we could make 6 sandwiches, with 8 slices of cheese we could make 8 sandwiches, and with 10 slices of bread we could make 5 sandwiches. How many sandwiches can we actually make? The answer is 5 because once we have made five sandwiches, we have run out of bread. We can't make any more sandwiches because we don't have any more bread. So we will end up with 5 sandwiches and leftover turkey (1 slice leftover) and leftover cheese (3 slices leftover). Using chemical terminology, we would say that the bread was our limiting reactant because it is the material we ran out of. The turkey and cheese are present in excess amounts. Notice that in this case, we couldn't just look at how many slices we had at the beginning in order to figure out what the limiting reactant was. We had more bread than anything else. BUT it took 2 slices of bread to make 1 sandwich. We could only make 5 sandwiches from 10 slices of bread. We had to take into account the recipe.

The same type of situation arises in chemistry. We do not always mix chemicals in exactly the right amounts. Often, we will have one material that we will run out of and other(s) that we will have in excess amount. In order to solve a limiting reactant problem, we will need first to get to our counting unit: moles. Once we do that, however, we will need to take into account our recipe. Just like we did in the sandwich example, we will figure out how many moles of product we can get from each of the reactants. One of the reactants will produce less product than the other(s). The one that produces the least amount of product is the limiting reactant because once we have formed that much product, we have run out of that reactant. We can make no more product without that reactant so that is the most amount of product we can get. We will have leftover amounts of the other reactant(s).

Let's work an example to show how to determine the limiting reactant.

Example 4-10: Suppose we mix together 10.0 g of iron and 8.00 g of oxygen and allow them to react to form iron(III) oxide. Which is the limiting reactant?

The first step in a stoichiometry problem is to write the balanced chemical equation.
$$4Fe\ (s) + 3O_2\ (g) \rightarrow 2Fe_2O_3\ (s)$$

We are given 10.0 g of Fe and 8.00 g of oxygen. In order to determine which is the limiting reactant, we shall figure out which will give the least amount of product. The one that gives the least amount of product is the limiting reactant because once this amount of product is formed, we will have run out of the limiting reactant. Let's start to fill in our stoichiometry table:

	$4Fe\ (s)$	+	$3O_2\ (g)$	$\rightarrow$	$2Fe_2O_3\ (s)$
measured	10.0 g		8.00 g		
conversion factor	55.85 g/mole		32.00 g/mole		
moles					

Let us first figure out how many moles of Fe_2O_3 we could get from 10.0 g of iron.
$$10.0\ g\ Fe\ x\ \frac{1\ mole\ Fe}{55.85\ g\ Fe} = 0.179\ moles\ Fe$$

$$0.179\ moles\ Fe\ x\ \frac{2\ moles\ Fe_2O_3}{4\ moles\ Fe} = 0.0895\ moles\ Fe_2O_3$$

	4Fe *(s)*	+	3O$_2$ *(g)*	→	2Fe$_2$O$_3$ *(s)*
measured	10.0 g		8.00 g		
conversion factor	55.85 g/mole		32.00 g/mole		
moles	0.179 moles				0.0895 moles (from Fe)

So if iron is our limiting reactant, then we could get 0.0895 moles of Fe$_2$O$_3$. What if O$_2$ is our limiting reactant? How many moles of product could we get from 8.00 g of O$_2$?

$$8.00 \text{ g O}_2 \text{ x } \frac{1 \text{ mole O}_2}{32.00 \text{ g O}_2} = 0.250 \text{ moles O}_2$$

$$0.250 \text{ moles O}_2 \text{ x } \frac{2 \text{ moles Fe}_2\text{O}_3}{3 \text{ moles O}_2} = 0.167 \text{ moles Fe}_2\text{O}_3$$

	4Fe *(s)*	+	3O$_2$ *(g)*	→	2Fe$_2$O$_3$ *(s)*
measured	10.0 g		8.00 g		
conversion factor	55.85 g/mole		32.00 g/mole		
moles	0.179 moles		0.250 moles		0.0895 moles (from Fe) OR 0.167 moles (from O$_2$)

So if iron is the limiting reactant, then we could get 0.0895 moles of iron(III) oxide and if oxygen is the limiting reactant then we could get 0.167 moles of iron(III) oxide. Which actually is the limiting reactant? The one that gives us the least amount of product is always the limiting reactant. In this case, the iron is the limiting reactant because it gives us less product. Once we have formed 0.0895 moles of iron(III) oxide, we have run out of iron. We can get no more product. We will have leftover oxygen.

	4Fe *(s)*	+	3O$_2$ *(g)*	→	2Fe$_2$O$_3$ *(s)*
measured	10.0 g		8.00 g		
conversion factor	55.85 g/mole		32.00 g/mole		
moles	0.179 moles LIMITING REACTANT		0.250 moles		0.0895 moles ~~(from Fe)~~ ~~OR~~ ~~0.167 moles~~ ~~(from O$_2$)~~

b) Determine the yield of a product based on the limiting reactant.

To determine the theoretical yield of product, we just need to finish the stoichiometry problem. We take the number of moles of product predicted from the limiting reactant and convert from moles of product to mass of product. Let's first just finish the last example before we work another example from start to finish.

Example 4-11: What mass of iron(III) oxide could we have obtained in the last example?

We are looking for the mass of iron(III) oxide. To determine this, we will need the molar mass of iron(III) oxide (159.7 g/mole).

	4Fe *(s)*	+	3O$_2$ *(g)*	→	2Fe$_2$O$_3$ *(s)*
measured	10.0 g		8.00 g		x g
conversion factor	55.85 g/mole		32.00 g/mole		159.7 g/mole
moles	0.179 moles LIMITING REACTANT		0.250 moles		0.0895 moles (from Fe) OR 0.167 moles (from O$_2$)

We just need to do the moles to mass calculation:

$$0.0895 \text{ moles Fe}_2\text{O}_3 \text{ x } \frac{159.7 \text{ g Fe}_2\text{O}_3}{1 \text{ mole Fe}_2\text{O}_3} = 14.3 \text{ g Fe}_2\text{O}_3$$

The final table would look like the following:

	4Fe *(s)*	+	3O$_2$ *(g)*	→	2Fe$_2$O$_3$ *(s)*
measured	10.0 g		8.00 g		14.3 g
conversion factor	55.85 g/mole		32.00 g/mole		159.7 g/mole
moles	0.179 moles LIMITING REACTANT		0.250 moles		0.0895 moles (from Fe) OR 0.167 moles (from O$_2$)

Example 4-12: A solution containing 1.34 g of calcium nitrate is mixed with another solution containing 2.42 g of sodium phosphate. The balanced equation for the reaction that occurs is
$$3Ca(NO_3)_2 \text{ *(aq)* } + 2Na_3PO_4 \text{ *(aq)* } \rightarrow Ca_3(PO_4)_2 \text{ *(s)* } + 6NaNO_3 \text{ *(aq)*}$$
What mass of calcium phosphate could be obtained from this reaction?

We already have a balanced chemical equation. Let's start filling in the table with the given and asked for information. Since all of the given was in terms of mass and since a mass is asked for, the conversion factors will all be molar masses.

	3Ca(NO$_3$)$_2$	+	2Na$_3$PO$_4$	→	Ca$_3$(PO$_4$)$_2$	+	6NaNO$_3$
measured	1.34 g		2.42 g		x g		
conversion factor	164.1 g/mole		163.9 g/mole		310.2 g/mole		
moles							

First, we shall figure out how many moles of calcium phosphate we can get if the calcium nitrate is the limiting reactant.

$$1.34 \text{ g Ca}(NO_3)_2 \text{ x } \frac{1 \text{ mole Ca}(NO_3)_2}{164.1 \text{ g Ca}(NO_3)_2} = 8.17 \text{ x } 10^{-3} \text{ moles Ca}(NO_3)_2$$

$$8.17 \times 10^{-3} \text{ moles Ca(NO}_3)_2 \times \frac{1 \text{ mole Ca}_3(PO_4)_2}{3 \text{ moles Ca(NO}_3)_2} = 2.72 \times 10^{-3} \text{ moles Ca}_3(PO_4)_2$$

	$3Ca(NO_3)_2$	+	$2Na_3PO_4$	$\rightarrow$	$Ca_3(PO_4)_2$	+	$6NaNO_3$
measured	1.34 g		2.42 g		x g		
conversion factor	164.1 g/mole		163.9 g/mole		310.2 g/mole		
moles	8.17×10^{-3} moles				2.72×10^{-3} moles (from Ca(NO$_3$)$_2$)		

Next, we shall figure out how many moles of calcium phosphate we can get if the sodium phosphate is the limiting reactant.

$$2.42 \text{ g Na}_3PO_4 \times \frac{1 \text{ mole Na}_3PO_4}{163.9 \text{ g Na}_3PO_4} = 1.48 \times 10^{-2} \text{ moles Na}_3PO_4$$

$$1.48 \times 10^{-2} \text{ moles Na}_3PO_4 \times \frac{1 \text{ mole Ca}_3(PO_4)_2}{2 \text{ moles Na}_3PO_4} = 7.38 \times 10^{-3} \text{ moles Ca}_3(PO_4)_2$$

	$3Ca(NO_3)_2$	+	$2Na_3PO_4$	$\rightarrow$	$Ca_3(PO_4)_2$	+	$6NaNO_3$
measured	1.34 g		2.42 g		x g		
conversion factor	164.1 g/mole		163.9 g/mole		310.2 g/mole		
moles	8.17×10^{-3} moles		1.48×10^{-2} moles		2.72×10^{-3} moles (from Ca(NO$_3$)$_2$) OR 7.38×10^{-3} moles from Na$_3$PO$_4$		

Calcium nitrate is the limiting reactant because it produces the smaller amount of product. The maximum amount of calcium phosphate we could produce is 2.72×10^{-3} moles. We then just need to convert from moles of calcium phosphate to grams of calcium phosphate.

$$2.72 \times 10^{-3} \text{ moles Ca}_3(PO_4)_2 \times \frac{310.2 \text{ g Ca}_3(PO_4)_2}{1 \text{ mole Ca}_3(PO_4)_2} = 0.844 \text{ g Ca}_3(PO_4)_2$$

	$3Ca(NO_3)_2$	+	$2Na_3PO_4$	$\rightarrow$	$Ca_3(PO_4)_2$	+	$6NaNO_3$
measured	1.34 g		2.42 g		0.844 g		
conversion factor	164.1 g/mole		163.9 g/mole		310.2 g/mole		
moles	8.17×10^{-3} moles LIMITING REACTANT		1.48×10^{-2} moles		2.72×10^{-3} moles ~~(from Ca(NO$_3$)$_2$)~~ ~~OR~~ ~~7.38×10^{-3} moles from Na$_3$PO$_4$~~		

Try Study Questions 19 in Chapter 4 of your textbook now!

We can also determine how much of the reactant in excess amount should remain at the end of the reaction.

Example 4-13: In the reaction in Example 4-12, how much sodium phosphate will remain at the end?

In Example 4-12, we determined that the calcium nitrate was the limiting reactant. There were 8.17×10^{-3} moles of calcium nitrate present in the reaction mixture. We can figure out how much sodium phosphate we need to react with this amount of calcium nitrate.

$$8.17 \times 10^{-3} \text{ moles Ca}(NO_3)_2 \times \frac{2 \text{ moles Na}_3PO_4}{3 \text{ moles Ca}(NO_3)_2} = 5.44 \times 10^{-3} \text{ moles Na}_3PO_4$$

$$5.44 \times 10^{-3} \text{ moles Na}_3PO_4 \times \frac{163.9 \text{ g Na}_3PO_4}{1 \text{ mole Na}_3PO_4} = 0.892 \text{ g Na}_3PO_4$$

Therefore, we only needed 0.892 g of Na_3PO_4. We started out with 2.42 g of Na_3PO_4. The quantity of Na_3PO_4 left over at the end of the reaction is 2.42 g – 0.892 g = 1.53 g of Na_3PO_4.

If we wanted to calculate the mass of sodium nitrate that could be formed, we could. We would start with the number of moles of calcium nitrate (our limiting reactant), figure out the number of moles of sodium nitrate that could be formed from this amount of calcium nitrate, and then calculate the mass of sodium nitrate this corresponds to by multiplying the number of moles of sodium nitrate by the molar mass of sodium nitrate.

Try Study Questions 25 in Chapter 4 of your textbook now!

• Calculate the theoretical and percent yields of a chemical reaction.

a) Explain the differences among actual yield, theoretical yield, and percent yield, and calculate percent yield (Section 4.5).

The theoretical yield is the maximum quantity of product that can be obtained from a chemical reaction. It is the mass of product that we calculate as the result of the stoichiometry problems we have been doing. Note that if the stoichiometry problem is a limiting reactant problem, the theoretical yield is the mass of product calculated based on the limiting reactant.

The actual yield for a reaction is the amount of product we actually obtain in the laboratory when someone actually runs the reaction. We cannot calculate what the actual yield for a reaction will be. It is something that must be determined in the laboratory. It is influenced by the extent to which the reaction actually proceeds as well as by experimental sources of error that lead to loss of product. The actual yield will never be greater than the theoretical yield.

The percent yield is a way to compare the actual and theoretical yields. The equation for percent yield is

$$\text{Percent Yield} = \frac{\text{Actual Yield}}{\text{Theoretical Yield}} \times 100$$

Example 4-14: In Example 4-12, we calculated that 0.844 g of calcium phosphate could be obtained. Let us suppose that someone in the lab actually carried out the reaction and obtained 0.802 g of calcium phosphate. What was the percent yield?

The actual yield is what was actually obtained in the laboratory: 0.802 g.
The theoretical yield is what our stoichiometry problem predicted: 0.844 g.

$$\text{Percent Yield} = \frac{\text{Actual Yield}}{\text{Theoretical Yield}} \times 100 = \frac{0.802 \text{ g}}{0.844 \text{ g}} \times 100 = 95.0\%$$

Try Study Questions 29 in Chapter 4 of your textbook now!

- ## Use stoichiometry to analyze a mixture of compounds or to determine the formula of a compound.

a) Use stoichiometry principles to analyze a mixture (Section 4.6).

Typically this type of problem involves a statement about procedures that were carried out to analyze a mixture. This type of problem can appear to be difficult 1) because each problem will be somewhat different due to different mixtures and procedures being used and 2) because there are typically a lot of words in the statement of the problem, and you will need to sort through the words to figure out exactly what was done. It is at heart a stoichiometry problem, and you need to follow each component of interest through its various transformations to get to the desired answer.

Example 4-15: You are given a mixture that contains sand, sodium chloride, and copper(II) carbonate. The initial mass of the mixture is 5.24 g. To this mixture you add some hydrochloric acid. The sodium chloride dissolves in the hydrochloric acid, and the copper(II) carbonate reacts with the hydrochloric acid according to the following reaction:

$$CuCO_3 \text{ (s)} + 2HCl \text{ (aq)} \rightarrow CuCl_2 \text{ (aq)} + H_2O \text{ (l)} + CO_2 \text{ (g)}$$

The aqueous solution is decanted from the sand that remains. The sand is rinsed multiple times, and these rinses are added to the aqueous solution. The sand is then dried. The mass of the dry sand is 1.24 g. The solution is neutralized and then aqueous potassium sulfide is added. The copper(II) chloride in the solution reacts with the potassium sulfide according to the following reaction:

$$CuCl_2 \text{ (aq)} + K_2S \text{ (aq)} \rightarrow CuS \text{ (s)} + 2KCl \text{ (aq)}$$

The reaction mixture is filtered. The mass of the solid copper(II) sulfide is 2.38 g. Calculate the percentage of the original mixture that was sand, that was sodium chloride, and that was copper(II) carbonate.

Clearly, this is a long problem. We need to sort through what takes place in the procedure. We start out with sand, sodium chloride, and copper(II) carbonate all together. When the hydrochloric acid is added, the sand stays solid, and the sodium chloride and copper(II) carbonate go into solution. The solid that is left is just sand. Its mass is 1.24 g. We can figure out the percent of the mixture that was sand:

$$\% \text{ sand} = \frac{1.24 \text{ g sand}}{5.24 \text{ g mixture}} \times 100 = 23.7\%$$

Let's move on to the solution. It contains sodium chloride. It also contains the copper(II) chloride that we got from the original copper(II) carbonate. We add to this the potassium sulfide. The sodium chloride does not react, but the copper(II) chloride does. The copper ions end up in copper(II) sulfide. We know the mass of the copper(II) sulfide.

From the mass of the copper(II) sulfide, we can calculate the number of moles of copper(II) sulfide present. Using the balanced equation for the reaction that formed the copper(II) sulfide, we can figure out how many moles of copper(II) chloride we had. From the number of moles of copper(II) chloride, we can figure out the number of moles of copper(II) carbonate, and from this we can figure out the mass of copper(II) carbonate. Let's now do this.

	$CuCl_2$ *(aq)*	+	K_2S *(aq)*	→	CuS *(s)*	+	$2KCl$ *(aq)*
measured					2.38 g		
conversion factor					95.62 g/mole		
moles							

$$2.38 \text{ g CuS} \times \frac{1 \text{ mole CuS}}{95.62 \text{ g CuS}} = 0.0249 \text{ moles CuS}$$

$$0.0249 \text{ moles CuS} \times \frac{1 \text{ mole CuCl}_2}{1 \text{ mole CuS}} = 0.0249 \text{ moles CuCl}_2$$

	$CuCl_2$ *(aq)*	+	K_2S *(aq)*	→	CuS *(s)*	+	$2KCl$ *(aq)*
measured					2.38 g		
conversion factor					95.62 g/mole		
moles	0.0249 moles				0.0249 moles		

Now we know how many moles of $CuCl_2$ we had in solution. We can work another stoichiometry problem to figure out the mass of copper(II) carbonate we started with.

	$CuCO_3$ *(s)*	+	$2HCl$ *(aq)*	→	$CuCl_2$ *(aq)*	+	H_2O *(l)* + CO_2 *(g)*
measured	x g						
conversion factor							
moles					0.0249 moles		

$$0.0249 \text{ moles CuCl}_2 \times \frac{1 \text{ mole CuCO}_3}{1 \text{ mole CuCl}_2} = 0.0249 \text{ moles CuCO}_3$$

$$0.0249 \text{ moles CuCO}_3 \times \frac{123.6 \text{ g CuCO}_3}{1 \text{ mole CuCO}_3} = 3.08 \text{ g CuCO}_3$$

	$CuCO_3$ *(s)*	+	$2HCl$ *(aq)*	→	$CuCl_2$ *(aq)*	+	H_2O *(l)* + CO_2 *(g)*
measured	3.08 g						
conversion factor	123.6 g/mole						
moles	0.0249 moles				0.0249 moles		

We therefore had 3.08 g of copper(II) carbonate in the original mixture. The percent copper(II) carbonate was

$$\% \text{ CuCO}_3 = \frac{3.08 \text{ g CuCO}_3}{5.24 \text{ g mixture}} \times 100 = 58.8\%$$

There was never a separate analysis done for the sodium chloride, but all of the percents must add to give 100%, so we can figure out the percent of sodium chloride in the mixture.
$$100\% - 23.7\% \text{ sand} - 58.8\% \text{ CuCO}_3 = 17.5 \% \text{ NaCl}$$

The mixture thus consisted of 23.7% sand, 58.8% $CuCO_3$, and 17.5% NaCl.

Once again, to solve this type of problem, spend some time reading the problem and sorting out what was done in each step of the analysis that was performed in the lab. Look for steps that deal with just one component of the mixture. Try to write balanced chemical equations for the steps in the analysis because it will be through the balanced chemical equations that you will be able to use the principles of stoichiometry we have learned.

Try Study Questions 35 in Chapter 4 of your textbook now!

b) Find the empirical formula of an unknown compound using chemical stoichiometry (Section 4.6).

This type of problem is a review of the empirical formula problems we covered in the last chapter, but the data will be given to you in a more realistic (and more complicated) way. Often, a combustion analysis will be performed, and the results reported will be the masses of the products of the combustion. From this data, you must then determine the empirical formula. In order to calculate an empirical formula we will need to take the combustion data and somehow determine the number of moles of each element in the compound. Most often, the type of compounds that you will face will be hydrocarbons or else compounds containing carbon, hydrogen, and oxygen.

Example 4-16: A particular hydrocarbon is analyzed by means of a combustion analysis. When 1.000 g of the compound undergo combustion, 2.743 g of carbon dioxide and 2.246 g of water are formed. What is the empirical formula of this compound?

Our goal in this problem is to obtain the empirical formula. In order to do this, we will need to determine how many moles of carbon there are and how many moles of hydrogen there are in this sample. The carbon dioxide will allow us to figure out the moles of carbon, and the water will allow us to figure out the moles of hydrogen. Let's start with the carbon dioxide. As in most problems in this chapter, we will go to moles using the molar mass.

$$2.743 \text{ g CO}_2 \times \frac{1 \text{ mole CO}_2}{44.010 \text{ g CO}_2} = 0.06233 \text{ moles CO}_2$$

The formula for CO_2 tells us that there is 1 mole of C for every mole of CO_2.
$$0.06233 \text{ moles CO}_2 \times \frac{1 \text{ mole C}}{1 \text{ mole CO}_2} = 0.06233 \text{ moles C}$$

Now, let's work on the hydrogen.
$$2.246 \text{ g H}_2\text{O} \times \frac{1 \text{ mole H}_2\text{O}}{18.015 \text{ g H}_2\text{O}} = 0.1247 \text{ moles H}_2\text{O}$$

The formula for H_2O tells us that there are 2 moles of H for each mole of H_2O.

$$0.1247 \text{ moles } H_2O \times \frac{2 \text{ moles H}}{1 \text{ mole } H_2O} = 0.2493 \text{ moles H}$$

Now we have the moles of C and the moles of H. The problem from here onward is the same as the empirical formula problems we worked in the last chapter. The smaller of these numbers is the moles of C. We will divide the number of moles of H by this number.

$$\text{Mole Ratio} = \frac{0.2493 \text{ moles H}}{0.06233 \text{ moles C}} = 4.000 \frac{\text{moles H}}{\text{mole C}}$$

The empirical formula is CH_4.

Try Study Questions 37 in Chapter 4 of your textbook now!

The problem is a little harder if instead of just being a hydrocarbon, the compound contains carbon, hydrogen, and oxygen.

Example 4-17: A compound with a formula of the type $C_xH_yO_z$ is analyzed by means of combustion analysis. When 1.500 g of the compound are analyzed, 3.410 g of carbon dioxide and 1.396 g of water are formed. What is the empirical formula of the compound?

We are asked once again for the empirical formula. To determine this, we will need to find the number of moles of each element. The problem starts out just as before to get the number of moles of C and of H. We get the number of moles of C by using the carbon dioxide information:

$$3.410 \text{ g } CO_2 \times \frac{1 \text{ mole } CO_2}{44.010 \text{ g } CO_2} = 0.07748 \text{ moles } CO_2$$

$$0.07748 \text{ moles } CO_2 \times \frac{1 \text{ mole C}}{1 \text{ mole } CO_2} = 0.07748 \text{ moles C}$$

We obtain the number of moles of H by using the water information:

$$1.396 \text{ g } H_2O \times \frac{1 \text{ mole } H_2O}{18.015 \text{ g } H_2O} = 0.07749 \text{ moles } H_2O$$

$$0.07749 \text{ moles } H_2O \times \frac{2 \text{ moles H}}{1 \text{ mole } H_2O} = 0.1550 \text{ moles H}$$

So now we have the moles of C and the moles of H. How do we get the moles of O? Oxygen appears in both the carbon dioxide and the water, and we had other oxygen enter into the reaction in the combustion besides that present in the compound. What can we do? We know the initial mass of the compound used. We also know the number of moles of C and the number of moles of H in that sample of the compound. From the number of moles of C and H, we can calculate the mass of C and H in the sample. If we subtract this from the original mass of the sample, then we will have the mass of O, which can be converted to moles of O.

$$0.07748 \text{ moles C} \times \frac{12.011 \text{ g C}}{1 \text{ mole C}} = 0.9306 \text{ g C}$$

$$0.1550 \text{ moles H} \times \frac{1.0079 \text{ g H}}{1 \text{ mole H}} = 0.1562 \text{ g H}$$

Mass Oxygen = 1.500 g – 0.9306 g – 0.1562 g = 0.4132 g O

$$0.4132 \text{ g O} \times \frac{1 \text{ mole O}}{15.999 \text{ g O}} = 0.02583 \text{ moles O}$$

We now have moles C, moles H, and moles O. The smallest of these is the number of moles of O, so we will divide all of the others by this number. We are thus setting the subscript for O to be 1 in the empirical formula.

$$\frac{C}{O} \text{ Mole Ratio} = \frac{0.07748 \text{ moles C}}{0.02583 \text{ moles O}} = 3.000 \frac{\text{moles C}}{\text{mole O}}$$

$$\frac{H}{O} \text{ Mole Ratio} = \frac{0.1550 \text{ moles C}}{0.02583 \text{ moles O}} = 6.001 \frac{\text{moles H}}{\text{mole O}}$$

The empirical formula is C_3H_6O.

Try Study Questions 41 in Chapter 4 of your textbook now!

Other Notes

1. Remember that a percent is calculated by dividing the part by the whole and then multiplying by 100.

$$\text{Percent} = \frac{\text{Part}}{\text{Whole}} \times 100$$

2. In a qualitative analysis, someone figures out what materials are present. In a quantitative analysis, one figures out how much of a material is present.

CHAPTER 5: Reactions in Aqueous Solution

Chapter Overview

In this chapter, you will learn about reactions in aqueous solutions. The key topics covered can be broken down into the following categories: 1) species in solution, 2) types of chemical reactions in aqueous solution, 3) writing chemical equations for reactions in aqueous solution, 4) solution concentration, and 5) solution stoichiometry. In order to understand what types of reactions occur in aqueous solutions, we must first understand what types of particles (species) are present in solution when we dissolve different compounds in water. We shall find that some materials produce ions in solution (ionic compounds, strong acids, and strong bases), some produce molecules in solution (most molecular compounds), and some produce both ions and molecules in solution (weak acids and weak bases). We next will turn our attention to four major types of reactions that occur in aqueous solution (precipitation, acid-base, gas-forming, and redox). The first three of these types are exchange reactions. You will learn how to predict the products for exchange reactions and how to write overall balanced chemical equations, total ionic equations, and net ionic equations for each type. You will learn that acids produce hydrogen ions (H^+) and bases produce hydroxide ions (OH^-) in solution. Redox reactions involve the transfer of electrons. We will define the terms oxidation (loss of electrons), reduction (gain of electrons), oxidizing agent, and reducing agent. We can recognize that a redox reaction has occurred if the oxidation numbers for some elements involved in the reaction change. In order to be able to do this, you will need to learn the rules for determining oxidation numbers. We will then get more quantitative by talking about solution concentration. The concentration unit used most often in chemistry is molarity (moles of solute/liter of solution). You will learn how to calculate the molarity of a solution and also how to calculate the pH ($pH = -\log[H^+]$) of a solution. You will then extend the arsenal of stoichiometry problems you can solve to those that involve solutions, including problems involving titrations.

Key Terms

In this chapter, you will need to learn and be able to use the following terms:

Acid: a substance that, when dissolved in water, increases the concentration of hydrogen ions (H^+) in solution.

Acid-base indicator: a dye that changes color at a particular pH.

Acidic oxide: a nonmetal oxide that can react with water to form an acidic solution.

Base: a substance that, when dissolved in water, increases the concentration of hydroxide ions (OH^-) in solution.

Basic oxide: a metal oxide that can react with water to form a basic solution.

Concentration: the amount of a material in a given amount of solvent or solution.

Electrode: a device such as a metal plate or wire for conducting electricity into and out of solutions in electrochemical cells.

Electrolyte: a compound whose aqueous solutions conducts electricity; these solutions contain ions.

Equivalence point: the point in a titration at which stoichiometrically equivalent amounts of the two reactants have been combined. In an acid-base titration, it is the point at which the amount of OH^- from the base exactly equals the amount of H^+ provided by the acid.

Exchange reaction: a reaction in which two substances switch which anion is connected to which cation: $MX + NY \rightarrow MY + NX$.

Molarity: the number of moles of solute per liter of solution; this is one of the most frequently used concentration units used in chemistry.

Net ionic equation: a chemical equation involving only those species undergoing chemical changes in the course of the reaction; it is the "action" equation.

Neutralization reaction: a reaction between an acid and a base that produces a salt and water.

Nonelectrolyte: a compound whose aqueous solutions do not conduct electricity.

Oxidation number: the charge an atom has, or appears to have, as determined by the guidelines for assigning oxidation numbers (when we assume that all bonds are ionic bonds).

Oxidation: the loss (or apparent loss) of electrons. When a material is oxidized, its oxidation number increases.

Oxidation-reaction reaction (redox reaction): a reaction that involves the transfer or apparent transfer of electrons.

Oxidizing agent: the substance in a redox reaction that causes another material to become oxidized. The oxidizing agent is itself reduced in the reaction.

Precipitate: an insoluble product of a reaction.

Precipitation reaction: an exchange reaction that produces an insoluble product.

Primary standard: an extremely pure material that can be weighed out exactly. Primary standards are used in standardizations.

Product-favored reaction: a reaction in which the reactants are completely or largely converted into products.

Reactant-favored reaction: a reaction that results in conversion of little if any of the reactants into products.

Reducing agent: the substance in a redox reaction that causes another material to become reduced. The reducing agent is itself oxidized in the reaction.

Reduction: the gain (or apparent gain) of electrons. When a material is reduced, its oxidation number decreases.

Salt: an ionic compound whose cation comes from a base and whose anion comes from an acid.

Solute: the component of a solution that is dissolved in another substance.

Solution: a homogenous mixture in which the components are evenly distributed in each other down to an atomic or molecular scale.

Solvent: the medium in which a solute is dissolved to form a solution.

Spectator ion: an ion that does not change during the course of a chemical reaction; it is present in solution at the beginning of the reaction and is still present in solution at the end of the reaction. Spectator ions are not shown in net ionic equations.

Standardization: the accurate determination of the concentration of an acid, base, or other reagent for use in a titration.

Strong acid: a compound that ionizes completely in water; for each molecule of acid placed in water, we get hydrogen ions in solution.

Strong base: a base that produces hydroxide ions in solution for each formula unit that dissolves in water.

Strong electrolyte: a compound that produces ions for each formula unit that dissolves in water; these solutions conduct electricity well.

Titration: a quantitative analysis in which one reactant is added incrementally to another reactant until a complete reaction has occurred between the quantities of the two reactants.

Volumetric flask: a flask that has been calibrated to hold a certain volume of solution when the solution level lines up with the single line that is marked at some point on its neck.

Weak acid: an acid that does not ionize completely in water; the acid will be present as both ions and molecules in solution.

Weak base: a base that does not ionize completely in water; the base will be present as both ions and molecules in solution.

Weak electrolyte: a compound that will be present as both ions and molecules in aqueous solutions; its solutions conduct electricity but not as well as those of a strong electrolyte at the same concentration.

Chapter Goals

By the end of this chapter you should be able to:

- **Understand the nature of ionic substances dissolved in water.**

  a) Explain the difference between electrolytes and nonelectrolytes and recognize examples of each (Section 5.1 and Figure 5.2).

  Compounds whose aqueous solutions conduct electricity are called electrolytes. In order to conduct electricity, these aqueous solutions must contain ions. Not surprisingly, ionic compounds that are soluble in water are electrolytes. The ions that are present in the ionic compound are separated from each other and surrounded by water molecules. For example, when the ionic compound sodium chloride dissolves in water, we get ions in solution (specifically, sodium ions and chloride ions).

  $$NaCl\ (s) \rightarrow Na^+\ (aq) + Cl^-\ (aq)$$

  Compounds whose aqueous solutions do not conduct electricity are called nonelectrolytes. The solute particles present in these aqueous solutions are molecules, not ions. Most molecular compounds that dissolve in water are nonelectrolytes. For example, when the molecular compound ethanol (C_2H_5OH) dissolves in water, each molecule of ethanol stays together as a single unit. We do not get charged particles in the solution.

  $$C_2H_5OH\ (l) \rightarrow C_2H_5OH\ (aq)$$

  Some molecular compounds (strong acids, weak acids, and weak bases), however, react with water to form ions and are thus electrolytes. An aqueous solution of the gas hydrogen chloride is called hydrochloric acid. Hydrogen chloride is a compound composed of only nonmetals, so we would predict that it would be molecular, but it reacts with the water in the solution to form ions. We can simplify our representation of this reaction as follows (we will consider a more complex representation later in the course):

  $$HCl\ (aq) \rightarrow H^+\ (aq) + Cl^-\ (aq)$$

Even though we started out with a molecular material, we ended up with ions in the solution; hydrochloric acid is an electrolyte.

We classify electrolytes as being strong or weak. For a strong electrolyte, we get ions in solution for each formula unit that dissolves. The solutions of strong electrolytes contain the solutes in the form of ions. Strong electrolytes are soluble ionic compounds and strong acids. Sodium chloride is a strong electrolyte. For each formula unit of sodium chloride that dissolves, we obtain one sodium ion and one chloride ion. Many strong bases, such as sodium hydroxide, are ionic compounds. They are strong electrolytes. For each formula unit of sodium hydroxide that dissolves we get one sodium ion and one hydroxide ion.

$$NaOH \ (s) \rightarrow Na^+ \ (aq) + OH^- \ (aq)$$

A strong acid such as hydrochloric acid is a molecular compound that reacts with water to form ions in solution. For strong acids, this reaction goes virtually to completion. Each molecule of HCl we start out with produces a hydrogen ion and a chloride ion in solution.

A weak electrolyte is a molecular substance in whose aqueous solutions some of the molecules reacts with water to form ions but where some of the molecules (usually most) remain as molecules. Examples of weak electrolytes are weak acids and weak bases. For example, the weak acid called acetic acid is a weak electrolyte.

$$CH_3COOH \ (aq) <=> H^+ \ (aq) + CH_3COO^- \ (aq)$$

The double arrow in this equation indicates an equilibrium. The forward reaction (acetic acid ionizing to form ions) occurs but the reverse reaction (the ions coming together to reform acetic acid molecules) also occurs. The acetic acid is present as both ions and molecules.

All of this information can be summarized in the following diagram:

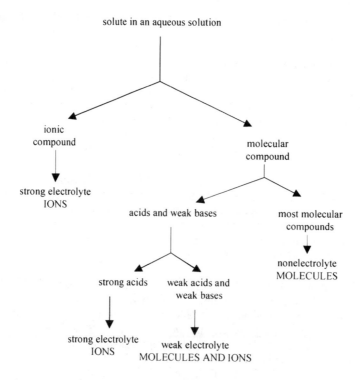

Example 5-1: Predict whether each of the following materials is a strong electrolyte, weak electrolyte, or nonelectrolyte.

      a. sucrose, $C_{12}H_{22}O_{11}$

This compound is composed of only nonmetals. It is therefore a molecular compound. It is not an acid or a base (we'll spend some more time talking about this in a little while). It is therefore a nonelectrolyte and present in aqueous solutions as molecules.

      b. LiBr

This compound is composed of a metal and a nonmetal. It is an ionic compound. It is therefore a strong electrolyte. When it dissolves in water, we get ions in solution.

      c. HNO_3

This compound is composed of only nonmetals, therefore it is a molecular compound. It is one of the strong acids listed in Table 5.2. It is a strong electrolyte. It will be present in aqueous solutions as ions.

      d. NH_3

This compound is composed of only nonmetals, therefore it is a molecular compound. It is one of the weak bases listed in Table 5.2. It is a weak electrolyte and will be present in aqueous solutions as both molecules and ions.

b) Predict the solubility of ionic compounds in water (Section 5.1 and Figure 5.3).

In order to predict the solubility of ionic compounds in water, you will need to learn the solubility rules listed in Figure 5.3. Once again, flash cards are probably the best way to help you do this. On one side, write an ion found in a salt, such as Cl^-. On the other side, write the general rule (soluble or insoluble) for salts containing that ion and any exceptions to the general rule. When you come to the card containing the chloride ion, for example, you should say, "Salts containing the chloride ion tend to be soluble with the exception of silver chloride, mercury(I) chloride, and lead(II) chloride." Once you have memorized the solubility rules, you will be able to predict the solubility of many compounds. To help you memorize the exceptions, note common exceptions. For the ions whose salts tend to be insoluble, the exceptions are always the alkali metal and ammonium salts. The hydroxide ions have one more exception to worry about. For the ions whose salts tend to be soluble, the ones for which there are no exceptions listed are easy. For the others, lead(II) salts tend to be exceptions and calcium, strontium, and barium salts are exceptions for a couple.

One ion that is not listed in the figure is the oxide ion. Salts containing the oxide ion tend to be insoluble with the exception of oxides containing the alkali metal cations.

Example 5-2: Predict whether each of the following salts is probably soluble or insoluble in water.

      a. $Ca(NO_3)_2$

From the solubility rules, you can see that salts containing the nitrate ion tend to be soluble in water. There are no exceptions listed so we would predict that calcium nitrate would be soluble.

      b. $Ca_3(PO_4)_2$

Salts containing the phosphate ion tend to be insoluble. Calcium ion is not listed as an exception. We would predict that this salt would be insoluble.

> c. Na_2CO_3
> Salts containing the carbonate ion tend to be insoluble. Sodium ion, however, is an alkali metal cation. The alkali metal cations are listed as one of the exceptions to the general rule so we would predict that this salt would be soluble.

Try Study Question 3 in Chapter 5 of your textbook now!

c) Recognize which ions are formed when an ionic compound or acid or base dissolves in water (Sections 5.1-5.3).

Ionic Compounds:

The ions formed in solution when an ionic compound dissolves in water are given by the name of the compound. For example, when lithium bromide dissolves in water, we get lithium ions and bromide ions in solution. This can be represented by the following equation.

$$LiBr \ (s) \rightarrow Li^+ \ (aq) + Br^- \ (aq)$$

We get one lithium ion and one bromide ion for each formula unit that dissolves.

When calcium bromide dissolves in water, we get calcium ions and bromide ions in solution.

$$CaBr_2 \ (s) \rightarrow Ca^{2+} \ (aq) + 2Br^- \ (aq)$$

As the formula predicts, we get one calcium ion and two bromide ions for each formula unit that dissolves. Notice that the bromide ions do not stay together as one unit Br_2^{2-} but instead separate from each other to form two separate particles in solution, $2Br^-$.

When sodium nitrate dissolves in water, just as the name predicts, we get sodium ions and nitrate ions in solution.

$$NaNO_3 \ (s) \rightarrow Na^+ \ (aq) + NO_3^- \ (aq)$$

Notice that the polyatomic ion does not break up into its constituent elements of nitrogen and oxygen. Instead, it stays together as one unit.

When calcium nitrate dissolves in water, we get calcium ions and nitrate ions in solution.

$$Ca(NO_3)_2 \ (s) \rightarrow Ca^{2+} \ (aq) + 2NO_3^- \ (aq)$$

In this case, once again, the polyatomic ion does not break apart into nitrogen and oxygen, but the two nitrate ions do separate from each other. We get one calcium ion and two nitrate ions in solution for each formula units of calcium nitrate that dissolves.

Acids:

Acids produce H^+ ions in solution. When a strong acid dissolves in water, each formula unit forms ions. We draw the ionization using a single arrow.

$$HCl \ (aq) \rightarrow H^+ \ (aq) + Cl^- \ (aq)$$
$$HNO_3 \ (aq) \rightarrow H^+ \ (aq) + NO_3^- \ (aq)$$

Weak acids also produce H^+ ions in solution, but the reaction does not go to completion. We end up with both ions and molecules in solution. There is an equilibrium, and we show this using double arrows.

$$CH_3COOH \ (aq) <=> H^+ \ (aq) + CH_3COO^- \ (aq)$$

Bases:

Bases produce hydroxide ions in aqueous solutions. Most of the strong bases we will run across are hydroxide salts of the group 1 cations. These are ionic compounds, so they are strong electrolytes, and their solutions contain the ions predicted by their names. For example, a solution of sodium hydroxide contains sodium ions and hydroxide ions.

$$NaOH \ (s) \rightarrow Na^+ \ (aq) + OH^- \ (aq)$$

A weak base is a weak electrolyte. An aqueous solution of a weak base will contain both molecules and ions. We will once again show an equilibrium situation by using the double arrow. The most common weak bases you will run across are ammonia and some compounds related to ammonia. We must explicitly show the reaction with water in order to show the ionization of a weak base.

$$NH_3\ (aq) + H_2O\ (l) <=> NH_4^+\ (aq) + OH^-\ (aq)$$

Try Study Question 5 in Chapter 5 of your textbook now!

- ## Recognize common acids and bases and understand their behavior in aqueous solutions (Sections 5.3 and Table 5.2).

    ### a) Know the names and formulas of common acids and bases.

    The definitions of acids and bases that we shall use for now is that acids increase the concentration of hydrogen ions in aqueous solutions and bases increase the concentration of hydroxide ions in aqueous solutions. We have already discussed strong and weak acids and bases as far as their behavior and the particles present in aqueous solutions for each. Some common acids and bases are listed in Table 5.2. You should memorize the names and formulas of the strong acids and strong bases in this table.

    ### b) Categorize acids and bases as strong or weak.

    We will assume, for the most part, that unless we tell you differently, an acid or base not listed as a strong acid or base in Table 5.2 is weak.

    Example 5-3: Identify each of the following as a strong or weak acid or base and write chemical equations for their dissociations or ionizations.

    a. $HClO_4$
    This is perchloric acid. It is one of the strong acids listed in Table 5.2 so it is a strong acid.
    $$HClO_4\ (aq) \rightarrow H^+\ (aq) + ClO_4^-\ (aq)$$

    b. KOH
    This is potassium hydroxide. It is a Group 1 hydroxide. It is a strong base.
    $$KOH\ (s) \rightarrow K^+\ (aq) + OH^-\ (aq)$$

    c. HNO_2
    This is nitrous acid. (It is not nitric acid, which was listed as a strong acid.) It is not listed as a strong acid, therefore we shall assume that it is weak (which indeed it is).
    $$HNO_2\ (aq) <=> H^+\ (aq) + NO_2^-\ (aq)$$

    Try Study Questions 13 and 15 in Chapter 5 of your textbook now!

- ## Recognize and write equations for the common types of reactions in aqueous solution.

    ### a) Predict the products of precipitation reactions (Section 5.2), which involve the formation of an insoluble reaction product by the exchange of anions between the cations of the reactants.

    In a precipitation reaction, we generally mix two soluble ionic compounds and form an insoluble material called a precipitate. This type of reaction falls into the category of chemical reactions called exchange reactions. In an exchange reaction, the ions switch partners: the cation of compound 1 ends up with the anion from compound 2, and the cation of compound 2 ends up with the anion from compound 1. This can be illustrated by the following equation in which M and N are cations and X and Y are anions:

$$MX + NY \rightarrow MY + NX$$

In a precipitation reaction, the driving force for the reaction is the formation of the solid material, which removes these ions from the solution.

In order to predict whether a precipitate will form, you will need to know the solubility rules. If the exchange reaction forms a compound that is not soluble then a precipitate will form.

Example 5-4: Predict whether a precipitate will form if aqueous solutions of calcium nitrate and sodium phosphate are mixed.

The reactants are calcium nitrate and sodium phosphate. We need to decide what the products would be. If a reaction occurs, the first cation (calcium) would end up with the second anion (phosphate). The second cation (sodium) would end up with the first anion (nitrate).

calcium nitrate + sodium phosphate → calcium phosphate + sodium nitrate

Now we will use our solubility rules to determine if any of the products are not soluble. Let's look at calcium phosphate. Phosphate salts tend to be insoluble. Calcium phosphate is not an exception to this rule. We would expect that the calcium phosphate would precipitate. This is a precipitation reaction. (For completion, let's look at sodium nitrate. All nitrate salts are soluble, so this compound is soluble.)

Example 5-5: Predict whether a precipitate will form if aqueous solutions of calcium nitrate and sodium chloride are mixed.

The reactants are calcium nitrate and sodium chloride. We need to decide what the products would be. If a reaction occurs, the first cation (calcium) would end up with the second anion (chloride), and the second cation (sodium) would end up with the first anion (nitrate).

calcium nitrate + sodium chloride → calcium chloride + sodium nitrate

What do the solubility rules predict? Let's look at calcium chloride. Chloride salts tend to be soluble, and calcium chloride is not one of our exceptions. This compound is soluble. Now let's look at sodium nitrate. All nitrate salts are soluble. This compound is soluble. Neither compound we would predict to form is insoluble. This is not a precipitation reaction. No reaction occurs in this case. We simply end up with a mixture of all four ions in solution.

---

b) Write net ionic equations and show how to arrive at such an equation for a given reaction (Sections 5.2 and 5.6).

There are three major types of equations we run across when we are dealing with reactions in aqueous solution: the overall balanced equation, the total ionic equation, and the net ionic equation. We will take a look at each of these using the reaction in Example 5-4 above:

calcium nitrate + sodium phosphate → calcium phosphate + sodium nitrate

The overall balanced equation is our normal balanced equation. We translate our word equation into chemical formulas and then balance the resulting equation.

$$3Ca(NO_3)_2 \ (aq) + 2Na_3PO_4 \ (aq) \rightarrow Ca_3(PO_4)_2 \ (s) + 6NaNO_3 \ (aq)$$

Notice that now you can determine the states for each of the compounds in the reaction (instead of us having to tell them to you) because you know that we started out with aqueous solutions and because you know the solubility rules.

The total ionic equation shows the exact species that are present in the reaction. For example, calcium nitrate is a soluble ionic compound. It will be present in solution as ions. Each formula unit of calcium nitrate will provide 1 Ca^{2+} ion and 2 NO_3^- ions in solution, so three formula units of calcium nitrate will provide 3 Ca^{2+} ions and 6 NO_3^- ions. Similarly, sodium phosphate is a soluble ionic compound. It will be present in solution as ions. Each formula unit of Na_3PO_4 will provide 3 Na^+ ions and 1 PO_4^{3-} ion, so two formula units of Na_3PO_4 will provide 6 Na^+ ions and 2 PO_4^{3-} ions in solution. On the right side of the equation, we come to $Ca_3(PO_4)_2$. This is our precipitate. It is a solid. It does not dissolve to give us ions in solution so it will stay together as one unit. Finally, we come to the sodium nitrate. It is a soluble ionic compound. It will be present in solution as ions. One formula unit of $NaNO_3$ provides one ion each of Na^+ and NO_3^-, so six formula units of $NaNO_3$ will provide 6 Na^+ ions and 6 NO_3^- ions. All of this is summarized in the total ionic equation for the reaction:

$$3Ca^{2+} \ (aq) + 6NO_3^- \ (aq) + 6Na^+ \ (aq) + 2PO_4^{3-} \ (aq) \rightarrow$$
$$Ca_3(PO_4)_2 \ (s) + 6Na^+ \ (aq) + 6NO_3^- \ (aq)$$

The net ionic equation shows us where the action is. It shows us which species actually change in the course of the reaction. Looking at the total ionic equation above, we can see that the nitrate ions start our as ions in solution and end up as ions in solution. They did not change, so we will not include them in our action equation. The sodium ions similarly start out as ions in solution and end up as ions in solution. They did not change, so we will not include them in our action equation. Ions such as the sodium ions and nitrate ions in this example which do not change in the course of the reaction are called spectator ions because they are there at the beginning, and they are there at the end. They simply "watched" what was going on. On the other hand, the calcium ions start out as ions in solution but end up as part of a solid. They did change, so we will include them in the action equation. The phosphate ions start out as ions in solution and end up as part of a solid, so they changed in the reaction. They too will be part of our action equation. The net ionic equation for this reaction is therefore

$$3Ca^{2+} \ (aq) + 2PO_4^{3-} \ (aq) \rightarrow Ca_3(PO_4)_2 \ (s)$$

The net ionic equation is a very powerful expression. It tells us that the heart of this reaction is to mix calcium ions and phosphate ions in solution, and they will form solid calcium phosphate. We did not have to pick calcium nitrate. We could have selected any soluble calcium salt and mixed it with any soluble phosphate salt, and we would have ended up forming solid calcium phosphate.

You may have noticed that if we know what the product is ($Ca_3(PO_4)_2$ in this case) for a precipitation reaction carried out by mixing two aqueous solution we can construct the net ionic equation simply by writing the ions that form this compound and then balancing the resulting equation.

Example 5-6: Write the overall balanced equation, the total ionic equation, and the net ionic equation for the reaction of aqueous solutions of potassium carbonate and calcium chloride.

The reactants are potassium carbonate and calcium chloride. The products of the exchange reaction would be potassium chloride and calcium carbonate. Potassium chloride is a soluble compound. Calcium carbonate is an insoluble compound. This is a precipitation reaction.

potassium carbonate + calcium chloride → potassium chloride + calcium carbonate

The equation showing the phases and properly balanced is
$$K_2CO_3 \ (aq) + CaCl_2 \ (aq) \rightarrow 2KCl \ (aq) + CaCO_3 \ (s)$$

In the total ionic equation, the potassium carbonate, calcium chloride, and potassium chloride will all be shown as dissociated ions, taking into account the subscripts and coefficient. The calcium carbonate will be shown as a unit because it is a solid.
$$2K^+ \ (aq) + CO_3{}^{2-} \ (aq) + Ca^{2+} \ (aq) + 2Cl^- \ (aq) \rightarrow 2K^+ \ (aq) + 2Cl^- \ (aq) + CaCO_3 \ (s)$$

The spectator ions are the potassium and chloride ions. The net ionic equation is
$$Ca^{2+} \ (aq) + CO_3{}^{2-} \ (aq) \rightarrow CaCO_3 \ (s)$$

Example 5-7: Write the net ionic equation for the precipitation reaction that results when aqueous solutions of silver nitrate and sodium sulfate are mixed.

We will do this example in a little faster way since only the net ionic equation is asked for. The reactants are silver nitrate and sodium sulfate. The possible products are silver sulfate and sodium nitrate. Silver sulfate is the precipitate. Sodium nitrate is soluble. The formula of the precipitate is Ag_2SO_4. To form this, we need to show a net ionic equation bringing together these ions:
$$2Ag^+ \ (aq) + SO_4{}^{2-} \ (aq) \rightarrow Ag_2SO_4 \ (s)$$

Try Study Question 11 in Chapter 5 of your textbook now!

c) Predict the products of acid-base reactions involving common acids and strong bases (Section 5.4).

Reactions between strong acids and strong bases are neutralization reactions. In general, acids react with strong bases to produce a salt and water. These reactions are exchange reactions. For example, nitric acid reacts with potassium hydroxide as follows:
$$HNO_3 \ (aq) + KOH \ (aq) \rightarrow KNO_3 \ (aq) + HOH \ (l)$$

You can recognize that HOH is simply H_2O, water. This equation is the equivalent of
$$HNO_3 \ (aq) + KOH \ (aq) \rightarrow KNO_3 \ (aq) + H_2O \ (l)$$

Try Study Question 21 in Chapter 5 of your textbook now!

d) Understand that the net ionic equation for the reaction of a strong acid with a strong base is $H^+ \ (aq) + OH^- \ (aq) \rightarrow H_2O \ (l)$ (Section 5.4).

Let's take a closer look at the reaction of nitric acid and potassium hydroxide. Nitric acid is a strong acid. It will be present in solution as H^+ ions and $NO_3{}^-$ ions. Potassium hydroxide is an ionic compound. It will be present in solution as K^+ ions and OH^- ions. On the products side, potassium nitrate is an ionic compound. It will be present in solution as K^+ ions and $NO_3{}^-$ ions. Water is a molecular compound. It will be present in solution (predominantly) as molecules. The total ionic equation for this reaction is therefore
$$H^+ \ (aq) + NO_3{}^- \ (aq) + K^+ \ (aq) + OH^- \ (aq) \rightarrow K^+ \ (aq) + NO_3{}^- \ (aq) + H_2O \ (l)$$

We can see that the spectator ions in this reaction are K^+ and $NO_3{}^-$. The hydrogen ions and hydroxide ions start out as ions in solution but end up as part of a molecular material. The net ionic equation is
$$H^+ \ (aq) + OH^- \ (aq) \rightarrow H_2O \ (l)$$

The net ionic equation for the reaction of any strong acid with any strong base will be the same. The hydrogen ions from the strong acid will react with the hydroxide ions from the strong base to form water molecules.

Example 5-8: Write the overall balanced equation, the total ionic equation, and the net ionic equation for the reaction of perchloric acid with sodium hydroxide.

Perchloric acid is a strong acid. Sodium hydroxide is a strong base.

Overall Balanced Equation:
$$HClO_4 \ (aq) + NaOH \ (aq) \rightarrow NaClO_4 \ (aq) + H_2O \ (l)$$

Total Ionic Equation:
$$H^+ \ (aq) + ClO_4^- \ (aq) + Na^+ \ (aq) + OH^- \ (aq) \rightarrow Na^+ \ (aq) + ClO_4^- \ (aq) + H_2O \ (l)$$

Net Ionic Equation:
$$H^+ \ (aq) + OH^- \ (aq) \rightarrow H_2O \ (l)$$

When we deal with weak acids, weak bases, or insoluble hydroxides, the net ionic equations become a bit more complicated for acid-base reactions because we do not have ions as the predominant species in solution. As such, they will be shown intact in the total ionic equation rather than broken up into ions. They will also appear in the net ionic equations.

Example 5-9: Write the overall balanced equation, the total ionic equation, and the net ionic equation for the reaction of acetic acid with potassium hydroxide.

Acetic acid is a weak acid, and potassium hydroxide is a strong base.

Overall Balanced Equation:
$$CH_3CO_2H \ (aq) + KOH \ (aq) \rightarrow KCH_3CO_2 \ (aq) + H_2O \ (l)$$

Because acetic acid is a weak acid, it will be present in solution as both ions and molecules, but the predominant form is as molecules. In the total ionic equation, therefore, we will show it not in the ionized form, but in the molecular form. The potassium hydroxide and potassium acetate are both ionic compounds and will be present in solution as ions.

Total Ionic Equation:
$$CH_3CO_2H \ (aq) + K^+ \ (aq) + OH^- \ (aq) \rightarrow K^+ \ (aq) + CH_3CO_2^- \ (aq) + H_2O \ (l)$$

In the net ionic equation only the potassium ions are spectator ions.

Net Ionic Equation:
$$CH_3CO_2H \ (aq) + OH^- \ (aq) \rightarrow CH_3CO_2^- \ (aq) + H_2O \ (l)$$

Example 5-10: Write the net ionic equation for the reaction of nitric acid with ammonia.

Nitric acid is a strong acid, and ammonia is a weak base.

The overall balanced chemical equation is
$$HNO_3 \ (aq) + NH_3 \ (aq) \rightarrow NH_4NO_3 \ (aq)$$

The nitric acid will be present as hydrogen ions and nitrate ions in solution, the ammonia will be present as ammonia molecules, and ammonium nitrate will be present as ammonium ions and nitrate ions. The only spectator ion is the nitrate ion. The net ionic equation is

H^+ *(aq)* + NH_3 *(aq)* → NH_4^+ *(aq)*

Try Study Question 25 in Chapter 5 of your textbook now!

e) Predict the products of gas-forming reactions (Section 5.5), the most common of which are those between a metal carbonate and an acid.

The major types of gas forming reactions are listed in Table 5.3 in your textbook. In these cases, you will start out looking at the reaction as a normal exchange reaction, but keep an eye out for predicting the formation of particular compounds. The ones to keep an eye out for are H_2CO_3, H_2SO_3, NH_4OH, and H_2S. By far, the most common type of gas-forming reaction you will run across is the reaction of a metal carbonate (or metal hydrogen carbonate) with an acid. The key compound to watch out for in that case is H_2CO_3.

Example 5-11: Write the overall balanced equation, total ionic equation, and net ionic equation for the reaction of potassium carbonate with hydrochloric acid.

In writing the overall equation, we start out as we would for a normal exchange reaction.

K_2CO_3 *(aq)* + 2HCl *(aq)* → 2KCl *(aq)* + H_2CO_3 *(aq)*

The key to this type of reaction is that H_2CO_3, carbonic acid, is not stable. It breaks down into water and carbon dioxide.

H_2CO_3 *(aq)* → H_2O *(l)* + CO_2 *(g)*

These two reactions can be added together to give the overall reaction

K_2CO_3 *(aq)* + 2HCl *(aq)* → 2KCl *(aq)* + H_2O *(l)* + CO_2 *(g)*

The carbon dioxide formed leaves the reaction vessel. So, whenever you would predict that H_2CO_3 would be a product of a reaction like this, keep in mind that it breaks down into water and carbon dioxide.

For the total ionic equation, the potassium carbonate will be present in solution as potassium ions and carbonate ions. The hydrochloric acid will be present as hydrogen ions and chloride ions, and the potassium chloride will be present as potassium ions and chloride ions. The water and the carbon dioxide are molecular compounds. They will each stay together as molecules.

$2K^+$ *(aq)* + CO_3^{2-} *(aq)* + $2H^+$ *(aq)* + $2Cl^-$ *(aq)* → $2K^+$ *(aq)* + $2Cl^-$ *(aq)* + H_2O *(l)* + CO_2 *(g)*

The potassium ions and chloride ions are spectator ions, and the net ionic equation is

CO_3^{2-} *(aq)* + $2H^+$ *(aq)* → H_2O *(l)* + CO_2 *(g)*

The reaction of a metal sulfite (or metal hydrogen sulfite) with an acid is very similar. For this type of reaction, we would predict the formation of sulfurous acid, H_2SO_3. Just like H_2CO_3 broke down into H_2O and CO_2, H_2SO_3 breaks down into H_2O and SO_2. For example, when potassium sulfite reacts with hydrochloric acid, we obtain the following overall balanced equation:

K_2SO_3 *(aq)* + 2HCl *(aq)* → 2KCl *(aq)* + H_2SO_3 *(aq)*

H_2SO_3 *(aq)* → H_2O *(l)* + SO_2 *(g)*

Overall: K_2SO_3 *(aq)* + 2HCl *(aq)* → 2KCl *(aq)* + H_2O *(l)* + SO_2 *(g)*

For the case of the reaction of an ammonium salt with a strong base, the compound to watch out for when you write the initial exchange reaction is NH_4OH. Once again, this is not a stable compound. It breaks down into NH_3 *(g)* and H_2O *(l)*. We could write the following equation when ammonium chloride reacts with potassium hydroxide.

NH_4Cl *(aq)* + KOH *(aq)* → KCl *(aq)* + NH_4OH *(aq)*

NH_4OH *(aq)* → NH_3 *(g)* + H_2O *(l)*

Overall: NH_4Cl *(aq)* + KOH *(aq)* → KCl *(aq)* + NH_3 *(g)* + H_2O *(l)*

The case of the reaction of a metal sulfide with an acid is the simplest case because the initial exchange reaction you would write would predict the formation of H_2S, which is already a gas. Potassium sulfide reacting with hydrochloric acid can be represented by the following equation.

K_2S *(aq)* + $2HCl$ *(aq)* → $2KCl$ *(aq)* + H_2S *(g)*

Try Study Question 27 in Chapter 5 of your textbook now!

f) Use the ideas developed in Sections 5.2-5.7 as an aid in recognizing four of the common types of reactions that occur in aqueous solution, and write balanced equations for such reactions (Section 5.6).

This is simply a matter of pulling together ideas we have already discussed. The different types of reactions are summarized in the following table:

Reaction Type	Driving Force
Precipitation	Formation of an insoluble compound
Acid-strong base	Formation of a salt and water
Gas-forming	Evolution of a water-insoluble gas such as CO_2
Oxidation-reduction	Transfer of electrons

Note that the first three of these reaction types involve the exchange of anions between the cations involved,

$$A^+B^- + C^+D^- → A^+D^- + C^+B^-$$

and so are called exchange reactions. The fourth (redox) involves the transfer of electrons.

Example 5-12: Write overall balanced chemical equations for each of the following reactions and classify each as a precipitation, an acid-base, or a gas-forming reaction.

a. copper(II) chloride + sodium hydroxide

The predicted products from the exchange reaction are copper(II) hydroxide and sodium chloride. Sodium chloride is a soluble compound, but copper(II) hydroxide is an insoluble compound (hydroxide salts tend to be insoluble, and copper(II) hydroxide is not one of the exceptions). The overall balanced equation is

$CuCl_2$ *(aq)* + $2NaOH$ *(aq)* → $Cu(OH)_2$ *(s)* + $2NaCl$ *(aq)*

This is a precipitation reaction.

b. hydrochloric acid and sodium hydroxide

Hydrochloric acid is a strong acid, and sodium hydroxide is a strong base.

HCl *(aq)* + $NaOH$ *(aq)* → $NaCl$ *(aq)* + H_2O *(l)*

This is an acid-base reaction.

> c. sodium hydrogen carbonate and nitric acid
> The predicted products are sodium nitrate and H_2CO_3. The H_2CO_3 should raise a red flag for us. It is one of the compounds to watch out for. It will break down into H_2O and CO_2.
>
> $NaHCO_3$ *(aq)* + HNO_3 *(aq)* → $NaNO_3$ *(aq)* + H_2CO_3 *(aq)*
> H_2CO_3 *(aq)* → H_2O *(l)* + CO_2 *(g)*
>
> The overall reaction is therefore
>
> $NaHCO_3$ *(aq)* + HNO_3 *(aq)* → $NaNO_3$ *(aq)* + H_2O *(l)* + CO_2 *(g)*
>
> This is a gas-forming reaction.

Try Study Question 29 in Chapter 5 of your textbook now!

g) Identify reactant- and product-favored reactions.

A reactant-favored reaction is a reaction that results in conversion of little if any of the reactants into products. A product-favored reaction is a reaction in which the reactants are completely or largely converted into products. We shall revisit and modify these definitions a little when we study chemical equilibrium later in the course.

Each of the reaction types we have studied so far is product-favored because they have driving forces that cause them to proceed. In the case of precipitation, acid-base, and gas forming reactions, the driving force is the removal of ions from the solution; this is accomplished by the formation of either a solid precipitate, water molecules, as a gas. In the case of redox reactions, the transfer of electrons occurs due to the difference in electrical potential energy between the materials involved (we shall examine this driving force in more detail later in the course).

Being able to predict whether given reactions will be reactant- or product-favored is one of the major goals of this course. We will revisit and expand upon this concept many times.

- **Recognize common oxidizing and reducing agents and identify oxidation-reduction reactions.**

a) Determine oxidation numbers of elements in a compound and understand that these numbers represent the charge an atom has, or appears to have, when the electrons of the compound are counted according to a set of guidelines (Section 5.7).

The oxidation number of an atom in a compound or polyatomic ion is the charge the atom has, or appears to have, when the electrons are counted according to a set of guidelines. In assigning oxidation numbers, we essentially make the assumption that all bonds are ionic. In cases where the compound actually is ionic, the oxidation number corresponds to the actual charge the ion has. If the compound is molecular, the oxidation number does not correspond to an actual charge, but it still can help us write the correct formula for the compound and to determine if a particular reaction is an oxidation-reduction reaction.

The rules for oxidation numbers are the following:

> 1. Each atom in a pure element has an oxidation number of zero. After all, if the atom is not combined with another element, it cannot have gained or lost electrons and thus has a charge of zero.

2. For monatomic ions, the oxidation number is equal to the charge on the ion. In this case, the material really is ionic so the oxidation number will be the same as the charge.

3. F always has an oxidation number of –1 in compounds with all other elements. Fluorine has the strongest pull for electrons. It always wins and gets the negative charge. Since it is in group 7A, its oxidation number is –1.

4. The oxidation number of O is –2 in most compounds except when combined with fluorine. Oxygen is the second strongest. It wins and gets the negative oxidation number in all cases except when it goes up against fluorine. The negative oxidation number for oxygen is –2 in most cases. (The exceptions are in peroxides O_2^{-2} where each oxygen would have an oxidation number of –1 and in superoxides O_2^{-} where each oxygen has an oxidation number of –1/2.)

5. The oxidation number of Cl, Br, and I is –1 except when combined with oxygen or fluorine. These come next. They get the negative oxidation number except when combined with oxygen or fluorine.

6. The oxidation number of H is +1 in most compounds. (The exception is in metal hydrides where it is –1.)

7. The algebraic sum of the oxidation numbers for the atoms in a neutral compound must be zero; in a polyatomic ion, the sum must be equal to the charge of the ion.

Example 5-13: Determine the oxidation number of each atom in the following species.

a. CaI_2

This compound is a binary ionic compound. The oxidation numbers will be the charges of the ions. The oxidation number of calcium is +2 because calcium is a Group 2A element. The oxidation number of each iodide ion is –1 because iodine is a Group 7A element.

b. HNO_3

This compound is not an ionic compound, but we can still determine the oxidation numbers. Start with the elements you know. Hydrogen has an oxidation number of +1 in most compounds (this is not a metal hydride). Oxygen has an oxidation number of –2 in most compounds (this is not a peroxide or superoxide). This just leaves the nitrogen. To figure out its oxidation number, we will need to use the fact that the sum of the oxidation numbers in the compound must add up to be zero. We have one +1 (from the hydrogen), three –2's (from the oxygens) and one nitrogen whose oxidation number we will call x in the equation below:

$$0 = 1(+1) + 1(x) + 3(-2)$$

We then solve for x.

$$0 = -5 + x$$
$$x = +5$$

The oxidation number of nitrogen in this compound is +5.

c. $NaClO_4$

This is an ionic compound. The sodium's oxidation number is +1 because sodium ions have a charge of 1+. The oxidation number of each oxygen will be –2. This leaves the Cl. Normally, it will have an oxidation number of –1, but in this case it is combined with oxygen so this will not necessarily be true here. We will determine its oxidation number by using the fact that the sum of the oxidation numbers of the elements in the compound will be zero.

$$0 = 1(+1) + 1(x) + 4(-2)$$
$$0 = -7 + x$$
$$x = +7$$

The oxidation number of chlorine in this compound is +7.

     d. MnO_4^-

The oxidation number of each oxygen will be –2. We need to calculate the oxidation number of the manganese. We have a polyatomic ion so the sum of the oxidation numbers will not add to equal zero but will add to equal the charge of the ion, in this case, –1.

$$-1 = 1(x) + 4(-2)$$
$$-1 = x - 8$$
$$x = +7$$

The oxidation number of the manganese in this ion is +7.

Try Study Question 35 in Chapter 5 of your textbook now!

b) Identify oxidation-reduction reactions (redox reactions) and identify the oxidizing and reducing agents and substances oxidized and reduced in the reaction (Section 5.7 and Tables 5.4 and 5.5).

Oxidation-reduction reactions involve the transfer (or apparent transfer) of electrons between species. Often, instead of saying "oxidation-reduction" reactions, we simply say "redox" reactions. Redox reactions can be broken down into two processes: oxidation and reduction. Oxidation is the loss (or apparent loss) of electrons. Reduction is the gain (or apparent gain) of electrons. There are two commonly used mnemonics used to help people remember which process is which. One is to picture a lion roaring and remember the saying "LEO the lion says GER." LEO stands for "Lose electrons oxidation." GER stands for "Gain electrons reduction." The other mnemonic is "OIL RIG." "Oxidation is loss; reduction is gain." If either of these helps you to remember which process is which, then use it.

In a redox reaction, both of these processes must occur. If something is gaining electrons, it must be getting the electrons from somewhere – from the substance that is being oxidized.

There are two other terms, which can sometimes be confusing. These are the terms oxidizing agent and reducing agent. The oxidizing agent is the material in the reaction that causes something else to get oxidized. In order to do this, it must itself be reduced in the process. Similarly, the reducing agent is the material that causes something else to get reduced. In order to do this, it must itself be oxidized. So the material that is oxidized is the reducing agent, and the material that is reduced is the oxidizing agent.

How can we tell if a particular material is oxidized or reduced? One of the best ways is to look at what happens to oxidation numbers. Oxidation numbers change in redox reactions.

When something is oxidized, its oxidation number goes up. When something is reduced, its oxidation number goes down. The easiest way to remember this is to remember the reduction case: the oxidation number gets reduced in the process of reduction.

Example 5-14: Determine if each of the following reactions is an oxidation-reduction reaction. If so, state which species is oxidized, which is reduced, which is the oxidizing agent, and which is the reducing agent.

a. Zn *(s)* + $CuCl_2$ *(aq)* → Cu *(s)* + $ZnCl_2$ *(aq)*

The first step is to determine the oxidation number of each atom in each species. Using the oxidation number rules we went over in the last goal, we can determine the following oxidation numbers.

$$\begin{array}{cccc} 0 & +2\ -1 & 0 & +2\ -1 \end{array}$$
Zn *(s)* + $CuCl_2$ *(aq)* → Cu *(s)* + $ZnCl_2$ *(aq)*

The zinc changed its oxidation number from 0 to +2. The oxidation number went up so it was oxidized. The copper changed its oxidation number from +2 to 0. Its oxidation number went down, so it was reduced.

The species that was oxidized was zinc metal. The material that was oxidized is the reducing agent so zinc metal is the reducing agent. The species that was reduced was $CuCl_2$; it is the oxidizing agent.

b. $2Na_2CrO_4$ *(aq)* + $3Na_2SO_3$ *(aq)* + $5H_2O$ *(l)* →
$$2Cr(OH)_3\ (s) + 3Na_2SO_4\ (aq) + 4NaOH\ (aq)$$

First, assign oxidation numbers.
$$\begin{array}{cccccc} +1\ +6\ -2 & +1\ +4\ -2 & +1\ -2 & +3\ -2\ +1 & +1\ +6\ -2 & +1\ -2\ +1 \end{array}$$
$2Na_2CrO_4$ *(aq)* + $3Na_2SO_3$ *(aq)* + $5H_2O$ *(l)* → $2Cr(OH)_3$ *(s)* + $3Na_2SO_4$ *(aq)* + $4NaOH$ *(aq)*

In this case, the Cr went from an oxidation number of +6 to an oxidation number of +3. The oxidation number went down so the Na_2CrO_4 is the material that is reduced. The S went from an oxidation number of +4 up to an oxidation number of +6, so the Na_2SO_3 is the material that is oxidized.

The Na_2CrO_4 is oxidized and is the reducing agent. The Na_2SO_3 is reduced and is the oxidizing agent.

c. $AgNO_3$ *(aq)* + $NaCl$ *(aq)* → $AgCl$ *(s)* + $NaNO_3$ *(aq)*

First assign oxidation numbers.
$$\begin{array}{cccc} +1\ +5\ -2 & +1\ -1 & +1\ -1 & +1\ +5\ -2 \end{array}$$
$AgNO_3$ *(aq)* + $NaCl$ *(aq)* → $AgCl$ *(s)* + $NaNO_3$ *(aq)*

No atoms changed their oxidation numbers, so this is not a redox reaction.

Try Study Question 39 in Chapter 5 of your textbook now!

With some experience, you may begin to recognize some common oxidizing and reducing agents such as those in Table 5.4 so that you do not need to go through the full process of assigning oxidation numbers in order to recognize redox reactions.

- ## Define and use molarity in solution stoichiometry.

a) Calculate the concentration of a solute in a solution in units of moles per liter (molarity), and use concentrations in calculations (Section 5.8).

Molarity is defined as the number of moles of *solute* per liter of *solution*.

Example 5-15: A solution is prepared by dissolving 10.0 g of barium chloride in enough water to make 500. mL of solution.

a. What is the concentration of this solution in units of molarity?
Molarity is moles of solute per liter of solution. The solute in this solution is the barium chloride. We are given the mass of barium chloride. Using the molar mass, we can calculate the number of moles.

$$10.0 \text{ g BaCl}_2 \times \frac{1 \text{ mole BaCl}_2}{208.2 \text{ g BaCl}_2} = 0.0480 \text{ moles BaCl}_2$$

We now need to divide this by the volume of solution in liters.

$$500. \text{ mL} \times \frac{1 \text{ L}}{1000 \text{ mL}} = 0.500 \text{ L}$$

$$c_{molarity} = \frac{0.0480 \text{ moles BaCl}_2}{0.500 \text{ L}} = 0.0961 \text{ M BaCl}_2$$

b. What is the concentration of each ion in this solution?
The concentration of $BaCl_2$ is 0.0961 M. When barium chloride dissolves in water, we get ions in solution. For each mole of $BaCl_2$, we get 1 mole of Ba^{2+} ions and 2 moles of Cl^- ions.

$$\frac{0.0961 \text{ moles BaCl}_2}{1 \text{ L sol'n}} \times \frac{1 \text{ mole Ba}^{2+}}{1 \text{ mole BaCl}_2} = 0.0961 \text{ M Ba}^{2+}$$

$$\frac{0.0961 \text{ moles BaCl}_2}{1 \text{ L sol'n}} \times \frac{2 \text{ moles Cl}^-}{1 \text{ mole BaCl}_2} = 0.192 \text{ M Cl}^-$$

One of the most important things for you to learn about molarity is that it gives us another way to calculate moles. Previously, the only way we had to get to moles was to use mass and the molar mass. Now we have a new way to get to moles, using the volume of a solution and the molarity.

This method of getting to moles is just as important as is the means of using the molar mass. Students sometimes get so stuck with the old way of using the mass and molar mass that they overlook this method. Don't fall into that trap. Let's work an example.

Example 5-16: How many moles of NaCl are there in 15.0 mL of a 0.342 M solution of NaCl?

We are given the volume of a solution in mL and its concentration in molarity. To use the molarity, we will need the volume in liters.

$$15.0 \text{ mL} \times \frac{1 \text{ L}}{1000 \text{ mL}} = 0.0150 \text{ L}$$

Now we can use the molarity as a conversion factor to get to moles. We could set up two conversion factors:

$$\frac{0.342 \text{ moles NaCl}}{\text{L sol'n}} \quad \text{and} \quad \frac{1 \text{ L sol'n}}{0.342 \text{ moles NaCl}}$$

In this case, we shall use the one with the moles on top and the volume on the bottom because we want to cancel L and end up with moles.

$$0.0150 \text{ L x } \frac{0.342 \text{ moles NaCl}}{\text{L sol' n}} = 5.13 \text{ x } 10^{-3} \text{ moles NaCl}$$

The definition of molarity involves moles of solute, volume of solution, and molarity. So far, we have seen how to solve for molarity and how to solve for moles. The third variable in the equation is volume. Let's work an example where we solve for it.

Example 5-17: What volume of a 0.342 M solution of NaCl contains 1.20 g of NaCl?

In this case, we are given the concentration of the solution and the mass of NaCl. We wish to solve for the volume of solution. We could do this if we knew the number of moles of NaCl. We can get this from the mass of NaCl.

$$1.20 \text{ g NaCl x } \frac{1 \text{ mole NaCl}}{58.44 \text{ g NaCl}} = 0.0205 \text{ moles NaCl}$$

This time, we will use our molarity to cancel moles and end up with L so we shall use the conversion factor with moles on the bottom and L on top.

$$0.0205 \text{ moles NaCl x } \frac{1 \text{ L sol' n}}{0.342 \text{ moles NaCl}} = 0.0600 \text{ L x } \frac{1000 \text{ mL}}{1 \text{ L}} = 60.0 \text{ mL}$$

Try Study Questions 41, 43, and 45 in Chapter 5 of your textbook now!

b) Describe how to prepare a solution of a given molarity from the solute and a solvent or by dilution from a more concentrated solution (Section 5.8).

Let us first consider how to prepare a solution of a given molarity from the solute and a solvent. In order to do this, we will need to weigh out the correct mass of solute and then add sufficient solvent to get the correct solution volume. A piece of glassware often used to help with this is a volumetric flask.

Example 5-18: Describe how to prepare 250. mL of a 0.300 M solution of NaCl starting with solid sodium chloride.

We need to figure out how much sodium chloride we need to weigh out. We are given the volume of the solution and its molarity. This is enough information to get us to moles. Once we have moles, we can calculate mass by using the molar mass.

$$250. \text{ mL sol' n x } \frac{1 \text{ L}}{1000 \text{ mL}} = 0.250 \text{ L sol' n NaCl}$$

$$0.250 \text{ L sol' n x } \frac{0.300 \text{ M NaCl}}{1 \text{ L sol' n}} = 0.0750 \text{ moles NaCl}$$

$$0.0750 \text{ moles NaCl x } \frac{58.44 \text{ g NaCl}}{1 \text{ mole NaCl}} = 4.38 \text{ g NaCl}$$

To prepare the solution, we would weigh out 4.38 g of NaCl and transfer this to a 250 mL volumetric flask. We would add some water to dissolve the NaCl and then add water until the solution level was exactly lined up with the line on the volumetric flask. The solution would be stoppered and shaken numerous times to ensure efficient mixing of the solution.

Try Study Questions 49 in Chapter 5 of your textbook now!

To prepare a solution of a given molarity by dilution, we start out with a more concentrated solution, take a portion of this more concentrated solution, and then add water to this portion until we reach the desired volume. The equation to use in working these problems is
$$c_c \bullet V_c = c_d \bullet V_d$$

If we know three of these quantities, we can solve for the fourth. The key to using this equation is to figure out what values to use for c_c, V_c, c_d, and V_d. Students are usually pretty good at figuring out which quantities correspond to c_c, c_d, and V_d. The point that they have the most difficulty with in using this equation is figuring out which volume to use for V_c. To keep this straight, always think back to what is done to make a dilution. We take a portion of the more concentrated solution and add water to it. For example, let's say that we have 2 L of a 1.0 M solution and take 10.0 mL of it to make a more dilute solution. The key volume is the volume of the portion of the more concentrated solution that we take to make the diluted solution. In this case, it is 10.0 mL. It does not matter what the total volume of the more concentrated solution is. We could have had 2 L or 1 L or a bucketful, or whatever volume we wanted of the more concentrated solution. What is important is that we took 10.0 mL of this solution to make the dilution.

Example 5-19: We have 250. mL of a 0.300 M solution of NaCl. We take 50.0 mL of this solution and dilute it to 500. mL. What is the concentration of the new solution?

This is a dilution problem. We will use the equation $c_c \bullet V_c = c_d \bullet V_d$. We need to identify each of these variables. c_c will be the concentration of the initial solution: 0.300 M. V_c will be the volume of the portion of the concentrated solution that we used to make the dilution: 50.0 mL. c_d is what we are looking for. V_d is the volume of the dilute solution that we made, 500. mL.
$$c_c \bullet V_c = c_d \bullet V_d$$
$$(0.300 \text{ M})(50.0 \text{ mL}) = c_d(500. \text{ mL})$$
$$c_d = 0.0300 \text{ M}$$

Example 5-20: What volume of a 0.200 M solution should we use to prepare 200. mL of a 0.0100 M solution?

Once again, this is a dilution problem. We will use the equation $c_c \bullet V_c = c_d \bullet V_d$. We need to identify each of these variables. c_c will be the concentration of the concentrated solution: 0.200 M. V_c will be the volume of the portion of the concentrated solution that we used to make the dilution; this is what we are asked for in this problem. c_d is the concentration of the diluted solution: 0.0100 M. V_d is the volume of the dilute solution that we will make, 200. mL.
$$c_c \bullet V_c = c_d \bullet V_d$$
$$(0.200 \text{ M})V_c = (0.0100 \text{ M})(200. \text{ mL})$$
$$V_c = 10.0 \text{ mL}$$

Therefore, we should transfer 10.0 mL of the 0.200 M solution, perhaps by means of a volumetric pipet, into a 200 mL volumetric flask and then add water to the flask until the solution level reaches the line on the flask. We would then stopper the flask and shake it numerous times.

Try Study Questions 51 in Chapter 5 of your textbook now!

c) Calculate the pH of a solution containing an acid or base and know what this means in terms of the relative amount of hydrogen ion in the solution. Calculate the hydrogen ion concentration of a solution from the pH (Section 5.9).

We define a solution as being acidic or basic relative to neutral water. The following are the concentrations of H^+ and OH^- ions in acidic, neutral, and basic solutions at 25°C.

Solution	Concentration of H^+	Concentration of OH^-
Acidic	Greater than 1.0×10^{-7} M	Less then 1.0×10^{-7} M
Neutral Water	1.0×10^{-7} M	1.0×10^{-7} M
Basic	Less then 1.0×10^{-7} M	Greater than 1.0×10^{-7} M

As the concentration of H^+ increases in an aqueous solution, the concentration of OH^- decreases, and as the concentration of OH^- increases, the concentration of H^+ decreases.

These numbers using scientific notation are not very convenient to use. Scientists have invented another way to express concentration information about H^+. This is the pH scale.
$$pH = -\log[H^+]$$

We can see that the pH of neutral water at 25°C is 7.
$$pH = -\log[H^+] = -\log[1.0 \times 10^{-7} \text{ M}] = 7.00$$

This is a more convenient number to use. There are a couple of peculiarities about the pH scale. The first is that the more acidic the solution is, the lower the pH. The more basic a solution is, the higher the pH.

Solution	pH Value (at 25°C)
Acidic	Less than 7.00
Neutral	7.00
Basic	Greater than 7.00

The second peculiarity is that pH is a logarithmic scale. A change of 1 pH unit implies a tenfold change in the hydrogen ion concentration. A solution with a pH of 5 is ten times more acidic than a solution with a pH of 6. A change of 2 pH units implies a one hundredfold $(10 \times 10 = 10^2)$ change in the hydrogen ion concentration. A change of 3 pH units implies a $10 \times 10 \times 10 = 10^3 = 1000$ fold change in the hydrogen ion concentration, and so forth.

Example 5-21: What is the pH of a 0.10 M solution of hydrochloric acid?

Hydrochloric acid is a strong acid. It ionizes completely. For each molecule of HCl, we end up with H^+ and Cl^- in solution. The concentration of H^+ in this solution is thus 0.10 M.
$$pH = -\log[H+] = -\log[0.10 \text{ M}] = 1.00$$

Notice that there are two decimal places after the decimal place in the answer. This is because there are two significant figures in the original concentration. As noted in your textbook's Example 5.11, you should have as many decimal places after the decimal place in a pH value as there are significant figures in the concentration used to calculate the pH.

Also note that we could solve this problem in which we gave the concentration of a strong acid because we could determine the concentration of hydrogen ions in the solution since strong acids ionize completely. We could not have worked this problem for a weak acid. To do that, we would need to know the extent to which the weak acid ionized. We will learn how to work problems like that much later in the course.

To go in the other direction, from pH to concentration of H+ ions, we use the equation
$$[H^+] = 10^{-pH}$$

Example 5-22: What is the hydrogen ion concentration of a solution whose pH is 4.50?

$$[H^+] = 10^{-pH} = 10^{-4.50} = 3.2 \times 10^{-5} \text{ M}$$

You should make sure you know how to use the log and the 10^x buttons on your calculator.

Try Study Questions 59 in Chapter 5 of your textbook now!

d) Solve stoichiometry problems using solution concentrations (Section 5.10).

With molarity, we have learned a new way to get to moles. We can modify our stoichiometry diagram to include this. In the following diagram, we can see that the central step has remained the same as it was before: we convert from moles of A to moles of B using the mole ratio from our balanced equation. All that has changed is that we have added a new way to get to moles or to come away from moles. We are no longer locked into using mass as our measured unit and molar mass as our conversion factor to get to moles; we can now use the volume of a solution as our measured unit and use molarity as our conversion factor.

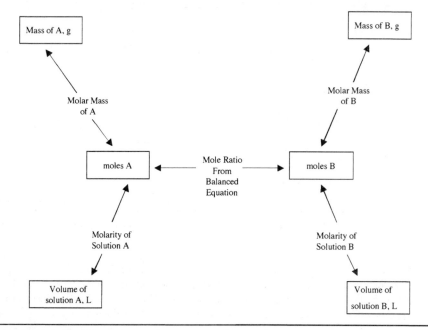

Example 5-23: What volume of a 0.100 M solution of hydrochloric acid do we need to react completely with 50.0 mL of a 0.500 M solution of sodium hydroxide?

This is a stoichiometry problem. We need the balanced chemical equation:
$$\text{HCl } (aq) + \text{NaOH } (aq) \rightarrow \text{NaCl } (aq) + \text{H}_2\text{O } (l)$$

We will again use a table to organize our information. For the examples in this chapter, we will not rewrite the table at each step, but simply show the initial table and the completed table. As you work a stoichiometry problem, you should fill in the table as you go along.

We are given the volume of the NaOH solution and the molarities of the two solutions. The molarities will be used as conversion factors to connect the volumes and moles.

	HCl *(aq)*	+	NaOH *(aq)*	→	NaCl *(aq)*	+	H$_2$O *(l)*
measured	x mL		50.0 mL				
conversion factor	0.100 M		0.500 M				
moles							

Using the volume of the sodium hydroxide solution and its molarity, we can calculate the number of moles of sodium hydroxide (keeping in mind that we need to convert from mL to L in order to use the molarity).

$$50.0 \text{ mL} \times \frac{1 \text{ L}}{1000 \text{ mL}} \times \frac{0.500 \text{ moles NaOH}}{1 \text{ L sol' n}} = 0.0250 \text{ moles NaOH}$$

Now that we have moles of NaOH, we can calculate the number of moles of HCl using the mole ratio. This is easy in this case since the two react in a 1:1 ratio. Finally, we can calculate the volume of HCl solution, using the molarity of the HCl solution as the conversion factor.

$$0.0250 \text{ moles NaOH} \times \frac{1 \text{ mole HCl}}{1 \text{ mole NaOH}} = 0.0250 \text{ moles HCl}$$

$$0.0250 \text{ moles HCl} \times \frac{1 \text{ L sol' n}}{0.100 \text{ moles HCl}} \times \frac{1000 \text{ mL}}{1 \text{ L}} = 2.50 \times 10^2 \text{ mL HCl sol' n}$$

The completed table looks like the following:

	HCl *(aq)*	+	NaOH *(aq)*	→	NaCl *(aq)*	+	H$_2$O *(l)*
measured	2.50 x 10^2 mL		50.0 mL				
conversion factor	0.100 M		0.500 M				
moles	0.0250 moles		0.0250 moles				

Example 5-24: What volume of a 0.150 M solution of copper(II) chloride solution is required to react with 1.00 g of zinc metal?

In this case, we are given the mass of zinc and are asked for the volume of the copper(II) chloride solution. Since we have a mass of zinc, the conversion factor we shall use to get to moles of zinc is the molar mass. Since we are asked for the volume of a solution, the conversion factor we shall use for the copper(II) chloride is the molarity of the solution.

	Zn *(s)*	+	CuCl$_2$ *(aq)*	→	Cu *(s)*	+	ZnCl$_2$ *(aq)*
measured	1.00 g		x mL				
conversion factor	65.39 g/mole		0.150 M				
moles							

$$1.00 \text{ g Zn} \times \frac{1 \text{ mole Zn}}{65.39 \text{ g Zn}} = 0.0153 \text{ moles Zn}$$

$$0.0153 \text{ moles Zn} \times \frac{1 \text{ mole CuCl}_2}{1 \text{ mole Zn}} = 0.0153 \text{ moles CuCl}_2$$

$$0.0153 \text{ moles CuCl}_2 \times \frac{1 \text{ L sol' n}}{0.150 \text{ moles CuCl}_2} \times \frac{1000 \text{ mL}}{1 \text{ L}} = 102 \text{ mL CuCl}_2 \text{ solution}$$

The completed table looks like the following:

	Zn (s)	+	CuCl$_2$ (aq)	→	Cu (s)	+	ZnCl$_2$ (aq)
measured	1.00 g		102 mL				
conversion factor	65.39 g/mole		0.150 M				
moles	0.0153 moles		0.0153 moles				

Try Study Questions 61 and 67 in Chapter 5 of your textbook now!

e) Explain how a titration is carried out, explain the procedure of standardization, and calculate concentrations or amounts of reactants from titration data (Section 5.10).

In a titration, two reactants are combined together by adding one reactant little by little to the other until we have added exactly enough of the reactants to get complete reaction. In an acid-base titration, for example, we start out with either an acid or base in a container and add the other until the amount of H$^+$ provided by the acid exactly equals the amount of OH$^-$ provided by the base. This point in the reaction is called the equivalence point. Often, we use a device called a buret to deliver the reactant that is added little by little. Of course, we need to have a way to detect when we have reached the equivalence point. For acid-base titrations, we will often use acid-base indicators to do this. These are compounds that change color at a particular pH. We select an acid-base indicator that will change color at the pH of the equivalence point.

To obtain quantitative information from the titration, we normally know enough information about one of the reactants to be able to calculate how many moles of it are present.

We can then use the mole ratio from the balanced equation to determine how many moles of the other material are present.

Problems involving titrations are really just stoichiometry problems.

Example 5-25: In a titration, 50.00 mL of a hydrochloric acid solution are titrated against 0.1004 M sodium hydroxide. If it takes 29.54 mL of the sodium hydroxide solution to reach the equivalence point, what was the concentration of the original hydrochloric acid solution?

In this case, we are given the volumes of both solutions and the concentration of the NaOH solution. We are asked for the concentration of the hydrochloric acid solution. This time, our unknown is the conversion factor in our table.

	HCl (aq)	+	NaOH (aq)	→	NaCl (aq)	+	H$_2$O (l)
measured	50.00 mL		29.54 mL				
conversion factor	x M		0.1004 M				
moles							

We need to get to moles, our counting unit. We can do this for the sodium hydroxide solution.

$$29.54 \text{ mL} \times \frac{1 \text{ L}}{1000 \text{ mL}} \times \frac{0.1004 \text{ moles NaOH}}{1 \text{ L sol' n}} = 2.966 \times 10^{-3} \text{ moles NaOH}$$

We can then use the mole ratio from the balanced equation to calculate how many moles of hydrochloric acid this corresponds to in this reaction.

$$2.966 \times 10^{-3} \text{ moles NaOH} \times \frac{1 \text{ mole HCl}}{1 \text{ mole NaOH}} = 2.966 \times 10^{-3} \text{ moles HCl}$$

At this point, we have the number of moles of HCl and the volume of the HCl solution. We can use this information to get the molarity of the solution.

$$c_{molarity} = \frac{2.966 \times 10^{-3} \text{ moles HCl}}{50.00 \text{ mL} \times \dfrac{1 \text{ L}}{1000 \text{ mL}}} = 0.05932 \text{ M HCl}$$

The completed table looks like the following:

	HCl *(aq)*	+	NaOH *(aq)*	→	NaCl *(aq)*	+	H$_2$O *(l)*
measured	50.00 mL		29.54 mL				
conversion factor	0.05932 M		0.1004 M				
moles	2.966 × 10^{-3} moles		2.966 × 10^{-3} moles				

WARNING: Some of you may have incorrectly learned an equation with which to solve titrations problems ($M_1V_1 = M_2V_2$) in a previous course. Do NOT use this equation. It only works for titration problems in which the mole ratio is 1:1, like it was in this example. There are titrations where the mole ratio will not be 1:1. In these cases, that equation will not provide the correct answer!

In order to work the last example, we had to start out by knowing the concentration of the NaOH solution. The only problem is that we cannot simply weigh out sodium hydroxide, dissolve it, and calculate the concentration. This is due, in part, to the fact that sodium hydroxide pellets pick up water from the air. When we weigh out sodium hydroxide pellets, we are weighing not only sodium hydroxide but this water as well. How could we know that the NaOH solution in this example is 0.1004 M? What we do is to perform a standardization of the sodium hydroxide solution. This is basically another titration. We can weigh out a known amount of a compound like potassium hydrogen phthalate (abbreviated KHP), a weak monoprotic acid. We can thus determine the number of moles of potassium hydrogen phthalate. We then titrate this against our sodium hydroxide solution. At the equivalence point, the number of moles of sodium hydroxide delivered will be equal to the number of moles of KHP we started with. If we have measured the volume of NaOH solution delivered, then we can determine the concentration of the NaOH solution by dividing this number of moles of NaOH by this volume (in liters). Now we know the concentration of the NaOH solution and can use the NaOH hydroxide solution in other titrations.

Try Study Questions 71 and 73 in Chapter 5 of your textbook now!

Other Notes

1. Electrons conduct electricity in a wire, and ions conduct electricity in a solution.

2. Chemical equations must be balanced for both mass and charge. For overall balanced equations, we sometimes do not worry about balancing charge explicitly because the formulas of the compounds ensure electrical neutrality. When we deal with net ionic equations, we must be careful that charge is balanced.

3. Common Properties of Acids and Bases

   Acids: sour taste (don't use this test in the lab), produce bubbles of CO_2 when added to a metal carbonate, produce H_2 gas when they are reacted with many metals, change the color of acid-base indicators (such as litmus), neutralize bases.

   Bases: bitter taste (don't use this test in the lab), slippery feel (don't use this test in the lab), change the color of acid-base indicators, neutralize acids.

4. Hydrogen ions do not exist as bare protons in solution. They combine with water molecules to form H_3O^+, hydronium ion, and other complexes containing even more molecules of water.

5. Acid/Base Behavior of Oxides

   Oxides of nonmetals can react with water to form acidic solutions. Sometimes, these oxides are called acidic oxides.
   Example: NO_2 *(g)* + H_2O *(l)* $\rightarrow$ HNO_3 *(aq)*

   Oxides of metals can react with water to form basic solutions. Sometimes, these oxides are called basic oxides.
   Example: Na_2O *(s)* + H_2O *(l)* $\rightarrow$ 2NaOH *(aq)*

Try Study Question 17 in Chapter 5 of your textbook now!

CHAPTER 6: Principles of Reactivity: Energy & Chemical Reactions

Chapter Overview

Most chemical and physical changes are accompanied by energy changes. In this chapter, we begin our study of thermodynamics (the science of heat and work) and its applications to chemical reactions. The first law of thermodynamics (the law of conservation of energy) states that energy is neither created nor destroyed in ordinary processes. There are two basic types of energy: kinetic energy and potential energy. The unit for energy most commonly used in chemistry is the joule (J). In analyzing a particular process, it is important to define the system (the part of the universe under study) and the surroundings (the rest of the universe). Due to the law of conservation of energy, any change in the energy of the system must come from or go to the surroundings and can be accounted for. A mathematical statement of the first law for processes in which heat (q) and work (w) are the only methods of energy transfer, is $\Delta E = q + w$. ΔE is the internal energy change for the system. You will learn the positive and negative sign conventions for q and w. If the only type of work that can be done is the expansion of a gas, and the process is carried out under conditions of constant volume, then the work will be equal to zero and ΔE will simply be q, $\Delta E = q_v$. Often, we will carry out processes under constant pressure rather than under constant volume. The heat transferred under conditions of constant pressure is defined to be the enthalpy change, $\Delta H = q_p$. ΔH will be negative for an exothermic process (heat transferred from the system to the surroundings) and will be positive for an endothermic process (heat transferred from the surroundings to the system).

The same amount of heat will cause the temperature of different materials to go up by different amounts. This is accounted for by the specific heat capacity (C). You will learn to use the equation: $q = C \cdot m \cdot \Delta T$ to calculate the amount of heat transferred in a given process that does not involve a change of state. In addition, you will learn that this equation is very useful in problems involving objects at different temperatures being brought into contact with each other. Heat transfer will occur between the two objects until they are at the same temperature. At this point, the heat lost by one object will be equal to the heat gained by the other object so that $q_1 + q_2 = 0$.

If a change of state does occur, then you will use another equation to calculate the heat transferred because during a change of state, even though heat is being added, no change of temperature occurs. For melting, $q = \Delta H_{fus} \cdot m$. For boiling, $q = \Delta H_{vap} \cdot m$.

We can determine the heat transferred in a chemical process by using a device called a calorimeter. There are two basic types of calorimeters. One is a constant pressure calorimeter, which will give a value of ΔH. The key equation usually used for this type of calorimeter is $q_{rxn} + q_{solution} = 0$. The other type of calorimeter is a constant volume calorimeter, of which the bomb calorimeter is the most common. This will lead to a value of ΔE. The key equation usually used for this type of calorimeter is $q_{rxn} + q_{bomb} + q_{water} = 0$.

Both E and H are state functions. This means that the values of ΔE and ΔH do not depend on the path chosen to go from the initial to the final state. All that matters is the value of the initial state and the final state. A consequence of ΔH being a state function is Hess's law, which states that if we add up a series of reactions, the total ΔH will be equal to the sum of the ΔH values for the reactions added. Sometimes, in order to get the given reactions to add up properly, we will need to modify the given reactions. If we reverse a chemical reaction, the sign of ΔH will be the opposite of what it was. If we multiply a reaction by a coefficient, then we will also multiply the ΔH value by the same coefficient. Another application of Hess's law involves using standard molar enthalpies of formation. The standard state for an element or compound is the most stable form of the substance in the physical state that exists at a pressure of 1 bar at a specified temperature. For solutes, the standard concentration is 1 M. The standard molar enthalpy of formation is defined to be the enthalpy change for the formation of one mole of a compound directly from its component elements in their standard states. We can calculate the enthalpy change for a reaction by using the equation:

$$\Delta H^{\circ}_{rxn} = \Sigma \left[\Delta H^{\circ}_f \left(products \right) \right] - \Sigma \left[\Delta H^{\circ}_f \left(reactants \right) \right]$$

You will also see that ΔH values can be used in stoichiometry calculations.

Key Terms

In this chapter, you will need to learn and be able to use the following terms:

Calorimeter: a device used for measuring the heats of chemical reactions.

Calorimetry: the experimental determination of the heats of chemical reactions.

Chemical potential energy: the potential energy associated with chemical bonding.

Electrical energy: the energy of electrons moving through a conductor.

Electrostatic energy: the potential energy that is associated with the separation of two charges.

Endothermic process: a process in which heat is transferred from the surroundings to the system.

Energy: the capacity to do work.

Enthalpy: the heat content of a substance at constant pressure. We abbreviate enthalpy by the letter H and changes in enthalpy by ΔH.

Exothermic process: a process in which heat is transferred from the system to the surroundings.

First law of thermodynamics: another name for the law of conservation of energy.

Gravitational energy: the potential energy that results from the attraction of any two masses.

Heat of fusion: the heat required to convert a solid at its melting point to a liquid, ΔH_{fus}.

Heat of vaporization: the heat required to convert a liquid at its boiling point to a gas, ΔH_{vap}.

Hess's law: if a reaction is the sum of two or more other reactions, ΔH for the overall process is the sum of the ΔH values of those reactions.

Internal energy: for a chemical system, this is the sum of the potential and kinetic energies of the atoms, molecules, or ions in the system. We abbreviate internal energy by the letter E and changes in internal energy by ΔE.

Kinetic energy: the energy of motion.

Law of conservation of energy: energy is neither created nor destroyed in ordinary processes.

Mechanical energy: the energy of a moving object at the macroscopic level.

Potential energy: energy that results from an object's position; stored energy.

Sound: a type of energy that results from the compression and expansion of the spaces between molecules.

Specific heat capacity: the quantity of heat required to raise the temperature of 1 gram of a substance by one kelvin.

Standard molar enthalpy of formation (ΔH_f^0): the enthalpy change for the formation of 1 mole of a compound directly from its component elements in their standard states.

Standard state: for an element or compound, this is the most stable form of the substance in the physical state that exists at a pressure of 1 bar (approximately 1 atmosphere) at a specified temperature. For a solute, the concentration is 1 M.

State function: a function whose value is determined only by the state of the system and not by the pathway by which that state was achieved.

Sublimation: the change of state in which a material changes directly from the solid state to the gaseous state without passing through the liquid state.

Surroundings: everything outside the system that can exchange energy with the system.

System: the object, or collection of objects, being studied.

Temperature: a property that determines the direction of heat transfer; heat transfers from something at a higher temperature to something at a lower temperature.

Thermal energy: heat energy due to the motion of atoms, molecules, and ions.

Thermal equilibrium: the condition at which two objects are at the same temperature.

Thermodynamics: the science of heat and work.

Work: work is done when a force is applied through a distance. Mathematically, w = F x d.

Chapter Goals

By the end of this chapter you should be able to:

- **Assess heat transfer associated with changes in temperature and changes of state.**

    a) Describe various forms of energy and the nature of heat and thermal energy transfer (Section 6.1).

    Energy is the capacity to do work. Work (w) is done when a force (F) is applied through a distance (d) and can be defined mathematically by the formula w = F x d.

    There are two major types of energy: kinetic energy and potential energy. Kinetic energy is the energy of motion. Thermal energy, mechanical energy, electrical energy, and sound are all types of kinetic energy.

    Potential energy results from an object's position; it is stored energy. Gravitational energy, chemical potential energy, and electrostatic energy are all types of potential energy.

    One form of energy may be converted into another form of energy. For example, the potential energy that a book has when it is held above the floor will be converted into kinetic energy if the person releases the book, letting it fall.

Just as there is a law of conservation of matter, there is also a law of conservation of energy. Energy can be neither created nor destroyed in ordinary changes. In other words, the energy of the universe is a constant for ordinary changes. Sometimes, this law is called the first law of thermodynamics. This law implies that if we carry out a procedure in which we convert one type of energy into others, then we should be able to account for all of the energy in some way at the end of the process. The energy has not been destroyed. We can trace the transformation of energy from one form to another.

As noted, one type of energy is thermal energy or heat. It is important to distinguish between heat and temperature. Heat and temperature are not the same thing. Heat is thermal energy. Temperature determines the direction of heat transfer; heat transfers from something at a higher temperature to something at a lower temperature. How much heat an object possesses depends not only on the temperature but also on what the material is made out of and on the amount of the material present.

When two objects at different temperatures are brought into contact, heat transfer will occur until the two objects are at the same temperature. If heat can only be transferred between these two objects, then the heat gained by the cooler object will be equal to the heat lost by the hotter object.

b) Use the most common energy unit, the joule, and convert between other energy units and joules (Section 6.1).

The most common energy unit used in science is the joule.

$$1 \text{ joule } = 1 \text{ J } = 1 \ \frac{\text{kg} \cdot \text{m}^2}{\text{s}^2}$$

There are a couple of other energy units that you should become familiar with as well. One is the calorie (cal). Notice that this calorie is written with a lowercase c. A calorie is the amount of heat required to raise the temperature of one gram of water 1°C (specifically, from 14.5°C to 15.5°C).

$$1 \text{ cal} = 4.184 \text{ J.}$$

The dietary Calorie (sometimes called the food Calorie) is also an important energy unit to know; it is the Calorie that is used as the energy unit on most food product containers in the United States. Notice that it is written with an uppercase C. One Calorie is equal to 1000 calories.

Example 6-1: The box for a particular type of cracker listed that three crackers had an energy content of 60. Calories. Calculate the number of calories, the number of joules, and the number of kilojoules to which this corresponds.

$$60. \text{ Cal x } \frac{1000 \text{ cal}}{1 \text{ Cal}} = 6.0 \text{ x } 10^4 \text{ cal}$$

$$6.0 \text{ x } 10^4 \text{ cal x } \frac{4.184 \text{ J}}{1 \text{ cal}} = 2.5 \text{ x } 10^5 \text{ J}$$

$$2.5 \text{ x } 10^5 \text{ J x } \frac{1 \text{ kJ}}{1000 \text{ J}} = 2.5 \text{ x } 10^2 \text{ kJ}$$

Try Study Question 5 in Chapter 6 of your textbook now!

c) Recognize and use the language of thermodynamics: the system and its surroundings; exothermic and endothermic reactions (Section 6.1).

The system in a thermodynamics experiment is the object, or collection of objects, being studied. The surroundings are everything outside the system that can exchange energy with the system. Technically, the surroundings would be the entire universe except for the system under study, but we usually do not have to go that far out from the system to account for just about all of the energy exchanged. Often, we will carry out a reaction in a container. We might consider the contents of the container and sometimes both the contents and the container as being the system. Everything else (the lab bench, the air in the room) would be considered to be the surroundings.

A process in which heat is transferred from the system to the surroundings is said to be exothermic. Heat is leaving the system and going into the surroundings. One way to remember this is that one leaves a building by means of an exit so the term that deals with heat leaving the system begins with ex-.

A process in which heat is transferred from the surroundings to the system is said to be endothermic. Heat is entering the system from the surroundings. The word enter begins with en-, so an endothermic process involves heat entering the system.

Example 6-2: Imagine you are boiling an egg. Define the system and surroundings and tell whether the process of cooking the egg is exothermic or endothermic.

There are a few ways to define the system and surroundings depending on exactly what we wanted to examine. Probably the easiest system to define is the egg with the surroundings being everything else (the water, the pan, the stove, etc.). For some cases, we might want to define the system to be the egg and the water and for yet other cases, we might want to define the system to be the egg, the water, and the pan. Let us consider the system to be the egg.

In order to cook the egg, we must provide heat. Heat enters the egg. This is an endothermic process.

Try Study Question 63 in Chapter 6 of your textbook now!

d) Use specific heat capacity in calculations of heat transfer and temperature changes (Section 6.2).

The same amount of heat added to different substances will cause different temperature changes. Adding the same amount of heat to a metal like aluminum *vs.* to water will cause different temperature changes in the two materials. A different temperature change will be caused even if we consider adding the same amount of heat to different amounts of the same material. The same amount of heat added to a teaspoon of water will cause a different change in temperature than it would for a whole pan of water. All of these considerations are taken into account in the specific heat capacity. The specific heat capacity (C) is the quantity of heat that is required to raise the temperature of 1 gram of a substance by one kelvin. Mathematically, it can be defined by the following equation:

$$\text{Specific Heat} = \frac{\text{heat}}{\text{mass} \bullet \text{change in temperature}}$$

$$C = \frac{q}{m \bullet \Delta T}$$

In using this and other equations, Δ means "change in." Δ indicates to take the final value of the quantity and subtract from it the initial value. Thus, ΔT means $T_{final} - T_{initial}$.

Because this equation deals with a *change* in temperature, it does not matter whether we use kelvins or degrees Celsius. The reason is that the size of a kelvin and a Celsius degree is the same. For example, in going from 0°C to 25°C, $\Delta T = 25°C - 0°C = 25°C$. Zero degrees Celsius corresponds to 273 K, and 25°C corresponds to 298 K. We would calculate the same ΔT in kelvin that we had in °C: 298 K – 273 K = 25 K.

Example 6-3: What is the specific heat of aluminum metal if 89.7 J of heat must be added to raise the temperature of 10.0 g of aluminum from 25.0°C to 35.0°C?

We are given q, m, T_f and T_i. We are asked to calculate C.

$$C = \frac{q}{m \cdot \Delta T}$$

$$= \frac{89.7 \text{ J}}{(10.0 \text{ g})(35.0°C - 25.0°C)}$$

$$= 0.897 \frac{\text{J}}{\text{g}°C}$$

Perhaps even more important than this form of the equation, we can rearrange the equation for specific heat capacity to solve for the heat, q.

$$q = C \cdot m \cdot \Delta T$$

Example 6-4: How much heat does it take to raise the temperature of a 35.0 g block of aluminum from 25.0°C to 100.0°C? The specific heat capacity of aluminum is 0.897 J/(g•K).

In this case, we are given C, m, T_f, and T_i, and are asked to solve for q.

$$q = C \cdot m \cdot \Delta T$$

$$= \left(0.897 \frac{\text{J}}{\text{g} \cdot \text{K}}\right)(35.0 \text{ g})(100.0°C - 25.0°C)$$

$$= 2.35 \times 10^3 \text{ J} \times \frac{1 \text{ kJ}}{1000 \text{ J}} = 2.35 \text{ kJ}$$

Example 6-5: To what temperature will a 25.0 g block of aluminum initially at 37.0°C rise if 548 J of heat are added? The specific heat capacity of aluminum is 0.897 J/(g•K)

In this case, we are given q, C, m, and T_i. We are asked to solve for T_i.

$$q = C \cdot m \cdot \Delta T$$

$$548 \text{ J} = \left(0.897 \frac{\text{J}}{\text{g} \cdot \text{K}}\right)(25.0 \text{ g})(T_f - 37.0°C)$$

We can distribute in the specific heat and the mass to obtain the following:

$$548 \text{ J} = 22.4 \frac{\text{J}}{\text{K}} T_f - 830. \text{ J}$$

We then collect the common terms and solve for the final temperature.

$$1378 \text{ J} = 22.4 \frac{\text{J}}{\text{K}} \text{ T}_f$$

$$\text{T}_f = 61.4°\text{C}$$

Try Study Questions 9 and 11 in Chapter 6 of your textbook now!

So far, we have considered cases in which we have dealt with heating or cooling only one substance. We shall now turn our attention to mixing together two substances that are at different temperatures. If this is done in an insulated container where heat can be transferred only between the two materials, then heat transfer will occur until the objects reach thermal equilibrium. At this point, two key things will be true: a) the final temperature of the two objects will be the same and b) the heat gained by one material will be equal to the heat lost by the other. This can be summarized by the following for two objects (1 and 2):

$$q_1 + q_2 = 0$$

Example 6-6: A 125 g block of copper at 65.0°C is dropped into an insulated container containing 250. g of water initially at 25.0°C. The final temperature of the mixture is 26.8°C. The specific heat capacity of water is 4.184 J/(g•K). What is the specific heat capacity of copper?

We start out with copper at a higher temperature and water at a lower temperature. They are mixed together. Heat will transfer between the two until they reach the same temperature. That temperature is 26.8°C. Due to the law of conservation of energy,

$$q_{Cu} + q_{H_2O} = 0$$

$$C_{Cu} \bullet m_{Cu} \bullet \Delta T_{Cu} + C_{H_2O} \bullet m_{H_2O} \bullet \Delta T_{H_2O} = 0$$

$$C_{Cu} \bullet m_{Cu} \bullet \left(T_{final_{Cu}} - T_{inital_{Cu}} \right) + C_{H_2O} \bullet m_{H_2O} \bullet \left(T_{final_{H_2O}} - T_{inital_{H_2O}} \right) = 0$$

We substitute in the values that we know and let *x* equal the specific heat capacity of copper.

$$x \bullet 125 \text{ g} \bullet (26.8°\text{C} - 65.0°\text{C}) + 4.184 \frac{\text{J}}{\text{g} \bullet \text{K}} \bullet 250. \text{ g} \bullet (26.8°\text{C} - 25.0°\text{C}) = 0$$

Notice that for each material, we calculate ΔT by taking the final temperature and subtracting the initial temperature. This results in a negative ΔT for copper, the material going down in temperature. It results in a positive ΔT for water, the material going up in temperature.

$$\left(-4.78 \times 10^3 \text{ g} \bullet \text{K}\right) x + 1.88 \times 10^3 \text{ J} = 0$$

$$x = 0.394 \frac{\text{J}}{\text{g} \bullet \text{K}}$$

The specific heat capacity of copper is thus 0.394 J/(g•K).

One of the harder problems algebraically is to solve the following type of problem for the final temperature.

Example 6-7: A 250. g piece of copper is heated to 75.0°C. It is then transferred to an insulated container containing 250. g of water at 20.0°C. The specific heat capacity of copper is 0.394 J/(g•K) and that of water is 4.184 J/(g•K). What is the final temperature of the mixture?

This problem starts out just like the last one. The sum of the heats for copper and water will be zero.

$$q_{Cu} + q_{H_2O} = 0$$

$$C_{Cu} \bullet m_{Cu} \bullet \Delta T_{Cu} + C_{H_2O} \bullet m_{H_2O} \bullet \Delta T_{H_2O} = 0$$

$$C_{Cu} \bullet m_{Cu} \bullet \left(T_{final_{Cu}} - T_{inital_{Cu}} \right) + C_{H_2O} \bullet m_{H_2O} \bullet \left(T_{final_{H_2O}} - T_{inital_{H_2O}} \right) = 0$$

Now we substitute in the given information; our unknown this time is the final temperature for the copper and the water. We know that heat transfer occurs until we reach thermal equilibrium, at which point the temperature of the copper and the water will be the same.

$$0.394 \frac{J}{g \bullet K} \bullet 250.\, g \bullet \left(T_f - 75.0°C \right) + 4.184 \frac{J}{g \bullet K} \bullet 250.\, g \bullet \left(T_f - 20.0°C \right) = 0$$

Once you have reached this point, you are finished with the chemistry. The rest of the problem is just algebra. We distribute in the specific heat capacities and masses and then combine like terms. We then solve for our unknown.

$$98.5 \frac{J}{°C} T_f - 7.39 \times 10^3\, J + 1.05 \times 10^3 \frac{J}{°C} T_f - 2.09 \times 10^4\, J = 0$$

$$1.14 \times 10^3 \frac{J}{°C} T_f - 2.83 \times 10^4\, J = 0$$

$$1.14 \times 10^3 \frac{J}{°C} T_f = 2.83 \times 10^4\, J$$

$$T_f = 24.7°C$$

The final temperature for both materials is 24.7°C. This example points out what we mean by specific heat capacity. The copper has lost the same amount of heat as the water gained. This heat loss caused the copper to drop in temperature from 75.0°C all the way down to 24.7°C, a 50.3°C difference in temperature. This same amount of heat caused the same mass of water to go up in temperature by only 24.7°C – 20.0°C = 4.7°C. The reason for this is that water has a much greater specific heat capacity; it takes much more heat to get water to go up in temperature than it does for copper.

Try Study Questions 15 and 17 in Chapter 6 of your textbook now!

e) Understand the sign conventions in thermodynamics (Section 6.2).

If heat and work are the only ways that a system can exchange energy with its surroundings, then the first law of thermodynamics, the law of conservation of energy, implies that we should be able to account for any energy change by taking into account the heat exchange and work that occurred. This can be summarized by the following statement of the first law:

$$\Delta E = q + w$$

We will better define ΔE in a later goal.

The sign conventions for heat (q) and work (w) for use in this equation are such that if the system ends up having a greater energy, then the sign is positive and that if the system ends up having less energy, then the sign is negative. It is useful to draw pictures to help you decide if q or w should be positive or negative.

Let's first consider heat, q. In an endothermic process, heat goes into the system from the surroundings.

The system ends up with a greater energy at the end than it had at the beginning. If we were to take the final energy of the system and subtract the initial energy of the system, then we would get a positive number. The sign for q for an endothermic process is thus positive.

Now let's consider an exothermic process. In this case, heat leaves the system and goes into the surroundings:

The system ends up with less energy at the end than it started with. If we take the final energy of the system and subtract the initial energy of the system, we will get a negative number. The sign for q for an endothermic process is thus negative.

Moving on to work, let us consider work being done on the system.

When work is done by the surroundings on the system, the system ends up with more energy at the end of the process. The sign for w for the surroundings doing work on the system is thus positive.

Finally, let us consider the system doing work on the surroundings.

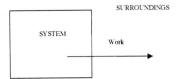

The system ends up with less energy at the end of the process than it had at the beginning. The sign for w when the system does work on the surroundings is negative.

Example 6-8: Calculate ∆E for a system if 35 kJ of work is done on the system, and the system gives off 45 kJ of heat.

According to the first law of thermodynamics, ∆E = q + w. In this case, heat is given off so q is negative. Work is done on the system, so w is positive.

$$\Delta E = -45 \text{ kJ} + 35 \text{ kJ}$$
$$\Delta E = -10 \text{ kJ}$$

Thus, the system ends up with 10 kJ less energy than it had at the beginning. Net, 10 kJ of energy have been transferred from the system to the surroundings.

f) Use heat of fusion and heat of vaporization to find the quantity of thermal energy involved in changes of state (Section 6.3).

Refer to Figure 6.11 in your textbook. It is important to know the basic form of this diagram because it often helps students to draw a sketch of this figure when they are solving problems in which a change of state may occur.

There are five regions in this figure. Starting from the left, we first have the solid region. In this region, as we add heat, the heat goes into raising the temperature of the solid. The equation that is used to calculate the heat in this region is the same as the equation for heat we have been using.

$$q_{solid} = C_{solid} \bullet m \bullet \Delta T$$

Following the solid region, we come to a flat region. In this region, we are changing the state of the material from being a solid to being a liquid. Students are often surprised by the fact that we get a flat line for a change of state. We are continuing to add heat, but the temperature does not change. Instead of going into raising the temperature, the added heat goes into changing the physical state of the material from a solid to a liquid, disrupting intermolecular attractions between the particles. The temperature at which this flat line occurs is the melting point of the solid. The temperature will level off until all of the solid has melted. The equation that is used to calculate the heat added in this region is

$$q_{melt} = \Delta H_{fus} \bullet m$$

In this equation, ΔH_{fus} is the heat of fusion of the solid. This is the heat required to change a solid to a liquid. The form of the equation shown above assumes that its units are J/g, although sometimes the units are given as J/mole.

Once the material has all melted, the temperature starts to go up again as we add heat. This is the liquid region. The equation for the heat added in this region is

$$q_{liquid} = C_{liquid} \bullet m \bullet \Delta T$$

After the liquid region, the graph levels off again. We are adding heat, but the heat is not going into changing the temperature but into changing the state of the material. We are boiling the liquid and changing it to the gaseous state. The equation that is used to calculate the heat added is similar to that we had for melting the solid, except that now we will use ΔH_{vap}, the heat of vaporization, instead of the heat of fusion. The heat of vaporization is the heat required to change a material from the liquid state to the gaseous state.

$$q_{boil} = \Delta H_{vap} \bullet m$$

Once all of the liquid has boiled, adding more heat once again raises the temperature of the material. The equation for the heat needed in this region is

$$q_{gas} = C_{gas} \bullet m \bullet \Delta T$$

Example 6-9: Calculate how much heat must be added to 50.0 g of liquid water at 25.0°C to steam at 125°C. The specific heat capacity of liquid water is 4.184 J/(g•°C) and that of water vapor is 2.03 J/(g•°C). The heat of vaporization of water is 2256 J/g.

As a first step, sketch Figure 6.11 and then mark off the regions that this problem involves. We are starting out with liquid water so at first we will add heat to increase the temperature from the initial temperature up to the boiling point temperature. After that, we will boil the water. Finally, we will add heat to raise the temperature of the gaseous water from the boiling point up to the final temperature. Note that for this problem, we do not need to worry about the solid region or the melting region.

Step 1: Raise temperature of liquid water from the initial temperature of 25.0°C to the boiling point of water, 100°C.
$$q_{liquid} = C_{liquid} \cdot m \cdot \Delta T$$
$$= 4.184 \frac{J}{g \cdot °C} \cdot 50.0 \text{ g} \cdot (100°C - 25.0°C)$$
$$= 1.6 \times 10^4 \text{ J}$$

Let your units help you in this type of problem. The unit for heat is J. Since the specific heat capacity has units of J/(g•°C), we know that we must multiply the specific heat capacity by a mass to cancel the g and by something involving temperature to cancel °C (or K). The second major thing to point out is that the final temperature we used was the boiling point of water, 100°C; we did not use the final temperature of the problem, 125°C. The reason for this is that the material remains a liquid only until we get to the boiling point.

Step 2: Boil the liquid.
$$q_{boil} = \Delta H_{vap} \cdot m$$
$$= 2256 \frac{J}{g} \cdot 50.0 \text{ g}$$
$$= 1.13 \times 10^5 \text{ J}$$

Once again, let your units help you. In this case, the units for the heat of vaporization are J/g. We still want to end up with J as our final unit, so we must multiply this only by the mass.

Step 3: Raise temperature of gaseous water from 100°C to 125°C.
$$q_{gas} = C_{gas} \cdot m \cdot \Delta T$$
$$= 2.03 \frac{J}{g \cdot °C} \cdot 50.0 \text{ g} \cdot (125°C - 100°C)$$
$$= 2.5 \times 10^3 \text{ J}$$

Step 4: Now we just need to add together all of the heats from the individual steps.
$$1.6 \times 10^4 \text{ J} + 1.13 \times 10^5 \text{ J} + 2.5 \times 10^3 \text{ J} = 1.31 \times 10^5 \text{ J} = 131 \text{ kJ}$$

Try Study Question 23 in Chapter 6 of your textbook now!
(Notice that in this problem, you will be decreasing the temperature, rather than increasing the temperature.)

- ## Apply the first law of thermodynamics.

   ### Understand the basis of the first law of thermodynamics (Section 6.4).

   The first law of thermodynamics is simply another way of stating the law of conservation of energy. If the internal energy of the system is changing, the energy must be coming from or else going into the surroundings. We should be able to account for these changes since energy is neither created nor destroyed in the universe. The energy change for the universe must be zero. If the only transfers that can occur between the system and the surroundings are heat and work, then if we account for any heat transferred to or from the system and any work done on or by the system, then we should know exactly how much energy the system has gained or lost, which will be equal to the amount of energy the surroundings lost or gained, respectively. This keeps the total energy in the universe a constant because

   $$\Delta E_{system} + \Delta E_{surroundings} = 0$$

- ## Define and understand the state functions enthalpy and internal energy.

   ### Recognize state functions whose values are determined only by the state of the system and not by the pathway by which that state was achieved (Section 6.4).

   We will now be a little more precise in what we mean by E. E is the internal energy. The internal energy in a chemical system is the sum of the potential and kinetic energies of the atoms, molecules, or ions in the system. We normally do not worry about the actual value of E for a material, but instead about the change in the internal energy in a process, ΔE. This can be evaluated by using the equation above, by taking into account the heat transferred between the system and the surroundings and also the work done on or by the system.

   One type of work that can often occur in chemical reactions is called P-V work. This is the work associated with a change in volume that occurs against a resisting external pressure. The equation for this type of work is

   $$w = -P \cdot \Delta V$$

   If we carry out a reaction under conditions of constant volume, then the P-V work will be zero because ΔV will be zero. Assuming that this was the only type of work possible, the equation for ΔE simplifies to

   $$\Delta E = q + 0$$
   $$\Delta E = q_v$$

   where the subscript v we wrote after q simply means that this is the heat transferred under conditions of constant volume.

   In chemistry, it is more usual to carry out processes under conditions of constant pressure than constant volume. In order to do this, we simply need to leave the reaction vessel in contact with the atmosphere, and the pressure will be essentially constant. For more precise work, we could set up things more carefully, but this suffices for many cases.

   The enthalpy, H, is defined to be the heat content of a substance under conditions of constant pressure. Just as was true for E, we are almost never interested in the actual value of H for a material, but instead in the enthalpy change, ΔH, for a process. ΔH is the heat change under conditions of constant pressure.

   $$\Delta H = q_p$$

   Because ΔH represents a heat change, the sign conventions for ΔH are the same as what we learned earlier for q:

Exothermic Reaction	Negative ΔH
Endothermic Reaction	Positive ΔH

So, under conditions of constant volume, $\Delta E = q_v$, and under conditions of constant pressure, $\Delta H = q_p$. We shall use these equations more when we deal with calorimetry. By far, we will use ΔH much more than ΔE in this course.

Enthalpy and internal energy are both state functions. A state function's value is determined only by the state of the system and not by the pathway by which that state was achieved. The bank balance example in the text is a very good one. Here is a variation of it. All that is important is the value that it currently has; it does not matter how it got there. A bank account starts at a value of zero. We could deposit $25, deposit another $25, deposit another $25 and then withdraw $35. The final state is $40 above zero. The change in the bank balance from the start to the end was $40 – $0 = $40. Alternatively, we could start again with a bank balance of zero and simply deposit $40. The final state is still $40 above zero. The change in the balance would be calculated the same way by taking the final state and subtracting the initial state and obtaining the same answer: $40 – $0 = $40. The pathway did not matter; all that mattered was the final and initial states.

We will see that the enthalpy change for a process will be the same regardless of whether we go from reactants to products in one step or in many steps. So long as we start at the same place and end at the same place, we will get the same value for ΔH.

Try Study Question 64 in Chapter 6 of your textbook now!
The answers to this question are not in the back of your textbook, so they are provided here.
64a. The volume of a balloon. This is a state function. The volume is what it is; it does not matter how it got to be that value.
b. The time it takes to drive from your home to your college or university. This is not a state function. Depending on the path taken and the traffic, this time will be a different value.
c. The temperature of the water in a coffee cup. This is a state function. It could have been heated from room temperature up to that temperature. It could have been cooled from boiling to that temperature. The pathway does not matter; all that matters is its current state.
d. The potential energy of a ball held in your hand. This is a state function. Once again, the previous history of the ball does not matter. All that matters is its current state.

- ## Calculate the energy changes occurring in chemical reactions and learn how changes are measured.

a) Recognize that when a process is carried out under constant pressure conditions, the heat transferred is the enthalpy change, ΔH (Section 6.5).

The heat transferred under conditions of constant pressure, q_p is equal to ΔH for the process. For more about this, see goal a under the previous bullet.

b) Describe how to measure the quantity of heat energy transferred in a reaction by using calorimetry (Section 6.6).

Calorimetry is the experimental determination of the heats of chemical reactions. There are two basic types of calorimetry experiments. Some are carried out at constant pressure, and some are carried out at constant volume.

In a constant pressure calorimetry experiment, our goal is to determine the heat of the reaction under conditions of constant pressure. This will give us ΔH for the reaction because $\Delta H = q_p$. In a constant volume calorimetry experiment, our goal is to determine the heat of the reaction under conditions of constant volume. This type of experiment will give us ΔE for the reaction because $\Delta E = q_v$.

Let us first consider constant pressure calorimetry. In this type of experiment, a reaction is often carried out in an insulated container that is exposed in some way to atmospheric pressure. Often, in general chemistry labs, the insulated device is a Styrofoam coffee cup.

Example 6-10: Suppose you mix together 50. mL of 1.00 M HCl with 60. mL of 1.00 M NaOH in a coffee cup calorimeter. The initial temperature of the solutions was 23.5°C. Following the reaction, the final temperature (extrapolated back to the time of mixing) was 29.5°C. Calculate ΔH for this reaction per mole of acid that reacts. Assume that the densities of all of the solutions are 1.0 g/mL and that their specific heat capacities are 4.20 J/(g•K).

Two changes in heat content take place within the system; the heat evolved in the reaction and the heat gained by the solution to increase its temperature. Because the reaction was carried out in an insulated container, the sum of these two heats must be zero.

$$q_{rxn} + q_{solution} = 0$$
$$q_{rxn} = -q_{solution}$$
$$= -C_{solution} \bullet m_{solution} \bullet \Delta T_{solution}$$

We know that the total volume of the solution will be approximately 110. mL since both of the solutions were aqueous solutions to begin with. Using the density of 1.0 g/mL, we obtain that the mass of the solution is 110. g.

$$q_{rxn} = -4.20\frac{J}{g \bullet K} \bullet 110.\ g \bullet (29.5°C - 23.5°C)$$
$$= -2.7 \times 10^3\ J = -2.7\ kJ$$

This is ΔH for the amounts of the reactants that actually reacted. The problem asks, however, for the value of ΔH per mole of acid that reacts.

	HCl *(aq)*	+	NaOH *(aq)*	→	NaCl *(aq)*	+	H$_2$O *(l)*
measured	50. mL		60. mL				
conversion factor	1.0 M		1.0 M				
moles	0.050 moles		0.060 moles				

Since this is a 1:1 reaction, we can see that the HCl is the limiting reactant and that the number of moles of HCl that reacts is 0.050 moles. (In this case, since the concentrations were equal, we could tell that the limiting reactant was HCl just by comparing the volumes mixed.)

$$\Delta H_{rxn} = \frac{-2.7\ kJ}{0.050\ moles\ acid} = -55\ \frac{kJ}{moles\ acid}$$

Try Study Question 35 in Chapter 6 of your textbook now!

In constant volume calorimetry, we will determine ΔE instead of ΔH. The device often used in constant volume calorimetry is a bomb calorimeter. In the case of a bomb calorimeter, the sum of the heats within the whole instrument will still equal zero, but the heat from the reaction gets transferred not only to the water but also to the device called the bomb. The key equation will therefore be

$$q_{rxn} + q_{bomb} + q_{H_2O} = 0$$

Example 6-11: A 1.50 g sample of glucose, $C_6H_{12}O_6$, is burned in a bomb calorimeter. The temperature rises from 21.35°C to 28.10°C. The calorimeter contains 600. g of water, the specific heat capacity of water is 4.184 J/(g•K), and the bomb has a heat capacity of 942 J/K. What is ΔE for the combustion of glucose per mole of glucose?

This problem involves a bomb calorimeter. It involves an experiment run under conditions of constant volume, and the heat will be related to ΔE. We start with the equation above:

$$q_{rxn} + q_{bomb} + q_{H_2O} = 0$$

We can expand the expressions for q_{bomb} and q_{water}.

$$q_{rxn} + C_{bomb} \bullet \Delta T + C_{H_2O} \bullet m_{H_2O} \bullet \Delta T = 0$$

We can then substitute in the values that we know and solve for q_{rxn}. Notice that our units help us know that we have set up the equation correctly.

$$q_{rxn} + 942\frac{J}{K} \bullet (28.10°C - 21.35°C) + \left(4.184\frac{J}{g \bullet K}\right)(600.\ g)(28.10°C - 21.35°C) = 0$$

$$q_{rxn} + 2.33 \times 10^4\ J = 0$$

$$q_{rxn} = -2.33 \times 10^4\ J$$

This is the amount of heat (ΔE) that is released for 1.50 g of glucose. The question asks for how much is released per mole of glucose.

$$1.50\ g\ C_6H_{12}O_6 \times \frac{1\ mole\ C_6H_{12}O_6}{180.2\ g\ C_6H_{12}O_6} = 8.32 \times 10^{-3}\ moles\ C_6H_{12}O_6$$

The heat released per mole can now be calculated:

$$\Delta E = \frac{-2.33 \times 10^4\ J}{8.32 \times 10^{-3}\ moles\ C_6H_{12}O_6} = -2.80 \times 10^6\ \frac{J}{mole} = -2.80 \times 10^3\ \frac{kJ}{mole}$$

Try Study Question 37 in Chapter 6 of your textbook now!

c) Apply Hess's law to find the enthalpy change for a reaction (Section 6.7).

Enthalpy is a state function. Its value is determined only by the state of the system and not by the pathway by which that state was achieved. In other words, so long as we start out with the initial state and end up with the final state, it does not matter how we get from the initial state to the final state. Thus, for a given chemical reaction, it does not matter if we carry out the process in a single step or in a series of steps that add up to the same overall reaction. This is the essence of Hess's Law: if a reaction is the sum of the two or more other reactions, ΔH for the overall process is the sum of the ΔH values of those reactions.

A typical Hess's law problem involves being given an overall equation for which we wish to know ΔH. What we do is to piece together other reactions whose ΔH values we do know in such a way that the reactions will add up to the desired chemical reaction. If we then add up the ΔH values for the individual steps, then we will obtain the ΔH for the overall process.

In working Hess's law problems, there are some key properties about ΔH that you will need to know:

1. If we reverse the direction in which we write an equation, then the sign of ΔH for the new equation will be the opposite of that of the original equation. This makes sense. If a process is exothermic in one direction, then it will be endothermic in the opposite direction.

$$H_2O \; (l) \rightarrow H_2O \; (g) \qquad \Delta H° = +44 \; kJ$$
$$H_2O \; (g) \rightarrow H_2O \; (l) \qquad \Delta H° = -44 \; kJ$$

2. If we multiply an equation by a coefficient, then we must multiply its ΔH value by the same coefficient.

$$H_2O \; (l) \rightarrow H_2O \; (g) \qquad \Delta H° = +44 \; kJ$$
$$2H_2O \; (l) \rightarrow 2H_2O \; (g) \qquad \Delta H° = 2(+44 \; kJ) = +88 \; kJ$$

In addition, to work an actual problem, it is suggested that you pay close attention to Problem Solving Tip 6.2 on p. 265 of your textbook. The key points are as follows:

1. Inspect the equation whose ΔH you wish to calculate, identifying the reactants and products, and locate those substances in the equations available to be added. Set up the given equations to get the reactants and products on the correct sides of the equations. To do this, you may need to reverse some of the given chemical equations.

2. Get the correct amount of the reactants and products on each side. To do this, you may need to multiply some of the given chemical equations by coefficients.

3. Make sure other substances in the given chemical equations that do not appear in the overall equation will cancel when the equations are added.

Example 6-12: Calculate the value of ΔH for the following reaction:

$$C_3H_8 \; (g) + 5O_2 \; (g) \rightarrow 3CO_2 \; (g) + 4H_2O \; (l) \qquad \Delta H = ?$$

given the following chemical equations for which we do know ΔH:

$$3C \; (s) + 4H_2 \; (g) \rightarrow C_3H_8 \; (g) \qquad \Delta H_1 = -104.7 \; kJ$$
$$C \; (s) + O_2 \; (g) \rightarrow CO_2 \; (g) \qquad \Delta H_2 = -393.5 \; kJ$$
$$2H_2 \; (g) + O_2 \; (g) \rightarrow 2H_2O \; (l) \qquad \Delta H_3 = -571.7 \; kJ$$

Step 1: C_3H_8 is a reactant in the desired equation. The only equation in which it appears in the known equations has it on the right side. We will need to reverse this equation. When we do this, we will change its sign. The CO_2 and the H_2O appear on the right side of the desired equation and that is where they are in the known equations so we can leave them written as they are. This will place the O_2 on the left side, which is where we want it.

$$C_3H_8 \; (g) \rightarrow 3C \; (s) + 4H_2 \; (g) \qquad \Delta H = -\Delta H_1 = +104.7 \; kJ$$
$$C \; (s) + O_2 \; (g) \rightarrow CO_2 \; (g) \qquad \Delta H_2 = -393.5 \; kJ$$
$$2H_2 \; (g) + O_2 \; (g) \rightarrow 2H_2O \; (l) \qquad \Delta H_3 = -571.7 \; kJ$$

Step 2: Now that we have our reactants and products on the correct sides of the equations, we will multiply by coefficients to get the correct amounts. We want only 1 C_3H_8 on the left, 3 CO_2 on the right, and 4 H_2O on the right so we will multiply the equations involving these by the coefficients necessary to get these amounts. The respective ΔH values will also be multiplied by these coefficients. Since O_2 is involved in more than one equation, we will simply double check that it comes out right when we use these coefficients.

$$C_3H_8 \; (g) \rightarrow 3C \; (s) + 4H_2 \; (g) \qquad \Delta H = +104.7 \text{ kJ}$$
$$3(C \; (s) + O_2 \; (g) \rightarrow CO_2 \; (g)) \qquad \Delta H = 3\Delta H_2 = 3(-393.5 \text{ kJ}) = -1180.5 \text{ kJ}$$
$$2(2H_2 \; (g) + O_2 \; (g) \rightarrow 2H_2O \; (l)) \qquad \Delta H = 2\Delta H_3 = 2(-571.7 \text{ kJ}) = -1143.4 \text{ kJ}$$

We get 3 oxygens from the CO_2 equation and 2 from the H_2O equation, giving us five total.

Step 3: We make sure that the other materials cancel out. In the top equation, we have 3C *(s)* on the right. In the second equation, we have 3C *(s)* on the left so these will cancel when we add the equations. In the top equation, we have $4H_2$ *(g)* on the right. In the bottom equation, we have $4H_2$ *(g)* on the left so they will cancel.

Now we simply add up the equations and the ΔH values.

$$C_3H_8 \; (g) \rightarrow 3C \; (s) + 4H_2 \; (g) \qquad \Delta H = +104.7 \text{ kJ}$$
$$3C \; (s) + 3O_2 \; (g) \rightarrow 3CO_2 \; (g)) \qquad \Delta H = -1180.5 \text{ kJ}$$
$$\underline{4H_2 \; (g) + 2O_2 \; (g) \rightarrow 4H_2O \; (l)) \qquad \Delta H = -1143.4 \text{ kJ}}$$
$$C_3H_8 \; (g) + 5O_2 \; (g) \rightarrow 3CO_2 \; (g) + 4H_2O \; (l)$$
$$\Delta H_{rxn} = +104.7 \text{ kJ} + -1180.5 \text{ kJ} + -1143.4 \text{ kJ} = -2219.2 \text{ kJ}$$

Try Study Question 43a in Chapter 6 of your textbook now!

d) Know how to draw and interpret energy level diagrams (Section 6.7).

You may find it useful to draw an energy level diagram to help you figure out a Hess's Law problem.

Example 6-13: Draw an energy level diagram to illustrate how to calculate ΔH_{rxn} for the problem in Example 6-12.

We wish to calculate the enthalpy change for going from C_3H_8 *(g)* + $5O_2$ *(g)* to $3CO_2$ *(g)* + $4H_2O$ *(l)*. This is ΔH_{rxn} on the diagram. ΔH_1 indicates that going from the elements to C_3H_8 *(g)* is an exothermic process. Thus on our energy diagram, we will write the elements 3C *(s)* + $4H_2$ *(g)* + $5O_2$ *(g)* above C_3H_8 *(g)* + $5O_2$ *(g)*.

To go from the elements to $3CO_2$ *(g)* we will need to carry out reaction 2 to form $3CO_2$ *(g)* three times. This is an exothermic process. We will place $3CO_2$ *(g)* + $4H_2$ *(g)* + $2O_2$ *(g)* below the elements on our diagram. The ΔH change corresponds to $3\Delta H_2$.

Next, we will combine the $4H_2$ *(g)* and the remaining $2O_2$ *(g)* to form the water molecules in a step that is also exothermic. This corresponds to $2\Delta H_3$.

We can see that the overall process in which we are interested (C_3H_8 *(g)* + $5O_2$ *(g)* $\rightarrow$ $3CO_2$ *(g)* + $4H_2O$ *(l)*) can be calculated as $3\Delta H_2 + 2\Delta H_3 - \Delta H_1$.

The completed energy diagram looks like the following:

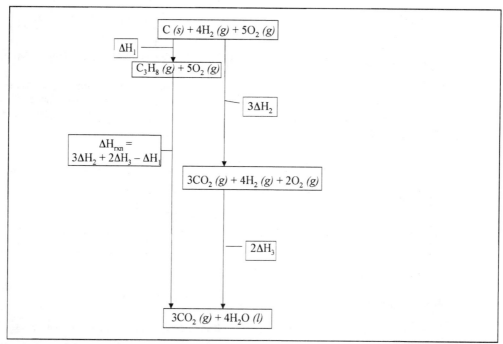

Try Study Question 43b in Chapter 6 of your textbook now!

e) Use standard molar enthalpy of formation, ΔH_f^o, to calculate the enthalpy change for a reaction, ΔH_{rxn}^o, (Section 6.8).

The standard state for an element or a compound is the most stable form of the substance in the physical state that exists at a pressure of 1 bar (approximately 1 atmosphere) and at a specified temperature. For a solute, the standard state is the solute present at a concentration of 1 M.

The standard molar enthalpy of formation is the enthalpy change for the formation of *one* mole of a compound directly from its *elements* in their standard states. Note in this definition that we form one mole of the compound and that we form the compound from its elements.

Example 6-14: Write the chemical equation that goes along with the standard molar enthalpy of formation of liquid water.

We must form one mole of water from the elements that go into making water.
$$H_2 \ (g) + 1/2 \ O_2 \ (g) \rightarrow H_2O \ (l)$$

Notice that we need to use a fractional coefficient for the oxygen. This is because we want to form only one mole of water.

Try Study Question 47 in Chapter 6 of your textbook now!

By definition, ΔH_f^o of an element in its standard state is zero.

We can use standard molar enthalpies of formation to calculate the enthalpy change for any chemical reaction:

$$\Delta H^{\circ}_{rxn} = \Sigma\left[\Delta H^{\circ}_f\left(products\right)\right] - \Sigma\left[\Delta H^{\circ}_f\left(reactants\right)\right]$$

In using this equation, we take into account not only the standard molar enthalpies of formation but also the stoichiometric coefficients.

Example 6-15: Calculate the enthalpy change for the following reaction using standard molar enthalpy of formation data:

$$C_3H_8 \ (g) + 5O_2 \ (g) \rightarrow 3CO_2 \ (g) + 4H_2O \ (l)$$

In order to use the equation above, we must know the standard molar enthalpies of formation. These are located in Table 6.2 as well as in Appendix L in your textbook.

$$\Delta H^{\circ}_{rxn} = \Sigma\left[\Delta H^{\circ}_f\left(products\right)\right] - \Sigma\left[\Delta H^{\circ}_f\left(reactants\right)\right]$$

$$\Delta H^{\circ}_{rxn} = \left[\left(3 \text{ moles}\right)\Delta H^{\circ}_f\left(CO_2(g)\right) + \left(4 \text{ moles}\right)\Delta H^{\circ}_f\left(H_2O(l)\right)\right]$$

$$- \left[\left(1 \text{ mole}\right)\Delta H^{\circ}_f\left(C_3H_8(g)\right) + \left(5 \text{ moles}\right)\Delta H^{\circ}_f\left(O_2(g)\right)\right]$$

$$\Delta H^{\circ}_{rxn} = \left[\left(3 \text{ moles}\right)\left(-393.509 \text{ kJ/mole}\right) + \left(4 \text{ moles}\right)\left(-285.83 \text{ kJ/mole}\right)\right]$$

$$- \left[\left(1 \text{ mole}\right)\left(-104.7 \text{ kJ}\right) + \left(5 \text{ moles}\right)\left(0 \text{ kJ/mole}\right)\right]$$

$$\Delta H^{\circ}_{rxn} = -2219.1 \text{ kJ}$$

Try Study Question 53 in Chapter 6 of your textbook now!

Other Notes

1. Enthalpy and Stoichiometry
Sometimes a chemical reaction will be written with an enthalpy or energy attached. For example,

$$P_4 \ (s) + 6Cl_2 \ (g) \rightarrow 4PCl_3 \ (l) \qquad \Delta H^{\circ} = -1279 \text{ kJ}$$

This enthalpy change is the enthalpy change to carry out 1 mole of this reaction, that is, for 1 mole of P_4 (s) to react with 6 moles of Cl_2 to form 4 moles of PCl_3. From this, we can write a number of different conversion factors. If the problem involves P_4, then we would use one of the following conversion factors:

$$\frac{-1279 \text{ kJ}}{1 \text{ mole } P_4} \quad \text{and} \quad \frac{1 \text{ mole } P_4}{-1279 \text{ kJ}}$$

If the problem involves Cl_2, we would use one of the following conversion factors:

$$\frac{-1279 \text{ kJ}}{6 \text{ moles } Cl_2} \quad \text{and} \quad \frac{6 \text{ moles } Cl_2}{-1279 \text{ kJ}}$$

If the problem involves PCl_3, then we would use one of these:

$$\frac{-1279 \text{ kJ}}{4 \text{ moles } PCl_3} \quad \text{and} \quad \frac{4 \text{ moles } PCl_3}{-1279 \text{ kJ}}$$

Example 6-16: Given the following balanced thermochemical equation, what would be the enthalpy change if 5.00 g of P_4 were reacted?

$$P_4 \ (s) + 6Cl_2 \ (g) \rightarrow 4PCl_3 \ (l) \qquad \Delta H^{\circ} = -1279 \text{ kJ}$$

This is a stoichiometry problem that involves P_4. We can set up a modified stoichiometry table. We were given the mass of phosphorus so we will need the molar mass of P_4 to get to moles. Notice that the ΔH given is used as a conversion factor in the table.

	P_4 (s)	+	$6Cl_2$ (g)	→	$4PCl_3$ (l)	$\Delta H° = -1279$ kJ
measured	5.00 g					x kJ
conversion factor	123.9g/mole					-1279 kJ/1 mole P_4
moles						

We can calculate the number of moles of P_4:

$$5.00 \text{ g } P_4 \text{ x } \frac{1 \text{ mole } P_4}{123.9 \text{ g } P_4} = 0.0404 \text{ moles } P_4$$

We can then calculate the enthalpy change for this amount of P_4:

$$0.0404 \text{ moles } P_4 \text{ x } \frac{-1279 \text{ kJ}}{1 \text{ mole } P_4} = -51.6 \text{ kJ}$$

The completed table looks like the following:

	P_4 (s)	+	$6Cl_2$ (g)	→	$4PCl_3$ (l)	$\Delta H° = -1279$ kJ
measured	5.00 g					-51.6 kJ
conversion factor	123.9 g/mole					-1279 kJ/1 mole P_4
moles	0.0404 moles					

Example 6-17: Given the following thermochemical equation, what mass of chlorine must have reacted if it was measured that 964 kJ of heat was released under conditions of constant pressure?

$$P_4 \text{ (s)} + 6Cl_2 \text{ (g)} \rightarrow 4PCl_3 \text{ (l)} \qquad \Delta H° = -1279 \text{ kJ}$$

We can set up our table. Notice that since this problem involves chlorine, we use the conversion factor involving the $\Delta H°$ and moles of chlorine.

	P_4 (s)	+	$6Cl_2$ (g)	→	$4PCl_3$ (l)	$\Delta H° = -1279$ kJ
measured			x g			-964 kJ
conversion factor			70.91 g/mole			-1279 kJ/6 moles Cl_2
moles						

We can use the given enthalpy change for the actual reaction carried out and use the relationship from our balanced thermochemical equation to calculate the number of moles of chlorine:

$$-964 \text{ kJ x } \frac{6 \text{ moles } Cl_2}{-1279 \text{ kJ}} = 4.52 \text{ moles } Cl_2$$

Now we just need to go from moles of Cl_2 to g of Cl_2 using the molar mass of Cl_2:

$$4.52 \text{ moles } Cl_2 \text{ x } \frac{70.91 \text{ g } Cl_2}{1 \text{ mole } Cl_2} = 321 \text{ g } Cl_2$$

The completed table looks like the following:

	P_4 (s)	+	$6Cl_2$ (g)	$\rightarrow$	$4PCl_3$ (l)	$\Delta H° = -1279$ kJ
measured			321 g			−964 kJ
conversion factor			70.91 g/mole			−1279 kJ/6 moles Cl_2
moles			4.52 moles			

Try Study Question 27 in Chapter 6 of your textbook now!

2. Many exothermic reactions are product-favored reactions. ΔH, however, is not the whole story for determining whether a reaction will be product-favored or reactant-favored because there are endothermic reactions that are also product favored. Another factor, to be studied later in the course, is needed to truly predict whether a reaction will be product-favored or not.

CHAPTER 7: Atomic Structure

Chapter Overview

In this chapter, we return to the issue of atomic structure. When last we left our model of the atom, we pictured that the protons and neutrons were located in the nucleus of the atom and that the electrons somehow were moving around outside the nucleus. In this chapter, we begin to look more closely at the arrangement of electrons in an atom. We begin by examining electromagnetic radiation. Some properties of electromagnetic radiation are best explained by treating it as if it were a wave. Other properties are best understood if we treat it as being composed of particles of energy called photons. The wavelength (λ) and the frequency (ν) of electromagnetic radiation are related by the equation $c = \lambda\nu$ where c is the speed of light. The energy of a single photon is given by the equation $E = h\nu$.

When white light is passed through a prism, we get a continuous spectrum containing all the frequencies of light. In contrast, when an element in the gas phase is excited and allowed to relax, it gives off light that produces a line spectrum when passed through a prism. Bohr was able to predict the line spectrum of hydrogen by assuming there were only certain energies an electron could have. He pictured the arrangement of electrons in an atom as being in fixed energy levels with the electrons traveling around the nucleus at set distances called orbits. His model worked very well for hydrogen but did not work well with atoms having more than one electron. De Broglie suggested that matter might have wave and particle properties. He derived an equation for the wavelength of a particle. This equation has been verified, and we now accept that matter has wave properties in addition to the more familiar particle properties. The modern view of the atom is derived using the wave properties.

In the modern quantum mechanical model, the electrons are pictured as being waves having only certain energies. The energies of electrons are thus quantized. Due to the Heisenberg uncertainty principle, if we know the energies very accurately, then there must be uncertainty in the positions. Rather than picturing the electrons orbiting the nucleus at fixed distances, we now picture the electrons as being in orbitals, regions of space where there is a high probability of finding the electrons. Associated with each orbital is a set of three quantum numbers. The principal quantum number (n) gives the energy level, the angular momentum quantum number (l) gives the subshell (s, p, d, or f), and the magnetic quantum number (m_l) gives the specific orbital within that subshell. The first energy level contains one 1s orbital. The second energy level, contains one 2s orbital and three 2p orbitals. The third energy level contains one 3s orbital, three 3p orbitals, and five 3d orbitals. The fourth energy level contains one 4s orbital, three 4p orbitals, five 4d orbitals, and seven 4f orbitals. Beyond the fourth energy level, each higher energy level has one s orbital, three p orbitals, five d orbitals, seven f orbitals, plus other orbitals that are not needed for any ground state configuration for any element now known. Lastly, the shapes of the different orbitals are discussed.

Key Terms

In this chapter, you will need to learn and be able to use the following terms:

Amplitude: the maximum height of a wave from its axis of propagation.

Angular momentum quantum number (l): an integer value from zero up to $n-1$, this quantum number indicates the type of subshell the electron is in.

Balmer series: the visible portion of the line emission spectrum of hydrogen; these lines correspond to the electron in hydrogen moving from higher energy levels down to the second energy level.

Electromagnetic radiation: radiation that consists of electric and magnetic fields in space, including light, microwaves, radio signals, and x-rays.

Excited state: an atom is in an excited state if any of its electrons are in a higher energy level than in the ground state.

Frequency: the number of waves that pass a given point in some unit of time. Frequency is abbreviated by the Greek letter nu (ν).

Ground state: an atom is in its ground state if all of its electrons are in the lowest possible energy levels.

Heisenberg uncertainty principle: we cannot know the exact position and the exact energy of an object at the same time; there must be a minimum uncertainty present.

Line emission spectrum: when the atoms of an element in the gas phase are excited, they emit electromagnetic radiation having only particular energies. When this light is passed through a prism, instead of getting a continuous spectrum, we obtain only certain lines in the spectrum.

Lyman series: a series of emission lines in the spectrum of hydrogen which fall in the ultraviolet region; these lines correspond to transitions of the electron in hydrogen from higher energy levels down to the first energy level.

Magnetic quantum number (m_l): an integer value from $-l$ up to $+l$, this quantum number indicates the orbital in a given subshell that the electron is in.

Nodal surface: in an orbital, a surface (often a plane) along which there is zero probability of finding an electron

Node: a point on a wave that has zero amplitude.

Orbital: a region of space in which we are likely to find an electron; each orbital corresponds to an allowed energy state of an electron in an atom or molecule.

Photoelectric effect: a phenomenon explained by Albert Einstein through assuming that electromagnetic radiation is quantized. The phenomenon involved the emission of electrons by metals. If light below a certain frequency is used, then no electrons are emitted, regardless the intensity of the light. Once a certain frequency is reached, the metal emits electrons, and the number of electrons emitted increases with increasing intensity of the light.

Principal quantum number (n): an integer value from one on up, this quantum number indicates the energy level (shell) of the electron.

Quantized: something is said to be quantized if it can have only particular values.

Quantum mechanics: also called wave mechanics, an approach to atomic behavior that describes the electron as being a wave. In this model of the electron, the energies of the electron are quantized.

Quantum numbers: a set of numbers that define the properties of an atomic orbital or identifies the spin state of an electron in the orbital.

Rydberg equation: an equation that predicts the wavelength of the lines in the emission spectrum of hydrogen.

Wave function: an equation, symbolized by the Greek letter ψ, which characterizes the electron as a wave and whose solutions give the energies of the electron.

Wavelength: the distance between one point on a wave and its repeat on the next waveform (for example, from one crest to the next crest). Wavelength is abbreviated by the Greek letter lambda (λ).

Wave-particle duality: the idea that both electromagnetic radiation and matter exhibit some properties like those of a wave and some properties like those of a particle.

Chapter Goals

By the end of this chapter you should be able to:

- **Describe the properties of electromagnetic radiation.**

    a) Use the terms wavelength, frequency, amplitude, and node (Section 7.1).

    It is important to know what each of these terms refers to. The wavelength (λ) is the distance from one point on a wave to its repeat on the next waveform. Often, it is measured from one crest to the next crest or from one trough to the next trough because these are the easiest points to locate accurately. The units for wavelength will be some unit of length such as nm, cm, or m.

    The frequency (ν) is the number of waves that pass a given point in a given period of time. The time interval is usually chosen as 1 second. Frequency is often measured in units of hertz (Hz). 1 Hz = 1/second

    The amplitude of a wave is the maximum height of a wave from its axis of propagation. A node is a point having an amplitude of zero.

    <u>Example 7-1</u>: Given the wave below,

    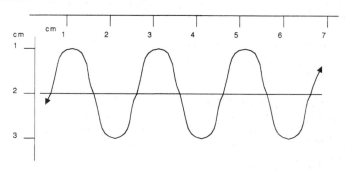

            a. determine the wavelength
    Using the horizontal ruler, the first crest is at 1.0 cm. The next crest is at 3.0 cm. The wavelength (the distance from one crest to the next) is thus 2.0 cm.

            b. determine the amplitude
    Using the vertical ruler, the axis of propagation lines up with the 2.0 cm mark on the ruler, and the crests lines up with the 1.0 cm mark. The amplitude is thus 1.0 cm.

            c. determine how many nodes are present in the portion of the wave shown
    A node occurs wherever the amplitude is zero. This occurs when the wave crosses the axis of propagation. There are thus 7 nodes displayed because the wave crosses the axis 7 times.

<u>Example 7-2</u>: 150 wave crests pass a given point in 5 seconds. What is the frequency of this wave?

The frequency is the number of waves passing a given point in a given period of time.
$$\nu = \frac{150}{5 \text{ s}} = 30 \text{ Hz}$$

---

b) Use Equation 7.1 ($c = \lambda\nu$), the relationship between the wavelength (λ) and frequency (ν) of electromagnetic radiation and the speed of light (c).

The wavelength and frequency of electromagnetic radiation are inversely proportional. As one gets larger, the other gets smaller. As the wavelength gets larger, the frequency gets smaller. As the wavelength gets smaller, the frequency gets larger.

The product of wavelength and frequency is a constant. For electromagnetic radiation, this constant is the speed of light, 2.998×10^8 m/s, which we symbolize with the letter c.
$$c = \lambda\nu$$

<u>Example 7-3</u>: Which has a longer wavelength, a microwave with a frequency of 3×10^{10} Hz or ultraviolet light with a frequency of 4×10^{14} Hz?

Frequency and wavelength are inversely proportional. As one gets bigger, the other gets smaller. The electromagnetic radiation with the larger wavelength will have the smaller frequency. 3×10^{10} Hz is smaller than 4×10^{14} Hz, thus the microwave radiation has the smaller frequency and the longer wavelength.

<u>Example 7-4</u>: What is the frequency of light that has a wavelength of 500. nm?

In this problem, we are given the frequency of light and the wavelength. They are related by the speed of light, c, through the equation
$$c = \lambda\nu$$

The value for the speed of light we are given is 2.998×10^8 m/s and the wavelength is 500. nm. When we solve this problem, we want to have the same unit of length in the speed of light and in the wavelength. We must either convert the speed of light to be in nm/s or else convert the wavelength to m; we shall do the latter.
$$500. \text{ nm} \times \frac{1 \text{ m}}{1 \times 10^9 \text{ nm}} = 5.00 \times 10^{-7} \text{ m}$$
We can now rearrange the equation to solve for ν by dividing each side by λ. We then substitute in the values and solve the problem.
$$c = \lambda\nu$$
$$\nu = \frac{c}{\lambda}$$
$$= \frac{2.998 \times 10^8 \text{ m/s}}{5.00 \times 10^{-7} \text{ m}}$$
$$= 6.00 \times 10^{14} \text{ Hz}$$

Try Study Question 3 in Chapter 7 of your textbook now!

c) Recognize the relative wavelength (or frequency) of the various types of electromagnetic radiation (Figure 7.3).

In Figure 7.3 in your text, the key thing to worry about is the order of the different types of electromagnetic radiation, rather than the actual values of the wavelengths and frequencies.

In order of increasing wavelength, the sequence is
$$\gamma \text{ rays} < \text{x-rays} < \text{UV} < \text{visible light} < \text{infrared} < \text{microwave} < \text{FM radio} <$$
$$\text{AM radio} < \text{long radio waves}$$

Within the category of visible light, in order of increasing wavelength, we have
$$\text{violet} < \text{indigo} < \text{blue} < \text{green} < \text{yellow} < \text{orange} < \text{red}$$

Because wavelength and frequency are inversely proportional, the order of the types of electromagnetic radiation according to increasing frequency is the reverse order:
$$\text{long radio waves} < \text{AM radio} < \text{FM radio} < \text{microwaves} < \text{infrared} < \text{visible light} <$$
$$\text{UV} < \text{x-rays} < \gamma \text{ rays}$$

The order for the visible light portion in order of increasing frequency is that which you may have learned using the mnemonic ROY G BIV
$$\text{red} < \text{orange} < \text{yellow} < \text{green} < \text{blue} < \text{indigo} < \text{violet}$$

As you will learn, frequency and energy are directly proportional, therefore the orders of the types of radiation according to increasing energy is exactly the same as that for increasing frequency.

Example 7-5: Which has the higher frequency, microwave radiation or radio waves?

From the order we went over above, we know that microwaves have a greater frequency than radio waves.

Try Study Question 1 in Chapter 7 of your textbook now!

d) Understand that the energy of a photon, a massless particle of radiation, is proportional to its frequency (Planck's equation, Equation 7.2). This is an extension of Planck's idea that energy at the atomic level is quantized (Section 7.2).

So far, we have been considering the wave properties of electromagnetic radiation. Electromagnetic radiation also has particle properties. When electromagnetic radiation delivers its energy, it delivers it in fixed amounts as if the light consists of packets or particles. These packets of energy are called photons. A photon is a massless particle of energy. The energy of a single photon of electromagnetic radiation is given by Planck's equation:
$$E = h\nu$$
where E is the energy of the photon, h is Planck's constant (6.626×10^{-34} J•s), and ν is the frequency of the electromagnetic radiation.

The energy of a photon is thus directly proportional to frequency. As the frequency increases, so does the energy. What is the relationship between energy and wavelength? Frequency and wavelength are inversely proportional. As the frequency increases, the wavelength decreases. Because frequency and energy are directly proportional, energy and wavelength must be inversely proportional. As the energy increases, the wavelength decreases. Using both the equation relating wavelength and frequency ($c = \lambda \nu$) and Planck's equation, we can derive an equation relating energy and wavelength:

$$E = \frac{hc}{\lambda}$$

Example 7-6: What is the energy of a photon of orange light having a frequency of 5.00 x 10^{14} Hz?

We are given the frequency and asked for the energy. We can simply use Planck's equation.

$$E = h\nu$$

$$= \left(6.626 \times 10^{-34} \text{ J} \bullet \text{s}\right)\left(5.00 \times 10^{14} \text{ Hz}\right)$$

$$= 3.31 \times 10^{-19} \text{ J}$$

Note that this is a very small amount of energy per photon. What is the energy for one mole of these photons?

$$3.31 \times 10^{-19} \ \frac{\text{J}}{\text{photon}} \ \text{x} \ \frac{6.022 \times 10^{23} \text{ photons}}{1 \text{ mole of photons}} \ \text{x} \frac{1 \text{ kJ}}{1000 \text{ J}} = 1.99 \times 10^2 \ \frac{\text{kJ}}{\text{mole}}$$

Try Study Question 7 in Chapter 7 of your textbook now!

- **Understand the origin of light from excited atoms and its relationship to atomic structure.**

a) Describe the Bohr model of the atom, its ability to account for the emission line spectra of excited hydrogen atoms, and the limitations of the model (Section 7.3).

In the Bohr model of the atom, electrons are pictured as moving in fixed circular orbits around the nucleus. Only orbits having particular radii are permitted, and the electrons can only have certain energies. Thus, according to Bohr, both the energy and distance of an electron from the nucleus are quantized.

According to the Bohr model, when an electron absorbs a photon of electromagnetic radiation, it gains the energy of the photon. In doing so, it moves from an orbit closer to the nucleus and having less energy up into an orbit further away from the nucleus and having more energy. Conversely, when an electron moves from a higher energy orbit further from the nucleus down to an orbit closer to the nucleus and having less energy, the electron gives off (emits) a photon of electromagnetic radiation that has the same energy as the energy difference between the two orbits. Because there are only particular energy levels, the photons that can be emitted can only have particular energies, corresponding to the energy differences between these levels. When the emission spectrum of an element in the gas phase is examined, it is found to be a line spectrum, consisting of electromagnetic radiation of particular energies. Having only particular energies in the spectrum is consistent with the Bohr model of the atom and the mechanism of absorption and emission of electromagnetic radiation we have been discussing.

Using the mathematical form of his model of the atom, Bohr was able to predict the Rydberg equation, an equation that predicts the wavelengths of the lines in the spectrum for hydrogen. The Bohr model of the atom does a good job at predicting the spectrum of hydrogen. However, it works well only for the element hydrogen (and other one electron systems). It does not work for other elements. While the Bohr model may have been the model of the atom that you learned in a previous course in middle school or high school, it is not the currently accepted model for the arrangement of electrons in an atom.

b) Understand that, in the Bohr model of the H atom, the electron can occupy only certain energy levels, each with an energy proportional to $1/n^2$ ($E = -Rhc/n^2$), where n is the principal quantum number (Equation 7.4, Section 7.3). If an electron moves from one energy state to another, the amount of energy absorbed or emitted in the process is equal to the difference in energy between the two states (Equation 7.5, Section 7.3).

According to the Bohr model, the actual energies of the energy levels in the hydrogen atom are given by the equation

$$E = -\frac{Rhc}{n^2}$$

where n is the principal quantum number. The principal quantum number gives the energy level the electron is in and has a value of 1 for the first energy level, 2 for the second energy level, etc.

Let's look at some of the implications of this formula. There is a negative sign in front of the equation. The zero for energy is defined to be a free electron at infinite distance from the nucleus. The negative sign means that the energies for an electron orbiting the nucleus are negative compared to this. This means that an electron orbiting the nucleus is at a lower energy than a free electron. As we move to greater and greater values of n, the energy goes up, approaching the zero point more and more closely. The second thing to notice is that n is in the denominator and is squared. The difference in energy between two energy levels varies according to which energy levels are involved. The energy also does not change linearly from energy level to energy level but goes as the square of the value. The energy for the second energy level is only one quarter as low as the energy for the first energy level. The energy for the third energy level is one ninth as low as the energy for the first level, and so forth. As we go to higher and higher energy levels, the values for the energies get closer and closer together. See Figure 7.10 in your textbook for an illustration of this.

As we have discussed before, for an electron to move from a lower energy level to a higher energy level, energy must be absorbed. When an electron moves from a higher energy level to a lower energy level, energy is emitted. The amount of energy absorbed or emitted is the difference in energy between the two levels. The energy difference between two energy levels is given by the equation:

$$\Delta E = E_f - E_i = -Rhc\left(\frac{1}{n_f^2} - \frac{1}{n_i^2}\right)$$

where f and i refer to the final and initial energy levels. A positive value from this equation indicates that this amount of energy is absorbed in moving from the initial to the final energy levels. A negative value indicates that this amount of energy is emitted in the move.

Example 7-7: Which transition will emit the higher energy photon, an electron in a hydrogen atom going from the fourth energy level to the third energy level or going from the third energy level to the second?

Both of these transitions involve the electron changing by one energy level. The difference in energy for the two transitions, however, will not be the same. The energies of the levels go as $1/n^2$, so the energy levels get closer and closer together as we move to higher and higher energy levels. The transition from the third energy level to the second energy level will involve a greater energy change and thus the emission of a more energetic photon than will the transition from the fourth energy level to the third energy level.

Example 7-8: Calculate the energy of a photon that excites an electron in a hydrogen atom from the first energy level to the fifth energy level.

We will use the following equation to solve this problem:

$$\Delta E = E_f - E_i = -Rhc\left(\frac{1}{n_f^2} - \frac{1}{n_i^2}\right)$$

In this equation, R, h, and c are constants we can look up: R is the Rydberg constant (1.0974×10^7 m^{-1}), h is Planck's constant (6.626×10^{-34} J•s), and c is the speed of light (2.998×10^8 m/s). In this problem, the final energy level is the fifth ($n = 5$), and the initial energy level is the first ($n = 1$).

$$\Delta E = -\left(1.0974 \times 10^7 \text{ m}^{-1}\right)\left(6.626 \times 10^{-34} \text{ J•s}\right)\left(2.998 \times 10^8 \frac{\text{m}}{\text{s}}\right)\left(\frac{1}{5^2} - \frac{1}{1^2}\right)$$

$$= 2.093 \times 10^{-18} \text{ J}$$

The energy of a photon that can produce this transition would be 2.093×10^{-18} J.

Try Study Question 17 in Chapter 7 of your textbook now!

- **Describe the experimental evidence for wave-particle duality.**

    Understand that in the modern view of the atom, electrons are described by the physics of waves (Section 7.4). The wavelength of an electron or any subatomic particle is given by de Broglie's equation (Equation 7.6).

    Many of the properties of light (electromagnetic radiation) can be explained by using a wave model. We have been using terms like wavelength and frequency that fit this model. In addition, light can be diffracted and exhibit other properties of waves. With their explanations of black body radiation and the photoelectric effect, Planck and Einstein proposed the revolutionary idea that light sometimes behaved as if it consisted of particles. Some experiments can thus be explained by assuming a wave model for light but others can be explained assuming a particle model for light. Which is light, a wave or a particle?

    When we try to understand something, we use as models for it things that make sense to us. These terms, wave and particle, are based on experiences in our daily lives. For macroscopic objects and phenomena, we understand what a particle is like by examining things like a baseball, and we understand what a wave is by looking at waves on a lake. Light goes beyond each of these terms. Light is light, something that goes beyond either of these terms. We can understand various aspects of light by noting that under certain conditions, it behaves *as if* it were a wave and that under other conditions, it behaves *as if* it were a particle. Light exhibits both wave and particle properties. We describe this by saying that there is a wave-particle duality.

    If light, something that traditionally was considered to be a wave, could have some particle properties, then perhaps matter, something that traditionally was viewed as being made of particles, could have some wave properties. This is what Louis de Broglie proposed. By combining Einstein's $E = mc^2$ equation and Planck's $E = h\nu$ equation and then assuming a speed other than the speed of light, he derived an equation that predicts a wavelength for a sample of matter having a mass m and a velocity v:

    $$\lambda = \frac{h}{mv}$$

This prediction was verified when it was discovered that electrons could be diffracted (a wave property) and behaved in this process as if they had the wavelength predicted by de Broglie's equation. Matter exhibits both wave and particle properties – a wave-particle duality. On our macroscopic level, we do not need to worry about the wave properties but at the subatomic level of the electron, the wave properties become a significant factor.

This is important because the modern view of the arrangement of electrons in an atom is based, not on the particle picture of the electron, but on the wave picture.

Example 7-9: Calculate the wavelength associated with a 73 kg (160 lb) object moving at a speed of 1.1 m/s (2.5 mi/h; a comfortable walking speed). This is a macroscopic object moving at a typical speed in our world.

To solve this problem, we need to use de Broglie's equation. For this equation, we want the mass to be in kg and the speed to be in m/s. We were nice and gave them to you already in these units, but in other problems, you might need to do conversions to get them into the proper units.

$$\lambda = \frac{h}{mv}$$

$$\lambda = \frac{6.626 \times 10^{-34} \text{ J} \cdot \text{s}}{(73 \text{ kg})(1.1 \text{ m/s})}$$

$$= 8.3 \times 10^{-36} \text{ m}$$

This wavelength is smaller than any measurable wavelength so we do not have to worry about the wave properties of this object. In the recommended exercise in your textbook, you will find, however, that the wavelength for the electron is measurable and significant.

Try Study Question 23 in Chapter 7 of your textbook now!

• Describe the basic ideas of quantum mechanics.

a) Recognize the significance of quantum mechanics in describing the modern view of atomic structure (Section 7.5).

Let's return now to developing our modern view of the atom. According to the Bohr model of the atom, the electrons travel around the nucleus in fixed energy levels that are at fixed distances from the nucleus. Werner Heisenberg showed that it is not possible to know both the exact position of an object and its exact energy at the same time. There is always a minimum amount of uncertainty associated with these. The more accurately we know one, the less accurately we know the other. At our macroscopic level, the amount of uncertainty does not cause problems, but at the level of electrons, the amount of uncertainty is significant. This raises problems for the Bohr model of the atom because it proposed fixed energy levels – known energies – and fixed distances from the nucleus – known positions. We cannot know both of these.

In our modern view of the atom, we have chosen to deal with the energies of an electron. If we are to know these accurately, then we must accept a large amount of uncertainty with regard to position. Instead of being able to state the exact position of the electron, we deal instead with the probability of being able to find the electron in a given region of space. Instead of speaking of an orbit of the electron at a fixed distance from the nucleus, we speak of an orbital, a region in space in which we are likely to find the electron. This approach to the arrangement of electrons is called quantum mechanics or wave mechanics.

b) Understand that an orbital for an electron in an atom corresponds to the allowed energy of that electron.

The quantum mechanical model of the atom, proposed by Schrödinger, involves complex mathematics that goes beyond the scope of this course. Nonetheless, we can deal with the results of these calculations in a descriptive fashion that can have meaning at this level. The solutions to the mathematical equations are called wave functions (ψ).

In this modern view of the atom, the energy of an electron is quantized – the electron can have only certain values of energy (this aspect of the Bohr model is retained). Now though, each of these energies is associated with a wave function ψ. The square of the wave function, ψ^2, is related to the probability of finding the electron within a given region of space. In the quantum mechanical model, an orbital corresponds to the allowed energy of the electron. We describe the location of the electron vaguely by defining a region in space in which there is a certain probability of finding the electron (such as a 90% or a 95% probability) when it has a particular energy value. Three quantum numbers (n, l, and m_l) arise as part of the mathematical solution of the Schrödinger equation. The values of these quantum numbers enable us to designate a particular orbital in an atom.

c) Understand that the position of the electron is not known with certainty; only the probability of the electron being at a given point of space can be calculated. This is the interpretation of the quantum mechanical model and embodies the postulate called the Heisenberg uncertainty principle.

This objective simply summarizes some of the key information we have been discussing. In the modern view of the atom, the energies of the electrons are quantized. Because we are specifying the energies exactly, we cannot specify the positions of the electrons accurately. Instead, we define regions of space in which we are likely to find an electron with a particular energy. These regions of space are called orbitals.

It is sometimes difficult for students to accept this view of the atom. In some earlier course, they have been taught the nice simplistic model of the electrons orbiting the nucleus at some fixed distance, like the planets orbiting the sun in the solar system. Some students do not want to give up this simplistic model for the more complex and less precise view of orbitals in which we do not know the exact positions of the electrons and in which we deal more with the wave properties of the electrons than with the particle properties. Nonetheless, this is the modern view of the atom, and you should strive to abandon the previous model and accept this new model (if you have not previously done so).

*   **Define the three quantum numbers (n, l, and m_l) and their relationship to atomic structure.**

a) Describe the allowed energy states of the electron in an atom using three quantum numbers n, l, and m_l (Section 7.5).

There are many different orbitals in an atom, corresponding to different possible solutions of the wave function. If we state the values of the three quantum numbers (n, l, and m_l), then we have specified a particular orbital in an atom. Each quantum number provides a different piece of information.

The principal quantum number, n, corresponds to the energy level (shell) of the electron. It can thus have any integer value from 1 on up. A value of $n = 1$ corresponds to the first energy level, $n = 2$ corresponds to the second energy level, etc.

A given energy level may be composed of one or more subshells. The angular momentum quantum number, l, designates the subshell the electron is in. Each value of l designates a different type of subshell. The values of l we shall deal with and their corresponding subshells are as follows:

Value of l	Corresponding Subshell Label
0	s
1	p
2	d
3	f

To determine which subshells are present in a given energy level, we use the following rule: for each value of n, l may have integer values going from $l = 0$ all the way up to $n - 1$.

For the first energy level, we start at $l = 0$ and go up to $n - 1$. In this case $n - 1 = 1 - 1 = 0$ so there is only one value of l possible: $l = 0$. Thus, the first energy level has only one type of subshell, an s subshell.

For the second energy level, we start at $l = 0$ and go up to $n - 1$. In this case, $n - 1 = 1$. There are two values of l possible: $l = 0$ and $l = 1$. The second energy level has two types of subshells: an s subshell and a p subshell.

For the third energy level, we start at $l = 0$ and go up to $n - 1$. In this case, $n - 1 = 2$. There are three values of l possible: $l = 0$, $l = 1$, and $l = 2$. The third energy level has three types of subshells: an s subshell, a p subshell, and a d subshell.

The fourth energy level has four types of subshells corresponding to $l = 0$, $l = 1$, $l = 2$, and $l = 3$. The fourth energy level has an s subshell, a p subshell, a d subshell, and an f subshell.

The magnetic quantum number, m_l, tells us the orbital within a given subshell in which we will find the electron. For a given value of l, m_l can have integer values from $-l$ up to $+l$. For an s sublevel, $l = 0$, therefore m_l may have only one value: 0. This means that there is only one s orbital in an s sublevel. For a p sublevel, $l = 1$, and m_l may have values of -1, 0, and $+1$. There are three values of m_l, so there are three orbitals in a p sublevel. For a d sublevel, $l = 2$, and m_l may have values of -2, -1, 0, $+1$, and $+2$. There are five different values of m_l, so there are five orbitals in a d sublevel. For an f sublevel, $l = 3$, and m_l may have values of -3, -2, -1, 0, $+1$, $+2$, and $+3$. There are seven different values of m_l, so an f sublevel has seven different orbitals.

Example 7-10: When $n = 3$, what are the possible values of l?

For each value of n, l may have values from 0 up to $n - 1$. $n - 1 = 3 - 1 = 2$. Therefore, l may have values of 0, 1, and 2.

Example 7-11: $l = 4$ corresponds to what is called a g subshell. How many orbitals are present in a g sublevel?

The number of orbitals in a subshell is given by how many values of m_l are possible for that subshell. m_l goes from $-l$ up to $+l$, therefore, m_l may have values of -4, -3, -2, -1, 0, $+1$, $+2$, $+3$, and $+4$. There would be 9 different g orbitals.

Example 7-12: Can there be an orbital with the following quantum numbers? $n = 2$, $l = 2$, m_l = −1

No, l may only have values from 0 up to n −1. In this case, n −1 = 2 −1 = 1. It is not possible to have $l = 2$ when $n = 2$.

Try Study Questions 27 and 33 in Chapter 7 of your textbook now!

We are now ready to pull all of this information together. This is the heart of this chapter. It is crucial for you to know the information that follows.

The first energy level has an s subshell. Within the s subshell, there is one s orbital. Thus, there is one 1s orbital. In this notation, the number refers to the energy level, and the letter refers to the subshell.

The second energy level has both an s subshell and a p subshell. An s subshell contains one s orbital, and a p subshell contains three p orbitals. Thus, there is one 2s orbital and three 2p orbitals.

The third energy level has an s subshell, a p subshell, and a d subshell. There is one 3s orbital, three 3p orbitals, and five 3d orbitals.

The fourth energy level has an s subshell, a p subshell, a d subshell, and an f subshell. There is one 4s orbital, three 4p orbitals, five 4d orbitals, and seven 4f orbitals.

Each of the higher energy levels have one s orbital, three p orbitals, five d orbitals, seven f orbitals, plus other orbitals that are not needed for any ground state configuration for any element now known.

Example 7-13: State which of the following is an incorrect designation for an orbital: 2p, 3s, 3f, 4d, 4f.

The second energy level contains one s orbital and three p orbitals, therefore 2p is fine.
The third energy level contains one s orbital, three p orbitals, and five d orbitals, therefore 3s is fine, but 3f is not correct. There are no 3f orbitals.
The fourth energy level contains one s orbital, three p orbitals, five d orbitals, and seven f orbitals, therefore 4d and 4f are correct.
The only incorrect designation is 3f.

We could also have solved this problem by using the quantum numbers. The correct orbitals all follow the rules gone over for quantum numbers. For the incorrect designation, however, $n = 3$. l must therefore may only have values of 0, 1, and 2. An f sublevel has $l = 3$. For $n = 3$, this is not an allowed value.

Try Study Question 39 and 41 in Chapter 7 of your textbook now!

b) Describe the shapes of the orbitals (Section 7.6).

An s orbital is spherical in shape. See Figure 7.14 in your text for an illustration of this.

A p orbital contains two regions of electron density. Separating these regions is a plane where there is no probability of finding the electron. This plane is called a nodal surface.

Examples of p orbitals are shown in Figure 7.16. Each p orbital can be pictured as being along one of the coordinate axes in space. We sometimes call these orbitals p_x, p_y, and p_z.

A d orbital contains two nodal surfaces. Most d orbitals contain four regions of electron density. There is one d orbital that looks different. This orbital has two regions of electron density with a "doughnut" shaped region of electron density around their middle. See Figure 7.15 in your text for drawings of all of the orbitals in an atom through the 3d orbitals.

An f orbital contains 3 nodal surfaces. Most f orbitals contain six regions of electron density, but there are some that have the "doughnut" type regions.

Other Notes

1. The Photoelectric Effect

The photoelectric effect deals with the ejection of electrons by a metal when light is shined on the metal. If light below a certain frequency is used, then no electrons are emitted. At these frequencies, it does not matter what the intensity of the light is; the metal simply does not emit electrons. Once a certain frequency is reached (the actual frequency depends on the metal), however, the metal begins to emit electrons. Once this frequency is reached, the greater the intensity of the light, the greater the number of electrons emitted.

Einstein explained this phenomenon. He assumed, like Planck, that electromagnetic radiation delivers its energy in packets (photons). It takes a certain amount of energy to get the metal to emit an electron. This energy must be provided by a single photon of electromagnetic radiation. The metal cannot store up energy from multiple photons to get to this energy. If a single photon of the electromagnetic radiation does not have sufficient energy, then the metal does not emit any electrons. It does not matter how many of these photons hit the metal; the metal will not emit electrons. This explains why lower frequency radiation does not cause any electrons to be emitted, no matter what the intensity is. At some frequency, however, a single photon does have enough energy to cause the metal to emit an electron. Once this frequency is reached, hitting the metal with more photons will cause more electrons to be emitted because each photon has the energy needed, thus the number of electrons emitted increases with increasing intensity of the radiation.

2. Nodal Surfaces

The number of nodal surfaces in an orbital is given by the *l* value. An s orbital ($l = 0$) has zero, a p orbital ($l = 1$) has one, a d orbital ($l = 2$) has two, and an f orbital ($l = 3$) has three.

The presence of nodal surfaces within orbitals further emphasizes that we are really dealing with the wave properties of the electron. If we consider the particle properties of an electron, the nodal surface does not make sense. How could a particle be present on each side of a nodal surface in a p orbital, for example, but never be at the nodal surface? Our particle view does not allow this, but a wave can certainly have zero amplitude right between a peak and a trough.

Example 7-14: How many nodal surfaces are present in a 3p orbital?

For a p orbital, $l = 1$, therefore there is one nodal surface.

Try Study Question 43 in Chapter 7 of your textbook now!

CHAPTER 8: Atomic Electron Configurations and Chemical Periodicity

Chapter Overview

In this chapter, we continue to develop our model of how electrons are arranged in an atom and then see how the electron arrangements influence various properties of the elements. We first discuss the property of electron spin. This leads us to a fourth quantum number called the electron spin quantum number, m_s. As part of this discussion, you will learn about paramagnetic materials and diamagnetic materials. A paramagnetic material is attracted into a magnetic field and contains unpaired electrons, whereas a diamagnetic material is slightly repelled by a magnetic field and has all of its electrons paired. According to the Pauli exclusion principle, no two electrons in an atom may have the same set of four quantum numbers. One of the major goals of this chapter is to teach you how to predict the electron configurations of the elements. You will learn how to write configurations using both orbital box diagrams and spdf notation. In both cases, we can abbreviate what we write by writing the preceding noble gas's symbol in brackets followed by the electrons that have been added beyond those in the noble gas. Another guiding principle is Hund's rule: in an atom in the ground state, if a sublevel that contains more than one orbital, each of these orbitals must be occupied by one electron before pairing occurs. You will also learn how to write the electron configurations of many ions.

Next, we will talk about periodic trends of the elements. One concept that helps with this is effective nuclear charge: the nuclear charge experienced by an outer level electron is smaller than the full nuclear charge due to shielding of the outer electrons by the inner electrons. The effective nuclear charge increases across a period. The size of atoms tends to increase down a group and decrease across a period. The first ionization energy, the energy to remove one electron from an atom in the gas phase, shows exactly the opposite trends: it decreases as we move down a group and increases as we move across a period. The removal of each successive electron from an atom requires more energy, with a large jump occurring once a noble gas configuration has been achieved. Electron affinity is the energy released when an electron is added to an atom in the gas phase. It tends to become less negative as we move down a group and more negative as we move across a period. The trends in ionization energy and electron affinity across a period are not smooth; many of the exceptions to the trends can be explained using arguments derived from the electron configurations of the elements involved. In addition to atomic sizes, we will also look at the sizes of ions. Monatomic cations are smaller than the atoms from which they are derived, whereas monatomic anions are larger. In general, metals tend to react by forming cations. The metals of Group 1A, Group 2A, and aluminum react by losing enough electrons to form cations with the same electron configuration as the preceding noble gas. When nonmetals form ions, they form anions having the same electron configuration as the next noble gas.

Key Terms

In this chapter, you will need to learn and be able to use the following terms:

Aufbau principle: the "building up" principle; as we move from element to element on the periodic table, the electrons are assigned to orbitals having the lowest energies possible, thus as we proceed, electrons are assigned to orbitals of increasingly higher energy.

Diamagnetic: when a diamagnetic material is placed in a magnetic field, it is slightly repelled by the magnet; a diamagnetic material contains no unpaired electrons.

Effective nuclear charge: the nuclear charge experienced by a particular electron in a multielectron atom, as modified by the presence of the other electrons.

Electron affinity: the energy of a process in which an electron is acquired by an atom in the gas phase.

Electron spin magnetic quantum number (m_s): a quantum number which may have either a value of $+1/2$ or $-1/2$; if an orbital contains one electron, the electron may have either value of m_s, but if the orbital contains two electrons, one electron will have $m_s = +1/2$, and the other electron will have $m_s = -1/2$.

Hund's rule: in a given subshell, each orbital is occupied by one electron before pairing occurs.

Ionization energy: the energy required to remove an electron from an atom in the gas phase.

Paramagnetic: when a paramagnetic material is placed into a magnetic field, it is attracted to the magnet; a paramagnetic material contains unpaired electrons.

Pauli exclusion principle: no two electrons in an atom can have the same set of four quantum numbers (n, l, m_l, and m_s).

Valence electrons: the electrons in an atom that go beyond the electron configuration of the previous noble gas.

Chapter Goals

By the end of this chapter you should be able to:

• **Understand the role magnetism plays in determining and revealing atomic structure.**

    a) Classify substances as paramagnetic (attracted to a magnetic field; characterized by unpaired electron spins) or diamagnetic (repelled by a magnetic field) (Section 8.1).

    When some materials are brought into a magnetic field, they are slightly repelled by the magnetic field; such materials are referred to as being diamagnetic. Other materials are attracted into the magnetic field; these materials are referred to as being paramagnetic.

    What causes some materials to be diamagnetic and others to be paramagnetic? The answer has to do with the arrangement of the electrons in the material. An electron behaves as though it has a spin. When an electron is placed in a magnetic field, its spin can either be aligned with the magnetic field or opposed to the field. We designate these two orientations using the electron spin magnetic quantum number, m_s. There are only two possible values for m_s, $+1/2$ and $-1/2$.

Each orbital in an atom can hold a maximum of two electrons. If there is only one electron in a given orbital, then the electron may have m_s equal to either +1/2 or –1/2. We say that the electron in that orbital is unpaired. If, however, the orbital has two electrons in it, one electron must have m_s equal to +1/2, and the other electron must have m_s equal to –1/2. We say that the electrons are paired in such an orbital.

When paired electrons are placed into a magnetic field, the magnetic effects of the two spinning electrons cancel each other. If we have a species in which all of the electrons are paired, therefore, all of the magnetic effects of the spinning electrons will cancel out. When this occurs, the material is slightly repelled by the magnet. A diamagnetic material thus contains all paired electrons.

If we have an unpaired electron in an orbital, the magnetic effect of the spinning electron is not canceled. This causes an attraction to the magnetic field. A paramagnetic material contains at least one unpaired electron.

The attraction of a paramagnetic material into a magnetic field is fairly small. The materials that we commonly refer to as being "magnetic", which are strongly attracted into a magnetic field, are better referred to as being ferromagnetic. Only a few metals and their alloys are ferromagnetic. Ferromagnetic materials are paramagnetic, but the effect is even stronger. A ferromagnetic material contains unpaired electrons just as a paramagnetic material does, but in a ferromagnetic material the unpaired electrons in a cluster of atoms all align in the same direction in the presence of a magnetic field, whereas for a material that is simply paramagnetic only some of the unpaired electrons align with magnetic field.

Example 8-1: Determine if each of the following materials is diamagnetic or paramagnetic.

    a. magnesium

In the diagram that follows, an electron with spin +1/2 is represented by an arrow pointing up, and an electron with spin –1/2 by an arrow pointing down. You will learn in this chapter that the electron configuration of magnesium is

↑↓	↑↓	↑↓ ↑↓ ↑↓	↑↓
1s	2s	2p	3s

All of the electrons are paired in this species so the material is diamagnetic.

    b. vanadium

You will learn in this chapter that the electron configuration of vanadium is

↑↓	↑↓	↑↓ ↑↓ ↑↓	↑↓	↑↓ ↑↓ ↑↓	↑ ↑ ↑ __ __	↑↓
1s	2s	2p	3s	3p	3d	4s

The electrons in the 3d orbitals are unpaired. Because there are unpaired electrons present, this material is paramagnetic.

Try Study Question 35 in Chapter 8 of your textbook now!

b) Recognize that each electron in an atom has a different set of the four quantum numbers, n, l, m_l, and m_s, where m_s, the spin quantum number has values of +1/2 or –1/2 (Section 8.2).

In the last chapter, we learned that with three quantum numbers (n, l, and m_l), we could specify a given orbital in an atom. The principal quantum number, n, gave the energy level. The angular momentum quantum number, l, gave the subshell (s, p, d, f). The magnetic quantum number, m_l, indicated the particular orbital within the subshell.

Each orbital can hold a maximum of two electrons. The two electrons in a filled orbital must have opposite values for the electron spin magnetic quantum number, m_s. Thus if we specify the values of all four quantum numbers, we have specified a given electron in the atom.

Each successive quantum number of the four narrows down the possible electrons until only one is possible. You can liken this to an address on a letter. Let's say that you share a room with one other person in a dormitory on your campus. We could specify the dorm, the floor of the dorm, the specific room on the floor, and finally which of the two people in the room the letter is for.

Example 8-2: What is the maximum number of electrons that can be identified with each of the following sets of quantum numbers?

       a. $n = 3$, $l = 2$, $m_l = -1$, $m_s = -1/2$

Four quantum numbers have been specified, and they are all allowed values. Only one electron in the atom can have all four of these quantum numbers.

       b. $n = 3$, $l = 2$, $m_l = -1$

In this case, we have specified the same first three quantum numbers, but not the spin quantum number. With these three quantum numbers, we have specified a given orbital in an atom. Each orbital can hold two electrons, so this set of three quantum numbers are the same for each of two electrons.

       c. $n = 3$, $l = 2$

Now, only two quantum numbers have been specified. We have identified the energy level (3rd) and the subshell (a d subshell) but we have not specified which d orbital we are talking about. There are five d orbitals in a d subshell, and each can hold two electrons. We have therefore only narrowed things down to one of ten possible electrons.

       d. $n = 3$

We have only specified the energy level. In the third energy level, there is an s orbital, three p orbitals, and five d orbitals. Each of these orbitals can hold two electrons. There are eighteen possible electrons with this value of n.

Try Study Questions 17 and 19 in Chapter 8 of your textbook now!

c) Understand that the Pauli exclusion principle leads to the conclusion that no atomic orbital can be assigned more than two electrons and that the two electrons in an orbital must have opposite spins (different values of m_s) (Section 8.2).

This objective simply formalizes some of what we have been talking about. The Pauli exclusion principle states that no two electrons in an atom can have the same set of four quantum numbers (n, l, m_l, and m_s). As we stated in the last chapter and in the last objective, the first three quantum numbers specify a given orbital. The Pauli exclusion principle states that the different electrons in an orbital must therefore have different values for m_s. Because m_s can have only two possible values, then each orbital can hold at most two electrons, and they must have opposite spins.

Specifying the four quantum numbers for an electron specifies one electron in the atom, just as we talked about in our discussion of the last goal.

- ## Understand effective nuclear charge and its role in determining atomic properties.

Understand effective nuclear charge, Z^*, and its ability to explain why different subshells in the same shell have different energies. Also, understand the role of Z^* in determining the properties of atoms (Sections 8.3 and 8.6).

The effective nuclear charge experienced by an electron in a multielectron atom is a value lower than that of the actual nuclear charge because the portion of the nuclear charge that an electron actually experiences is affected by the presence of the other electrons in the atom.

In a hydrogen atom, there is one proton and one electron. The electron feels the full effect of the +1 charged nucleus. In an atom with more electrons, the outer electrons do not feel the full attractive force of the nucleus because of the presence of the other electrons. We shall work through the same case as your textbook does: lithium. There are three protons in the nucleus, and there are three electrons. The first two electrons fill the 1s orbital. The other electron is in the 2s orbital. How much of the nuclear charge will the electron in the 2s orbital experience? If there were no other electrons present, the electron would experience the full +3 charge of the nucleus, but there are other electrons. The electrons in the 1s orbital affect how much of the nuclear charge the outer energy level electron will experience. On average, the electron in the 2s orbital will be further away from the nucleus than the lower energy 1s electrons. The 1s electrons can be thought of as shielding the outer electron from the full attractive force of the nucleus because we have negative charges in between the negative 2s electron and the nucleus. If the 2s electron were always further away from the nucleus than the 1s electrons and if each 1s electrons were 100% effective in shielding ability, the 2s electron would experience an effective nuclear charge of +3 (from the protons) − 2 (from the inner electrons) = +1. The 2s electron, however, is not always further away. It penetrates the 1s orbital some. As a 2s electron moves closer to the nucleus in this area of penetration, the effective nuclear charge approaches that which the electron would experience if no other electrons were present (+3 for lithium). In lithium, the average effective nuclear charge experienced by the 2s electron is close to 1 but is a little larger due to the penetration of the 2s orbital into the 1s orbital; the average turns out to be 1.28.

Effective nuclear charge can be used to explain why different subshells within a given energy level have different energies. The reason has to do with the amount that the orbitals in these subshells can penetrate the inner orbitals. The penetration of an s subshell into the inner orbitals is the greatest, thus the effective nuclear charge experienced by an electron in an s subshell will be greater than that of electrons in other subshells within that energy level. The greater effective nuclear charge leads to a greater attraction to the nucleus, which leads to a lower energy. Because it has the lowest energy, an s subshell will fill before any of the other subshells. The order of penetrating ability is s > p > d > f. The effective nuclear charges experienced by an electron thus go in the same order: s > p > d > f. The higher the effective nuclear charge, the lower the energy of an electron in a given energy level so the order of electron energies is s < p < d < f.

The key trends to remember dealing with effective nuclear charge are the following:
- o As one moves across a period, the effective nuclear charge experienced by the outermost electrons tends to increase.
- o Within a given energy level, the effective nuclear charge experience by an electron goes in the order: s > p > d > f. This explains why in a given energy level in a multielectron atom, the s orbital is at a lower energy than the p orbitals which are lower than the d orbitals, which are lower than the f orbitals.

We shall explore the role effective nuclear charge has in determining some of the periodic properties of atoms when we discuss those properties.

Example 8-3: Place the following elements in order of increasing effective nuclear charge felt by the outermost p electrons: N, O, F.

These three elements are all in the same period. The element furthest to the right will have the greatest effective nuclear charge, and the one furthest to the left will have the least. The order is N < O < F.

• Write the electron configuration for elements and monatomic ions.

a) Using the periodic table as a guide, depict electron configurations of the elements and monatomic ions using orbital box or the spdf notation. In both cases, configurations can be abbreviated with the noble gas notation (Sections 8.3 and 8.4).

This is one of the most important goals for this chapter. In writing electron configurations, we will assign the electrons by starting with the lowest energy electrons and then go up in energy until we have assigned all of the electrons in the atom. Let the periodic table help you. See Figure 8.7 in your textbook. This figure indicates which set of orbitals is filling in the different regions of the periodic table. Notice that in the s and p regions, the energy level filling is the same as the row number. Thus in the fourth row of the periodic table, we fill the 4s and the 4p orbitals. In the d block, the energy level lags one behind the row number, thus d-block elements in the fourth row are filling the 3d orbitals. The f block elements lag behind two energy levels from the row number. Thus the 4f orbitals get filled when we are in the sixth row of the periodic table.

There are two major types of notation that we shall use to show electron configurations. The first is called orbital box notation. This involves drawing a box (or line) to represent an orbital and then representing the electrons by arrows. An arrow pointing up will represent an electron with $m_s = +1/2$ and an arrow pointing down will indicate an electron with $m_s = -1/2$. The first electron to occupy an orbital could be shown either with an arrow up or down, but for consistency, we will usually show the first arrow pointing up. If there are two electrons in an orbital, then we must have one arrow pointing up and one arrow pointing down.

The second type of notation simply lists the sublevels that are fully or partially occupied and indicates the number of electrons in each using superscripts. We shall call this type of notation spdf notation.

In both types of notation, we can abbreviate the configurations by writing in brackets the last noble gas before the element in question and then writing the configuration for the electrons that have been added since that noble gas.

We will now begin to take a walk through the periodic table, showing the electron configurations for various elements.

The first element is hydrogen, which has one electron. This electron will go into the lowest energy orbital, the 1s orbital. We will indicate this by showing one arrow going up in the 1s orbital for the orbital box notation. In spdf notation, we will write 1s for the energy level and subshell and a superscript of 1 to indicate that there is one electron in this subshell.

H   ↑ $1s^1$
      1s

With helium, we finish the first energy level.

He   ↑↓ $1s^2$
      1s

With lithium, the first energy level is filled. One more electron remains. It will start the second energy level. Lithium is in the second row on the periodic table, is in the s block and is in the first column in this block so the last part of its electron configuration will be $2s^1$. We also could write the notations using the noble gas core; we have shown this for the spdf notation.

Li   ↑↓   ↑ $1s^2 2s^1$ $[He]2s^1$
     1s   2s

With beryllium, we finish the 2s orbital.

Be   ↑↓   ↑↓ $1s^2 2s^2$ $[He]2s^2$
     1s   2s

Let's now take a look at boron. Tracing through the periodic table, we filled the 1s orbital with hydrogen and helium, the 2s orbital with lithium and beryllium. Boron is in the first column of the p block in the second row on the periodic table. Its last electron will be indicated by $2p^1$.

B   ↑↓   ↑↓   ↑ _ _ $1s^2 2s^2 2p^1$ $[He]2s^2 2p^1$
   1s   2s   2p

With carbon, we are faced with a choice. In the p orbitals, do we get two electrons in one orbital or do we get one electron in each of two orbitals? According to Hund's rule, in a given sublevel, each orbital must be occupied by one electron before any pairing occurs. The author once heard a high school teacher refer to this as being like a fair dessert rule: "Everyone must get a slice of pie before anyone can have seconds." In carbon, therefore we end up with the two p orbitals each with one electron.

C   ↑↓   ↑↓   ↑ ↑ _ $1s^2 2s^2 2p^2$ $[He]2s^2 2p^2$
   1s   2s   2p

We know from the periodic table that the final entry for nitrogen will be $2p^3$ because nitrogen is in the p block, in the second row of the periodic table, and in the third column of the p block. Following Hund's rule, we put each of the electrons in the p orbitals into separate orbitals.

N   ↑↓   ↑↓   ↑ ↑ ↑ $1s^2 2s^2 2p^3$ $[He]2s^2 2p^3$
   1s   2s   2p

When we get to oxygen, we have no choice but to begin pairing up electrons in the 2p orbitals.

O   ↑↓   ↑↓   ↑↓ ↑ ↑ $1s^2 2s^2 2p^4$ $[He]2s^2 2p^4$
   1s   2s   2p

Fluorine will have five electrons in the 2p orbitals. We know this because fluorine is in the fifth column in the p block in the second row of the periodic table.

F   ↑↓   ↑↓   ↑↓ ↑↓ ↑ $1s^2 2s^2 2p^5$ $[He]2s^2 2p^5$
   1s   2s   2p

At neon, we are at the next noble gas and have filled the 2p orbitals.

Ne   ↑↓   ↑↓   ↑↓ ↑↓ ↑↓ $1s^2 2s^2 2p^6$
    1s   2s   2p

The next element is sodium. We are back to being in the first column of the s block but this time, we are in the third row of the periodic table. Also, the last noble gas is now neon instead of helium.

Na   ↑↓   ↑↓   ↑↓ ↑↓ ↑↓   ↑ $1s^2 2s^2 2p^6 3s^1$ $[Ne]3s^1$
     1s   2s   2p   3s

From this point on, we shall only go over some elements' electron configurations. At phosphorus, we are at the third column of the p block in the third row of the periodic table. We once again follow Hund's Rule.

P   ↑↓   ↑↓   ↑↓ ↑↓ ↑↓   ↑↓   ↑ ↑ ↑ $1s^2 2s^2 2p^6 3s^2 3p^3$
     1s   2s   2p   3s   3p

                                                     $[Ne]3s^2 3p^3$

You can see that the orbital box notation is taking up more and more space. From now on, we will only show the spdf notation and the spdf configuration using the noble gas core.

Calcium is located in the second column in the s block of the fourth row on the periodic table, so our final entry for it will be $4s^2$. We can trace through on the periodic table all that must be there. Starting at hydrogen, we have gone through the 1s orbital, the 2s orbital, the 2p orbitals, the 3s orbital, the 3p orbitals, and now the 4s orbital. The preceding noble gas is argon.

Ca $1s^2 2s^2 2p^6 3s^2 3p^6 4s^2$ $[Ar]4s^2$

With scandium, we are in a d block for the first time. We are in the first column of the d block in the fourth row on the periodic table. The energy level for the d block elements lags one behind the energy level. Our final entry will be $3d^1$. At first, write down the configuration in the order that you go through the periodic table:

Sc $1s^2 2s^2 2p^6 3s^2 3p^6 4s^2 3d^1$ $[Ar]4s^2 3d^1$

For the final answer, however, it is customary to group together all of the subshells from the same energy level, so our final configuration would be

Sc $1s^2 2s^2 2p^6 3s^2 3p^6 3d^1 4s^2$ $[Ar]3d^1 4s^2$

For the elements scandium through titanium, we fill the 3d orbitals. If we were to show the orbital box notation, we would follow Hund's rule, filling all five 3d orbitals with one electron before we would begin pairing the electrons.

For the elements from gallium through krypton, we are filling the 4p orbitals. As an example, we shall consider selenium. As our first attempt, we would write

Se $1s^2 2s^2 2p^6 3s^2 3p^6 4s^2 3d^{10} 4p^4$ $[Ar]4s^2 3d^{10} 4p^4$

We would then rewrite it in the order

Se $1s^2 2s^2 2p^6 3s^2 3p^6 3d^{10} 4s^2 4p^4$ $[Ar]3d^{10} 4s^2 4p^4$

For Rb and Sr, we fill the 5s orbital, for yttrium through cadmium the 4d orbitals, for indium through xenon, the 5p orbitals, and for cesium and barium, the 6s orbital.

When we get to lanthanum, we would have the following

La $1s^2 2s^2 2p^6 3s^2 3p^6 4s^2 3d^{10} 4p^6 5s^2 4d^{10} 5p^6 6s^2 5d^1$ $[Xe]6s^2 5d^1$

which would be rewritten as

La $1s^2 2s^2 2p^6 3s^2 3p^6 3d^{10} 4s^2 4p^6 4d^{10} 5s^2 5p^6 5d^1 6s^2$ $[Xe]5d^1 6s^2$

Following lanthanum, we jump down to the lanthanide series at the bottom of the periodic table. This group of elements involves filling f orbitals for the first time. The f block lags two energy levels behind the row number of the periodic table. Since the lanthanide series falls in the middle of the sixth row on the periodic table, we are filling the 4f orbitals. For praseodymium, we are in the second column of the f block so we would predict two electrons in the 4f orbitals. We would first write the following

Pr $1s^22s^22p^63s^23p^64s^23d^{10}4p^65s^24d^{10}5p^66s^25d^14f^2$ $[Xe]6s^25d^14f^2$

This would be rewritten as

Pr $1s^22s^22p^63s^23p^63d^{10}4s^24p^64d^{10}4f^25s^25p^65d^16s^2$ $[Xe]4f^25d^16s^2$

Let's now do one final element: meitnerium (element 109). This is in the seventh column of the d block in row 7 of the periodic table so the final entry we will first write will be $6d^7$.

Mt $1s^22s^22p^63s^23p^64s^23d^{10}4p^65s^24d^{10}5p^66s^24f^{14}5d^{10}6p^67s^25f^{14}6d^7$

         $[Rn]7s^25f^{14}6d^7$

This would be rewritten in order as

Mt $1s^22s^22p^63s^23p^63d^{10}4s^24p^64d^{10}4f^{14}5s^25p^65d^{10}5f^{14}6s^26p^66d^77s^2$

         $[Rn]5f^{14}6d^77s^2$

There are some exceptions to the configurations we have been discussing. We will not worry about most of these in this course, but you should be aware of some. The elements chromium and molybdenum in Group 6B on the periodic table are exceptions as are copper, silver, and gold in Group 2B. For each of these, instead of two electrons in the s orbital we get only one; the other electron that would have gone into the s orbital ends up the d subshell instead. You should probably know these exceptions in the transition metals. The following are the configurations for chromium and copper:

	Predicted	Actual
Cr	$[Ar]3d^44s^2$	$[Ar]3d^54s^1$
Cu	$[Ar]3d^94s^2$	$[Ar]3d^{10}4s^1$

In the f block, there are many exceptions. One common motif besides that we have gone over (in which the electrons beyond the s subshell are distributed with one in the d subshell and the remainder in the f subshell) is having all of the electrons beyond the s subshell go into the f subshell. Europium falls into this category:

	Predicted	Actual
Eu	$[Xe]4f^65d^16s^2$	$[Xe]4f^76s^2$

Don't worry about memorizing the exceptions in the f block; simply be aware that they exist.

Example 8-4: Write the spdf configuration for gallium.

Gallium (Ga) is in the first column in the p block of the fourth row of the periodic table. Our final entry will be $4p^1$. We trace through the periodic table until we get to $4p^1$. We first write:

$1s^22s^22p^63s^23p^64s^23d^{10}4p^1$

and then rewrite it in order of increasing energy levels

$1s^22s^22p^63s^23p^63d^{10}4s^24p^1$

Example 8-5: Write the noble gas core configuration for osmium (Os).

Osmium is in the sixth column of the d block of the sixth row. The d block lags behind one energy level from the row number. The final entry will be $5d^6$. The last noble gas was Xe. Since Xe, we have filled the 6s orbital and the 4f orbitals (remember the energy level for the f orbitals lags behind the row number by two), and then started to fill the 5d sublevel. We first write

$[Xe]6s^24f^{14}5d^6$

In the proper order, this becomes

$[Xe]4f^{14}5d^66s^2$

Be careful when you are dealing with elements beyond lanthanum. Students often forget to include the electrons in the f orbitals.

Try Study Questions 5, 7, and 9 in Chapter 8 of your textbook now!

To form a cation from a neutral atom, one or more electrons are lost. Metals tend to form cations. The cations of the metals in Groups 1A, 2A, and aluminum have the same electron configuration as the previous noble gas. A neutral atom of magnesium has the electron configuration

Mg $1s^2 2s^2 2p^6 3s^2$ $[Ne]3s^2$

Magnesium will lose the two electrons in the 3s orbital to form an ion with a 2+ charge: Mg^{2+}, which has the same electron configuration as the previous noble gas, Ne:

Mg^{2+} $1s^2 2s^2 2p^6$ $[Ne]$

Many of the metals further down in Groups 3A through 5A form cations by losing their highest energy level p electrons to obtain the same configuration as the last transition metal in Group 2B. For example, the element tin has the electron configuration

Sn $[Kr]4d^{10}5s^2 5p^2$

Tin forms ions with a 2+ charge by losing the two electrons in the 5p subshell and obtaining the configuration of cadmium:

Sn^{2+} $[Kr]4d^{10}5s^2$

When the transition metals form ions, they first lose electrons from the ns orbital, after which they lose electrons from the $(n-1)$ d sublevel, where n is the row number on the periodic table. For example, a neutral atom of cobalt has the electron configuration

Co $[Ar]3d^7 4s^2$

Cobalt forms ions with a 2+ charge and a 3+ charge. For the 2+ ion, the two electrons in the 4s orbital are lost. For the 3+ ion, the two electrons in the 4s orbital and one of the 3d electrons are lost.

Co^{2+} $[Ar]3d^7$
Co^{3+} $[Ar]3d^6$

The fact that the s electrons are lost first helps to explain why so many transition metals have ions with a 2+ charge.

<u>Example 8-6:</u> Predict the electron configuration of the following ions:

a. potassium ion

Potassium is a main group element in Group 1A. The electron configuration of neutral potassium is

K $1s^2 2s^2 2p^6 3s^2 3p^6 4s^1$

To form the ion, the one electron in the fourth energy level will be lost, leaving the electron configuration of argon

K^+ $1s^2 2s^2 2p^6 3s^2 3p^6$

b. Ti^{2+}

Titanium is a transition metal with the electron configuration

Ti $[Ar]3d^2 4s^2$

To form a 2+ ion, two electrons must be lost. These will come from the 4s orbital. The electron configuration of the ion is

Ti^{2+} $[Ar]3d^2$

c. Bi^{3+}

Bismuth is a main group element in Group 5A. It has the electron configuration
$$[Xe]4f^{14}5d^{10}6s^26p^3$$
To form the ion with a 3+ charge, three electrons must be lost; these will be the electrons in the 6p sublevel.
$$[Xe]4f^{14}5d^{10}6s^2$$

Try Study Question 13 in Chapter 8 of your textbook now!

Nonmetals tend to form ions by gaining electrons to attain the same electron configuration as the next noble gas; the resulting ions are negatively charged (anions). For example, fluorine is in Group 7A. It is one electron away from having the noble gas configuration of neon.

F $1s^22s^22p^5$ $[He]1s^22p^5$

When it forms an ion, therefore, it gains one electron and ends up with the same electron configuration as neon. The fluoride ions has the electron configuration

F^- $1s^22s^22p^6$ $[He]1s^22p^6 = [Ne]$

Example 8-7: What is the electron configuration of the sulfide ion?

Sulfur is in Group 6A on the periodic table. It has the following electron configuration:
     S $[Ne]3s^23p^4$
It is two electrons away from the noble gas configuration of argon. It will thus gain two electrons, forming an anion with a charge of 2–.
     S^{2-} $[Ne]3s^23p^6 = [Ar]$

Try Study Question 11 in Chapter 8 of your textbook now!

b) Recognize that electrons are assigned to the subshells of an atom in order of increasing subshell energy. In the H atom the subshell energies increase with increasing *n*, but, in a many-electron atom, the energies depend on both *n* and *l* (see Figure 8.4).

In the hydrogen atom, the energy of the electron is determined solely by the energy level it is in. In other words, all of the subshells within an energy level have exactly the same energy *in hydrogen*. In hydrogen, therefore, the 2s and the 2p orbitals have exactly the same energy. The 3s, 3p, and 3d orbitals all have the same energy as each other. This is true only for one electron systems like the hydrogen atom.

In a multielectron atom, however, the situation is more complicated. The subshells within a given energy level have different energies. An s subshell will be lower in energy than a p subshell which will be lower than a d subshell which will be lower than an f subshell. Thus, the 2s subshell is lower in energy than the 2p subshell. This is why we filled he 2s orbital before we began to fill the 2p orbitals when we were doing our walk through the periodic table. As we saw, it even gets more complicated because the energy levels overlap. Remember, we filled the 4s orbital before we filled the 3d orbitals. Beyond just looking at the periodic table, is there any way to predict this order? Yes, there is. Electrons are assigned to subshells in order of increasing *n* + *l*, and for two subshells with the same *n* + *l* value, electrons are assigned first to the subshell of lower *n*.

Example 8-8: Predict the order in which the 3s, 3p, 3d, 4s, and 4p subshells will fill (without relying on the periodic table).

The following table lists the $n + l$ value for the 3s, 3p, 3d, 4s, and 4p orbitals.

Orbital	$n + l$ value
3s	$3 + 0 = 3$
3p	$3 + 1 = 4$
3d	$3 + 2 = 5$
4s	$4 + 0 = 4$
4p	$4 + 1 = 5$

We can see that the 3s orbital has the lowest $n + l$ value, so it will be the first of these orbitals to fill. There are two sublevels with $n + l = 4$, the 3p and the 4s orbitals. In choosing between these, we select to fill first the one with the lowest n, so the 3p subshell will fill before the 4s subshell. Following this, we will move on to the $n + l = 5$ subshells, the 3d and the 4p subshells. In selecting between these, the 3d will fill before the 4p because it has the lower value for n. The order is thus 3s, 3p, 4s, 3d, 4p just as we would have predicted from our work with the periodic table.

The order of filling for all of the subshells used for the ground states of the known elements (and a few more) is given in Figure 8.5 in your text.

c) When assigning electrons to atomic orbitals, apply the Pauli exclusion principle and Hund's rule (Section 8.3 and 8.4).

We used these principles when we were doing our walk through the periodic table. Reviewing, the Pauli exclusion principle states that no two electrons in an atom may have the same set of four quantum numbers. This comes into play in writing electron configurations using orbital box notation because in any filled orbital, one electron will be represented by an arrow pointing up and the other electron by an arrow pointing down.

According to Hund's rule, in a given subshell containing more than one orbital, each orbital must get occupied by one electron before pairing occurs. We used this in writing electron configurations using orbital box notation in that each orbital in a given subshell received one arrow before we drew any orbital in that subshell containing two arrows.

- **Understand the fundamental physical properties of the elements and their periodic trends.**

  a) Predict how properties of atoms – size, ionization energy (IE), and electron affinity (EA) – change on moving down a group or across a period of the periodic table (Section 8.6). The general periodic trends for these properties are as follows:

  i) Atomic size decreases across a period and increases down a group.

  These trends in atomic size are best observed in the main group elements. Why do these trends occur? Atomic size increases as we move down a group. As we move down a group, we progressively place electrons into higher and higher energy levels. The higher energy electrons tend to spend more of their time farther from the nucleus so the size of the atoms increases. This part of the trend is probably in line with what you would have predicted.

The other trend, the size decreasing as we move across a period from left to right, might not line up with your initial gut reaction. Why would the size decrease as we add more electrons? The answer goes back to the idea of effective nuclear charge. As we move across a period, the effective nuclear charge increases. The outer electrons feel a greater attraction for the nucleus and are pulled in tighter. The size thus gets smaller as we move across the period.

The periodic trends for atomic size of the transition metals are largely the same as those of the main group elements with some differences. The size usually does increase as we move down a group, but the change from row to row is not as great. As we move across a period, the size initially decreases, just as we found for the main group elements, but then levels off and finally increases toward the end of the transition metals due to electron-electron repulsions within the d subshell.

Example 8-9: Place the following elements in order of increasing atomic size: Rb, Sr, and Cs.

Atomic size increases as we move down a group. Cesium is larger than rubidium. Atomic size decreases as we move across a period. Rubidium is larger than strontium. If rubidium is larger than strontium and if cesium is larger than rubidium, then cesium must also be larger than strontium. The correct sequence, in order of increasing atomic size, is strontium < rubidium < cesium.

Try Study Question 25 in Chapter 8 of your textbook now!

ii) IE increases across a period and decreases down a group.

Ionization energy is the energy required to remove an electron from an atom in the gas phase. Energy is required to remove an electron from an atom; this is an endothermic process. An electron is associated with an atom by means of electrostatic attraction between the negative electron and the positive nucleus. To remove an electron, therefore, energy must be supplied to overcome this electrostatic force.

The most common ionization energy to discuss is the first ionization energy, the energy required to remove one electron from an atom in the gas phase. Sometimes, we simply refer to this as the ionization energy. The first ionization energy tends to increase across a period and decrease down a group. This can be rationalized by thinking about atomic size. Atomic size increases as we move down a group. The outermost electrons are, on average, further from the nucleus. The electrostatic attraction between two charged bodies decreases with increasing distance, so it takes less energy to remove an electron from the larger atoms. The first ionization energy thus decreases as we move down a group. On the other hand, atomic size decreases as we move across a period due to the increased effective nuclear charge. The electrons are held closer to the nucleus; there is a greater attraction between the nucleus and an outer electron. It is therefore harder to remove such an electron, and the first ionization energy increases as we move across a period on the periodic table.

The trend for ionization energy across a period is not entirely smooth; instead of consistently going up, there are some cases where the ionization energy decreases on moving from one element to the next (see Figure 8.13 in your text). Let's look at the second period elements; the other periods have similar cases. We start out the second period as expected: in moving from lithium to beryllium, the ionization energy increases. When we go to boron, however, the ionization energy actually decreases instead of increasing. Why would this be? With boron, we start the 2p subshell. The increase in effective nuclear charge in moving from beryllium to boron is not enough to counteract the fact that the electron in question is in the slightly higher energy 2p subshell. The ionization energy actually goes down. Following this, our trend of increasing across a period returns for carbon and nitrogen, but then the ionization energy goes down when we move from nitrogen to oxygen. With oxygen, we get the first pairing of electrons in the 2p subshell. Electron-electron repulsions in the paired 2p orbital cause the ionization energy to come out lower. Following this, our usual trend returns, and the ionization energy increases as we move to fluorine and then to neon.

Recapping therefore, the first ionization energy tends to decrease as we move down a group, and increase as we move across a period. We get exceptions to the trend across a period in group 3A where we start each p subshell and in group 6A where we start to pair electrons in each p subshell.

If we remove more than one electron from an atom, we find that with each electron removed, the ionization energy increases. This is because we are removing a negative electron from a particle that is increasingly positive in charge. We also find that there is a very large increase in ionization energy when we go to remove an electron from an ion that has an electron configuration equivalent to that of a noble gas. This explains why metals form cations by losing enough electrons to obtain the same electron configuration as the previous noble gas. Going beyond this simply takes too much energy for ordinary processes. Thus, for an element in Group 1A, we can remove one electron with a reasonable amount of energy. At this point, we have formed an ion with a noble gas configuration. After this, the energy required to remove a second electron is very large, so lithium tends to form only a 1+ ion in ordinary chemical reactions. For an element in Group 2A, we can remove one electron and then remove another electron using some more energy at which point we have formed an ion with a noble gas configuration. The energy required to remove a third electron is very large, so magnesium tends to form a 2+ ion in ordinary chemical reactions but does not form ions with a higher positive charge.

Example 8-10: Place the following elements in order of increasing first ionization energy: B, C, Al.

The first ionization energy of boron will be larger than that of aluminum because first ionization energy decreases as we go down a group. The first ionization energy of carbon will be greater than that of boron because it is further to the right and moving from Group 3A to Group 4A is not a point where one of our exceptions comes in. If carbon's first ionization energy is greater than that of boron and boron's is greater than that of aluminum, then carbon's will also be larger than that of aluminum. The sequence in order of increasing ionization energy is aluminum < boron < carbon.

Try Study Questions 31a and c and 33 a and b in Chapter 8 of your textbook now!

iii) The affinity for an electron generally increases across a period (the value of EA becomes more negative) and decreases down a group.

We speak of an atom having an affinity or "liking" for electrons. The electron affinity of an atom is the energy of a process in which an electron is acquired by the atom in the gas phase. Some of the terminology associated with electron affinity is sometimes confusing. The process of an atom gaining an electron in the gas phase is usually an exothermic one. The actual values of the electron affinities are thus usually negative numbers. An atom with a more negative electron affinity has a greater "liking" for electrons than one with a more positive value. An atom with a more negative electron affinity is said to have a greater affinity for the electron being added. Let's say we are comparing two atoms, one with a –400 kJ/mole electron affinity and one with a –100 kJ/mole electron affinity. The atom with the –400 kJ/mole electron affinity has a more negative electron affinity and thus has a greater affinity or "liking" for an electron. The element with the greater affinity for an electron has the more negative electron affinity.

As we move down a group, the electron being added will need to go into an energy level further out, where the electrostatic attraction will be less. As we move down a group on the periodic table, therefore, the electron affinity tends to get less negative; as we move down a group the affinity for the electron decreases. As we move across a period on the periodic table, the effective nuclear charge increases. An electron coming in will be attracted by a greater effective charge. It makes sense, therefore, that the electron affinity tends to get more negative as we move across a period; as we move across a period, the atom has a greater affinity for an electron being added.

There are some irregularities to these trends. As we move from the second period down to the third, the normal trend of the electron affinity getting less negative as we move down a group does not hold for many cases. This is because the atoms of the second period are very small and electron-electron repulsions in these small atoms cause their electron affinities not to be as negative as we might have expected.

The other type of exception is similar to those that we encountered with ionization energy but shifted over by one group. Let's move across the second period looking for exceptions, just as we did with ionization energy. We start out with an exception to our rule. As we move from lithium to beryllium, the electron affinity gets less negative instead of more negative. Why does beryllium have a less negative electron affinity? The electron coming into a beryllium atom would need to start a new p subshell. Adding an electron to a new subshell is not as favorable as adding one to a subshell that already has electrons in it. After beryllium, the trend goes as expected for awhile. The electron affinity of boron is more negative than that of beryllium and that of carbon is more negative than that of boron. After that, however, the electron affinity again becomes less negative at nitrogen. Why? With nitrogen, we have all three p orbitals with one electron in them. Adding one more electron causes us to have to begin pairing electrons in the p orbitals. This is not as favorable. The electron affinities then become more negative as we move from nitrogen to oxygen and then to fluorine. It then gets less negative for neon. In this case, the extra electron would need to start a whole new energy level. This is not favorable so the electron affinity is less negative than for fluorine. The key exceptions thus occur as we move from Group 1A to Group 2A, as we move from Group 4A to Group 5A, and as we move from Group 7A to Group 8A.

Example 8-11: Which atom in each pair has a more negative electron affinity?

   a. Na or K
K is below Na on the periodic table. Electron affinity tends to become less negative as we move down a group, therefore we would expect Na to have the greater electron affinity.

   b. Al or Si
The general trend is for electron affinity to become more negative as we move from left to right. Si is further right than Al, and they are not in our critical groups for exceptions, so we would expect that the electron affinity of Si would be more negative than that of Al.

   c. Si or P
Si is in Group 4A, and P is in Group 5A. This is one of the critical transitions where we get exceptions to the general trend that electron affinities become more negative as we move from left to right. We therefore should predict that the general trend will be broken and that silicon will have a more negative electron affinity than phosphorus.

Try Study Questions 31b and 33 c in Chapter 8 of your textbook now!

The periodic trends are summarized in the following table:

As we move	Atomic Size	First Ionization Energy	Electron Affinity
Down a Group	Increases	Decreases	Becomes Less Negative
Across a Period	Decreases	Increases	Becomes More Negative

b) Recognize the role that ionization energy and electron affinity play in the chemistry of the elements (Section 8.7).

Group 1A and 2A metals and aluminum generally form cations with an electron configuration equivalent to that of the previous noble gas. When a nonmetal forms an ion, it usually acquires enough electrons to form an anion with the electron configuration of the next higher noble gas. These two statements can be rationalized using what we know about ionization energy and electron affinity.

Group 1A and 2A metals and aluminum tend to have a fairly low first ionization energy; this means that it does not take a great amount of energy to cause them to lose an electron. Successive ionization energies get increasingly larger. Once a noble gas configuration has been reached, there is a very large jump in ionization energy. The consequence of this is that once a noble gas configuration has been achieved, these metals do not tend to lose any more electrons. The metals also tend to have less negative electron affinities, so they do not tend to form anions readily. These metals thus tend to react by losing electrons, and they will lose enough electrons to obtain a noble gas configuration but no more.

On the other hand, the nonmetals have high first ionization energies. This means that it takes a lot of energy to get them to lose an electron. On the other hand, they have more negative electron affinities. Nonmetals thus tend to gain electrons and not lose them. They will gain sufficient electrons to attain the next noble gas's electron configuration.

An element such as carbon has a higher ionization energy than the Group 1A and 2A metals but not as negative an electron affinity as the nonmetals that readily form anions. Instead of forming an ion, carbon usually reacts by sharing electrons.

Other Notes

1. The valence electrons in an atom are the electrons that have been added since the last noble gas. For the main group elements, they are the s and p electrons of the highest energy level. For the transition metals, they are the s and p electrons of the highest energy level, and the d electrons belonging to the previous energy level. For example, in vanadium, the valence electrons are the two 4s electrons and the two 3d electrons.

2. Similarities in properties of the elements in a vertical column on the periodic table are the result of similar valence shell electron configurations.

3. The radius of a monatomic cation is smaller than that of the atom from which it is derived. The radius of a monatomic anion is larger than that of the atom from which it is derived. Within a series of isoelectronic ions, the one with the largest number of protons will have the smallest radius and the one with the smallest number of protons will have the largest radius.

<u>Example 8-12:</u> Identify the species in the following pairs that is larger.

      a. Na or Na$^+$

Cations are smaller than the atoms from which they are derived, so the sodium atom is larger. In a sodium atom, there are 11 protons and 11 electrons. In a sodium ion, there are still 11 protons but only 10 electrons. The 11 protons can pull the 10 electrons in more tightly.

      b. Br or Br$^-$

Anions are larger than the atoms from which they are derived, so the bromide ion is larger. In a bromine atom, there are 35 protons and 35 electrons. In a bromide ion, there are still 35 protons, but now there are 36 electrons. The 35 protons cannot hold the 36 electrons as tightly due to electron-electron repulsions.

      c. S^{2-} or Cl$^-$

Both of these ions have an electron configuration that is the same as that of argon; they have the same number of electrons. Sulfide has 16 protons, whereas chloride has 17 protons. The ion with the smaller number of protons will be larger, so S^{2-} is larger than Cl$^-$.

Try Study Questions 33d and 51 b and e in Chapter 8 of your textbook now!

CHAPTER 9: Bonding and Molecular Structure: Fundamental Concepts

Chapter Overview

In this chapter, you will learn more about how atoms combine together to form compounds. The electrons involved in chemical bonding are the valence electrons in an atom. For main group elements, these are the electrons in the highest energy level's s and p orbitals, and the total number of valence electrons in a main group element is given by the group number on the periodic table. Many elements follow the octet rule when they react by gaining, losing, or sharing enough electrons to obtain a noble gas configuration with filled s and p sublevels in their outer shells.

There are two main types of bonding we shall study in this chapter: ionic bonding and covalent bonding. Ions are formed by the transfer of electrons between atoms. In an ionic compound in the solid state, the ions are held together in a crystal lattice by electrostatic interactions. The force of the electrostatic interaction depends on the magnitude of the charges and on the distance between the ions. The lattice energy, the difference in energy between separated ions in the gas phase and the same ions in a solid crystal lattice, can be calculated using a Born-Haber cycle.

We can represent a molecular compound or a polyatomic ion using Lewis structures. These show the connectivity between the atoms and any lone pairs present in the valence shells of the atoms. You will learn how to draw Lewis structures for many compounds. Sometimes, more than one possible Lewis structure can be drawn for the same molecule or polyatomic ion. In such cases, the real structure will be somewhere in between the possible Lewis structures. These different Lewis structures are called resonance structures. A Lewis structure does not provide us with the actual shape of a molecule or polyatomic ion; it is, however, the first step in determining the shape. You will learn how to use valence shell electron-pair repulsion theory (VSEPR) to determine these shapes. Based on the number of groups surrounding the central atom, we can determine the electron-pair geometry. Based on where the atoms end up in this electron-pair geometry, we can determine the molecular geometry. You will also learn how to calculate the formal charge of an atom in a molecule or polyatomic ion.

Electronegativity is a measure of an atom's ability to attract electrons to itself in a chemical bond. If two atoms with very different electronegativities bond, then an ionic bond is formed. If two atoms with equal electronegativities bond, then a nonpolar covalent bond is formed. If two atoms with different electronegativities (but not as different as for ionic bonding) bond, a polar covalent bond is formed. In a polar covalent bond, one atom gets the shared electrons more of the time and so has a partial negative charge; the other atom does not get the shared electrons as much and so has a partial positive charge. Just because a molecule has polar bonds does not guarantee that the overall molecule will be polar. You will learn how to predict whether a molecule will be polar or not. The bond order is the number of electron pairs shared by two atoms. The bond length decreases as the bond order between two atoms increases and as the size of the atoms decreases. The bond dissociation energy is the energy required to break a bond in the gas phase. Breaking bonds takes energy. Making bonds in the gas phase releases energy. You will learn how to use bond dissociation energies to estimate ΔH for chemical reactions.

Key Terms

In this chapter, you will need to learn and be able to use the following terms:

Axial: in the trigonal bipyramidal electron-pair geometry, the two positions directly across from each other.

Bent: the molecular geometry that results when two atoms and one lone pair are connected to a central atom in the trigonal planar electron-pair geometry and when two atoms and two lone pairs are connected to a central atom with the tetrahedral electron-pair geometry.

Bond angle: the angle formed between two atoms that are connected to the same central atom.

Bond dissociation energy (D): the enthalpy change for breaking a bond in a molecule with the reactants and products in the gas phase.

Bond length: the distance between the nuclei of two bonded atoms.

Bond order: the number of bonding electron pairs shared by two atoms in a molecule or polyatomic ion.

Bonding pair: a pair of electrons that is shared between two atoms.

Bonding: describes the forces that hold adjacent atoms together.

Born-Haber cycle: an energy cycle in which the formation of one mole of a solid crystalline ionic compound from its elements is pictured as going through the following steps: conversion of the elements to gaseous monatomic atoms, the loss of electrons by the metal atoms, the gain of electrons by the nonmetal atoms, and finally the formation of a solid crystal lattice from the isolated ions in the gas phase; the overall sum of this cycle will equal the enthalpy of formation of the solid crystalline ionic compound.

Chemical bond: an interaction between the valence electrons in two atoms that results in an attractive force between the two atoms.

Coordinate covalent bond: a covalent bond in which both electrons in the bond were contributed by one atom.

Core electrons: all of the electrons in an atom except for the valence electrons.

Covalent bond: a force of attraction between two atoms that results from the sharing of electrons in the valence shells of two atoms.

Dipole moment (μ): the product of the magnitude of the partial charges on a molecule and the distance by which they are separated. The dipole moment is a vector quantity having both a magnitude and a direction.

Double bond: a double bond exists when four electrons are shared by two atoms; two pairs of electrons are shared.

Electronegativity (χ): a measure of the ability of an atom in a molecule to attract electrons to itself.

Electroneutrality principle: electrons in a molecule or polyatomic ion will be distributed in such a way that the charges on all atoms are as close to zero as possible.

Electron-pair geometry: the geometry taken up by the valence electron pairs around a central atom.

Equatorial: in the trigonal bipyramidal electron-pair geometry, the three positions in a plane midway between the two axial positions.

Formal charge: the charge calculated for an atom by comparing the number of valence electrons in an atom in a molecule or polyatomic ion (in which lone pairs are assigned solely to the atom and bonding pairs are evenly split between the two atoms involved in the bond) to the number of valence electrons that a free atom of that element would have. It is calculated by the equation: Formal charge = group number − [Lone pair electrons + 1/2 Bonding electrons]

Free radical: a chemical species containing an unpaired electron; these tend to be very reactive.

Ionic bond: the electrostatic attraction between two oppositely charged ions. Ion formation results from the transfer of electrons from one atom to another.

Isoelectronic: containing the same number and arrangement of valence electrons.

Isostructural: having the same structure.

Lattice energy: the energy for the formation of one mole of a solid crystalline ionic compound from its constituent ions in the gas phase.

Lewis electron dot symbol: in a Lewis dot symbol for an element, the chemical symbol is used to represent the nucleus and core electrons of an atom; the valence electrons are represented by dots placed around the symbol.

Lewis structure: a representation of a covalently bonded species in which the chemical symbol is used to represent the nucleus and core electrons of each atom; a lone pair belonging solely to one atom is represented by two dots, and a pair of electrons that is shared between two atoms is represented by a line connecting the two atoms.

Linear: the electron-pair geometry having two groups connected to a central atom, and the molecular geometry that results when both groups connected central atom in the linear electron-pair geometry are atoms and when two atoms and three lone pairs are connected to a central atom in the trigonal bipyramidal electron-pair geometry.

Lone pair: a pair of electrons that belongs solely to one atom in a covalent species; these are nonbonding electrons.

Molecular geometry: the arrangement of the central atom and the atoms attached to it.

Nonbonding electrons: electrons that are not shared between atoms in a covalent species; they belong solely to one atom.

Nonpolar covalent bond: a bond in which the bonding electrons are shared equally between the two bonded atoms.

Octahedral: the electron-pair geometry when six groups are connected to a central atom and the molecular geometry when all six groups in an octahedral electron-pair geometry are atoms.

Octet rule: the atoms of many elements lose, gain, or share electrons to obtain eight electrons in the highest energy level.

Octet: an electron configuration having completed s and p sublevels in the highest energy level. All of the noble gases (except helium) have this configuration.

Polar covalent bond: a bond that involves sharing electrons but in which the two atoms do not share the electrons equally. One of the atoms will be partially positive, and one will be partially negative.

Resonance hybrid: the actual structure of a molecule or ion that exhibits resonance; it is somewhere in between, a composite of, the resonance structures that can be drawn.

Resonance structures: these are used to represent bonding in a molecule or ion when a single Lewis structure fails to describe accurately the actual electronic structure.

Seesaw: the molecular geometry when four atoms and one lone pair are connected to a central atom with the trigonal bipyramidal electron-pair geometry.

Square planar: the molecular geometry when four atoms and two lone pairs are connected to a central atom with the octahedral electron-pair geometry.

Square pyramidal: the molecular geometry when five atoms and one lone pair are connected to a central atom with the octahedral electron-pair geometry.

Structure: the way atoms are arranged in space.

Tetrahedral: the electron-pair geometry having four groups connected to a central atom and the molecular geometry when all four groups in a tetrahedral electron-pair geometry are atoms.

Trigonal bipyramidal: the electron-pair geometry when five groups are connected to a central atom and the molecular geometry when all five groups in a trigonal bipyramidal electron-pair geometry are atoms.

Trigonal planar: the electron-pair geometry having three groups connected to a central atom, and the molecular geometry that results when all three groups connected to a central atom with the trigonal planar electron-pair geometry are atoms.

Trigonal pyramidal: the molecular geometry when three atoms and one lone pair are connected to a central atom with the tetrahedral electron-pair geometry.

Triple bond: a triple bond exists when six electrons are shared by two atoms; three pairs of electrons are shared.

T-shaped: the molecular geometry when three atoms and two lone pairs are connected to a central atom with the trigonal bipyramidal electron-pair geometry.

Valence electrons: for a main group element, the valence electrons are the s and p electrons of the highest energy level. For transition metals, they are the electrons in the highest energy level s orbital as well as the electrons in the partially filled d subshell of the next highest energy level.

Valence shell electron-pair repulsion (VSEPR) model: bonding and lone pair electron pairs in the valence shell of an element repel each other and seek to be as far apart as possible.

Chapter Goals

By the end of this chapter you should be able to:

- ## Understand the difference between ionic and covalent bonds.

  a) Describe the basic forms of chemical bonding – ionic and covalent – and the differences between them (Section 9.2).

  Ion formation involves the transfer of one or more electrons between two atoms. One particle ends up positively charged, and the other ends up negatively charged. The ionic bond is the electrostatic force of attraction that results between two oppositely charged ions.

  Example 9-1: Draw the Lewis symbols showing the formation of the ionic compound made from calcium atoms and chlorine atoms.

  Calcium is in Group 2A, so it starts out with two valence electrons. It will react by losing these two electrons to obtain the noble gas configuration of neon. Chlorine is in group 7A and has seven valence electrons. It will react by gaining one electron to achieve the noble gas configuration of argon. Because calcium has two electrons to lose and each chlorine will gain only one electron, we will need two chlorine atoms per atom of calcium. The resulting particles are calcium ions with a 2+ charge and chloride ions each with a 1– charge.

Notice that in the case of ionic bonding, we form separate particles (in this case, one calcium ion and two chloride ions) that are held together by electrostatic forces.

In covalent bonding, atoms do not fully transfer electrons from one to the other. Instead, they share pairs of valence electrons in covalent bonds.

Example 9-2: Draw the Lewis structure for water, a covalent compound.

The formula for water is H_2O. Neither the hydrogen nor the oxygen is able to completely pull the electrons away from the other so the atoms end up sharing electrons.

The oxygen has two pairs of electrons that solely belong to it and has a share in two other pairs with the hydrogens. The oxygen has achieved an octet through sharing electrons. The hydrogens each have one pair of electrons that they share with oxygen. Each hydrogen, through sharing electrons, has ended up with an electron configuration like that of the noble gas helium. Notice that in the case of covalent bonding, we end up forming a single particle, a molecule of water. Often, we represent a shared pair of electrons using a line:

b) Predict from the formula whether a compound has ionic or covalent bonding, based on whether a metal is part of the formula (Section 9.2).

In many cases, we can predict whether a compound will be ionic or covalent simply by looking at the formula. In general, if a binary compound is composed of a metal combined with a nonmetal, then the compound will be ionic, though there are some exceptions. If a binary compound is composed of a nonmetal combined with another nonmetal, then the compound will be covalent. A polyatomic ion has covalent bonding within it, but the group of atoms has a charge and is thus an ion. Compounds containing polyatomic ions are ionic except when the polyatomic ion is combined only with hydrogen in an acid.

Example 9-3: Predict whether each of the following compounds will be ionic or covalent.

   a. LiF
This compound is composed of the metal lithium and the nonmetal fluorine. It will be ionic.

   b. N_2O
This compound is composed of two nonmetals, nitrogen and oxygen. It will be covalent.

c. $Ca(NO_3)_2$
This compound consists of a metal and a polyatomic ion. It will be ionic.

d. HNO_3
This compound is composed of the nitrate ion combined with hydrogen. It is covalent.

c) Write Lewis symbols for atoms (Section 9.1).

In the Lewis symbols for main group elements, the symbol of the element represents the nucleus and the core electrons. The valence electrons of the main group elements (those in the s and p orbitals of the highest energy level) are written as dots around the symbol. Each side of the symbol (top, bottom, left, and right) can hold a maximum of two dots. We write a dot on each side before we begin pairing up dots.

Example 9-4: Write the Lewis symbol for each of the following elements:

a. aluminum

Aluminum is in group 3A, so there are three valence electrons, corresponding to the two electrons in the 3s subshell and one electron in the 3p subshell. We write the symbol for aluminum, Al, and then place three unpaired dots around the symbol:

$$\cdot \overset{\textstyle \cdot}{Al} \cdot$$

b. nitrogen

Nitrogen is in group 5A of the periodic table, so there are five valence electrons. We write the symbol for nitrogen, N, and then place five dots around it. We first write four unpaired dots. At this point, we still have one more electron to place, so we pair up the dots on one side:

$$\cdot \overset{\textstyle \cdot\cdot}{\underset{\textstyle \cdot}{N}} \cdot$$

Notice that in the case of a noble gas (other than helium), there will be two electrons on each side of the symbol giving a total of eight electrons. We have stated previously that many atoms react to obtain a noble gas electron configuration. Eight valence electrons is a very stable configuration. This configuration is sometimes referred to as an octet. The fact that many atoms react to achieve this noble gas configuration of eight valence electrons is sometimes referred to as the octet rule.

Example 9-5: Give the number of bonds that oxygen is expected to form if it obeys the octet rule.

Oxygen is in Group 6A of the periodic table. Its Lewis symbol is

$$\overset{\textstyle \cdot\cdot}{\underset{\textstyle \cdot}{:O}} \cdot$$

This atom is two electrons away from having an octet. It is expected to form two bonds in order to obtain an octet.

Try Study Question 3 in Chapter 9 of your textbook now!

d) Describe the basic ideas underlying ionic bonding and explain how such bonds are affected by the sizes and charges of the ions (Section 9.3).

As we have gone over, ions are formed by the transfer of electrons between atoms. In forming ionic compounds, main group atoms often gain or lose electrons to obtain the same electron configuration as a noble gas.

The ions in an ionic compound are held together by electrostatic attraction between the oppositely charged ions. There are two main factors that influence the relative strength of this attraction between ions:

> 1) The magnitude of the ion charge. The greater the ion charges, the greater the attraction will be.

> 2) The distance between the ions. This is determined by the size of the ions. The larger the size, the smaller the attraction will be.

Example 9-6: For each pair of ionic compounds, predict which will be held together by a greater electrostatic attraction.

   a. NaCl or NaBr

In these two compounds, sodium ions are combined with either chloride or bromide. Chloride and bromide both have the same charge, 1–. Bromine is below chlorine on the periodic table. The bromide ion should thus be larger than the chloride ion. The larger the size of the ions, the smaller the attraction so there should be a greater electrostatic attraction in NaCl than in NaBr.

   b. KCl or CaS

In potassium chloride, we have a 1+ ion combined with a 1– ion. In calcium oxide, we have a 2+ ion combined with a 2– ion. The greater charges in CaS lead to a greater attraction between the ions in CaS than in KCl. (The size factor is not as important here, and it does not work consistently for these two compounds. The potassium ion will be larger than the calcium ion but the chloride ion will be smaller than the sulfide ion.)

Try Study Question 5 in Chapter 9 of your textbook now!

e) Understand lattice energy and know how lattice energies are calculated (Born-Haber cycle); recognize trends in lattice energy and how melting points of ionic compounds are correlated with lattice energy (Section 9.3).

The lattice energy of an ionic compound is the energy for the formation of one mole of a solid crystalline ionic compound from its constituent ions in the gas phase. Lattice energies can be calculated by means of a Born-Haber cycle. In a Born-Haber cycle, we picture the formation of one mole of a solid crystalline ionic compound from its elements as occurring through the following steps:

> 1) conversion of the elements to gaseous monatomic atoms,

> 2) the loss of electrons by the metal atoms (corresponding to one or more ionization energies of the metal),

> 3) the gain of electrons by the nonmetal atoms (corresponding to one or more electron affinities of the nonmetal), and finally

4) the formation of a solid crystal lattice from the isolated ions in the gas phase (corresponding to the lattice energy).

The overall sum of this cycle will equal the enthalpy of formation of the solid crystalline ionic compound. The Born-Haber cycle for the formation of sodium chloride is shown in Figure 9.4 in your textbook. So long as we know all of the energies pictured in the cycle except for one, we can calculate the one that we do not know. Often the unknown value is the lattice energy.

<u>Example 9-7</u>: Calculate the lattice energy for one mole of calcium fluoride.

The equation that deals with the formation of calcium fluoride is
$$Ca\ (s) + F_2\ (g) \rightarrow CaF_2\ (s) \quad \Delta H = -1219.6\ \text{kJ/mole}$$
This value for the enthalpy change was obtained by looking look up the standard enthalpy of formation for this compound in Appendix L in your text. All of the steps of the Born-Haber cycle must add up to give this value.

In the first step of the cycle, we need to take calcium metal and convert it to calcium gas. This is an endothermic step.
$$Ca\ (s) \rightarrow Ca\ (g) \qquad \Delta H_f^o = +178.2\ \text{kJ/mole}$$

Likewise, we need to take fluorine gas and convert it to fluorine atoms in the gas phase. Appendix L lists the ΔH to make one mole of F atoms; twice this value will give us ΔH to make two moles of F atoms. This is an endothermic step.
$$2[1/2\ F_2\ (g) \rightarrow F\ (g)] \qquad \Delta H = 2\ \text{moles}(+78.99\ \text{kJ/mole}) = +157.98\ \text{kJ}$$

Next, we need to convert the gaseous molecules to the ions. In the case of calcium, we need to form a 2+ ion. This will be the sum of the first two ionization energies for calcium. The first ionization energy of calcium is listed in Appendix F. The second ionization of energy of calcium is 1148 kJ/mole. This too is an endothermic step.
$$Ca\ (g) \rightarrow Ca^{2+}\ (g) + 2e^- \quad \text{First Ionization Energy + Second Ionization Energy}$$
$$= 599\ \text{kJ/mole} + 1148\ \text{kJ/mole} = +1747\ \text{kJ/mole}$$

Each electron must be gained by one of the fluorine atoms. A total of two moles of fluorine atoms will each gain one electron. This will correspond to two times the electron affinity for fluorine. The electron affinity for fluorine is found in Appendix F. Finally, we have a step that is exothermic.
$$2[F\ (g) + e^- \rightarrow F^-\ (g)] \qquad 2(\text{Electron Affinity}) = 2\ \text{moles}(-328\ \text{kJ/mole}) = -656\ \text{kJ}$$

At this point, the only step left in the cycle is the lattice energy, the energy to go from the gaseous ions to the solid compound. This is what we are trying to determine. All of the steps in this cycle must add up to give the same value as the ΔH^o_f listed in Appendix L, because both processes have the same starting and finishing points:
$$-1219.6\ \text{kJ} = +178.2\ \text{kJ} + 157.98\ \text{kJ} + 1747\ \text{kJ} + -656\ \text{kJ} + \text{lattice energy}$$
$$\text{lattice energy} = -2647\ \text{kJ}$$

Try Study Question 8 in Chapter 9 of your textbook now!
The answer to this question is not listed in the back of your textbook. In this case, instead of being asked for the lattice energy, you are asked to calculate the standard molar enthalpy of formation of the compound. (The lattice energy for lithium fluoride is given in Table 9.3 in your textbook.) The answer is –607 kJ/mole.

We find the same trends in lattice energy that we have already discussed for the amount of electrostatic attraction between ions. As the magnitude of the charges increases, the lattice energy becomes more negative. Likewise, as the size of the ions decreases, the lattice energy becomes more negative.

In general, the greater the lattice energy, the higher the melting point tends to be. This makes sense because we need to pull apart the ions. The more tightly they are held together, the harder it will be to pull them apart, so a higher temperature is needed to melt the compound.

• Draw Lewis electron dot structures for small molecules and ions.

a) Draw Lewis structures for molecular compounds and ions (Section 9.4).

As we have seen, we can draw Lewis structures for molecular compounds. In these structures, the chemical symbols represent the nuclei and core electrons. We use dots to represent the valence electrons. A pair of electrons belonging solely to one atom is referred to as a lone pair of electrons; these are nonbonding electrons. A shared pair of electrons is a bonding pair and is often represented by a line connecting two atoms.

Many atoms react to obtain eight electrons (an octet) surrounding them. For some simple cases, we can use the Lewis symbols for the atoms and simply pair up any electrons that are unpaired by having them become part of a shared pair of electrons with another atom, as we did with our example for forming water above. On the other hand, this simple method is *not* usually the best method to use to draw Lewis structures. Instead, a more systematic method that will work in many more cases is the preferred method. The key steps are summarized below:

1. Determine the arrangement of atoms within a molecule. In this step, you will determine which atom is the central atom and which atoms are attached to the central atom. Usually, the central atom will be the one with the lowest electronegativity (this is also usually the one with the lowest electron affinity). Hydrogen is never (well, almost never) a central atom. The halogens are very rarely the central atom. Sometimes, there is more than one central atom in a species.

2. Determine the total number of valence electrons in the species. This is the sum of the valence electrons in each atom. If the species has a negative charge, you will need to add as many electrons as the charge is negative. If the species has a positive charge, you will need to subtract as many electrons as the charge is positive.

3. Place one pair of electrons between each pair of bonded atoms to form single bonds.

4. Use any remaining electrons as lone pairs around each terminal atom (except H, which is satisfied with only a duet of electrons) so that each terminal atom is surrounded by eight electrons.

5. At this point, one of three cases will occur:

a. You will have used all of the valence electrons and each atom has satisfied the octet rule or its exception to the octet rule. If this is the case, you are finished.

b. You will have satisfied the octet rule for each terminal atom (or its exception) and you have electrons that have not been used. Place these on the central atom. If you satisfy its octet and still have electrons left, check that the central atom is from period 3 or higher on the periodic table. If it is not, you have made a mistake. If it is from period 3 or higher, then place the remaining electrons around the central atom, expanding beyond an octet.

c. You have used up all the electrons, but not all of the atoms have satisfied the octet rule (or their exception). Such a case means that our assumption of only single bonds in the species is not correct. Form one or more multiple bonds (double or triple bonds) so that each atom satisfies its octet (or exception to the octet rule). To do this, move one or more of the lone pairs on the terminal atoms so that they are now bonding pairs.

Example 9-8: Draw the Lewis structure for water, H_2O.

1. Determine the central atom and the terminal atoms.
In this case, oxygen must be the central atom, even though it has a higher electronegativity than hydrogen. This is because hydrogen cannot be the central atom.

2. Determine the number of valence electrons.
We obtain the number of valence electrons in each atom by looking at the group number.
$$2H = 2 \times 1 = \quad 2$$
$$1O = 1 \times 6 = \quad \underline{6}$$
$$8$$

3. Connect the central atom to the terminal atoms by single bonds.
$$H\text{——}O\text{——}H$$

4. Use any remaining electrons as lone pairs on the terminal atoms.
Because the terminal atoms are hydrogens, we cannot do this.

5. We have an example of case b. We have satisfied the duets of hydrogen and still have four electrons left. We place these on the central atom.

$$H\text{——}\overset{\bullet\bullet}{\underset{\bullet\bullet}{O}}\text{——}H$$

At this point, everything has an octet. We are finished with this Lewis structure.

Example 9-9: Draw the Lewis structure for I_3^-.

1. The central atom must be iodine.

2. Number of valence electrons:
$$3I = 3 \times 7 = \quad 21$$
$$\text{neg. charge} = \quad \underline{+1}$$
$$22$$

3. Use all single bonds.
$$I\text{——}I\text{——}I$$

4. Complete octets of terminal atoms.

5. This structure uses 16 of the 22 electrons. Completing the octet on the central iodine will take us up to 20 electrons used. There are still 2 electrons that need to be placed. Iodine is beyond period 3 on the periodic table. It can expand its octet. We will have a total of 10 electrons around the central iodine in this structure. We place the Lewis structure for an ion in square brackets and place the overall charge of the ion outside the brackets.

Example 9-10: Draw the Lewis structure for formaldehyde, CH_2O.

1. The central atom will be carbon.

2. Number of valence electrons:

$$1C = 1 \times 4 = \quad 4$$
$$2H = 2 \times 1 = \quad 2$$
$$1O = 1 \times 6 = \quad \underline{6}$$
$$\qquad\qquad\qquad 12$$

3. Use all single bonds.

4. Complete octets of terminal atoms.

5. We have used all the electrons, but the central carbon does not have an octet. It needs another pair of electrons. We will complete its octet by moving one of the lone pairs to become a second bonding pair between carbon and oxygen.

Try Study Questions 11 and 13 in Chapter 9 of your textbook now!

b) Understand and apply the octet rule; recognize exceptions to the octet rule (Sections 9.4 and 9.6).

The octet rule states that many elements will react so as to get eight electrons (four pairs) in the highest energy level. This corresponds to completed s and p subshells in the highest energy level, which is the same configuration as a noble gas. The octet rule is not a hard and fast rule; there are many exceptions.

The exceptions you should know are the following:

1. Hydrogen atoms get only two electrons (one pair).

2. Beryllium in covalent compounds gets only four electrons (two pairs) in the highest energy level.

3. Boron sometimes gets only six electrons (three pairs) in the highest energy level, but sometimes it does get eight (by means of a coordinate covalent bond).

4. Elements in the third period and above can get more than eight electrons in the highest energy level.

5. There are species that contain an odd number of valence electrons. While many fewer in number than those with an even number of valence electrons, some of these compounds do exist. It is impossible to achieve octets around each atom in such a species; there will be one unpaired electron on one of the atoms (probably the central atom). Species with an unpaired electron are called free radicals.

c) Write resonance structures, understand what resonance means, and know how and when to use this means of representing bonding (Section 9.5).

In drawing the Lewis structure for ozone (O_3), we obtain as our first attempt:

In this structure, we have used up all of the valence electrons, but the middle oxygen does not have a completed octet. Following our usual rule we move one of the lone pairs in to form a double bond. With this structure, we have used all of the valence electrons and have completed octets on each atom.

On the other hand, we could just as easily have constructed the following Lewis structure:

Which is correct? The first Lewis structure with all octets would indicate that we have a single bond going from the middle oxygen to the left oxygen and that we have a double bond going from the middle oxygen to the right oxygen. The second Lewis structure with all octets would indicate the opposite. Both indicate that we have a single bond in one position and a double bond in the other. Bond length is related to how many bonds are present between the atoms. Scientists measured the bond length in ozone and discovered that the bond lengths are all the same and have a value in between that expected for a single bond and that for a double bond. This indicates that the bonding is the same for the two positions and that it is neither a single bond nor a double bond, but something about halfway in between, one and a half bonds. We can't really draw that very well, so what we do is to draw both Lewis structures and connect them with a double-headed arrow to indicate that the real situation is somewhere in between these two extremes.

We describe this situation by saying that ozone exhibits resonance. A material is said to exhibit resonance when a single Lewis structure fails to describe accurately the actual electronic structure. The real case lies somewhere in between the different Lewis structures that can be drawn; we say that the true species is a resonance hybrid, something in between (a composite or average of) the resonance structures that we can draw. As we showed for ozone, we write all of the resonance structures and connect them with a double-headed arrow.

When should you recognize that different resonance structures are possible? If, in drawing a Lewis structure, you have a choice to make as to where to make a multiple bond, then you need to consider resonance structures in which you make the multiple bond using each of the equivalent atoms.

Here are some key things to keep in mind:

1. Resonance exists when there are different possible arrangements of the electrons in a molecule or ion.

2. Resonance structures will involve changing multiple bonds: double or triple bonds.

3. In looking for resonance, we consider moving electrons around, not atoms.

4. The resonance structures that we draw do not represent the real structure of the species; the real structure will be something in between what we can draw.

Example 9-11: Draw the Lewis structure for the nitrate ion (NO_3^-). If any resonance structures exist, write them.

1. The N will be the central atom.

2. # of valence electrons
   $1N = 1 \times 5 =$ 5
   $3O = 3 \times 6 =$ 18
   Neg. Charge = $\underline{1}$
                    24

3. Connect by single bonds.

4. Complete the octets of the terminal atoms.

5. With this structure, the nitrogen does not have an octet, but we have used all of the valence electrons. We resolve this situation by moving a lone pair into being a bonding pair.

6. In the original structure that we drew with no double bonds, all of the oxygens were equivalent. We arbitrarily chose to make the double bond to the top oxygen. We could just as easily have made the double bond with either the left or the right oxygens. We will therefore draw all three structures and connect them by double-headed arrows. The real polyatomic ion will be somewhere in between these three extremes. The bonding between the nitrogen and oxygen turns out to be equivalent to about one and a third bonds because the bond we are drawing as a double bond is distributed over three equivalent regions.

Try Study Question 15 in Chapter 9 of your textbook now!

- ## Use the valence shell electron-pair repulsion theory (VSEPR) to predict the shapes of simple molecules and ions and to understand the structure of more complex molecules.

Predict the shape or geometry of molecules and ions of main group elements using VSEPR theory (Section 9.7). Table 9.11 summarizes the relation between valence electron pairs, electron pair and molecular geometry, and molecular polarity.

The model that we shall use to determine molecular geometry is the valence shell electron-pair repulsion (VSEPR) model. According to this model, the electron pairs (bonding pairs and nonbonding electron pairs) in the valence shell of an atom repel each other and seek to be as far apart as possible. It will be important in the following discussion to distinguish between two terms: electron-pair geometry and molecular geometry.

The electron-pair geometry tells us where the different groups around a central atom end up. For these purposes, we will define a group to be either an atom or a lone pair. The atoms correspond to the presence of bonding electrons, and the lone pairs correspond to the presence of nonbonding electrons. An atom corresponds to 1 group, whether it is connected to the central atom by a single, double, or triple bond. Likewise, each lone pair corresponds to 1 group. These groups will get as far apart as possible in the electron-pair geometry. This electron-pair geometry will serve as the template upon which the molecule or polyatomic ion will be built. Different numbers of groups around the central atom lead to different electron-pair geometries. You should learn the names and bond angles and be able to draw sketches of each of these. They are summarized in the following table:

# of groups around central atom	Electron-Pair Geometry	Bond Angles	Sketch
2	Linear	180°	X——M——X
3	Trigonal planar	120°	

4	Tetrahedral	109.5°	
5	Trigonal bipyramidal	Axial-axial: 180° Equatorial-equatorial: 120° Axial-Equatorial: 90°	
6	Octahedral	Groups next to each other: 90° Groups across from each other: 180°	

In the sketches, a regular line represents a group in the plane of the paper, a solid wedge represents a group coming out at you from the paper, and a dotted line represents a group going away from you.

There are two different types of positions in the trigonal bipyramidal electron-pair geometry. To help keep these straight, keep in mind some terms from geography about the earth. The two positions that are directly across from each other are like the axis of the earth, so they are referred to as being axial. The three positions in a plane midway between the axial positions are in the region that is like the equator of the earth, so they are referred to as being equatorial.

The molecular geometry tells us where the *atoms* end up in the molecule or polyatomic ion. To determine the molecular geometry, we first look at the electron-pair geometry in which we consider all the groups on the central atom and then cover up the lone pairs to see where the atoms end up. It is as if we are saying that the lone pairs are invisible. Their effects are felt and help influence where the atoms end up, but in naming the molecular geometry, we only consider where the atoms are. If you see the term geometry or shape without a descriptor indicating whether the person is referring to the electron-pair or molecular geometry, the person usually is referring to the molecular geometry.

The bond angles in the molecule or polyatomic ion will largely be determined by the electron-pair geometry. This is because the lone pairs exert an influence even though we treat them as being invisible in naming the molecular geometry.

There are a couple more considerations about lone pairs:

1. Lone pairs take up more than their fair share of the space. This tends to push the atoms closer together. The bond angles in a molecule or polyatomic ion are largely determined by the electron-pair geometry, but if lone pairs are present, the bond angles will not be exactly those predicted by the electron-pair geometry. Often they are a bit smaller.

2. It does not matter where we place lone pairs in the linear, trigonal planar, and tetrahedral electron-pair geometries. For a trigonal bipyramidal electron-pair geometry, lone pairs preferentially go to the equatorial positions. For an octahedral electron-pair geometry, it does not matter where the first lone pair goes, but if we have two lone pairs, they will go to positions that are opposite one another.

The steps to follow in determining molecular geometry are the following:

1. Draw the Lewis structure for the molecule or polyatomic ion.

2. Count the number of groups surrounding the central atom. Remember that a group is either an atom or a lone pair.

3. Based on the number of groups, select the correct electron-pair geometry. This is the template upon which the molecule or ion will be built.

4. Sketch the electron-pair geometry, showing all of the atoms and lone pairs.

5. Based on where the atoms end up in this sketch, determine the molecular geometry. It may help to draw a new sketch in which you leave all of the atoms exactly where they were in the sketch of the electron-pair geometry but in which you do not draw the lone pairs.

6. The bond angles in the molecule or ion will largely be determined by the electron-pair geometry.

Example 9-12: Determine the molecular geometry and bond angles of methane, CH_4.

1. Hopefully, at this point, you can determine that the correct Lewis structure for this molecule is

2. There are four atoms attached to the central atom, so there are four groups around the central atom.

3. The electron-pair geometry for four groups is tetrahedral.

4. Sketch:

5. In this case, all of the groups are atoms; there are no lone pairs around the central atom. The molecular geometry is thus exactly the same as the electron-pair geometry: tetrahedral.

6. The bond angles will be 109.5°.

Example 9-13: Determine the molecular geometry and bond angles of formaldehyde, CH_2O.

1. We have seen that the Lewis structure for formaldehyde is

2. There are three atoms and no lone pairs around the central atom, so there are three groups. Notice that it does not matter that we have a double bond connecting the carbon and oxygen; the oxygen atom still counts as just one group.

3. The electron-pair geometry is trigonal planar.

4. Sketch:

5. Just like in the last example, there are no lone pairs around the central atom, so the molecular geometry will be the same as the electron-pair geometry: trigonal planar.

6. The bond angles will be around 120°.

Example 9-14: Determine the molecular geometry and bond angles in water, H_2O.

1. We have seen that the Lewis structure for water is

$$H——\overset{\bullet\bullet}{\underset{\bullet\bullet}{O}}——H$$

2. There are four groups around the central atom: two atoms and two lone pairs.

3. The electron-pair geometry will be tetrahedral.

4. Sketch:

5. We have lone pairs on the central atom, so the molecular geometry will not be the same as the electron-pair geometry. We now cover up the lone pairs and look at where the atoms are.

The molecular geometry is bent.

6. The bond angles will be around but less than 109.5°.

Example 9-15: Determine the molecular geometry and bond angles in I_3^-.

1. We have seen that the Lewis structure is

$$\left[\quad \overset{\bullet\bullet}{\underset{\bullet\bullet}{\overset{\bullet}{\,}I}}——\overset{\bullet\;\bullet\bullet}{\underset{\bullet\bullet}{I}}——\overset{\bullet\bullet}{\underset{\bullet\bullet}{I\,}\overset{}{\,}} \quad \right]^{-}$$

2. There are two atoms and three lone pairs around the central I.

3. The electron pair geometry for five groups is trigonal bipyramidal.

4. In drawing the electron-pair geometry, keep in mind that lone pairs preferentially go to the equatorial positions in the trigonal bipyramidal template. There are three lone pairs, so all three equatorial positions will be occupied by lone pairs.

5. We now cover up the lone pairs.

This is a linear molecular geometry.

6. The iodines are at the axial positions, so we would expect the bond angle to be 180°.

If we need to determine the molecular geometry for a species that exhibits resonance, we need only consider one of the resonance structures. All of the resonance structures will have the same molecular geometry. This is a result of the fact that we do not move the atoms around when we draw resonance structures.

The following table summarizes the different molecular geometries that are possible for the different electron-pair geometries:

# of groups	Electron-Pair Geometry	Composition of Groups	Molecular Geometry
2	Linear	2 atoms	Linear
3	Trigonal planar	3 atoms	Trigonal planar
		2 atoms, 1 lone pair	Bent
4	Tetrahedral	4 atoms	Tetrahedral
		3 atoms, 1 lone pair	Trigonal pyramidal
		2 atoms, 2 lone pairs	Bent
5	Trigonal bipyramidal	5 atoms	Trigonal bipyramidal
		4 atoms, 1 lone pair	Seesaw
		3 atoms, 2 lone pairs	T-shaped
		2 atoms, 3 lone pairs	Linear
6	Octahedral	6 atoms	Octahedral
		5 atoms, 1 lone pair	Square pyramidal
		4 atoms, 2 lone pairs	Square planar

Try Study Questions 19, 21, and 23 in Chapter 9 of your textbook now!

Sometimes a molecule or polyatomic ion will have more than one central atom. In such cases, we simply follow all of these steps for each central atom.

Example 9-16: Determine the molecular geometry and bond angles around each central atom in methanol, CH_3OH.

1. Lewis Structure:

2. There are four groups around the carbon (four atoms) and four groups around the oxygen (two atoms and two lone pairs).

3. The electron-pair geometry around carbon is tetrahedral and that around the oxygen is also tetrahedral.

4. Sketch:

5. For the molecular geometry, we obtain the following sketch:

The molecular geometry is tetrahedral around the carbon and bent around the oxygen.

6. The bond angles around both the carbon and oxygen are around 109.5°.

Try Study Question 27 in Chapter 9 of your textbook now!

- **Use electronegativity and formal charge to predict the charge distribution in molecules and ions and to define the polarity of bonds.**

  a) Calculate formal charges for atoms in a molecule based on the Lewis structure (Section 9.8).

  The formal charge of an atom in a molecule or polyatomic ion is the charge calculated by comparing the number of valence electrons around the atom in the molecule or ion (according to a particular set of rules) compared to the number of valence electrons in a neutral atom of the element. In determining the number of valence electrons present around an atom in the molecule or polyatomic ion, we picture lone pairs belonging solely to the atom on which they are written and bonding pairs being shared equally between the atoms that are bonded together. In a bonding pair, one electron is assigned to each atom. Formal charge is sometimes referred to as the covalent limit because we assume 100% equal sharing of electrons. We can summarize all of this with the equation:

  Formal charge = group number – [LPE +1/2 BE]

The formal charges of all the atoms in a molecule or polyatomic ion will add up to the overall charge of the molecule or ion.

Example 9-17: Calculate the formal charge of each atom in NH_3BF_3.

With a little work, you can hopefully figure out that the Lewis structure for this species is

Each hydrogen is identical in this structure, so we will just consider the case for one H. Each hydrogen is involved in one bond to N. We divide each bond down the middle so in the molecule each hydrogen has one valence electron associated with it. A neutral atom of hydrogen has one valence electron. This is the same number, so the formal charge of hydrogen is zero. Showing this mathematically, we obtain

Formal charge = group number – [LPE +1/2 BE] = 1 – [0 + 1/2(2)] = 0

Each fluorine is identical so we will just consider the case for one F. Each one has three lone pairs. Thus it has six electrons from the lone pairs. It is also involved in one bond to B. Dividing this bond between the F and B, the bond contributes one electron to the F, giving a total of 6+1 = 7 valence electrons around the F in the molecule. An atom of fluorine has 7 valence electrons. This is the same number, so the formal charge is zero. Mathematically,

Formal charge = group number – [LPE +1/2 BE] = 7 – [6 + 1/2(2)] = 0

Now, let's look at the N. It has no lone pairs. There are four bonds, so these contribute four electrons to N. An atom of nitrogen starts out with 5 valence electrons. In effect, it has lost one electron in going from a free atom to being in this molecule. Its formal charge is +1. Mathematically,

Formal charge = group number – [LPE +1/2 BE] = 5 – [0 + 1/2(8)] = +1

Finally, let us consider the case of the boron atom. It has no lone pairs. There are four bonds to it, so these contribute four electrons to it. A neutral boron atom has three valence electrons. In effect, it has gained one electron, so its formal charge is –1. Using the equation, we obtain

Formal charge = group number – [LPE +1/2 BE] = 3 – [0 + 1/2(8)] = –1

Let's now double check that the formal charges add up to the overall charge. This is a neutral molecule so the overall charge is zero. The hydrogens and fluorines each have formal charges of zero. The nitrogen has a formal charge of +1, and the boron has a formal charge of –1. The sum of all these is zero.

Example 9-18: Calculate the formal charge on each atom in the acetate ion, $CH_3CO_2^-$.

The Lewis structure for acetate is

In both of these resonance structures, we calculate the formal charges of the hydrogens and the two carbons to be zero:

Formal charge of hydrogen = $1 - [0 + 1/2(2)] = 0$
Formal charge of left carbon = $4 - [0 + 1/2 (8)] = 0$
Formal charge of right carbon = $4 - [0 + 1/2 (8)] = 0$

In each resonance structure, we calculate the double bonded oxygen to have a formal charge of zero and the single bonded oxygen to be -1:

Formal charge of double bonded oxygen = $6 - [4 + 1/2 (4)] = 0$
Formal charge of single bonded oxygen = $6 - [6 + 1/2 (2)] = -1$

In the left resonance structure, the top oxygen has a formal charge of zero, and in the right structure, it has a formal charge of -1. The net formal charge of this oxygen in the resonance hybrid will be the average of these two: $-1/2$. In the same way, we determine that the right oxygen also has a net formal charge of 1/2 because it has a formal charge of -1 in the left resonance structure and of 0 in the right resonance structure.

The sum of all the formal charges will once again equal the total charge of the species, -1. We can verify that this is true. The hydrogens and the carbons have zero formal charge. Each of the two oxygens has a formal charge of $-1/2$, so the total charge is -1, as it should be for the acetate ion.

Try Study Questions 29 and 31 in Chapter 9 of your textbook now!

b) Define electronegativity and understand how it is used to describe the unequal sharing of electrons between atoms in a bond (Section 9.8).

Electronegativity is a measure of the ability of an atom in a molecule to attract electrons to itself. The higher the electronegativity, the greater will be the ability of the atom to attract electrons to itself. The element with the highest electronegativity is fluorine. Electronegativity tends to decrease down a group and increase across a period on the periodic table.

If two identical atoms are bonded together, the atoms will have identical electronegativities. Neither one can attract the electrons in the bond any better than the other. The electrons are shared equally. On the other hand, if two atoms are bonded together that have different electronegativities, the one that has the greater electronegativity will attract the electrons more strongly than the other. The electrons spend more time around the atom that has the greater electronegativity. There is thus an unequal sharing of the electrons. The atom with the greater electronegativity gets the electrons more than the other one and so ends up with a partial negative charge (δ^-). The atom with the lower electronegativity gets the shared electrons less of the time and ends up with a partial positive charge (δ^+). We refer to a bond such as this one in which the bonding electrons are shared but are not shared equally as a polar covalent bond.

The difference in electronegativity between two atoms that are bonded together determines how polar a bond is. The greater this difference, the greater the polarity is. We can think about this in terms of a tug of war for the electrons. Let's picture a tug of war between two people. If the two people are equally strong, neither one will win the tug of war. We could have two very weak people in the tug of war. Neither one will win. Similarly, we could have two very strong people in the tug of war. Once again, no one will win. This corresponds to a situation like that in a nonpolar covalent bond. The two atoms have an equal ability to attract the electrons and so share them equally. If, however, we set up a tug of war between two people with different strengths, then the one who is stronger will get a greater amount of the rope. In the case of a molecule, if we have two atoms with different electronegativities, the one with the greater electronegativity gets the electrons in the bond more of the time and becomes partially negative; the one with the smaller electronegativity gets the electrons in the bond less of the time and becomes partially positive. In this case, we have a polar covalent bond. If we set up a tug of war between two people with very different strengths, then the stronger person might be able to pull the rope completely away from the other person. In the case of two atoms, if one has a much greater electronegativity than the other, then it might be able to pull one or more electrons completely away from the one with the lower electronegativity; in this case, we end up with ion formation.

In reality, rather than a sharp cutoff between ionic and covalent bonding, we find in nature a continuum from bonds that are largely covalent and just a little ionic to those that are largely ionic and just a little covalent.

Example 9-19: Which of the following bonds would be more polar?

    a. N-H or O-H?
Both of these bonds involve hydrogen. Nitrogen and oxygen both have electronegativities greater than that of hydrogen so both bonds will be polar with the nitrogen or oxygen having partial negative charges and the hydrogen having a partial positive charge. The question is which has a greater electronegativity difference, N-H or O-H? Both nitrogen and oxygen are in the second row of the periodic table but oxygen is further to the right. The difference between oxygen and hydrogen will be greater than that between nitrogen and hydrogen, so the O-H bond will be more polar.

    b. O-F or Li-F
Both of these bonds involve fluorine. Oxygen and fluorine are both nonmetals on the right side of the periodic table. Lithium is a metal on the left side. The electronegativities of oxygen and fluorine are closer to each other than those of lithium and fluorine, so the electronegativity difference between lithium and fluorine will be greater. The Li-F bond will be more polar; in fact it is largely ionic, as we have noted before.

Try Study Question 33 in Chapter 9 of your textbook now!

c) Combine formal charge and electronegativity to gain a perspective on the charge distribution in covalent molecules and ions (Section 9.8).

So far, we have been concerned with equivalent resonance structures in which we moved around a multiple bond to different equivalent atoms. These structures were also equivalent in that the formal charges that resulted were the same in each resonance structure. In such a case, we said that the actual structure was a hybrid of all of the structures with each structure having an equal weight in determining the actual electron distribution in the resonance hybrid.

There are also nonequivalent resonance structures in which the hybrid will be more like one (or more) of the resonance structures than the other(s). In these cases, the atoms that are involved in multiple bonding are not equivalent. How will we determine which is(are) the more important structure(s)? There are two guiding principles we shall use:

    1. The atoms will be distributed in such a way that the formal charges on all atoms are as close to zero as possible.

2. The more important structure(s) will have a negative charge residing on a more electronegative atom and a positive charge on a less electronegative atom.

Example 9-20: For NO_2^+, determine which resonance structures the actual species will more closely resemble.

Here are the possible Lewis structures, along with the formal charges on each atom:

$$\left[\; \overset{+1}{:O} \equiv \overset{+1}{N} - \overset{-1}{\underset{\cdot\cdot}{\overset{\cdot\cdot}{O}}} : \; \right]^+ \longleftrightarrow \left[\; \overset{-1}{:\underset{\cdot\cdot}{\overset{\cdot\cdot}{O}}} - \overset{+1}{N} \equiv \overset{+1}{O} : \; \right]^+ \longleftrightarrow \left[\; \overset{0}{\underset{\cdot\cdot}{\overset{\cdot\cdot}{O}}} = \overset{+1}{N} = \overset{0}{\underset{\cdot\cdot}{\overset{\cdot\cdot}{O}}} \; \right]^+$$

The first two structures are not very good. Compared to the third structure, there are many more atoms with a formal charge. In addition, in the first two structures, we have a formal charge of +1 on an oxygen, the second most electronegative atom. In the structure on the right, we have less formal charge and the only atom with a positive charge is the nitrogen. While nitrogen has a high electronegativity, it is not as high as that of oxygen, so it will take a positive charge before oxygen. The third structure is the most important resonance structure.

Another issue that sometimes arises is that the charges predicted by formal charge in a resonance structure do not always line up with what we know from electronegativity.

Example 9-21: Considering both formal charges and bond polarities, predict on which atom or atoms the negative charge resides in the following species: NH_4^+.

The Lewis structure for NH_4^+ is

$$\left[\begin{array}{c} H \\ | \\ H - \overset{+1}{N} - H \\ | \\ H \end{array} \right]^+$$

In this Lewis structure, there is a formal charge of +1 on the nitrogen and of zero on each of the hydrogens. On the other hand, we know nitrogen is more electronegative than hydrogen and that each bond should be polarized so that the nitrogen is partially negative and the hydrogen is partially positive. The predictions based on electronegativity and formal charge are working in opposite directions. The electronegativity predictions are closer to reality. The positive charge, instead of being located on the nitrogen, is spread out among the hydrogens.

Try Study Question 39 in Chapter 9 of your textbook now!

- ## Predict the polarity of molecules.

a) Understand why some molecules are polar whereas others are nonpolar (Section 9.9). See Table 9.8.

A molecule is said to be polar if the distribution of electrons is such that there is an accumulation of electron density on one side of the molecule, giving it a partial negative charge, and that there is a corresponding lack of electron density on the other side, giving it a partial positive charge. The quantity that measures the extent of the polarity of a molecule is the dipole moment. This is the product of the partial charges and the distance by which they are separated. The dipole moment is a vector quantity, having both a magnitude and a direction. It is drawn as an arrow from the partially positive end to the partially negative end.

If a molecule contains only nonpolar covalent bonds, then it will be a nonpolar molecule because the separation of charge must be zero. There will be a uniform distribution of electrons throughout. On the other hand, if a molecule contains polar covalent bonds, then it may end up either as a polar molecule or as a nonpolar molecule, depending on how the atoms are distributed in space and on whether the constituents are all the same or not. This is because the effects of polar bonds may cancel out in the overall molecule.

Let's look at some examples of this. First let's consider the case of carbon dioxide. The Lewis structure is

There are two atoms attached to the central atom. The electron pair geometry will be linear, and because both groups are atoms, so will the molecular geometry. Each carbon-oxygen double bond will be polar with the oxygen being partially negative and the carbon being partially positive. The dipole moment for each of these bonds would point from the carbon to the oxygen. The two dipole moments will be pointing in opposite directions and have equal magnitudes. They will cancel out each other, giving the molecule a net zero dipole moment. The molecule will thus be nonpolar, even though it contains polar bonds.

Now, let's take a look at water. We determined earlier that it has a bent molecular geometry. The O-H bonds will be polar with the oxygen being partially negative and the hydrogen being partially positive. In this case, the dipole moments do not cancel out, so this molecule is polar.

b) Predict the polarity of a molecule (Section 9.9 and Examples 9.13 and 9.14).

How can you predict whether a molecule will be polar or not? The key is whether the distribution of charge is symmetrical or not. For molecules with only one central atom, it is easiest to remember the situations where a nonpolar molecule will result. All of the other situations will be polar.

Electron-pair geometry	Nonpolar molecule
Linear	Both groups the same
Trigonal planar	All groups the same
Tetrahedral	All groups the same
Trigonal bipyramidal	All groups the same OR Both axial groups the same and all three equatorial groups the same but the axial and equatorial groups may be different
Octahedral	All groups the same OR All groups across from each other the same but the different pairs may be different

Let's consider the carbon dioxide and water examples again. Carbon dioxide has a linear electron-pair geometry. Both groups are the same, so this is a nonpolar molecule. Water has a tetrahedral electron-pair geometry. Two of the groups are hydrogen atoms, and two of the groups are lone pairs. The groups are not all the same, so this is a polar molecule.

Another method we can use to determine if a molecule is polar is to imagine the more electronegative atom in each bond pulling the molecule in the direction of the dipole moment. Ask yourself the question, "If we take into account all of the bonds, will the molecule move?" If so, the molecule is polar. If not, the molecule is nonpolar. Note that we are only imagining the molecule moving or not. The dipole moments do not pull the molecule so that it will move; this is simply a trick to help us figure things out.

In the case of CO_2, we picture the oxygens pulling on the carbon with equal strengths in opposite directions. The molecule will not move, so it is a nonpolar molecule. For H_2O, the pulls would not cancel out. The molecule would end up moving, so it is a polar molecule.

Try Study Question 45 in Chapter 9 of your textbook now!

- ## Understand the properties of covalent bonds and their influence on molecular structure.

  ### a) Define and predict trends in bond order, bond length, and bond dissociation energy (Section 9.10).

  The bond order is the number of bonding electron pairs shared by two atoms in a molecule. A bond order of 1 indicates a single bond, of 2 indicates a double bond, etc. When we have a molecule with resonance, we may obtain fractional bond orders. The bond order connecting atoms X and Y in a molecule can be calculated using the following equation:

  $$\text{Bond order} = \frac{\text{number of shared pairs linking X and Y}}{\text{number of X- Y links in the molecule or ion}}$$

Example 9-22: What is the bond order for each bond in the following molecules or ions?

a. nitrogen

The Lewis structure of nitrogen is

:N≡≡N:

We can see that the bond order is 3 without doing any calculations because there is a triple bond and no resonance structures. (The calculation is: bond order = 3/1 = 3.)

b. nitric acid

The Lewis structure for nitric acid is

The bonds that do not change in the two resonance structures are the easiest because we can simply look at the structures and determine the bond orders:

O-H	Bond order is 1
N-O of the OH group	Bond order is 1

The two oxygens that do not have a hydrogen attached are different in the two resonance structures. For these, we will do the calculation. There are three bonding electron pairs involved with these two N-O links
> N-O with no H attached   Bond order = 3/2

Try Study Question 47 in Chapter 9 of your textbook now!

Bond length is the distance between the nuclei of two bonded atoms. The major determinant of bond length is the size of the atoms involved in the bond. In addition, the bond order also affects bond length. The greater the bond order, the shorter the bond length connecting two atoms will be.

Example 9-23: Place the following bonds in order of increasing bond length: C–O, C=O, C–N.

C–O and C=O both connect carbon and oxygen. The double bond will be shorter than the single bond.

In comparing the C-O and the C-N bonds, both involve carbon, so the difference in size will come about from the oxygen and nitrogen. Recall that atomic size tends to decrease as we move from left to right across a period of the periodic table, so oxygen will be smaller than nitrogen. The C–O bond will be shorter than the C–N bond.

The C=O bond will be the shortest, followed by the C–O bond, followed by the C–N bond.

Try Study Question 49 in Chapter 9 of your textbook now!

The bond dissociation energy (D) is the enthalpy change for breaking a bond in a molecule with the reactants and products in the gas phase.

Breaking bonds in a molecule is endothermic. It takes energy to break a bond. Making bonds from atoms or radicals in the gas phase is exothermic.

The higher the bond order the greater the energy required to break a bond. A triple bond has a higher bond dissociation energy than a double bond, and a double bond has a greater bond dissociation energy than a single bond.

b) Use bond dissociation energies, D, in calculations (Section 9.10 and Example 9.15)

Average bond dissociation energies for many bonds are listed in Table 9.10 in your textbook. When using this table, keep in mind that these values are for reactions carried out in the gas phase only and that they are average values. The actual bond dissociation energy in a particular compound may be different, but the average values are nonetheless good for getting an estimate of the enthalpy change in a reaction.

To work a problem, we break all of the bonds that need breaking. The enthalpy change for this is calculated by adding up all of their bond dissociation energies, taking into account how many of these bonds are broken. We then form all of the bonds that need forming. This is summarized by the following equation.

$$\Delta H^o_{rxn} = \Sigma\, D(\text{bonds broken}) - \Sigma\, D(\text{bonds formed})$$

Example 9-24: Estimate the enthalpy change for the combustion of methane.

The balanced chemical equation is

$$CH_4 \ (g) + 2O_2 \ (g) \rightarrow CO_2 \ (g) + 2H_2O \ (g)$$

Notice that the water in this equation must be in the gaseous state if we are going to use bond dissociation energies.

It often helps in working this type of problem to draw out the Lewis structures for each species.

Next, determine the bonds broken and the enthalpy required to do this. In this case, all of the bonds on the reactant side must be broken.

4 C-H bond	= 4 moles x 413 kJ/mole =	1652 kJ
2 O=O bonds	= 2 moles x 498 kJ/mole =	996 kJ
		2648 kJ

This corresponds to the energy we must put in to break the bonds in the reactants.

Now determine the bonds made in the products and the enthalpy change to do this. This corresponds to the energy given off when the new bonds are made.

2 C=O bonds	= 2 moles x 745 kJ/mole =	1490 kJ
4 O-H bonds	= 4 moles x 463 kJ/mole =	1852 kJ
		3342 kJ

Notice that there are four O–H bonds. Each molecule has two O–H bonds and there are two molecules in the balanced equation, so there are 2 x 2 = 4 O–H bonds.

For the entire reaction, therefore,

$$\Delta H^{\circ}_{rxn} = \Sigma \ D(\text{bonds broken}) - \Sigma \ D(\text{bonds formed})$$

$$= 2648 \ kJ - 3342 \ kJ$$

$$= -694 \ kJ$$

Try Study Question 55 in Chapter 9 of your textbook now!

Other Notes

1. Chemical reactions involve the loss, gain, or rearrangement of the valence electrons. For a main group element, the valence electrons are the s and p electrons of the highest energy level; for these main group elements, the number of valence electrons is equal to the group number. (Using the IUPAC system, this is still true for groups 1 and 2, but for the other main group elements (groups 13-18), the number of valence electrons is obtained by covering up the 1 in the group number. Group 13 thus has 3 valence electrons.) For transition metals, they are the electrons in the highest energy level s orbital as well as the electrons in the partially filled d subshell of the next highest energy level. All of the other electrons in an atom are referred to as core electrons.

All elements in a group on the periodic table have the same number of valence electrons. This similarity in electron structure is the reason why the elements in the same group have similar properties.

<u>Example 9-25:</u> How many valence electrons are there in an atom of each of the following elements?

   a. P

Phosphorus is a main group element in Group 5A on the periodic table. There are thus 5 valence electrons. Specifically, these electrons are two 3s electrons and the three 3p electrons.

   b. V

Vanadium is a transition metal with the electron configuration $[Ar]3d^3 4s^2$. There are thus 5 valence electrons.

   c. Sn

Tin is a main group element in Group 4A on the periodic table. There are thus 4 valence electrons. Notice that in this case, we do not count the 4d electrons. This is because the 4d subshell is a filled subshell in an inner energy level.

Try Study Question 1 in Chapter 9 of your textbook now!

2. Species that contain the same number and arrangement of valence electrons are said to be isoelectronic.

CHAPTER 10: Bonding and Molecular Structure: Orbital Hybridization and Molecular Orbitals

Chapter Overview

In this chapter, we will explore covalent bonding in much more detail. You will learn about the two major theories of covalent bonding: valence bond theory and molecular orbital theory. In valence bond (VB) theory, we picture a chemical bond as arising from the overlap of atomic orbitals on different atoms. As two isolated atoms approach each other and their orbitals start to overlap, the potential energy of the system goes down. This continues until a minimum is reached. After this point, the energy goes back up again. The reason the potential energy at first goes down is that more electron density is placed between the two positively charged nuclei, decreasing the electrostatic repulsion between the two nuclei. Eventually, the nuclei get close enough that the electrostatic repulsion of the two nuclei for each other and of the electrons for each other cannot be compensated for, and so the potential energy begins to go up again. The distance between the nuclei at the point of the minimum potential energy is the average bond length for the bond.

We introduce the idea of hybridization of atomic orbitals on an atom in order to achieve the correct electron-pair geometries that we studied in the last chapter. Each of the electron-pair geometries is associated with a particular hybridization of the orbitals on the central atom. We picture the required atomic orbitals on the atom mixing together to form the new hybrid orbitals that are involved in bonding. In forming hybrid orbitals, we always end up with the same number of hybrid orbitals as atomic orbitals that were mixed together. You will learn about two different types of covalent bonds: sigma (σ) bonds resulting from head-to-head overlap of two atomic orbitals and pi (π) bonds resulting from the sideways overlap of unhybridized p orbitals. In a σ bond, the region of increased electron density is along the bond axis. In a π bond, it is above and below the bond axis. A single bond is always a σ bond, a double bond consists of a σ bond and a π bond, and a triple bond consists of a σ bond and two π bonds. There is free rotation around a single bond but not around a double or triple bond at room temperature.

Molecular orbital (MO) theory is a second theory of covalent bonding. In MO theory, we picture the atomic orbitals on an atom coming together to form new orbitals that belong to the molecule as a whole. When atomic orbitals combine, they can be pictured as combining in an additive way or in a subtractive way. When they combine in an additive way, they form MOs that are lower in energy than the original atomic orbitals; these MOs are called bonding MOs. When they combine in a subtractive way, the resulting MOs are higher in energy than the original atomic orbitals; these molecular orbitals are called antibonding MOs. Electrons in the bonding MOs contribute to the bonding between two atoms, and electrons in the antibonding MOs take away from the bonding. In combining orbitals in MO theory, there is orbital conservation: the same number of molecular orbitals is produced as there were atomic orbitals in the first place. When completing MO diagrams, we follow the Pauli exclusion principle and Hund's rule. MO theory was able to predict successfully that oxygen is a paramagnetic molecule.

Key Terms

In this chapter, you will need to learn and be able to use the following terms:

Antibonding molecular orbital: a molecular orbital in which the energy of the orbital is higher than that of the parent atomic orbitals.

Bond axis: a line connecting the nuclei of two atoms that are covalently bonded.

Bonding molecular orbital: a molecular orbital in which the energy of the orbital is lower than that of the parent atomic orbitals.

Delocalized: spread out

HOMO: the highest occupied molecular orbital in a molecule (or polyatomic ion).

Hybrid orbital: an orbital formed by mixing two or more atomic orbitals.

Isomers: compounds that have the same formula but different structures.

LUMO: the lowest unoccupied molecular orbital in a molecule (or polyatomic ion).

Molecular orbital theory: a theory of covalent bonding that pictures the atomic orbitals coming together and forming new orbitals that belong to the resulting molecule.

Orbital hybridization: the combination of atomic orbitals to form a set of equivalent hybrid orbitals.

Pi (π) bond: a covalent bond that results from the sideways overlap of unhybridized p orbitals on two different atoms. A pi bond concentrates electron density above and below the bond axis.

Sigma (σ) bond: a covalent bond that results from the head-to head overlap of atomic orbitals on two different atoms. A sigma bond concentrates electron density along the bond axis.

Valence bond theory: a theory of covalent bonding that pictures chemical bonds as being formed by the overlap of atomic orbitals in the valence shells of the combining atoms.

Chapter Goals

By the end of this chapter you should be able to:

*   **Understand the differences between valence bond theory and molecular orbital theory.**

    a) Describe the main features of valence bond theory and molecular orbital theory, the two commonly used theories for covalent bonding (Section 10.1).

    The basis of valence bond theory is the idea that chemical bonds are formed by the overlap of atomic orbitals. A covalent bond is pictured as forming when two atomic orbitals overlap. The emphasis is on the specific atomic orbitals that overlap to form a particular bond and on the type of bond that results, a sigma bond or a pi bond (see below). In order to account for the molecular geometries that are observed, the concept of orbital hybridization (see below) was introduced. Each electron-pair geometry that we studied in Chapter 9 has a particular type of hybridization required. This theory provides a nice visual picture of what is going on in bonding and fits in well with the electron-pair ideas we have been studying.

    In molecular orbital theory, we imagine that the atomic orbitals in the atoms combine to form new orbitals that belong to the molecule as a whole. We call these new orbitals "molecular orbitals" because they belong to the molecule. When orbitals combine to form molecular orbitals, they will form both bonding and antibonding molecular orbitals. Placing electrons into a bonding molecular orbital places electron density between the two nuclei and contributes to the bonding in the molecule. Placing electrons into an antibonding molecular orbital, on the other hand, reduces the electron density between the two nuclei and lessens the bonding that there will be in the molecule. This theory works better if we need a more quantitative picture of what is going on and is necessary in discussions of excited states of molecules.

b) Recognize that the premise for valence bond theory is that bonding results from the overlap of atomic orbitals. By virtue of the overlap of orbitals, electrons are concentrated (or localized) between two atoms (Section 10.2).

The main points of valence bond theory are

1. Orbitals overlap to form a bond between two atoms.

2. Two electrons of opposite spin can be accommodated in the overlapping orbitals. Usually one electron is supplied by each of the two bonded atoms.

3. Because of orbital overlap, the bonding electrons have a higher probability of being found within a region of space influenced by both nuclei. Both electrons in the bond are simultaneously attracted to both nuclei.

Why does a chemical bond arise? First, imagine bringing together two hydrogen nuclei with no electrons present. Each nucleus contains one proton and is positively charged. Because they are both positive, they would repel each other, and the amount of this repulsion would get larger the closer the nuclei got to each other. Now, let's picture two hydrogen atoms coming together to form a hydrogen molecule. This is pictured in Figure 10.1 in your textbook. Starting at the right side of this diagram, as the two isolated atoms approach, their electron clouds begin to overlap. As they do so, the energy goes down, forming a more stable species. This is very different from the case with the bare nuclei, where the energy went up. Why does the energy go down? It must be because of the electrons. In the overlap region, there is greater electron density in between the two nuclei. This electron density shields the nuclei from experiencing the repulsion that they would if the electrons were not there. The system is at a lower energy than the isolated atoms. As the nuclei get closer, the energy keeps going down until we get to a minimum that occurs at a particular distance between the nuclei. If the nuclei approach any closer than this, the energy goes up rather sharply. The electrostatic repulsions between the two nuclei and between the electrons simply get too great. The distance between the nuclei at which the minimum energy is reached is the bond length, and the energy that must be applied to break the bond so that we go all the way back up the curve to isolated atoms is the bond energy. A chemical bond forms because the bonded atoms are at a lower energy than they would be if they were separate particles.

The key idea is that bonding occurs when we concentrate electron density between two atoms.

c) Distinguish how sigma (σ) and pi (π) bonds arise. For σ bonding, orbitals overlap in a head-to-head fashion, concentrating electrons along the bond axis. Sideways overlap of p atomic orbitals results in π bond formation, with electrons above and below the molecular plane (Section 10.2).

A σ bond arises when two orbitals overlap in a head-to-head fashion to form a bond. In such a bond, the electrons are concentrated along the bond axis. Figure 10.2 in your textbook shows various σ bonds.

A π bond arises when two p orbitals overlap in a sideways fashion above and below the bond axis. To get a π bond, there must be unhybridized p orbitals on the atoms that are bonding. Figure 10.10c in the textbook illustrates a π bond. If two atoms are joined by a single bond, it is a σ bond. π bonds occur when multiple bonds are involved. A double bond consists of one σ bond and one π bond. A triple bond consists of one σ bond and two π bonds.

Example 10-1: How many σ bonds and how many π bonds are in the following molecule?

All of the single bonds are σ bonds as is 1 bond in each of the multiple bonds. There are 10 σ bonds (6 C–H bonds, 2 C–C bonds, 1 bond in the double bond, and 1 bond in the triple bond). There are 3 π bonds (1 bond in the double bond and 2 bonds in the triple bond).

Try Study Question 37b in Chapter 10 of your textbook now!

d) Understand how molecules having double bonds can have isomeric forms.

If two atoms are connected by a single bond, the two atoms can rotate freely at room temperature. If two atoms are connected by a double or a triple bond, however, there is not free rotation of the bonded atoms. This lack of free rotation must be caused by the π bond(s). In order to form a π bond, the unhybridized p orbitals involved in the bond must be lined up with each other in order to get maximum overlap. If we try to rotate one atom relative to the other, the p orbitals will not line up as well. The bonding will not be as strong. At the extreme point, where the two p orbitals are at right angles to each other, there cannot be any overlap. Rotating the atoms thus causes the π bond to be broken. It requires quite a bit of energy to do this, more than is available at room temperature.

A consequence of the lack of rotation about a double bond is *cis-trans* isomerism. In *cis-trans* isomerism, we have two compounds that have the same formula, except that the groups around a double bond are distributed differently. It arises when each carbon atom in the double bond has two different groups attached to it. The easiest case to see is one such as that pictured on p. 455 in your textbook. Each carbon has attached to it one hydrogen and one chlorine. Because we cannot rotate about the double bond at room temperature, the substituents are locked into place. The case in which the two chlorines are on the same side of the double bond is referred to as the *cis-* isomer. That in which the two chlorines are on opposite sides of the double bond is called the *trans-* isomer. You might be able to remember this because a transatlantic flight goes from one side of the Atlantic Ocean to the other side.

Example 10-2: For each molecule pictured, determine if a *cis-trans* isomer would exist. If so, label the name with the proper prefix (*cis-* or *trans-*). Draw and name the other isomer.

a.

1, 1, 2- trichloroethylene

This compound does not meet the criteria for *cis-trans* isomerism. It does not have two
different groups connected to each carbon atom. The carbon on the right does have two
different groups attached to it (H and Cl), but the carbon on the left has both groups being the
same (Cl atoms). If we were to reverse the order of the two substituents on the left carbon,
we would have the same compound.

b.

2-pentene

This compound does meet the criteria for *cis-trans* isomerism; each carbon has two different
groups attached to it. In this case, the two carbon-containing groups are on the same side of
the double bond, so this is *cis*-2-pentene.

The other isomer would be

trans-2-pentene

Try Study Question 13 in Chapter 10 of your textbook now!

• **Identify the hybridization of an atom in a molecule or ion.**

Use the concept of hybridization to rationalize molecular structure (Section 10.2).

For some molecules, we can picture bond formation by the overlap of the atomic orbitals
with which we are already familiar. For example, the HF molecule shown in Figure 10.2b in
your textbook can be pictured as arising from the overlap of the 1s orbital on the hydrogen
and one of the 2p orbitals on the fluorine. On the other hand, some molecular geometries
cannot be predicted using the orbitals with which we are familiar.

For example, consider methane, CH_4. We know from VSEPR theory (and experimental
evidence) that this molecule has a tetrahedral electron-pair geometry and a tetrahedral
molecular geometry. There are four equivalent single bonds with bond angles of 109.5°. We
also know that the ground state electron configuration of carbon is $[He]2s^22p^2$. We have
paired electrons in the 2s orbital and two electrons in the 2p subshell. We would predict,
therefore, that each of the p electrons could become part of a shared pair with another atom
to form two single bonds. This is not the four that we know form. Let's say that we excite the
atom and move one of the s electrons into the p subshell, giving $[He]2s^12p^3$. This is a little
better because now we have four unpaired electrons. We could form four bonds, but the p
orbitals would form bonds at 90° to each other, and we could expect that the bond involving
the 2s electron would be different from the bonds involving the 2p electrons. This does not
correspond to getting four equivalent single bonds each at 109.5°. Linus Pauling suggested

the idea of orbital hybridization to resolve this dilemma. The 2s orbital and the three 2p orbitals mix together to form four new orbitals that have some properties of an s orbital and some properties of a p orbital. Because there were three p orbitals and only one s orbital, the new orbitals will each be 3/4 like p orbitals and 1/4 like an s orbital. We call the resulting hybrids sp^3 orbitals to indicate that one s and 3 p orbitals were combined. These are in the second energy level so they are sometimes called $2sp^3$ orbitals. These new hybrid orbitals are all equivalent and have the desired bond angles.

Each of the electron-pair geometries we studied has a corresponding set of hybrid orbitals.

# of Groups Around Central Atom	Electron-Pair Geometry	Atomic Orbitals Used	Number of Hybrid Orbitals	Hybrid Orbitals
2	Linear	s + p	2	sp
3	Trigonal Planar	s + p + p	3	sp^2
4	Tetrahedral	s + p + p + p	4	sp^3
5	Trigonal Bipyramidal	s + p + p + p + d	5	sp^3d
6	Octahedral	s + p + p + p + d + d	6	sp^3d^2

In each case, we always end up with the same number of hybrid orbitals being formed as the number of atomic orbitals that were mixed. In the case of the sp^3 hybrid orbitals, we mix together four atomic orbitals (one s and three p), and we end up with four sp^3 hybrid orbitals.

To form the hybrid orbitals for the trigonal bipyramidal and octahedral electron-pair geometries, we must have d orbitals available. This is why we could expand the octet of a central atom only if it came from the third period or higher on the periodic table. For the first two periods, we are dealing with the first and second energy levels, which do not have d orbitals. Starting with the third period, we are filling energy levels that have d orbitals.

In the case of atoms with sp^2 hybrid orbitals, there is a p orbital that does not hybridize. In those with sp hybrid orbitals there are two p orbitals that do not hybridize. These leftover p orbitals are the ones that engage in π bonding to form double or triple bonds, if necessary.

Example 10-3: State the hybridization of each atom in the following molecule.

None of the hydrogens are hybridized. They bond using their 1s orbitals. Each carbon in the C=C double bond has three groups around it, so each of these is sp^2 hybridized. The next carbon has four groups around it, so it is sp^3 hybridized. Each carbon in the triple bond has two groups around it, so each of these is sp hybridized.

Example 10-4: State the hybridization of the nitrogen in ammonia, NH_3.

The Lewis structure for ammonia is

There are four groups around the nitrogen (3 atoms and 1 lone pair). The electron-pair geometry is tetrahedral. This indicates that the hybridization will be sp^3.

Try Study Questions 3, 5, and 7 in Chapter 10 of your textbook now!

- **Understand the differences between bonding and antibonding molecular orbitals.**

a) Understand molecular orbital theory (Section 10.3), in which atomic orbitals are combined to form bonding orbitals, nonbonding orbitals, or antibonding orbitals that are delocalized over several atoms. In this description, the electrons of the molecule or ion are assigned to the orbitals beginning with the one at lowest energy, according to the Pauli exclusion principle and Hund's rule.

In molecular orbital theory, we picture that the atomic orbitals on the bonded atoms cease to exist because they come together to form new orbitals belonging to the molecule as a whole.

The major interactions among atomic orbitals to form molecular orbitals are usually between orbitals of similar energy. Thus, we consider a 1s orbital on one atom combining with the 1s orbital on another atom, a 2p orbital on one atom combining with a 2p orbital on another, etc.

Atomic orbitals combine to form molecular orbitals in both additive and subtractive fashions. When atomic orbitals interact in an additive fashion, electron density is increased between the two nuclei. The resulting molecular orbital is called a bonding molecular orbital. A bonding molecular orbital is lower in energy than the original atomic orbitals that went into forming it. Electrons in bonding molecular orbitals contribute to bonding between the atoms because these electrons are at a lower energy than in the original atoms.

When atomic orbitals interact in a subtractive fashion, the electron density is less between the two nuclei than in the original orbitals. The resulting molecular orbital is called an antibonding molecular orbital. Electrons in antibonding molecular orbitals take away from the bonding between the atoms because these electrons are at a higher energy than in the original atoms. We represent that an orbital is an antibonding molecular orbital by including a superscript asterisk as part of the symbol for the orbital. Thus, the antibonding sigma orbital formed from two 1s orbitals is referred to as the "sigma star 1s orbital", σ^*_{1s}.

In the diatomic molecules that we shall study, there will be only bonding and antibonding molecular orbitals. There is a third type of molecular orbital that you will run across in triatomic and higher molecules, a nonbonding molecular orbital. These have the same energy as the original atomic orbitals. Electrons in nonbonding molecular orbitals do not affect the energy of the molecule compared to the separated atoms.

In valence bond theory, we pictured the formation of σ and π bonds. In molecular orbital theory, we picture the formation of σ and π molecular orbitals. σ MOs have increased electron density along the bond axis whereas π MOs have increased electron density above and below the bond axis.

There are some key principles to keep in mind about molecular orbitals:

   1) The total number of molecular orbitals is always equal to the total number of atomic orbitals contributed by the atoms that have combined.

2) The bonding molecular orbital is lower in energy than the parent orbitals, and the antibonding orbital is higher in energy.

3) Electrons of the molecule are assigned to orbitals of successively higher energy according to the Pauli exclusion principle and Hund's rule.

4) Atomic orbitals combine to form molecular orbitals most effectively when the atomic orbitals are of similar energy.

Using molecular orbital theory, the bond order can be calculated as follows:

$$\text{Bond Order} = \frac{1}{2}\left(\text{\# of } e^- \text{ in bonding MOs} - \text{\# of } e^- \text{ in antibonding MOs}\right)$$

Example 10-5: Using molecular orbital theory, predict the electron configuration and bond order for Li_2.

The electron configuration of each uncombined lithium atom is $1s^2 2s^1$. Only the 1s and 2s orbitals are occupied, so if we combine two lithium atoms, then we will have a total of four atomic orbitals to combine into four molecular orbitals. The s orbitals will combine into σ_{1s}, σ^*_{1s}, σ_{2s}, and σ^*_{2s} molecular orbitals. Each lithium atom has three electrons, so we will need to place six electrons into the molecular orbitals that have formed. Each molecular orbital can hold two electrons of opposite spin. We start with the orbital with lowest energy and progressively move to higher energy orbitals.

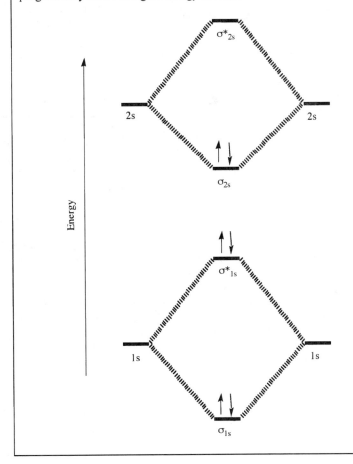

The bond order is

$$\text{Bond Order} = \frac{1}{2}\left(\text{\# of e}^- \text{ in bonding MOs} - \text{\# of e}^- \text{ in antibonding MOs}\right)$$

$$= \frac{1}{2}(4 - 2) = 1$$

Sometimes, we ignore the effects of the core electrons and only show the valence electrons in MO diagrams. The reason for this is that the core electrons will have completely filled bonding and antibonding MOs, offsetting each other and leading to zero bonding. Any bonding present in the molecule will result from what is going on with the valence electrons. In the example above, we could have shown just the σ_{2s} and $\sigma*_{2s}$ orbitals.

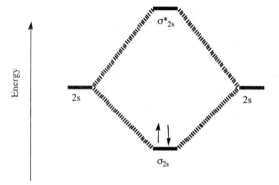

The bond order would still be calculated to be 1:

$$\text{Bond Order} = \frac{1}{2}(2 - 0) = 1$$

Try Study Question 15 in Chapter 10 of your textbook now!

b) Use molecular orbital theory to explain the properties of O_2 and other diatomic molecules.

Oxygen gas (O_2) is paramagnetic. It is attracted into a magnetic field. This means that it must have unpaired electrons. Valence bond theory predicts that all of the electrons in oxygen would be paired, so valence bond theory does not adequately explain the bonding in oxygen. Molecular orbital theory, however, does correctly predict that oxygen has unpaired electrons. This success of molecular orbital theory greatly enhanced its acceptance as an alternative bonding theory. Each oxygen atom has six valence electrons, so there are twelve total valence electrons in O_2. The molecular orbital diagram for the valence electrons in oxygen is shown in the figure that follows. The order of the σ_{2p} and π_{2p} molecular orbitals is different in this diagram than in the one pictured in Figure 10.22 in your textbook. For the reasons noted in the box on p. 464 in your text, the order shown in your textbook is that used for B_2, C_2, and N_2. The order shown here is used for O_2, F_2, and Ne_2.

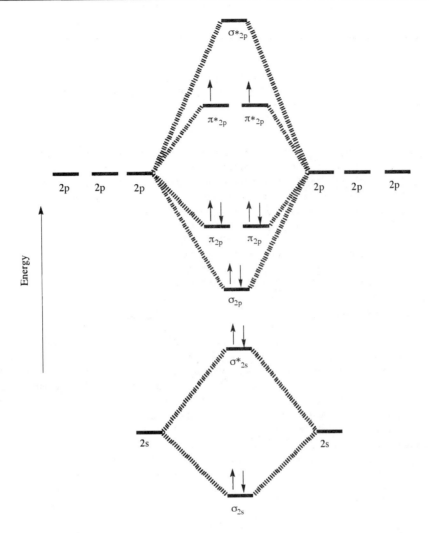

$$\text{Bond Order} = \frac{1}{2}\left(\text{\# of e}^- \text{ in bonding MOs} - \text{\# of e}^- \text{ in antibonding MOs}\right)$$

$$= \frac{1}{2}(8 - 4) = 2$$

A double bond is predicted. MO theory and VB theory agree on this point.

Try Study Question 17 in Chapter 10 of your textbook now!

Sometimes chemists speak of the HOMO or LUMO in a molecule. The HOMO is the highest occupied molecular orbital in the molecule. The LUMO is the lowest unoccupied molecular orbital. In our oxygen example, the HOMO is the π^*_{2p} MO. The LUMO is σ^*_{2p}.

Try Study Question 19 in Chapter 10 of your textbook now!

CHAPTER 11: Carbon: More Than Just Another Element

Chapter Overview

Organic chemistry is the study of compounds containing carbon (except the oxides of carbon, carbides, and carbonates). Millions of different carbon compounds are known. There are so many because carbon atoms can form four bonds, can form single, double, and triple bonds, and can link together to form chains and rings. Isomers are compounds with the same molecular formulas but which are different compounds. For a given molecular formula, there are often many different isomers. You will learn about structural isomers and stereoisomers. Within the class of stereoisomers, you will learn about geometric isomers and optical isomers.

Hydrocarbons are compounds that consist of only carbon and hydrogen. Alkanes are hydrocarbons that have the general formula C_nH_{2n+2} and contain only carbon-carbon single bonds. Alkenes contain at least one carbon-carbon double bond. Those containing one double bond have the general formula C_nH_{2n}. Compounds containing a carbon-carbon triple bond are alkynes. Those having only one triple bond have the general formula C_nH_{2n-2}. You will learn how to name and draw the structures for each of these types of compounds. Aromatic compounds are yet another class of hydrocarbons. The aromatic compounds you will study contain a benzene ring. Alkanes react with halogens by substitution reactions. Alkenes react with many compounds by addition reactions. Aromatic compounds react with some compounds by a different kind of substitution reaction.

Organic compounds containing elements other than carbon and hydrogen can be treated as hydrocarbons in which one or more hydrogen has been replaced by a functional group containing other elements. These functional groups are the reactive parts of the molecule. You will study the following classes of compounds containing functional groups: alcohols, ethers, amines, aldehydes, ketones, carboxylic acids, esters, and amides. You should learn the structure of each functional group, how to name and write formulas for their compounds, and their typical reactions.

Polymers are giant molecules made by chemically joining together many small molecules called monomers. Addition polymers are formed by linking molecules containing carbon-carbon double bonds by addition reactions. Condensation polymers are formed by the reaction of two different reactant molecules, each containing two functional groups. In this type of reaction, a small molecule (such as water) is eliminated. Many biological compounds and modern materials are polymers.

Key Terms

In this chapter, you will need to learn and be able to use the following terms:

Addition polymers: polymers formed by the addition reactions of monomers containing carbon-carbon double bonds.

Addition reaction: a compound having a double or triple bond reacts with a material with the general formula X–Y such that X ends up attached to one of the atoms that was originally involved in the double or triple bond and Y ends up attached to the other.

Alcohol: an organic compound containing a hydroxyl (–OH) functional group (without a carbonyl group being attached to the same carbon); an alcohol can also be thought of as a water molecule in which one of the hydrogens is replaced by a hydrocarbon chain.

Aldehyde: a carbonyl compound in which an organic group (–R) and a hydrogen atom are attached to the carbon of the carbonyl group.

Alkane: a hydrocarbon in which the carbon atoms are connected by only C–C single bonds and that does not contain a ring of carbon atoms. The general formula for an alkane is C_nH_{2n+2}.

Alkene: a hydrocarbon containing a carbon-carbon double bond. The general formula for an alkene containing one double bond is C_nH_{2n}.

Alkyne: a hydrocarbon containing a carbon-carbon triple bond. The general formula for an alkyne containing one triple bond is C_nH_{2n-2}.

Amide: a carbonyl compound in which an organic group (–R) and an amino group (–NR'$_2$) are attached to the carbon of the carbonyl group.

Amine: a compound that can be thought of as an ammonia molecule in which one or more of the hydrogens has been replaced by a hydrocarbon chain(s).

Aromatic compounds: compounds characterized by the presence of a benzene ring or related structure.

Carbohydrate: a compound with the general formula $C_x(H_2O)_y$, carbohydrates contain many –OH groups and also either an aldehyde or ketone when written in their open chain form.

Carbonyl group: a carbon double bonded to an oxygen (C=O).

Carboxylic acid: a carbonyl compound in which an organic group (–R) and a hydroxyl (–OH) are attached to the carbon of the carbonyl group.

Chiral: a term used to describe molecules (and other objects) that have non-superimposable mirror images.

Condensation polymer: a polymer formed in a condensation reaction of two different reactant molecules, each containing two functional groups.

Condensation reaction: a chemical reaction in which two molecules react by splitting out, or eliminating, a small molecule.

Copolymer: a polymer formed by the polymerization of two or more different monomers.

Cycloalkane: a hydrocarbon in which the carbon atoms are connected by only C–C single bonds and that contain a ring of carbon atoms. The general formula for a cycloalkane is C_nH_{2n}.

Disaccharide: a carbohydrate containing two monosaccharide units bonded together.

Elastomer: a material that goes back to its initial shape after stretching and then releasing it.

Enantiomers: a pair of molecules that are non-superimposable mirror images of each other.

Ester: a carbonyl compound formed from the reaction of an alcohol (R'OH) and a carboxylic acid (RCO$_2$H). The general formula of an ester is RCO$_2$R'.

Esterification reaction: the reaction of an alcohol and a carboxylic acid to form an ester.

Fat: a triester of glycerol and three long chain carboxylic acids; a fat is a solid at room temperature.

Functional group: an atom or group of atoms attached to a carbon atom in an organic compound.

Geometric isomers: stereoisomers that differ according to the arrangement in space of groups that are attached to two different atoms that cannot rotate.

Hydrocarbon: a compound composed of only carbon and hydrogen.

Hydrogenation reaction: an addition reaction in which H_2 reacts with a compound containing a double or triple bond.

Hydrolysis reaction: a reaction with water; in organic chemistry, this reaction often breaks apart a compound.

Isomers: compounds that have identical elemental compositions but are different compounds.

Ketone: a carbonyl compound in which two organic groups (–R) are attached to the carbon of the carbonyl group.

Monomer: the molecules that are joined together to form a polymer.

Monosaccharide: a carbohydrate containing only one sugar ring or chain.

Monounsaturated fatty acid: a long chain carboxylic acid that contains one carbon-carbon double bond.

Oil: a triester of glycerol and three long chain carboxylic acids, an oil is a liquid at room temperature.

Optical isomers: stereoisomers that are non-superimposable mirror images.

Organic chemistry: the chemistry of compounds containing carbon (except the oxides of carbon, carbides, and carbonates).

Plane-polarized light: this is produced when light of only one wavelength is passed through a polarizing filter. The resulting light vibrates in only one direction instead of in all directions.

Polymer: a giant molecule made by chemically joining many small molecules together.

Polyunsaturated fatty acid: a long chain carboxylic acid that contains more than one carbon-carbon double bond.

Primary alcohol: an alcohol in which the –OH group is attached to a carbon that is itself attached to only one other carbon.

Primary amine: an amine in which one of the hydrogens in NH_3 has been replaced by a hydrocarbon chain.

Saponification: the hydrolysis of a fat or oil using aqueous NaOH or KOH.

Saturated compound: a hydrocarbon containing only single bonds.

Saturated fatty acid: a long chain carboxylic acid that does not contain any carbon-carbon multiple bonds.

Secondary alcohol: an alcohol in which the –OH group is attached to a carbon that is itself attached to two other carbons.

Secondary amine: an amine in which two of the hydrogens in NH_3 have been replaced by hydrocarbon chains.

Stereoisomers: compounds with the same formula and in which there is a similar attachment of atoms but in which the atoms have different orientations in space.

Structural isomers: compounds having the same elemental compositions but in which the atoms are linked together in different ways.

Substitution reaction: one atom (or group of atoms) takes the place of another atom (or group of atoms) in a compound.

Tertiary alcohol: an alcohol in which the –OH group is attached to a carbon that is itself attached to three other carbons.

Tertiary amine: an amine in which all three of the hydrogens in NH_3 have been replaced by hydrocarbon chains.

Thermoplastics: polymers that soften and flow when they are heated and harden when they are cooled.

Thermosetting plastics: polymers that set (become a solid) when heated; once they have set, they cannot be resoftened.

Unsaturated compound: a hydrocarbon containing double or triple bonds.

Chapter Goals

By the end of this chapter you should be able to:

• **Classify organic compounds based on formula and structure.**

Understand the factors that contribute to the large numbers of organic compounds and the wide array of structures (Section 11.1).

A carbon atom has four valence electrons in its outer shell. In order to achieve an octet, it needs to form four bonds. Carbon atoms can form single, double, and triple bonds. A carbon atom can thus achieve an octet in many ways: 1) by forming four single bonds, 2) by forming a double bond and two single bonds, 3) by forming two double bonds, and 4) by forming a triple bond and a single bond.

The large number of organic compounds that are possible is caused mainly by 1) the many different ways that carbon can bond and 2) the ability of carbon atoms to bond to other carbon atoms to form chains and rings of carbon atoms.

Try Study Question 91 in Chapter 11 of your textbook now!

• **Recognize and draw structures of structural isomers and stereoisomers for carbon compounds.**

Recognize and draw structures of geometric isomers and optical isomers (Section 11.1).

Isomers are compounds that have the same elemental composition but that are different compounds. Structural isomers are isomers in which the atoms are hooked together in different ways. For example, the compounds n-butane and 2-methylpropane are structural isomers.

| butane | 2-methylpropane |

Both of these compounds have the same exact molecular formula, C_4H_{10}, but they are different compounds. In butane, all four carbon atoms are in the same chain, whereas in 2-methylpropane, there is a three carbon chain with a one carbon group hanging off the middle carbon.

To draw structural isomers, move around the atoms, but make sure that you follow all of the rules for drawing Lewis structures you learned previously. Keep in mind that carbon will form four bonds, hydrogen will form one bond, and oxygen will usually form two bonds. One way to know if you have formed a structural isomer is to see if the compounds would have different names according to the rules you will learn later in this chapter.

Example 11-1: Draw all of the possible isomers for C_5H_{12}.

One possible isomer would be to place all five carbons in a row. The name of this structure according to the rules you will learn is pentane.

pentane

What we do now is to move atoms around, still following our Lewis structure rules to see if we end up with new compounds. We could try having a four carbon chain with a one carbon chain hanging off it and a three carbon chain with two one carbon chains hanging off it. According to the nomenclature rules you will learn, the name of the compound with a chain of four carbons is 2-methylbutane. The name of the compound with the three carbon chain is 2,2-dimethylpropane.

2-methylbutane

2,2-dimethylpropane

These three compounds are the only structural isomers of C_5H_{12}. There are some other drawings that students sometimes think might be other isomers but are not. Let's take a look at a couple of these and see why they are not new structural isomers.

In the structure on the left, we still have a five carbon chain. If you place a pencil on the end carbon and start drawing, you can connect all five carbons without needing to lift the pencil. This is just another representation of pentane. The structure on the right is another representation of 2-methylbutane; we have just written the molecule in the opposite direction where carbon 1 is on the right instead of on the left.

The only three structural isomers are pentane, 2-methylbutane, and 2,2-dimethylpropane.

Stereoisomers are compounds that have the same formula and in which there is a similar attachment of atoms but in which the atoms have different orientations in space. One class of stereoisomers is geometric isomers. These are isomers that differ according to the arrangement in space of groups that are attached to two different atoms that cannot rotate.* The *cis-trans* isomers we discussed in the last chapter are geometric isomers.

Example 11-2: Identify whether the following has a *cis-* or *trans-* isomer and write the structural formula for the other.

Each carbon of the double bond has two different groups attached to it besides the other C of the double bond. This is a case where we will have *cis-trans* isomerism. In this structure, one of the carbon groups is above the double bond, and the other is below it. This is the *trans*-isomer. The *cis*- isomer would have both carbon groups on the same side of the double bond:

Optical isomers are stereoisomers that are non-superimposable mirror images of each other. Molecules having non-superimposable mirror images are said to be chiral, and the two mirror image compounds are called enantiomers. Most properties of enantiomers are identical; one property in which they differ is the way in which they interact with plane-polarized light. Plane-polarized light is produced when light of one wavelength is passed through a polarizing filter. The resulting light vibrates in only one direction instead of in all directions. When such light is passed through a solution containing an optically active compound, the plane of vibration shifts. The two enantiomers rotate the plane in opposite directions.

The most common type of chiral compound has four different groups attached to the same tetrahedral carbon atom.

Example 11-3: Draw the mirror image of the compound shown. Determine whether the new structure is an enantiomer of the original compound.

To draw the mirror image, imagine a mirror to the right of the molecule. The things that are close to the mirror in the given compound will be close to the mirror in the mirror image as well. Here is the pair of mirror images:

To determine if the two drawings are of the same compound or different compounds, we line up two of the groups and see if the other two groups line up or not. In this case, let us rotate the molecule so that the -CH₃ and -CH₂CH₃ groups are aligned.

The –H and –OH groups end up at different places in the two diagrams. These are two different compounds. Even before we began to draw anything, we could have predicted that the mirror images would be non-superimposable because there are four different groups attached to the same C (-H, -OH, -CH₂CH₃, and -CH₃).

- ## Name and draw structures of common organic compounds.

a) Draw structural formulas and name simple hydrocarbons, including alkanes, alkenes, alkynes, and aromatic compounds (Section 11.2).

Alkanes are hydrocarbons in which the carbon atoms are connected only by C–C single bonds and which do not contain any rings of carbon atoms. The general formula of an alkane is C_nH_{2n+2}. Alkanes contain only single bonds. Such compounds are referred to as being saturated. There is a systematic way to name alkanes.

1. *Identify the longest carbon chain in the molecule.* Be careful. The longest carbon chain may not be in a straight horizontal line in the drawing. For example, in the following drawing, you might be tempted to think that the longest carbon chain is three carbons long because there are three carbons in a horizontal row across the middle of the drawing.

CH₃—CH₂
CH₃—C——CH₃
CH₂—CH₂—CH₃

The longest carbon chain, however, is six carbons long as highlighted in the following drawing.

2. *The molecule will be named as a derivative of the single chain alkane containing the number of carbon atoms in the longest chain.* The single chain alkanes are listed in Table 11.2 in your book. You should memorize the names for those containing 1-10 carbons. The compound in this example will be named as a derivative of hexane.

3. *Number the longest chain so that the first substituent you come to will have the lowest number possible.* In our example, if we begin numbering with the top carbon, then the carbon with substituents is carbon 3. If we begin numbering with the bottom right carbon, then the substituents are on carbon 4. We number in the direction that gives the smaller number, so we will start at the top carbon.

```
  1      2
CH₃—CH₂
        | 3
CH₃—C——CH₃
        |
      CH₂—CH₂—CH₃
       4    5    6
```

4. *Each substituent is identified by a name and the position of substitution in the carbon chain.* The name of a substituent consisting of a hydrocarbon chain is based on the name of the hydrocarbon carbon containing the same number of carbon atoms, but replacing the –ane ending with –yl. The CH₃- group is named based on the hydrocarbon methane; its name is methyl. More than one of a particular group is indicated by prefixes: di-, tri-, etc. In our example, we have two methyl groups attached to carbon 3. The name is 3,3-dimethylhexane.

Example 11-4: Name the following compound:

```
                        CH₃
                        |
                        CH₂   CH₃
                        |     |
CH₃—CH₂—CH₂——CH₂—CH——CH——CH₂—CH₃
```

The longest carbon chain is the one going horizontal across the page. It contains eight carbons, so this compound will be named as a derivative of octane. If we start numbering at the left carbon, the substituents would be at positions 5 and 6. If we start numbering at the right carbon, the substituents would be at positions 3 and 4. We always want the first substituent to have the lowest possible number, so we will start numbering with the carbon on the right. We have a one carbon group, a methyl group, at position 3. We have a two carbon group, an ethyl group, at position 4. We will list these in alphabetical order (ignoring any prefixes). The name of this compound is 4-ethyl-3-methyloctane.

To draw an alkane from its name, first write the longest carbon chain, then attach at the correct positions the correct number of carbons for each carbon chain branching off the main chain. Finally, add hydrogens to the carbons so that each carbon has four bonds.

Example 11-5: Draw the structural formula for 3-ethyl-2,2-dimethylhexane.

"Hexane" indicates that the longest carbon chain is six carbons long.

$$C\!-\!C\!-\!C\!-\!C\!-\!C\!-\!C$$

There is an ethyl group at position 3 and two methyl groups at position 2.

Finally, add hydrogens so that each carbon has four bonds.

Try Study Questions 5 and 7 in Chapter 11 of your textbook now!

In cycloalkanes, like in alkanes, the carbon atoms are connected by C–C single bonds. In cycloalkanes, however, there is a ring of carbon atoms. The general formula for a cycloalkane is C_nH_{2n}. It is two less than for the alkane with the same number of carbon atoms because we can imagine removing a hydrogen atom from each of the terminal carbon atoms of the alkane and then joining together the two ends of the chain to form the ring of the cycloalkane.

Example 11-6: Draw the structural formula for cyclooctane.

This compound has eight carbon atoms in a ring.

Sometimes, we simplify our drawings even further by letting a vertex in the polygon represent a carbon atom with however many hydrogen atoms are needed to make the carbon have four bonds. This simplified picture of cyclooctane is

Each carbon in this drawing is connected to two other carbons, therefore each carbon must also have two hydrogens attached to it to bring the total number of bonds per carbon atom up to four.

Alkenes contain one or more carbon-carbon double bonds. The general formula for an alkene containing one double bond is C_nH_{2n}. We have already discussed *cis-trans* isomerism in alkenes. Alkenes are named similarly to alkanes with the following differences:

1. The ending for the name is –ene instead of –ane.

2. The carbon chain that forms the basis for the name is the longest carbon chain that contains the double bond.

3. The position of the double bond is stated in the name, and the carbon chain is numbered so as to give this position the lowest possible number.

4. If *cis-trans* isomers are possible, we must state which isomer is actually present.

Example 11-7: Name the following compound.

$$CH_3$$
$$H_3C \quad\quad CH_2CHCH_3$$
$$C=C$$
$$H \quad\quad H$$

The longest carbon chain that contains the double bond is six carbons long; the compound will be named as a hexene. The chain is numbered to give the double bond the smallest possible number, so the double bond is at position 2; this is a 2-hexene. There is a methyl group at position 5 in the chain, so this is 5-methyl-2-hexene. Finally, the arrangement around the double bond is *cis*. The name of this compound is *cis*-5-methyl-2-hexene.

Example 11-8: Draw the structural formula for trans-5-methyl-3-heptene.

The longest carbon chain containing the double bond will be seven carbons long. There will be a double bond at position 3, and the configuration around the double bond will be trans. This gives the following carbon skeleton for this chain:

$$C-C$$
$$C=C$$
$$C-C-C$$

There is a methyl group at position 5.

Finally, we fill in the hydrogens so that each carbon has four bonds.

Try Study Question 15 in Chapter 11 of your textbook now!

Alkynes contain a carbon-carbon triple bond. The general formula for an alkyne containing one triple bond is C_nH_{2n-2}. Both alkenes and alkynes are unsaturated compounds because they contain double or triple bonds. These double or triple bonds could be broken and more hydrogens added to the carbons.

Alkynes are named similarly to alkenes, except that the ending is –yne and *cis-trans* isomers are not possible around a triple bond.

Example 11-9: Name the following compound.

The longest chain containing the triple bond is five carbons long, so this will be a pentyne. The triple bond is at position 1 in the chain. There are two methyl groups attached to carbon 3. The name of this compound is 3,3-dimethyl-1-pentyne.

Aromatic compounds are characterized by the presence of a benzene ring or related structure. Benzene has the formula C_6H_6. In benzene, the carbon atoms are arranged in a ring. Each carbon is sp^2 hybridized forming σ bonds to a hydrogen atom and two other carbon atoms. Each carbon also has an unhybridized p orbital. Two different resonance structures of alternating single and double bonds between the carbon atoms can be drawn.

The true structure of benzene lies in between these two extremes of the resonance structures. There is resonance delocalization of the electrons in the p orbitals to form a π network over the entire ring.

Naming substituted benzene rings in which other groups have replaced two or more of the hydrogens, we number the benzene ring so as to give the substituents the lowest possible numbers. We then state the position number and name of the substituents followed by the word "benzene". If there are only two substituents, then we may use an older system of nomenclature using the letters *o*, *m*, and *p*, which stand for *ortho*, *meta*, and *para*. The relationships for these positions are shown in the following figure:

Example 11-10: Name the following compound.

This compound has two 2-carbon chains at adjacent positions on the benzene ring. One of the positions will be position 1, and the other will be position 2. This compound's name is 1,2-diethylbenzene. Alternatively, we could say that the two positions are *ortho* to one another, so this compound could be named *o*-diethylbenzene.

Example 11-11: Draw the structure for *p*-chloropropylbenzene.

The *p* means that the two substituents will be directly across the benzene ring from each other. At one position will be a chlorine atom, and at the other will be an alkyl group that is three carbons long.

Try Study Question 25 in Chapter 11 of your textbook now!

On pp. 492-493, there are some compounds containing the benzene ring that have special names. You should check with your instructor to see if you need to memorize any of these.

b) Identify possible isomers for a given formula (Section 11.2).

To identify the isomers for a given formula, keep in mind the following things: 1) the types of isomers that are possible, and 2) the fact that carbon will form four and only four bonds.

Beyond that, we simply play around with the structures, trying out whatever variations we can come up with, double-checking that we haven't broken any rules and that we haven't repeated a previous structure.

Example 11-12: Draw the possible alkynes with the formula C_5H_8.

Probably the easiest isomer to come up with involves having all five carbons in a row and the triple bond between the first and second carbons. This isomer is called 1-pentyne.

H——C≡≡C——CH₂—CH₂—CH₃

We can imagine moving the triple bond to another position, between carbons 2 and 3. This is 2-pentyne.

CH₃—C≡≡C——CH₂—CH₃

If we were to move the triple bond further down the chain, we would repeat 2-pentyne and 1-pentyne, so these would not add to the number of isomers. This takes care of all of the isomers with five carbons in a row. Let's now consider whether there are any isomers with four carbons in the main chain. We can have the triple bond between carbons 1 and 2. We cannot place the methyl group at positions 1 or 4 because this would make a 5 carbon chain again. We cannot have the methyl group at position 2 because carbon can only have four bonds. We can place the methyl group at position 3. This isomer is 3-methyl-1-butyne.

CH₃
|
H——C≡≡C——CH——CH₃

Try Study Questions 9 and 17 in Chapter 11 of your textbook now!

c) Name and draw structures of alcohols and amines (Section 11.3).

Alcohols can be thought of as either a hydrocarbon chain in which a hydrogen has been replaced by a hydroxyl (–OH) functional group or as a water molecule in which one of the hydrogens has been replaced by a hydrocarbon chain.

They are named as derivatives of the corresponding hydrocarbons. The longest chain containing the carbon with the hydroxyl group is used as the main chain in the molecule. The hydrocarbon chain is numbered so that the –OH group has the smallest possible number. The position of the hydroxyl group is indicated, and the final –e in the name is replaced by –ol.

In a primary alcohol, the –OH group is attached to a carbon that is attached to one carbon atom. In a secondary alcohol, the –OH group is attached to a carbon that is attached to two carbon atoms. In a tertiary alcohol, the –OH group is attached to a carbon that is attached to three carbon atoms.

Example 11-13: Name the following compounds and for parts a and b, determine if they are primary, secondary, or tertiary alcohols.

a.

The longest chain containing the hydroxyl group is four carbons long, so this will be named based on butane. This is an alcohol, so we replace the final –e with –ol. This is a butanol. We number the chain from the side closest to the hydroxyl. This would be the right side, so the hydroxyl is at position 2. The full name of this compound is 2-butanol. The carbon to which the –OH is attached is connected to two carbon atoms. This is a secondary alcohol.

b.

Even though the chain that runs from left to right has more carbons, we must select the longest chain that contains the hydroxyl group. This chain starts with the top carbon, moves down to the horizontal chain and goes to the right. It contains five carbons. The hydroxyl is at position 1. There is an ethyl group going off to the left at position 2 in the main chain. The name of this compound is 2-ethyl-1-pentanol. The –OH is attached to a carbon that is connected to only one other carbon atom. This is a primary alcohol.

c.

The chain in this molecule is five carbons long. There are two hydroxyls; we will leave in place the final –e in the name of the hydrocarbon chain, but end the name with -diol to indicate the presence of two alcohol groups. These hydroxyls are at positions 2 and 3 in the chain. The name of this compound is 2,3-pentanediol. Both hydroxyl groups are secondary.

Example 11-14: Write the structure for the following alcohol: 3-pentanol.

The hydrocarbon chain will be five carbons long. There will be one hydroxyl at position 3.

Try Study Question 31 in Chapter 11 of your textbook now!

Amines can be thought of as ammonia (NH_3) molecules in which one or more hydrogen has been replaced by a hydrocarbon chain(s). A primary amine has one hydrocarbon chain attached to the nitrogen, a secondary amine has two attached to the nitrogen, and a tertiary amine has three attached to the nitrogen. They are named by listing the hydrocarbon chains in alphabetical order followed by the ending "amine".

Example 11-15: Name the following compounds and specify whether the amine is a primary, secondary, or tertiary amine.

    a.

$$CH_3—CH_2—NH_2$$

This compound has a two carbon chain attached to the nitrogen atom. The name is ethylamine. Because there is only one of the hydrogens of NH_3 replaced by a hydrocarbon chain, this is a primary amine.

    b.

$$CH_3—CH_2—NH$$
$$|$$
$$CH_3$$

This compound has a two carbon chain and a one carbon chain attached to the nitrogen. We list these in alphabetical order, so this is ethylmethylamine. It is a secondary amine.

Example 11-16: Draw the structure for diethylamine.

This name indicates that there are two chains consisting of two carbons attached to a nitrogen atom. This is a secondary amine, so two of the three hydrogen atoms of NH_3 have been replaced by hydrocarbon chains, leaving just one hydrogen.

$$CH_3—CH_2—NH$$
$$|$$
$$CH_2$$
$$|$$
$$CH_3$$

This could also be written as $(CH_3CH_2)_2NH$ or $(C_2H_5)_2NH$.

Try Study Question 33 in Chapter 11 of your textbook now!

d) Name and draw structures of carbonyl compounds – aldehydes, ketones, acids, esters, and amides (Section 11.4).

These compounds all contain a carbon atom connected to an oxygen atom by a double bond. The C=O function group is called a carbonyl group.

Aldehydes
General Formula:

$$O$$
$$\|$$
$$R—C—H$$

where R stands for hydrogen or a hydrocarbon chain.

Naming: In the systematic nomenclature, the main chain is the longest chain containing the carbonyl carbon. Replace the final –e in the name of the corresponding alkane with –al. The carbonyl carbon is carbon 1.

Common Names to Know: The aldehyde containing one carbon is called formaldehyde. That with two carbons is called acetaldehyde.

Example 11-17: Name the following compound.

$$CH_3-\underset{\underset{CH_3}{|}}{CH}-CH_2-\underset{\underset{}{\overset{\overset{O}{\|}}{C}}}-H$$

The longest chain is four carbons long. There is a methyl group at position 3. The name of this compound is 3-methylbutanal.

Example 11-18: Draw the structure for pentanal.

This compound has five carbons in a row. The terminal carbon must have a carbonyl group and a hydrogen atom attached.

$$H-\underset{\underset{H}{|}}{\overset{\overset{H}{|}}{C}}-\underset{\underset{H}{|}}{\overset{\overset{H}{|}}{C}}-\underset{\underset{H}{|}}{\overset{\overset{H}{|}}{C}}-\underset{\underset{H}{|}}{\overset{\overset{H}{|}}{C}}-\underset{\underset{}{\overset{\overset{O}{\|}}{C}}}-H$$

Ketones
General Formula:

$$R-\underset{\underset{}{\overset{\overset{O}{\|}}{C}}}-R'$$

Naming: In the systematic nomenclature, the main chain is the longest chain containing the carbonyl carbon. Replace the final –e in the name of the corresponding alkane with –one.
The position of the carbonyl carbon is specified (other than in propanone where there is only one possibility) and is given the smallest number possible.

Common Names to Know: The common name for propanone is acetone.

Example 11-19: Name the following compound.

$$CH_3-CH_2-CH_2-\underset{\underset{}{\overset{\overset{O}{\|}}{C}}}-CH_2-CH_3$$

The longest chain is six carbons long. We number from the right in order to give the carbonyl group the lowest possible number, which is 3. The name is 3-hexanone.

Example 11-20: Draw the structure for 3-methyl-2-butanone.

There are four carbons in the main chain. There is a carbonyl group at position 2 and a methyl group at position 3.

$$H-\underset{\underset{H}{|}}{\overset{\overset{H}{|}}{C}}-\underset{\underset{}{\overset{\overset{O}{\|}}{C}}}-\underset{\underset{\underset{\underset{\underset{H}{|}}{H-\overset{\overset{H}{|}}{C}-H}}{|}}{|}}{\overset{\overset{H}{|}}{C}}-\underset{\underset{H}{|}}{\overset{\overset{H}{|}}{C}}-H$$

Carboxylic Acids
General Formula:

O
‖
R——C——O——H

Naming: The main chain is the longest chain containing the carbonyl carbon. Replace the final –e in the name of the corresponding alkane with –oic and add the word acid. The position of the carbonyl carbon is numbered as position 1.

Common Names to Know: The common name for the carboxylic acid with one carbon is formic acid. That with two carbons is acetic acid.

Example 11-21: Name the following compound.

O
‖
CH₃—CH₂—C——O——H

The longest chain containing the carbonyl group is three carbons long. The name of this acid is propanoic acid.

Example 11-22: Draw the structure for pentanoic acid.

There are five carbons in the chain. At the terminal position is a carboxyl group.

H   H   H   H   O
| | | | ‖
H——C——C——C——C——C——O——H
| | | |
H   H   H   H

Esters
General Formula:

O
‖
R——C——O——R'

Naming: There are two parts to the name of an ester. The first part is the name of the hydrocarbon chain attached to the O that does not include the carbonyl group (R' in the preceding figure). This hydrocarbon chain is named as an alkyl group. The second part is the name of the hydrocarbon chain containing the carbonyl carbon. We remove the final –e of the hydrocarbon name and replace it with –oate.

Common Names to Know: Instead of methanoate, a one carbon chain that contains the carbonyl is called formate. Instead of ethanoate, a two carbon chain containing the carbonyl carbon is called acetate.

Example 11-23: Name the following compound.

O
‖
H₃C——CH₂—C——O——CH₂——CH₃

The hydrocarbon chain that does not contain the carbonyl group has two carbon atoms; it is an ethyl group. The hydrocarbon chain containing the carbonyl group has three carbons. The name of this compound is ethyl propanoate.

Example 11-24: Draw the structure for methyl acetate.

There is one C in the chain that does not contain the carbonyl and two in the one that does.

```
        H    O         H
        |    ||        |
  H——C——C——O——C——H
        |              |
        H              H
```

Amides
General Formula:

```
        O
        ||
  R——C——NR'₂
```

Naming: A hydrocarbon chain attached to the N and not part of the chain with the carbonyl is named by writing N- and the name of this chain. This is followed by the name of the chain with the carbonyl. The final –e of the corresponding alkane's name is changed to –amide.

Common Names to Know: Instead of methanamide and ethanamide, carbonyl parts containing one and two carbons are called formamide and acetamide, respectively.

Example 11-25: Name the following compounds.

a.

```
               O
               ||
  CH₃—CH₂—C——NH₂
```

There are no hydrocarbon chains attached to the N other than the one containing the carbonyl group. This chain has three carbon atoms. The name of this compound is propanamide.

b.

```
        O        CH₃
        ||       /
  H——C——N
                 \
                 CH₃
```

Besides the chain containing the carbonyl, there are two methyl groups attached to the N. There is one C in the chain containing the carbonyl. The name is N,N-dimethylformamide.

Example 11-26: Draw the structure for N-ethylbutanamide.

The chain containing the carbonyl has four carbons. Attached to the nitrogen is a two carbon chain. The remaining bond to the nitrogen is to a hydrogen.

```
      H    H    H    O         H    H
      |    |    |    ||        |    |
  H——C——C——C——C——N——C——C——H
      |    |    |    |         |    |
      H    H    H    H         H    H
```

Try Study Questions 39, 41, and 51 in Chapter 11 of your textbook now!

• Know the common reactions of organic functional groups.

This goal applies specifically to the reactions of alkenes, alcohols, amines, aldehydes and ketones, and carboxylic acids.

Addition Reaction to an Alkene
In an addition reaction, a molecule with the general formula X–Y (such as hydrogen, halogens, hydrogen halides, and water) adds across the carbon-carbon double bond such that X ends up attached to one carbon and Y to the other.

Example 11-27: Predict the product from the reaction of ethene (ethylene) with water.

One –H from the water will end up attached to one carbon of the double bond, and the –OH will end up attached to the other.

Oxidation of an Alcohol
A primary alcohol can be oxidized first to an aldehyde and then to a carboxylic acid.

A secondary alcohol can be oxidized to a ketone.

Usual oxidizing agents used for these reactions are $KMnO_4$ and $K_2Cr_2O_7$.

Example 11-28: Predict the product of the oxidation of 2-propanol by potassium dichromate.

The structure of 2-propanol is

This is a secondary alcohol because the carbon to which the –OH group is attached is itself attached to two other carbons. The oxidation will yield a ketone.

Reduction of Aldehydes, Ketones, and Carboxylic Acids

The reverse reaction of oxidation is reduction. Carboxylic acids can be reduced to aldehydes and eventually to primary alcohols. Ketones can be reduced to secondary alcohols. Common reducing agents for these reactions are $NaBH_4$, $LiBH_4$, and H_2.

Example 11-29: Predict the product of the reduction of propanal by $NaBH_4$.

Propanal is an aldehyde. The reduction of an aldehyde will yield a primary alcohol.

Esterification: Reaction of a Carboxylic Acid with an Alcohol

An acid reacts with an alcohol to form an ester.

Example 11-30: Predict the products of the reaction of acetic acid with 1-propanol.

The products are propyl acetate and water.

Hydrolysis of an Ester

This is the reverse reaction to an esterification. An ester reacts with water to form an acid and an alcohol.

This reaction is usually performed using a base instead of water. The carboxylate anion is then converted to the acid by reacting it with an acid.

Synthesis and Hydrolysis of an Amide

Just like a carboxylic acid can react with an alcohol to form an ester, it can also react with an amine to form an amide. An amide can undergo a hydrolysis reaction to reform the carboxylic acid and the amine.

Example 11-31: Predict the products of the hydrolysis of N,N-dimethylacetamide.

N,N-dimethylacetamide is an amide. A hydrolysis reaction is a reaction with water. The products will be a carboxylic acid and an amine. The –OH of the water will end up as part of the acid, and the other hydrogen will end up connected to the nitrogen of the amine.

The products are acetic acid and N,N-dimethylamine.

Reaction of an Amine with an Acid

Amines are basic, so they will react with acids. The H^+ of the acid will attach to the lone pair on the nitrogen of the amine.

Example 11-32: Predict the products of the reaction of dimethylamine and hydrochloric acid.

HCl is an acid; the amine is a base. H$^+$ will be transferred from HCl to the amine.

Reaction of a Carboxylic Acid with a Base
Carboxylic acids are acidic, so they will react with bases. The hydrogen attached to the –OH group in the acid will be removed.

Example 11-33: Predict the products of the reaction of formic acid with sodium hydroxide.

Formic acid is an acid; NaOH is a base.

Try Study Questions 19, 43, 45, 47, and 63 in Chapter 11 of your textbook now!

- ## Relate properties to molecular structure.

a) Describe the physical and chemical properties of the various classes of hydrocarbon compounds (Section 11.2).

Physical Properties: Hydrocarbons are colorless. They tend to be insoluble in water and soluble in nonpolar solvents. The smaller alkanes (methane, ethane, propane, and butane), alkenes and alkynes are gases at room temperature and pressure. The aromatics and higher molecular weight alkanes, alkenes, and alkynes are liquids and solids.

Chemical Properties:

Alkanes: Overall, alkanes exhibit relatively low chemical reactivity; there are only a couple types of chemical reactions in which they participate. 1) Alkanes burn in air to produce CO_2 and H_2O in exothermic reactions. 2) Alkanes also react with the halogens in a reaction that replaces hydrogen atoms in the alkanes with halogen atoms; these reactions are initiated by visible or ultraviolet light.

Alkenes and Alkynes: Complete oxidation yields CO_2 and H_2O. Alkenes and alkynes also participate in addition reactions with hydrogen, halogens, hydrogen halides, and water.

Aromatics: Complete oxidation yields CO_2 and H_2O. Even though they are unsaturated, aromatic compounds do not undergo addition reactions. Instead, they undergo substitution reactions in which one or more hydrogens is replaced by another group. These reactions require the presence of reagents such as H_2SO_4, $AlCl_3$, or $FeCl_3$.

Try Study Question 13 in Chapter 11 of your textbook now!

b) Recognize the connection between the structures and the properties of alcohols (Section 11.3)

An alcohol has two distinct regions: a polar part due to the –OH group and a nonpolar part due to the hydrocarbon chain. Both regions play important roles in the properties of alcohols.

An alcohol has a higher boiling point than the corresponding hydrocarbon. This is due to the presence of the –OH group. This group can engage in strong intermolecular forces (hydrogen bonding) with other molecules of the alcohol. The longer the hydrocarbon chain is, the higher the boiling point will be. This is due to the general trend that the higher the molecular weight of a material is, the higher the boiling point.

The smaller alcohols are soluble in water. This is caused by the ability of the polar –OH group to interact favorably with polar water. As the chain length of the hydrocarbon part increases, the alcohol becomes more like the hydrocarbon because the –OH group represents a smaller fraction of the molecule. The solubility in water decreases. The presence of more than one –OH group in alcohols further increases the boiling point and solubility in water.

Example 11-34: Place the following compounds in order of increasing solubility in water:
$$CH_3CH_2OH \qquad CH_3(CH_2)_3OH \qquad CH_3(CH_2)_5OH$$

The solubility in water decreases as the length of the hydrocarbon chain increases. The compound with the smallest solubility will have the longest chain length. In order of increasing solubility, the order is
$$CH_3(CH_2)_5OH < CH_3(CH_2)_3OH < CH_3CH_2OH$$

---

c) Know the structures and properties of several natural products, including carbohydrates (Section 11.4) and fats and oils (Section 11.4).

Carbohydrates (sugars) contain many –OH groups and either an aldehyde or ketone when written in their open chain form. The general formula of a carbohydrate is $C_x(H_2O)_y$. Many carbohydrates are chiral due to the presence of carbons with four different groups attached to them. Many carbohydrates have ring forms in addition to the open chain form. These form when one of the –OH groups reacts with the aldehyde or ketone carbon. The oxygen of the reacting –OH becomes part of the ring, and the aldehyde or ketone oxygen is converted to a new hydroxyl group. In the case of glucose, it is the –OH of carbon 5 that reacts with the aldehyde carbon. This is shown in the following figure. The difference between the α and β ring forms is the orientation around carbon 1, whether the –OH is up or down in the figure.

α-D-glucose   D-glucose   β-D-glucose
                               (open-chain)

Glucose is soluble in water. Glucose is a monosaccharide because it has one sugar ring or chain. Table sugar is the compound sucrose ($C_{12}H_{22}O_{11}$). It is a disaccharide, meaning that it contains two monosaccharide units that have joined together. Sucrose is formed by the combination of glucose and fructose. Its structure is shown in the following figure.

Fats and oils are based on glycerol (1,2,3-propanetriol). In fats and oils, each of the hydroxyl groups of glycerol is esterified to a long chain carboxylic acid. Because they are present in fats, these long chain carboxylic acids are sometimes called fatty acids. They typically have chain lengths of 12 to 18 carbon atoms. The distinguishing physical difference between a fat and an oil is simply the physical state at which it is found at room temperature. A fat is solid, whereas an oil is a liquid.

glycerol   a fat or oil

A saturated fatty acid is one that does not contain any double bonds between the carbon atoms. A monounsaturated fatty acid contains one double bond between carbon atoms somewhere in the chain. A polyunsaturated fatty acid contains more than one double bond between carbon atoms in the chain. The greater the number of double bonds in the chain, the greater is the chance that the resulting triester will be a liquid at room temperature.

When a fat or oil is hydrolyzed using aqueous NaOH, glycerol and the sodium salts of the fatty acids are produced. This is shown in the following figure. The sodium salts of the fatty acids are soaps. This process is sometimes called saponification.

•   **Identify common polymers.**

a) Write equations for the formation of addition polymers and condensation polymers, and describe their structures (Section 11.5).

Polymers are giant molecules made by chemically joining many small molecules together. The small molecules that are joined together to form a polymer are called monomers.

The monomers of addition polymers are alkenes.

Example 11-35: Predict the structure of the polymer formed by the addition polymerization of propylene $CH_2=CHCH_3$.

Draw the structure of the monomer. Look for the carbon-carbon bond in the structure. This is the area of action in the reaction. It will be converted to a single bond with all of the groups still around it and single bonds going out to the other links in the chain.

This polymer is called polypropylene because it was formed by combining propylene monomers. It still uses the name propylene even though it no longer contains a double bond.

Try Study Question 65a in Chapter 11 of your textbook now!

A condensation reaction is one in which two molecules react by splitting out, or eliminating a small molecule. A condensation polymer is one formed by the condensation reaction of two different reactant molecules, each containing two functional groups. For example, ethylene glycol (with two alcohol groups) can react with propanedioc acid (with two carboxylic acid groups) to form a polymer. One alcohol can react with one carboxyl group to form an ester. In doing so, a water molecule is eliminated as the ester linkage is formed. This leaves an alcohol group on one side that can react with another acid and also leaves an acid group free on the other side that can react with another alcohol. These reactions occur, forming a long-chained molecule, with alternating parts derived from the diacid and the diol.

Common condensation polymers are polyesters and polyamides.

Try Study Questions 65b and 79 in Chapter 11 of your textbook now!

b) Relate properties of polymers to their structures (Section 11.5).

Polyethylene made up of long linear chains has a higher density than that made up of chains containing branches. This is because the linear chains can pack more tightly together. The higher density polyethylene is called high density polyethylene (HDPE) and is used for making items such as milk bottles. Low density polyethylene (LDPE), containing the branched chains, is softer and more flexible. It is used in plastic wrap and sandwich bags.

Sometimes, we can predict the properties of polymers made from substituted ethylene ($CH_2=CHX$) monomers based on the properties of the X group. For example, polymers without polar X groups (such as polystyrene) often dissolve in organic solvents. On the other hand, polyvinyl alcohol contains an –OH group in each monomer group; this makes the polymer have an affinity for water.

Other Notes

1. The bond enthalpies for carbon-carbon single, double, and triple bonds as well as for carbon-hydrogen bonds are large in comparison to many other types of bonds. This means that it takes quite a bit of energy to break these bonds.

2. Even though we often draw the carbon ring in cycloalkanes as a flat ring, those with more carbon atoms than cyclobutane have rings that are not flat. Because a six-membered ring is so common, we often show the actual conformation of the three-dimensional shape. There are two key conformations: the chair form and the boat form. You should become used to seeing the chair form:

3. An ether can be thought of as a water molecule in which both of the hydrogen atoms have been replaced by an organic group. The general formula is ROR'.

4. Alkynes also undergo addition reactions. Addition of two moles of chlorine to acetylene is an example.

5. Natural rubber is a polymer. The monomer for natural rubber is 2-methyl-1,3-butadiene, also called isoprene. In the polymer, there is a double bond between carbons 2 and 3 of the monomer unit, and the geometry around this double bond is *cis-*.

isoprene   natural rubber

CHAPTER 12: Gases & Their Properties

Chapter Overview

With Chapter 12, we begin a more in-depth study of the states of matter. Gases are in some ways the easiest to understand because we can often ignore intermolecular forces and assume the size of the particles is so small that we don't have to take it into account when determining the volume available for them to move in. We begin by examining what we mean by the pressure of a gas and by defining various units associated with pressure and showing how to convert between them. We then look at some two-variable gas laws. Boyle's law states that the pressure and volume of a gas are inversely proportional (at constant T and n). The mathematical equation you should learn for this gas law is

$$P_1 V_1 = P_2 V_2$$

Charles's law states that the volume and temperature of a gas are directly proportional (at constant P and n). The mathematical equation you should learn for this gas law is

$$\frac{V_1}{T_1} = \frac{V_2}{T_2}$$

The general gas law combines these two gas laws into one (at constant n):

$$\frac{P_1 V_1}{T_1} = \frac{P_2 V_2}{T_2}$$

Avogadro's hypothesis states that equal volumes of gases under the same conditions of temperature and pressure have equal numbers of particles. The ideal gas law is

$$PV = nRT$$

You will learn how to use this equation to solve many problems. R is the gas constant. R = 0.082057 L atm/(mole K) = 8.3145 J/(mole K). Because one of the terms involved in the ideal gas law is the number of moles, it can be useful in solving stoichiometry problems involving gases. Dalton's law states that the total pressure of a mixture of gases is equal to the sum of the partial pressures of the gases in the mixture. In dealing with Dalton's law, the concentration unit called the mole fraction is introduced. The mole fraction of component A in a mixture is calculated as follows:

$$X_A = \frac{\text{moles of component A}}{\text{total moles}}$$

The theory we use to explain gas behavior is the kinetic-molecular theory. It proposes that gases consist of small particles (atoms or molecules) in motion. We can use it to predict the relationships in the ideal gas law. In a gas, the particles do not all move at the same speed; there is a distribution of speeds. If we increase the temperature, the kinetic energy of the particles increases; more particles move at faster speeds. If we have two gases at the same temperature, the particles in the one with the larger molar mass move with a smaller average speed. Graham's law of effusion is

$$\frac{\text{Rate of effusion of gas 1}}{\text{Rate of effusion of gas 2}} = \sqrt{\frac{\text{molar mass of gas 2}}{\text{molar mass of gas 1}}}$$

Under conditions of high pressure and/or low temperature, some assumptions of the ideal gas law fall apart. Under these circumstances, the van der Waals equation sometimes gives more accurate results. This equation has correction terms for intermolecular forces (a) and molecular volume (b).

Key Terms

In this chapter, you will need to learn and be able to use the following terms:

Avogadro's hypothesis: under conditions of constant temperature and pressure, equal volumes of gases contain equal numbers of particles.

Boyle's law: the pressure and the volume of a gas are inversely proportional (under conditions of constant temperature and moles of gas). This results in the equation:

$$P_1V_1 = P_2V_2$$

Charles's law: the volume and the absolute temperature of a gas are directly proportional (under conditions of constant pressure and moles of gas). This results in the equation:

$$\frac{V_1}{T_1} = \frac{V_2}{T_2}$$

Compressibility: a measure of how much a material's volume changes when the pressure is changed. Gases have a high compressibility.

Dalton's law of partial pressures: the total pressure of a mixture of gases is equal to the sum of the partial pressures of the various gases in the mixture.

Diffusion: the mixing of the molecules of two or more substances by random molecular motion.

Effusion: the movement of a gas through a tiny opening in a container into another container where the pressure is lower.

General gas law (combined gas law): a gas law relating two conditions of pressure, temperature, and volume for the same sample of a gas; it combines both Boyle's law and Charles's law:

$$\frac{P_1V_1}{T_1} = \frac{P_2V_2}{T_2}$$

Graham's law: an equation that relates the rates of effusion (or diffusion) to the molar masses of the gases being compared. The equation is

$$\frac{\text{Rate of effusion of gas 1}}{\text{Rate of effusion of gas 2}} = \sqrt{\frac{\text{molar mass of gas 2}}{\text{molar mass of gas 1}}}$$

Ideal gas law constant (R): the proportionality constant in the ideal gas law.
R = 0.082057 L atm/(mole K) = 8.3145 J/(mole K).

Ideal gas law: $PV = nRT$

Kinetic-molecular theory of gases: a theory to explain the behavior of gases, it postulates that gases consist of small particles (atoms or molecules) that are separated from each other by a great distance in comparison to their sizes, that these molecules are in motion and collide with each other with no loss of energy, and that the average kinetic energy of a gas is proportional to the temperature of the gas.

Maxwell-Boltzmann distribution curve: a plot showing the relationship between the number of molecules and their speed (or energy).

Partial pressure: the pressure exerted by one of the components in a mixture of gases.

Pressure: force per area

Root-mean-square (rms) speed: a special type of average speed

$$\text{rms speed} = \sqrt{\mu^2} = \sqrt{\frac{3RT}{M}}$$

This equation is sometimes called the Maxwell equation. To use this equation, the molar mass (M) must be in kg/mole, and the value of R used is 8.3145 J/(mole K.)

Standard molar volume: the volume occupied by 1 mole of a gas at STP. The value is 22.4 L.

Standard temperature and pressure (STP): 0°C and 1 atm.

van der Waals equation: a variation of the ideal gas law in which there are correction factors for intermolecular forces (a) and molecular volume (b)

$$\left(P + a\left[\frac{n}{V}\right]^2\right)(V - bn) = nRT$$

Chapter Goals

By the end of this chapter you should be able to:

- ## Understand the basis of the gas laws and know how to use those laws.

  ### a) Describe how pressure measurements are made and work with the units of pressure, especially atmospheres (atm) and millimeters of mercury (mm Hg) (Section 12.1).

  Pressure is force per area. Pressures are often measured using a barometer. A barometer can be made using a tube sealed at one end by completely filling it with a liquid (usually mercury), covering the open end of the tube, placing the open end in a container that holds the same liquid, and uncovering the open end below the surface of the liquid in the container. The liquid inside the tube will then fall until the pressure exerted by the liquid in the tube exactly balances the pressure exerted by the atmosphere on the liquid in the container. We measure the height from the surface of the liquid in the container to the surface of the liquid in the tube. This height is proportional to the pressure. Look at Figures 12.2 in your text and be sure that you understand the set-up of this type of barometer.

  The mercury in this type of barometer will rise to a height of about 760 mm at sea level. One standard atmosphere (1 atm) is defined to be exactly 760 mm Hg. You may sometimes also run across the unit torr. One torr is the same thing as 1 mm Hg.

     1 atm = 760 mm Hg = 760 torr

  In chemistry, we do not usually use the SI unit for pressure, but it is the pascal (Pa).

     1 atm = 101300 Pa

  Yet another unit for pressure is the bar. This unit of pressure is the standard pressure for the field of thermodynamics and is a little less than 1 atm.

     1 bar = 100000 Pa = 0.9872 atm

  You need to be able to convert between these different pressure units.

  Example 12-1: Convert a pressure of 99.3 kPa to atm, bars, and mm Hg.

  In order to carry out these conversions, we use the conversion factors listed above. As usual, we will let our units help us in determining which values go on the top and which on the bottom of our conversion factors.

  $$99.3 \text{ kPa} \times \frac{1000 \text{ Pa}}{1 \text{ kPa}} \times \frac{1 \text{ bar}}{100000 \text{ Pa}} = 0.993 \text{ bars}$$

  $$99.3 \text{ kPa} \times \frac{1000 \text{ Pa}}{1 \text{ kPa}} \times \frac{1 \text{ atm}}{101300 \text{ Pa}} = 0.980 \text{ atm}$$

$$0.980 \text{ atm} \times \frac{760 \text{ mm Hg}}{1 \text{ atm}} = 745 \text{ mm Hg}$$

Try Study Question 1 in Chapter 12 of your textbook now!

b) Understand the origins of the gas laws (Boyle's law, Charles's law, and Avogadro's hypothesis) and know how to apply them (Section 12.2).

There are four major variables that affect gases: pressure (P), volume (V), temperature (T), and the number of moles (n) of the gas. We will first look at some relationships in which we allow only two of these variables to change, and we keep the other two constant.

First, we will consider the relationship between pressure and volume when we keep the temperature and number of moles constant. Boyle's law deals with this. If we have a gas in a sealed syringe and we push down on the plunger of the syringe, the volume of the gas will go down. In other words, as we increase the pressure, the volume decreases. Volume and pressure go in opposite directions. Mathematically, we say that pressure and volume are inversely proportional (under conditions of constant T and n) and represent this as

$$P \propto \frac{1}{V}$$

We can convert a proportionality into an equality by introducing a proportionality constant. Let us call this constant c_B.

$$P = c_B \frac{1}{V}$$

We can solve this equation for the constant by multiplying both sides of the equation by V.

$$PV = c_B$$

What this implies is that the pressure of a gas times its volume is equal to a constant. Thus, if we have two sets of conditions, the pressure times the volume under one set of conditions will be equal to the pressure times the volume under the other set of conditions (provided that the temperature and the number of moles of the gas are the same). We can represent this by an equation for Boyle's law that you should memorize:

$$P_1 V_1 = P_2 V_2$$

Example 12-2: Suppose that 2.00 L of helium at 875 mm Hg is allowed to expand to occupy a volume of 2.50 L. What is the new pressure of the helium gas?

In solving this type of problem, it is useful to set up a table.

Original Conditions	Final Conditions
P_1 = 875 mm Hg	P_2 = x
V_1 = 2.00 L	V_2 = 2.50 L

In this case, we are given P_1, V_1, and V_2, and we are trying to solve for P_2. The temperature and the number of moles of the gas are constant. This is a Boyle's Law problem.

$$P_1 V_1 = P_2 V_2$$

We want to solve this equation for P_2. That means that we want to get P_2 in the numerator on one side of the equation, and we want to get it by itself. It is already in the numerator. The thing that is keeping it from being by itself is the V_2. P_2 is multiplied by V_2. To undo this and get P_2 by itself, we need to divide both sides of the equation by V_2. When we do this, we end up with

$$P_2 = \frac{P_1 V_1}{V_2}$$

$$= \frac{(875 \text{ mm Hg})(2.00 \text{ L})}{2.50 \text{ L}}$$

$$= 7.00 \times 10^2 \text{ mm Hg}$$

Try Study Question 5 in Chapter 12 of your textbook now!

Charles's Law deals with volume and temperature (under conditions of constant pressure and number of moles of gas). If we heat a gas, it expands. In other words, as the temperature increases, so does the volume. Volume and temperature go in the same direction. The volume and the temperature in kelvin are directly proportional (under conditions of constant P and n).

$$V \propto T$$

We could introduce a proportionality constant, solve for the constant, and then consider two sets of conditions, like we did when we derived the form of Boyle's law you should memorize. Skipping to the chase, the equation to memorize for Charles's law is

$$\frac{V_1}{T_1} = \frac{V_2}{T_2}$$

The trickiest thing in a Charles's law problem is that the temperatures MUST be in kelvin. Student often miss points on exams because they forget to switch temperatures in °C to K (and of course, we professors of chemistry usually write problems with the temperatures in °C to check whether students have learned this point).

Example 12-3: A 10.0 L sample of nitrogen gas at 10.°C is heated to a new temperature of 20.°C. Assuming that the pressure remains the same, what is the new volume of the gas.

Once again, we will set up a table to help us organize our thinking.

Original Conditions	Final Conditions
$T_1 = 10.°C$	$T_2 = 20.°C$
$V_1 = 10.0 \text{ L}$	$V_2 = x$

We are given T_1, V_1, and T_2, and we are asked to solve for V_2. This is a Charles's law problem. First, we must convert the temperatures to K!

$$T_1 = (10.°C + 273°C)\frac{1 \text{ K}}{1°C} = 283 \text{ K}$$

$$T_2 = (20.°C + 273°C)\frac{1 \text{ K}}{1°C} = 293 \text{ K}$$

Now, we can use our Charles's law equation.

$$\frac{V_1}{T_1} = \frac{V_2}{T_2}$$

We need to solve this for V_2. V_2 is already in the numerator. What is keeping it from being by itself is that it is divided by T_2. To undo this division, we must multiply both sides by T_2.

$$V_2 = \frac{V_1 T_2}{T_1} = \frac{(10.0 \text{ L})(293 \text{ K})}{283 \text{ K}} = 10.4 \text{ L}$$

Had we forgotten to convert from °C to K, we would have ended up with the wrong answer of 20.0 L, a very different answer! CONVERT THE TEMPERATURES TO KELVIN!

Try Study Question 7 in Chapter 12 of your textbook now!

Boyle's law and Charles's law can be combined into the general gas law (combined gas law). This gas law relates pressure, volume, and absolute temperature assuming constant number of moles of a gas. The equation for this gas law is

$$\frac{P_1 V_1}{T_1} = \frac{P_2 V_2}{T_2}$$

This equation is very useful in solving problems where all three variables (P, T, and V) can change. In addition, notice that if only two variables change, you can obtain any two-variable gas law by covering up the variable that is not involved in the problem. For example, a Boyle's law problem is done under conditions of constant temperature, so we would cover up the temperature terms and obtain the Boyle's law equation:

$$P_1 V_1 = P_2 V_2$$

If you do not like memorizing, therefore, you could just memorize the general gas law equation and derive the others we have discussed so far.

Example 12-4: A sample of argon gas initially at 0°C occupies a volume of 2.24 L and exerts a pressure of 760. mm Hg. If the gas is heated to 100.°C and allowed to expand to 3.00 L, what is the new pressure of the gas?

Initial Conditions	Final Conditions
P_1 = 760 mm Hg	P_2 = x
V_1 = 2.24 L	V_2 = 3.00 L
T_1 = 0°C	T_2 = 100.°C

In this case, we are given information about P, T, and V, and all three are allowed to vary. We will use the general gas law to solve this problem.

$$\frac{P_1 V_1}{T_1} = \frac{P_2 V_2}{T_2}$$

Just like for Charles's law, the temperatures must be in kelvin.

$$T_1 = (0°C + 273°C)\frac{1 \text{ K}}{1°C} = 273 \text{ K}$$

$$T_2 = (100.°C + 273°C)\frac{1 \text{ K}}{1 °C} = 373 \text{ K}$$

Next, we rearrange the general gas law equation for the desired variable. We want to solve for P_2. It is already in the numerator. To get it by itself, we need to undo dividing by T_2 and multiplying by V_2. Thus we will multiply both sides by T_2 and divide each side by V_2.

$$P_2 = \frac{P_1 V_1}{T_1} \cdot \frac{T_2}{V_2} = \frac{(760 \text{ mm Hg})(2.24 \text{ L})}{273 \text{ K}} \cdot \frac{373 \text{ K}}{3.00 \text{ L}} = 775 \text{ mm Hg}$$

Try Study Question 13 in Chapter 12 of your textbook now!

The final relationship for this section is Avogadro's hypothesis, which states that under the same conditions of temperature and pressure, equal volumes of gases contain equal numbers of particles. Thus, 2 liters of hydrogen gas contains the same number of molecules as does 2 liters of nitrogen gas. Another consequence is that 4 liters of hydrogen gas should contain twice as many molecules as does 2 L of hydrogen gas at the same temperature and pressure. Mathematically, we can state that under conditions of constant P and T

$$V \propto n$$

Example 12-5: Hydrogen and oxygen gases combine to form water according to the equation
$$2H_2 \ (g) + O_2 \ (g) \rightarrow 2H_2O \ (l)$$
If 436 mL of hydrogen gas at a certain temperature and pressure is used in this reaction, how much oxygen gas at the same temperature and pressure should be required?

The balanced equation indicates that twice as many moles of hydrogen gas are needed than of oxygen gas. Because volume and the number of particles are proportional, twice the volume of hydrogen gas should be required as oxygen gas.

$$436. \ \text{mL} \ H_2 \ \times \ \frac{1 \ \text{mL} \ O_2}{2 \ \text{mL} \ H_2} \ = \ 218 \ \text{mL} \ O_2$$

Try Study Question 15 in Chapter 12 of your textbook now!

• **Use the ideal gas law.**

a) Understand the origin of the ideal gas law and know how to use the equation (Section 12.3).

Boyle's law states that volume is inversely proportional to pressure, Charles's law states that volume is directly proportional to temperature, and Avogadro's hypothesis states that volume is directly proportional to the number of moles of the gas. Putting these together,

$$V \propto \frac{nT}{P}$$

This proportionality can be converted to an equality by introducing a proportionality constant. We shall call this constant R.

$$V = R\frac{nT}{P}$$

This equation can be rearranged into the form in which we usually find it:
$$PV = nRT$$
You should definitely memorize this equation! This is the ideal gas law. This equation works very well for gases under normal pressures and temperatures. Under conditions of very high pressure and/or low temperature, the equation does not work so well, and the gases are said to deviate from ideality.

R is the same for all gases; its value depends upon the units used. In most gas law problems, the value of R we use is

$$R = 0.082057 \ \frac{L \ atm}{mole \ K}$$

In order to use this value of R, the pressure must be expressed in atm, the volume in L, and the temperature in K. If a problem gives P, V, or T in other units than these, then we must convert them into these units before we can use the ideal gas law.

Example 12-6: What pressure should 1.00 g of oxygen exert if it is in a 250. mL vessel at a temperature of 15°C?

The ideal gas law is PV = nRT. We are looking for the pressure, so this will be our unknown. The volume is given to us, but is in units of mL. We will need to convert this to L.

$$250. \text{ mL} \times \frac{1 \text{ L}}{1000 \text{ mL}} = 0.250 \text{ L}$$

We are not given the number of moles directly, but we are given the mass of the gas; we can use the molar mass of oxygen gas to determine the number of moles. The molar mass of O_2 is 2 x 16.00 = 32.00 g/mole.

$$1.00 \text{ g } O_2 \times \frac{1 \text{ mole } O_2}{32.00 \text{ g } O_2} = 0.0313 \text{ moles } O_2$$

We are given the temperature in °C; we will need to convert this to K.

$$(15°C + 273°C)\frac{1 \text{ K}}{1°C} = 288 \text{ K}$$

We can now use the ideal gas law.

$$PV = nRT$$

$$P = \frac{nRT}{V}$$

$$P = \frac{(0.0313 \text{ moles})\left(0.082057 \dfrac{\text{L atm}}{\text{mole K}}\right)(288 \text{ K})}{0.250 \text{ L}}$$

$$= 2.95 \text{ atm}$$

Try Study Question 15 in Chapter 12 of your textbook now!

b) Calculate the molar mass of a compound from a knowledge of the pressure of a known quantity of a gas in a given volume at a known temperature (Section 12. 3).

The molar mass (M) of a compound can be calculated by taking the mass of the compound in the sample and dividing by the number of moles of the compound present:

$$\text{Molar Mass} = \frac{\text{mass}}{\text{moles}}$$

$$M = \frac{m}{n}$$

The ideal gas law gives us a new way to calculate the number of moles of a gas. If we know the pressure, volume, and temperature of a gas, we can calculate the number of moles. If we also know the mass of the gas present, then we can calculate the molar mass.

Example 12-7: Calculate the molar mass of a gas if 1.63 g of the gas placed in a 500. mL flask at 25°C exerts a pressure of 942 mm Hg.

We are asked to solve for the molar mass. M = m/n. We are given m, 1.63 g. All we need to do is to find a way to get n, and then we could calculate the molar mass. We can use the ideal gas law to determine n.

$$PV = nRT$$

$$n = \frac{PV}{RT}$$

We are given the pressure, the volume, and the temperature, and we know the value of R. Before we can substitute into this equation, we must adjust some of the units in the given information to be in the form it needs to be for our units of R. The pressure must be in atm.

$$P = 942 \text{ mm Hg} \times \frac{1 \text{ atm}}{760 \text{ mm Hg}} = 1.24 \text{ atm}$$

The volume must be in L.

$$V = 500 \text{ mL} \times \frac{1 \text{ L}}{1000 \text{ mL}} = 0.500 \text{ L}$$

The temperature must be in K.

$$T = (25°C + 273 \ °C)\frac{1 \text{ K}}{1°C} = 298 \text{ K}$$

Now we can substitute into the equation:

$$n = \frac{PV}{RT}$$

$$= \frac{(1.24 \text{ atm})(0.500 \text{ L})}{\left(0.082057 \ \frac{\text{L atm}}{\text{mole K}}\right)(298 \text{ K})}$$

$$= 0.0253 \text{ moles}$$

All that is left is to calculate the molar mass:

$$M = \frac{m}{n}$$

$$= \frac{1.63 \text{ g}}{0.0253 \text{ moles}}$$

$$= 64.3 \text{ g/mole}$$

Try Study Question 29 in Chapter 12 of your textbook now!

- ## Apply the gas laws to stoichiometric calculations (Section 12.4).

  ### a) Stoichiometric calculations involving gases.

  As we have seen, the ideal gas law provides us with another method to calculate the number of moles of a species.

  $$n = \frac{PV}{RT}$$

  Because we can calculate moles by this method, we can use it in solving some stoichiometry problems. The following diagram builds on the one we used earlier for mass-mass and solution stoichiometry.

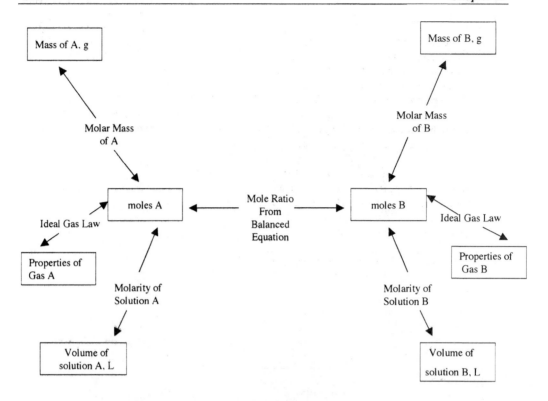

Example 12-8: Sodium chloride can be prepared from the reaction of sodium metal and chlorine gas. What mass of sodium chloride could be prepared from 290. mL of chlorine gas at a pressure of 752 mm Hg and a temperature of 25°C?

We are asked for the mass of sodium chloride and given information about the chlorine gas. This is a stoichiometry problem. The first step is to write a balanced chemical equation and to set up our table with the given information.

	2Na *(s)*	+	Cl₂ *(g)*	→	2NaCl *(s)*
Measured			P = 752 mm Hg V = 290. mL T = 25°C		x g
Conversion Factor					
Moles					

The connecting link between properties of a gas and moles is the ideal gas law. Before we can use the ideal gas law with the value of R that we know, we must first convert the pressure to be in atm, the volume to be in L, and the temperature to be in K.

$$P = 752 \text{ mm Hg} \times \frac{1 \text{ atm}}{760 \text{ mm Hg}} = 0.989 \text{ atm}$$

$$V = 290. \text{ mL} \times \frac{1 \text{ L}}{1000 \text{ mL}} = 0.290 \text{ L}$$

$$T = (25°C + 273°C)\frac{1 \text{ K}}{1°C} = 298 \text{ K}$$

Now, we can use the ideal gas law to determine the number of moles of chlorine gas.

$$PV = nRT$$

$$n = \frac{PV}{RT} = \frac{(0.989 \text{ atm})(0.290 \text{ L})}{\left(0.082057 \dfrac{\text{L atm}}{\text{mole K}}\right)(298 \text{ K})} = 0.0117 \text{ moles Cl}_2$$

	2Na *(s)*	+	Cl$_2$ *(g)*	→	2NaCl *(s)*
Measured			P = 752 mm Hg V = 290. mL T = 25°C		x g
Conversion Factor			$n = \dfrac{PV}{RT}$		
Moles			0.0117 moles		

According to the balanced equation, there are two moles of sodium chloride produced for each mole of chlorine gas.

$$0.0117 \text{ moles Cl}_2 \times \frac{2 \text{ moles NaCl}}{1 \text{ mole Cl}_2} = 0.0235 \text{ moles NaCl}$$

We are asked for the mass of sodium chloride that could be produced. The connecting link between mass and moles is the molar mass. Hopefully, you can calculate that the molar mass of sodium chloride is 58.44 g/mole.

$$0.0235 \text{ moles NaCl} \times \frac{58.44 \text{ g NaCl}}{1 \text{ mole NaCl}} = 1.37 \text{ g NaCl}$$

The completed stoichiometry table is as follows:

	2Na *(s)*	+	Cl$_2$ *(g)*	→	2NaCl *(s)*
Measured			P = 752 mm Hg V = 290. mL T = 25°C		1.37 g
Conversion Factor			$n = \dfrac{PV}{RT}$		58.44
Moles			0.0117 moles		0.0235 moles

Try Study Questions 31 and 33 in Chapter 12 of your textbook now!

b) Use Dalton's law of partial pressures (Section 12.5).

Imagine that we have a mixture of different gases (A, B, C, ...). What will the total pressure be? Dalton's law states that the total pressure will be the sum of the pressures that each gas would exert if it were alone under the same conditions.

$$P_{total} = P_A + P_B + P_C + ...$$

where P_A, P_B, P_C, etc. represent the pressures that each individual gas (A, B, C, ...) would exert on its own. These pressures are called the partial pressures of the gases. The partial pressure of a gas in a mixture is the pressure due to that particular gas.

Dalton's law indicates that the presence of other gases does not affect the pressure that each gas exerts. This is because each gas molecule is moving independently.

One consequence of this is that the total pressure is dependent solely on the number of molecules that are present in the gas mixture; it does not matter what the molecules are.

$$P_{total}V = n_{total}RT$$

Another consequence is that we can determine the partial pressure of a gas if we know what fraction of the molecules in the mixture is the gas of interest. If half the molecules are of one gas, the partial pressure due to that gas will be half the total pressure. Mathematically stated,

$$P_A = X_A P_{total}$$

where X_A is the mole fraction of component A. The mole fraction is the fraction of the molecules (moles) in a mixture that is the component of interest.

$$X_A = \frac{\text{moles of component A}}{\text{total moles}}$$

A mole fraction is set up like a percent – the part over the whole – except that we don't multiply by 100. Just like the total of all the percents in a mixture must be 100%, the sum of all the mole fractions in a mixture must be 1. If we have 3 moles of N_2 and 1 mole of O_2 in a mixture, the mole fraction of nitrogen will be 3/4 = 0.75. The partial pressure of the nitrogen will be 0.75 times the total pressure. The mole fraction of oxygen must be $1 - 0.75 = 0.25$.

Example 12-9: What is the pressure of 5.00 L of a gas mixture consisting of 0.30 moles of N_2, 0.20 moles of O_2, and 0.15 moles of CO_2? The temperature is 298 K.

Each gas behaves independently. The total number of moles of gas is equal to
$$n_{total} = 0.30 \text{ moles} + 0.20 \text{ moles} + 0.15 \text{ moles} = 0.65 \text{ moles}$$

We can then use this number of moles in the ideal gas law, where the number of moles is the total number of moles of gas in our mixture.

$$P_{total}V = n_{total}RT$$

$$P_{total} = \frac{n_{total}RT}{V} = \frac{(0.65 \text{ moles})\left(0.082057 \, \frac{L \text{ atm}}{\text{mole K}}\right)(298 \text{ K})}{5.00 \text{ L}} = 3.2 \text{ atm}$$

Example 12-10: What is the partial pressure of each gas in the last example?

To solve this problem, we can either solve the ideal gas law for each gas or use the equation
$$P_A = X_A P_{total}$$
We shall do the latter.

First we calculate the mole fractions:
$$X_{N_2} = \frac{\text{moles of } N_2}{\text{total moles}} = \frac{0.30 \text{ moles}}{0.65 \text{ moles}} = 0.46$$
$$X_{O_2} = \frac{\text{moles of } O_2}{\text{total moles}} = \frac{0.20 \text{ moles}}{0.65 \text{ moles}} = 0.31$$
$$X_{CO_2} = \frac{\text{moles of } CO_2}{\text{total moles}} = \frac{0.15 \text{ moles}}{0.65 \text{ moles}} = 0.23$$

Then we calculate the partial pressures.
$$P_{N_2} = X_{N_2} P_{total} = (0.46)(3.2 \text{ atm}) = 1.5 \text{ atm}$$
$$P_{O_2} = X_{O_2} P_{total} = (0.31)(3.2 \text{ atm}) = 0.98 \text{ atm}$$
$$P_{CO_2} = X_{CO_2} P_{total} = (0.23)(3.2 \text{ atm}) = 0.74 \text{ atm}$$

Try Study Question 37 in Chapter 12 of your textbook now!

• **Understand kinetic-molecular theory as it is applied to gases, especially the distribution of molecular speeds (energies) (Section 12. 6).**

a) **Apply the kinetic molecular theory of gas behavior at the molecular level (Section 12.6).**

So far, we have been considering macroscopic properties of gases such as pressure and volume. As often is the case, chemists try to explain macroscopic behavior in terms of submicroscopic particles (atoms and molecules). The theory that we use to explain the behavior of gases is called the kinetic-molecular theory. The assumptions of the kinetic-molecular theory are as follows:

> 1. Gases consist of particles (molecules or atoms). The separation between these particles is greater than the size of the particles themselves.
>
> 2. The particles of a gas are in continual, random, and rapid motion. As they move, they collide with one another and with the walls of their container, but they do so without loss of energy.
>
> 3. The average kinetic energy of a gas is proportional to the gas temperature. All gases have the same average kinetic energy at the same temperature.

The ideal gas law, solved for pressure, is

$$P = \frac{nRT}{V}$$

The pressure exerted by a gas is thus directly proportional to the number of moles of the gas and to the temperature but is inversely proportional to the volume. Let us see how the kinetic-molecular theory predicts these very relationships.

If we increase the number of molecules in a sample of a gas, we increase the number of collisions that will occur with the container walls. The pressure will go up. Pressure is directly proportional to the number of molecules (thus moles) of the gas.

If we decrease the size of the container, there will be more collisions per unit area. The pressure thus increases. Decreasing the volume increased the pressure. Pressure and volume are inversely proportional.

If we increase the temperature, we increase the average kinetic energy of the gas. The particles thus collide with the wall of the container with a greater force. Because they are moving more quickly, there are also more collisions with the container wall. The pressure will go up. Pressure and temperature are directly proportional.

By assuming that gases consist of these very small particles called molecules (or atoms), we are able to explain the macroscopic observations that we made earlier regarding the relationship between the pressure of a gas and the volume, temperature, and number of moles of the gas.

There are some other features about molecular speeds and energies you should be familiar with. The molecules in a sample of a gas do not all move at the same speed. Some molecules are moving more slowly, and some are moving more rapidly. Most are moving at somewhere in between. There is a distribution of speeds in the sample. A typical distribution is shown in Figure 12.14 in your text. The maximum in this curve is the most probable speed. When we increase the temperature, a greater fraction of the molecules move with a faster speed. The graph gets extended to the right and also gets squashed down in the middle.

The kinetic energy of a particle is given by

$$K.E. = \frac{1}{2}mv^2$$

For a collection of gas molecules, the average kinetic energy of a molecule is given by

$$\overline{K.E.} = \frac{1}{2}m\overline{\mu}^2$$

where $\overline{\mu}$ is the average speed of the molecules in the sample.

One of the assumptions of the kinetic-molecular theory is that all gases, regardless of molar mass, have the same kinetic energy. If we have two gases at the same temperature, the one with the greater molar mass will have the smaller average speed. You can rationalize this in the following way: if I throw a golf ball and a bowling ball with the same energy, the bowling ball will move with a slower speed than the golf ball. You should look at Figure 12.15 in your text. This figure shows that (at the same temperature) the average speed of the different molecules decreases with increasing molar mass.

Example 12-11: Place the following molecules in order of increasing average speed at the same temperature: O_2, CO_2, SO_2, Ar.

The speed will decrease as the molar mass increases. So, we need to calculate the molar mass of each gas and then put the gases in the reverse order of their molar masses.

The molar masses are as follows:
        O_2: 32.0 g/mole, CO_2: 44.0 g/mole, SO_2: 64.1 g/mole, Ar: 39.9 g/mole

The order of these gases from the slowest to the fastest is SO_2, CO_2, Ar, and O_2.

Try Study Question 45 in Chapter 12 of your textbook now!

b) Understand the phenomena of diffusion and effusion and know how to use Graham's law (Section 12.7).

Thomas Graham studied two similar sounding processes: diffusion and effusion. Diffusion is the gradual mixing of the molecules of two or more substances by random molecular motion. The odor of perfume is spread throughout a room by means of diffusion. Effusion is the movement of a gas through a tiny opening a container into another container where the pressure is lower. A latex balloon filled with helium deflates due to effusion of the helium through pores in the latex.

The rates of diffusion or effusion of two gases are related by the equation:

$$\frac{\text{Rate of effusion of gas 1}}{\text{Rate of effusion of gas 2}} = \sqrt{\frac{\text{molar mass of gas 2}}{\text{molar mass of gas 1}}}$$

The two key things to note in this equation are that we use the *square roots* of the molar masses and that the gas on top on one side of the equation is on the bottom on the other side.

The greater the molar mass of a gas, the slower it will diffuse or effuse.

Example 12-12: An unknown gas effuses through a barrier 5.6 times more slowly than does hydrogen. What is the molar mass of the unknown gas?

This problem involves effusion. Graham's law for effusion states that

$$\frac{\text{Rate of effusion of gas 1}}{\text{Rate of effusion of gas 2}} = \sqrt{\frac{\text{molar mass of gas 2}}{\text{molar mass of gas 1}}}$$

$$\frac{\text{Rate of effusion of } H_2}{\text{Rate of effusion of unknown gas}} = \sqrt{\frac{\text{molar mass of unknown gas}}{\text{molar mass of } H_2}}$$

In this case, we are told that the unknown gas effuses 5.6 times more slowly does hydrogen. Thus, the rate of effusion of hydrogen over that of the unknown gas is 5.6. The molar mass of hydrogen is 2.02 g/mole.

$$5.6 = \sqrt{\frac{\text{molar mass of unknown gas}}{2.02}}$$

$$5.6 = \frac{\sqrt{\text{molar mass of unknown gas}}}{1.42}$$

$$8.0 = \sqrt{\text{molar mass of unknown gas}}$$

$$(8.0)^2 = \text{molar mass of unknown gas}$$

$$\text{molar mass of unknown gas} = 64 \text{ g/mole}$$

Try Study Questions 47 and 49 in Chapter 12 of your textbook now!

- **Recognize why real-world gases do not behave like ideal gases.**

    Appreciate the fact that gases usually do not behave as ideal gases (Section 12.9). Deviations from ideal behavior are largest at high pressure and low temperature.

    The ideal gas law works extremely well for a large number of gases under ordinary conditions around normal atmospheric pressure and room temperature. When we go to more extreme conditions of higher pressure or lower temperature, significant deviations from ideality can occur. This is caused by a breakdown of some of our assumptions. Our ideal gas model assumes that the particles act entirely independently of each other. In other words, we are assuming that there are no intermolecular forces between the particles. We also assume that the particles have the entire volume to move around in, that we do not have to subtract out any volume for the space taken up by the physical particles themselves. Neither assumption is true. The particles do have intermolecular forces, and they do take up some space themselves. Under normal conditions, however, ignoring these factors does not introduce very much error. When we go to higher pressures and/or lower temperatures, these factors become significant, and we must correct for them.

    Johannes van der Waals tried one method to correct for these effects. He introduced two correction factors into the ideal gas law, one to correct for intermolecular force effects (a) and one to correct for the molecular volume effect (b).

$$\left(P + a \left[\frac{n}{V} \right]^2 \right) (V - bn) = nRT$$

The van der Waals equation has the same basic form as the ideal gas law in that we have a pressure term times a volume term being equal to nRT. The van der Waals equation, however, is more complicated than the ideal gas law both in the way that it looks and in the fact that each gas will have different values for a and b.

Students sometimes get the mistaken view that the ideal gas law is fake, and the van der Waals equation is true. The ideal gas law works remarkably well under normal conditions. The van der Waals equation is another model, which works better under some circumstances, but it also runs into difficulty in yet other circumstances.

<u>Example 12-13:</u> Compare the pressure predicted for CO_2 gas by the ideal gas law and by the van der Waals equation when 1.00 mole of gas is held in a container with a volume of 2.00 L at a temperature is 298 K.

First let's solve the ideal gas law problem.

$$PV = nRT$$

$$P = \frac{nRT}{V}$$

$$P = \frac{(1.00\ \text{mole})\left(0.082057\ \dfrac{\text{L atm}}{\text{mole K}}\right)(298\ \text{K})}{2.00\ \text{L}}$$

$$= 12.2\ \text{atm}$$

Now, let's use the van der Waals equation. From Table 12.2 in your text, we find that the value of a is 3.59 atm L^2/mol^2 and the value of b is 0.0427 L/mol for CO_2.

$$\left(P + a\left[\frac{n}{V}\right]^2\right)(V - bn) = nRT$$

$$P + a\left[\frac{n}{V}\right]^2 = \frac{nRT}{V - bn}$$

$$P = \frac{nRT}{V - bn} - a\left[\frac{n}{V}\right]^2$$

$$= \frac{(1.00\ \text{mole})\left(0.082057\ \dfrac{\text{L atm}}{\text{mole K}}\right)(298\ \text{K})}{2.00\ \text{L} - \left(0.0427\ \dfrac{\text{L}}{\text{mole}}\right)(1.00\ \text{mole})} - \left(3.59\ \dfrac{\text{atm L}^2}{\text{mole}^2}\right)\left[\dfrac{1.00\ \text{mole}}{2.00\ \text{L}}\right]^2$$

$$= 11.6\ \text{atm}$$

Try Study Question 51 in Chapter 12 of your textbook now!

Other Notes

1. Standard temperature and pressure (STP) for gases is defined to be 0°C and 1 atm.

2. At STP, 1 mole of an ideal gas occupies 22.4 L. This quantity is called the standard molar volume of a gas.

3. The density of a gas may be obtained by using the equation:

$$d = \frac{PM}{RT}$$

where M is the molar mass of the gas.

CHAPTER 13: Intermolecular Forces, Liquids, and Solids

Chapter Overview

In this chapter, we examine the liquid and solid states. Unlike gases, we must consider intermolecular forces when we study these. Ionic compounds engage in ion-ion interactions. Polar covalent compounds engage in dipole-dipole interactions. Compounds containing a H atom bonded to a F, O, or N atom can engage in a strong dipole-dipole interaction called a hydrogen bond. Nonpolar covalent compounds engage in induced dipole/induced dipole interactions. A mixture of an ionic compound and a polar compound involves ion-dipole interactions, and a mixture of a polar compound and a nonpolar compound involves dipole/induced dipole interactions. In general, for species of similar molar mass, the strength of the various forces goes in the order: ion-ion > ion-dipole > dipole-dipole > dipole/induced dipole > induced dipole/induced dipole. Many of water's properties, such as the fact that is a liquid at room temperature and that its solid form is less dense than its liquid form, are a result of its extensive hydrogen bonding.

You will learn about the different types of phase changes including which are exothermic and which are endothermic. We will discuss liquid-vapor equilibria in some detail. If a liquid is placed in a sealed container, a dynamic equilibrium will be set up between the liquid and gas states. The pressure that the vapor of the liquid exerts when equilibrium is established is called the equilibrium vapor pressure. The boiling point is the temperature at which the vapor pressure of a liquid is equal to the external pressure. The stronger the intermolecular forces are, the higher the boiling point will be. The vapor pressure, molar enthalpy of vaporization, and absolute temperature are related by the Clausius-Clapeyron equation. The critical temperature for a substance is the temperature beyond which we cannot obtain a liquid, no matter what pressure is exerted. The surface tension is the energy required to break through the surface of a liquid. It is dependent, in part, upon the strength of the intermolecular forces in the liquid. The viscosity of a liquid is a measure of the resistance of a liquid to flow; it too is dependent, in part, upon the strength of the intermolecular forces. Capillary action is the movement of a liquid up a tube. Both capillary action and the shape of a liquid's meniscus are dependent upon the adhesive forces between the liquid and the container and also on the cohesive forces between the molecules of the liquid itself.

Most solids have a regular structure, a crystal lattice. The smallest repeating unit in a crystal lattice is called the unit cell. We will focus on the cubic unit cells: simple cubic (sc), body-centered cubic (bcc), and face-centered cubic (fcc). A sc unit cell contains one atom per unit cell, a bcc cell contains 2 atoms, and a fcc cell contains 4 atoms. The lattices of many ionic compounds are built by placing one of the ions at the lattice points in a unit cell and placing the other ions at holes in the lattice. The empirical formula for an ionic compound can be determined from its unit cell. Network solids, such as diamond, graphite, and silicates, are composed of a three-dimensional array of covalently bonded atoms. Amorphous solids, such as glass and many polymers, do not have any long-range regularity.

A phase diagram for a material shows the state of matter present at combinations of pressure and temperature. Lines on a phase diagram represent combinations of pressure and temperature where two states are in equilibrium and phase changes occur. At the triple point, the three states of matter are in equilibrium. You should be able to use a phase diagram and identify the triple point and critical point.

Key Terms

In this chapter, you will need to learn and be able to use the following terms:

Adhesive forces: forces of attraction between two different materials.

Amorphous solids: solids with no long-range regularity.

Boiling point: the temperature at which the equilibrium vapor pressure of a liquid is equal to the external pressure.

Capillary action: the spontaneous movement of a liquid up a tube.

Cohesive forces: forces of attraction between molecules of the same material.

Condensation: the conversion of a material from the gaseous state into the liquid state.

Critical point: the point at which the vapor pressure *vs.* temperature curve for a material ends.

Critical pressure: the pressure at the critical point.

Critical temperature: the temperature above which a material cannot be liquefied, no matter how much pressure is applied.

Crystal lattice: the solid regular array of atoms, ions, or molecules in a crystalline solid.

Dipole/induced dipole attraction: forces of attraction between a polar molecule and a nonpolar molecule; this force is generated when the presence of a polar molecule causes the electrons in the nonpolar molecule to shift, setting up a temporary dipole on the nonpolar molecule.

Dipole-dipole attraction: forces of attraction between two polar molecules; the positive part of one will line up with the negative part of the other.

Dynamic equilibrium: a situation in which two reverse processes are occurring at the same rate.

Energy of hydration: the solvation energy when the solvent is water.

Equilibrium vapor pressure: the pressure of the vapor of a substance in contact with its liquid (or solid) phase in a sealed container.

Freezing point: the temperature at which a liquid freezes

Freezing: the process of converting a liquid into a solid.

Hydrogen bond: a particularly strong type of dipole-dipole interaction; it arises between a small very electronegative atom (usually F, O, or N) and a hydrogen that is covalently bonded to another small very electronegative atom (usually F, O, or N).

Induced dipole/induced dipole attraction (London dispersion forces): temporary correlations of electron movements on different molecules that result in temporary dipoles being established and attracting the molecules to one another; while all molecules possess this type of intermolecular force, this is the only type of intermolecular force possible between two nonpolar molecules.

Intermolecular forces: interactions between molecules, ions, or molecules and ions.

Ion-dipole attraction: forces of attraction between an ion and a polar molecule; the ion and the portion of a polar molecule that has the opposite charge will line up and be attracted to one another.

Ionic solids: solids made up of positive and negative ions.

Lattice points: the corners of a unit cell in a crystal lattice.

Melting point: the temperature at which a solid melts.

Melting: the process of converting a solid into a liquid.

Molecular solids: solids in which molecules are the fundamental repeating unit.

Network solids: solids in which the atoms are held together in infinite two- or three-dimensional networks by means of covalent bonds.

Normal boiling point: the boiling point under an external pressure of 760 mm Hg.

Octahedral hole: in an ionic solid, an ion is in an octahedral hole if it is surrounded by six ions of the other element arranged in an octahedral geometry.

Phase diagram: a graph showing the state of matter present for a substance under different combinations of temperature and pressure.

Polarizability: the ease with which the electron cloud of an atom or a molecule can be distorted.

Polarization: the process of inducing a dipole.

Solvation energy: the energy associated with surrounding a material with molecules of a solvent.

Standard molar enthalpy of crystallization (ΔH^o_{cryst}): the enthalpy change accompanying the transformation of 1 mole of a liquid into a solid under standard conditions.

Standard molar enthalpy of fusion (ΔH^o_{fusion})**:** the enthalpy change accompanying the transformation of 1 mole of a solid into a liquid under standard conditions.

Standard molar enthalpy of sublimation ($\Delta H^o_{sublimation}$)**:** the enthalpy change accompanying the transformation of 1 mole of a solid directly into the gaseous state under standard conditions.

Standard molar enthalpy of vaporization (ΔH^o_{vap})**:** the enthalpy change accompanying the conversion of 1 mole of a liquid into the gaseous state under standard conditions.

Sublimation: the process of converting a solid into a gas without passing through the liquid state.

Surface tension: the energy required to break through the surface of a liquid or to disrupt a drop of the liquid and spread the material out as a film.

Tetrahedral hole: in an ionic solid, an ion is in a tetrahedral hole if it is surrounded by four ions of the other element arranged in a tetrahedral geometry.

Triple point: the one combination of temperature and pressure where the solid, liquid, and gaseous phases of a material are all in equilibrium with each other.

Unit cell: the smallest repeating unit of a crystal lattice.

Vaporization (evaporation): the conversion of a material from the liquid state into the gaseous state.

Viscosity: a measure of the resistance of liquids to flow; more viscous materials do not flow easily.

Volatility: the tendency of the molecules of a liquid to escape from the liquid phase and enter the vapor phase.

Chapter Goals

By the end of this chapter you should be able to:

•   **Describe intermolecular forces and their effects.**

   a) Describe the various intermolecular forces found in liquids and solids (Sections 13.2 and 13.4).

   In the last chapter, we studied gases. Our study was made easier because we could assume in most cases that the particles were moving with sufficient kinetic energy that we could ignore any forces of attraction or repulsion between the particles. With liquids and solids, however, we cannot make this assumption. The particles are held in the liquid and solid states due to the intermolecular forces between the particles. The kinetic energy of the particles is not sufficient to overcome these forces completely. We begin this chapter by examining these forces. Intermolecular forces are interactions between particles: between molecules, between ions, or between molecules and ions.

The strength of the intermolecular forces between the particles of a material have many consequences on the properties of the materials. When the particles are held together by stronger forces, it will take more energy to disrupt the forces. Materials with stronger intermolecular forces have higher melting points and boiling points than those with weaker intermolecular forces. Intermolecular forces also play a role in determining whether a material will dissolve in a particular solvent.

All intermolecular forces have their origin in electrostatic forces of attraction and repulsion.

Ion-Ion Forces. The strongest forces of attraction between two particles are between a positively charged ion and a negatively charged ion as in an ionic compound. These forces are in the range of 700 – 1100 kJ/mole.

Ion-Dipole Forces. If the interaction between two ions is the greatest interparticle force, next in line would be one where we replace one of the ions by a polar molecule, a molecule with a region of partial positive charge and a region of partial negative charge. This type of interaction is called an ion-dipole interaction. We run across these when we mix an ionic compound and a polar covalent compound; the most common example is in the solution of an ionic material in water. When an ionic compound dissolves in a polar solvent, the ions get surrounded by molecules of the polar compound; the ions are said to be solvated (when water is the solvent, we say that the ions are hydrated). We get forces of attraction between the positive ions and the negative ends of the polar compound and also between the negative ions and the positive ends of the polar compound. Associated with this process is the enthalpy of solvation (hydration). In general, this becomes more exothermic as the ionic radius gets smaller and as the ion charge increases.

Example 13-1: In each pair, indicate which compound will have the greater enthalpy (heat) of hydration.

a. $Mg(NO_3)_2$ or $Ca(NO_3)_2$

In this case, the nitrate ions are the same, so the difference in the energy of hydration will come from the metal ions. The magnesium and calcium ions both have a 2+ charge, so the difference will come from the size of the ions. The smaller the ion, the greater the energy will be. Magnesium is further up Group 2A on the periodic table, so it is the smaller ion. $Mg(NO_3)_2$ will have the greater energy of hydration.

b. $Mg(NO_3)_2$ or $Al(NO_3)_3$

In this case, magnesium and aluminum are in the same row of the periodic table. $Al(NO_3)_3$ will have the greater energy of hydration because the aluminum ion has both the smaller size and also the larger charge.

Try Study Question 9 in Chapter 13 of your textbook now!

Dipole-Dipole Forces. We started our discussion of interparticle forces with ion-ion interactions. We then replaced one of the ions with a polar molecule and dealt with ion-dipole interactions. Continuing in this way, we now replace the other ion with a polar molecule and consider the interactions between two polar molecules. These are dipole-dipole interactions. The positive side of one molecule lines up with the negative side of another.

(There is a particularly strong type of dipole-dipole interaction called hydrogen bonding that we shall consider a little later). Polar compounds tend to have higher enthalpies of vaporization and higher boiling points than nonpolar compounds of a similar size. This is because molecules of polar compounds are attracted to each other by dipole-dipole interactions, which are stronger than the intermolecular forces that hold together molecules of nonpolar compounds.

Intermolecular forces also affect solubility. Polar molecules are likely to dissolve in a polar solvent, and nonpolar molecules are likely to dissolve in a nonpolar solvent. It is unlikely that polar molecules will dissolve in nonpolar solvents or that nonpolar molecules will dissolve in polar solvents. These statements are sometimes summarized as "like dissolves like." Notice that we said "likely" and "unlikely." These tendencies have some notable exceptions. Perhaps the statement with the greatest number of exceptions is that of nonpolar solutes not dissolving in polar solvents. There are a number of nonpolar materials that have a significant solubility in water, for instance.

Example 13-2: Which should have the higher boiling point: butane ($CH_3CH_2CH_2CH_3$) or acetone (CH_3COCH_3)?

These two compounds have the same molar masses (58 g/mole). Butane is a hydrocarbon; hydrocarbons are nonpolar. Acetone is a polar molecule; its molecular shape is shown in the following figure.

Because they are polar, acetone molecules can interact with each other by means of dipole-dipole interactions. The nonpolar butane molecules will not have dipole-dipole interactions. Because of the stronger intermolecular forces present in acetone, acetone should have the higher boiling point.

Example 13-3: Which should be more soluble in water: butane or acetone?

Water is a polar solvent. Like dissolves like. The polar acetone should be more soluble in the polar water. The nonpolar butane should not be as soluble in the polar water.

Dipole/Induced Dipole Forces. When we mix together a nonpolar compound and a polar compound, there can be some attraction between their molecules. These forces are called dipole/induced dipole interactions. The polar molecule has a permanent dipole; it is the "dipole" part of this force. The nonpolar molecule contributes the "induced dipole" part. What is an induced dipole and how does it form?

The electron cloud is symmetrically distributed in a nonpolar molecule; that's what makes it nonpolar. When we bring a polar molecule close to a nonpolar molecule, something happens. Let us picture the negative portion of the polar molecule approaching the nonpolar molecule as in Figure 13.6a in your textbook. The electrons on the nonpolar molecule will respond to this approaching negative portion of the polar molecule. Because they are also negative, they will be repelled (like charges repel) and spend more time on the other side of the molecule.

There will thus be a greater electron density on one side of the originally nonpolar molecule than on the other. It will have become slightly polar. Of course, once the polar molecule is removed, the electrons will go back to the way they were. We call the temporary dipole an induced dipole because the presence of the polar molecule caused it to occur.

The process of inducing a dipole is called polarization. It is easier to cause some molecules to get an induced dipole than other. In other words, different molecules have different polarizabilities. In general, the larger the molecule, the more polarizable because the electron cloud extends further out from the nuclei. We can get a feel for the polarizability of similar molecules by looking at their molar masses. In general, the higher the molar mass, the greater the polarizability of the molecule will be. For example, I_2 is more polarizable than Cl_2.

Induced dipole/induced dipole forces. Imagine a nonpolar molecule. Over time, the electron cloud is symmetrically distributed over the whole molecule. At a given instant in time, there may be a greater electron density on one side than on the other. For this moment of time, there is a temporary dipole on the molecule. Now imagine that there is another nonpolar molecule nearby. The temporary dipole on one molecule will induce a temporary dipole on the other molecule. There will be an attraction between the two molecules. These momentary attractions and repulsions between the molecules are called induced dipole/induced dipole forces. Sometimes, they are also called London dispersion forces. Each individual induced dipole/induced dipolar force is not that large, but over the length of a large molecule, they can add up to become very significant. This type of correlation of electron motions occurs in all molecules. With nonpolar molecules, however, it is the only kind of force available.

In general, the strength of intermolecular forces goes in the order that we have studied them:
ion-ion > ion-dipole > dipole-dipole > dipole/induced dipole > induced dipole/induced dipole

b) Tell when two molecules can interact through a dipole-dipole attraction and when hydrogen bonding may occur. The latter occurs most strongly when H is attached to O, N, or F (Sections 13.2 and 13.3).

A dipole-dipole interaction occurs when we have polar molecules interacting with other polar molecules. The molecules of a polar compound interact with each other by means of dipole-dipole forces. In some cases, a polar molecule is large enough that it can bend back on itself, and set up dipole-dipole interactions between two polar regions on the same molecule.

There is a particularly strong type of dipole-dipole interaction called a hydrogen bond. The word "bond" can be a little misleading. A hydrogen bond is not a covalent bond but is simply a very strong dipole-dipole intermolecular force. In order to get a hydrogen bond, we must have a molecule that has within it a hydrogen atom that is covalently bonded to a small, very electronegative atom (usually F, O, or N); the hydrogen will be partially positive. This hydrogen atom then interacts by means of the dipole-dipole interaction (the hydrogen bond) with the lone pair on another small, very electronegative atom (usually F, O, or N).

X and Y are both small very electronegative
atoms (usually F, O, or N)

c) Identify instances in which molecules interact by induced dipoles (dispersion forces) (Section 13.2).

All molecules interact by means of induced dipole/induced dipole forces. For nonpolar compounds, this is the only type of interaction available between their molecules.

Example 13-4: Determine the type of intermolecular forces that must be overcome in converting each of the following from a liquid to a gas.

Because dispersion forces will be present for all molecules, we could list these for each of the following cases. Usually, we will just list them if this is the only type of force available. The key to solving this problem is to determine whether the materials are polar or nonpolar. If they are polar, we must also determine if hydrogen bonding is present.

a. ethanol (CH_3CH_2OH)
The Lewis structure for ethanol is

We can already tell that this will be a polar molecule, so there will be dipole-dipole forces. In addition, we have a hydrogen atom connected to an oxygen, so these will be hydrogen bonds.

b. acetone (CH_3COCH_3)
The Lewis structure for acetone is

We need to determine the molecular shape to tell if this will be polar or not. The arrangement around the two end carbons is tetrahedral; that for the central carbon is trigonal planar.

This is a polar molecule. The molecules will interact by means of dipole-dipole forces. There is no hydrogen attached to an N, O, or F. There will not be hydrogen bonding.

c. propane ($CH_3CH_2CH_3$)
This is a hydrocarbon. It will be nonpolar. The molecules can only interact by means of induced dipole/induced dipole forces.

Try Study Questions 3 and 7 in Chapter 13 of your textbook now!

- ## Understand the importance of hydrogen bonding.

  ### Explain how hydrogen bonding affects the properties of water (Section 13.3).

  Hydrogen bonding is very important in many compounds. Hydrogen bonding between the different nitrogenous bases in DNA is the key that allows the genetic code to work. Hydrogen bonding is also responsible for helping to determine the structures of proteins.

  In addition, hydrogen bonding helps to determine the properties of one of the most abundant compounds on Earth, water. Water possesses some properties that are different from those of similar compounds, largely due to the vast hydrogen bonding network that exists in water.

  A water molecule can form four hydrogen bonds: each of the hydrogen atoms and each of the lone pairs on the oxygen can be involved in a hydrogen bond. In ice, we get all four of these hydrogen bonds occurring, locking the atoms into place. In liquid water, there are still around four hydrogen bonds per molecule, but the molecules can move around so a given water molecule is constantly changing the other water molecules to which it is hydrogen bonded.

  Water boils at a high temperature for a molecule with its molar mass. At room temperature, some molecules with larger molar masses such as N_2, O_2, CO_2, and even other molecules with similar formulas containing elements from group VIA (H_2S, H_2Se, and H_2Te) are already gases; they have already boiled. Yet water, with its puny molar mass of only 18 g/mole is still a liquid and won't boil until we get to 100°C under normal atmospheric pressure. The higher boiling point indicates that water has stronger intermolecular forces between its molecules than these other materials.

  Another unusual property of water is that its solid form is less dense than its liquid form; solid ice floats in liquid water. This is the opposite from what is true for most materials. The reason why this is true again has to do with water's ability to hydrogen bond. In order to get those four hydrogen bonds forming at the optimal angles, the water molecules in the solid arrange themselves in hexagonal rings. (These six-sided rings are also the underlying reason why snowflakes have six sides.) There is open space in the middle of each ring. When the ice melts, some of the hydrogen bonds break, and the water molecules can approach each other more closely, moving into this space. In the liquid, we get the same amount of matter packed into a smaller space, so the liquid is more dense than the solid. Water actually has its maximum density at 4°C.

  Water has a larger than usual heat capacity, as well as large heats of fusion and vaporization. This is because of the hydrogen bonds holding the water molecules together.

- ## Understand the properties of liquids.

  ### a) Explain the processes of evaporation and condensation, and use the enthalpy of vaporization in calculations (Section 13.5).

  Evaporation (vaporization) is the process in which a substance in the liquid state becomes a gas. Molecules escape from the liquid surface and enter the gaseous state. In order for a molecule to do this, the molecule must possess sufficient kinetic energy to overcome the intermolecular forces to the other molecules of the liquid, be at the surface, and be moving in the proper direction. Vaporization is an endothermic process because energy must be added to each molecule to overcome the intermolecular forces so that the molecule can escape into the gas phase. The enthalpy change for evaporating one mole of a liquid is called the standard molar enthalpy of vaporization, ΔH°_{vap}. (In Chapter 6, we dealt with the enthalpy of vaporization, but on a per gram basis.)

$$\text{Liquid} \xrightarrow{\text{vaporization}} \text{Vapor} \quad \Delta H = \Delta H^{\circ}_{vap}$$

Condensation is the reverse of vaporization. In this case, a molecule goes from being in the gaseous state to being in the liquid state. Because condensation and vaporization are reverse processes, ΔH°_{cond} and ΔH°_{vap} will have equal magnitudes but opposite signs.

$$\text{Vapor} \xrightarrow{\text{condensation}} \text{Liquid} \quad \Delta H = \Delta H^{\circ}_{cond} = -\Delta H^{\circ}_{vap}$$

Your body takes advantage of the fact that vaporization is an endothermic process. The reason that you sweat is so that the energy from your body can be transferred to the water. This energy is used in the endothermic vaporization of the water. The water molecules then carry this energy away from your body as they evaporate and leave the surface of your body.

<u>Example 13-5:</u> How much heat is required to evaporate 35.0 g of liquid bromine? ΔH°_{vap} of bromine is 30.0 kJ/mole.

We are given the mass of Br_2 and the molar enthalpy of vaporization of Br_2. We are asked for the amount of heat required to vaporize this mass of bromine. The enthalpy of vaporization is on a per mole of bromine basis, so in order to use it, we must determine how many moles of bromine we have. We are given the mass of Br_2; we can use the molar mass to get to moles.

$$35.0 \text{ g } Br_2 \times \frac{1 \text{ mole } Br_2}{159.8 \text{ g } Br_2} = 0.219 \text{ moles } Br_2$$

We can then use the molar enthalpy of vaporization to go from moles to kJ.

$$0.219 \text{ moles } Br_2 \times \frac{30.0 \text{ kJ}}{1 \text{ mole } Br_2} = 6.57 \text{ kJ}$$

Try Study Question 11 in Chapter 13 of your textbook now!

b) Define the equilibrium vapor pressure of a liquid, and explain the relationship between the vapor pressure and boiling point of a liquid (Section 13.5).

If you put some water in an open container, it will eventually evaporate completely. If, however, you put the water into a tightly sealed container, at first, some of it will evaporate. After that, it looks like the evaporation stops and that nothing more happens. It turns out that quite a lot is happening. Water is still evaporating. Why then does it look like nothing is happening? Water is also condensing. The two processes are occurring at the same rate.

Rate of evaporation = Rate of condensation

This type of situation is called a dynamic equilibrium. A dynamic equilibrium occurs when two reverse processes occur at the same rate. We indicate the presence of a dynamic equilibrium by using two arrows, one pointing to the left and one pointing to the right:

Liquid <=> Vapor

Once the liquid-vapor equilibrium has been established, there will be a certain pressure exerted by the water molecules in the vapor phase in the space above the liquid. This pressure is called the equilibrium vapor pressure (or just the vapor pressure). Every liquid will have its own value for the equilibrium vapor pressure at a given temperature. Your textbook shows a graph of the equilibrium vapor pressure for three different substances *vs.* temperature in Figure 13.18. One thing to notice is that the vapor pressure goes up with increasing temperature. This is because at higher temperatures more molecules will have sufficient kinetic energy to go into the gas phase.

Volatility is the tendency for the molecules of a liquid to go into the gas phase. At a given temperature, a greater fraction of the molecules of a more volatile material will be in the gas phase than those of a less volatile material. Volatility is related to intermolecular forces. For molecules of similar molar mass, the relationships in the following table hold:

If the molecules of a liquid have ...	Ease to convert from liquid to gas	Volatility	Equilibrium vapor pressure	Boiling Point (b.p.)
Stronger intermolecular forces	Harder	Less volatile	Lower vapor pressure	Higher b.p.
Weaker intermolecular forces	Easier	More volatile	Higher vapor pressure	Lower b.p.

The table above mentions boiling point. The boiling point of a liquid is the temperature at which its vapor pressure is equal to the external pressure. (If the external pressure is 760 mm Hg, the boiling point is referred to as the normal boiling point.) At a temperature below the boiling point, a bubble of the vapor that forms inside the liquid will get crushed and not make it to the surface. Once the boiling point is reached, the vapor pressure inside the bubble is equal to the external pressure so a bubble can form and not be crushed but instead make it to the surface to release the vapor into the air.

If the external pressure is increased, the liquid will boil at a higher temperature. If the external pressure is reduced, the liquid will boil at a lower temperature. At high altitudes, the atmospheric pressure is lower. High altitude cooking directions therefore call for boiling food longer because the water is boiling at a lower temperature. It will take longer to cook at the lower temperature.

Example 13-6: Using Figure 13.18 in the textbook,

    a. Determine the normal boiling point of diethyl ether.
The boiling point is the temperature at which the vapor pressure equals the external pressure. The normal boiling point is the boiling point at 760 mm Hg. We simply go to a pressure of 760 mm Hg, find the corresponding point on the curve, and read the temperature off the temperature axis. We could estimate this value to be 35°C for diethyl ether. (The text gives it more precisely as 34.6°C.)

    b. At what pressure could we get water to boil at a temperature of 90°C?
In this case, we find the temperature of 90°C on the temperature axis, go up to the curve, and read the pressure off the pressure axis. We can estimate this to be about 500 mm Hg.

Try Study Questions 17 and 19 in Chapter 13 of your textbook now!

c) Describe the phenomena of the critical temperature, T_C, and critical pressure, P_C, of a substance (Section 13.5).

Under normal circumstances, if we apply enough pressure to a gas, the gas will eventually liquefy. Every material, however, has a temperature beyond which it cannot be liquefied, no matter how much pressure is applied; this will occur at different temperatures for different materials. The vapor pressure *vs.* temperature curve ends at this temperature. The combination of temperature and pressure at this point is called the critical point. The value of the pressure at this point is called the critical pressure, P_C.

The material that exists at temperatures and pressures above TC and PC is called a supercritical fluid. The density of a supercritical fluid is like that of a liquid because the high pressure pushes the molecules together, but the kinetic energy of the molecules is high enough that they can overcome the intermolecular forces and not "stick together" to form a liquid. A supercritical fluid thus behaves like a gas but has a density like that of a liquid.

Example 13-7: The critical point of water is $T_C = 374.0°C$ and $P_C = 217.7$ atm. Can water be liquefied at a temperature of 500°C?

The temperature of 500°C is higher than the critical temperature, so water cannot be liquefied at this temperature.

Try Study Question 37 in Chapter 13 of your textbook now!

d) Describe how intermolecular interactions affect the cohesive forces between identical liquid molecules, the energy necessary to break through the surface of a liquid (surface tension), and the resistance to flow, or viscosity, of liquids (Section 13.5).

A molecule in the middle of a liquid will feel attractions to other molecules in all directions. A molecule at the surface of a liquid, however, will not feel intermolecular forces above it.

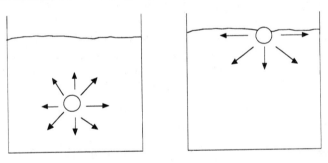

The effect of this is that there is a net inward force of attraction on the surface molecules. This pulls them closer together contracting the surface and making the liquid behave as though it had a skin. The surface tension is a measure of the toughness of this skin. It is the energy required to break through the surface or to disrupt a liquid drop and spread the material out as a film. The surface tension is, in part, dependent on the strength of the intermolecular forces. The greater the intermolecular forces are, the greater the surface tension is.

Viscosity is a measure of the resistance of a liquid to flow. Molasses is more viscous than water. It might not surprise you at this point to learn that intermolecular forces also play a role in determining the viscosity of a material. The greater the intermolecular forces are, the more viscous a material will be. In considering this, we must consider all of the intermolecular forces over the whole molecule. Longer chains have greater intermolecular forces than smaller chains due to increased London dispersion forces. Long chains can also get tangled with each other thus causing an increase in viscosity.

When a thin tube of glass is placed in water, the water climbs up into the tube. You may have seen this if you have had a blood test in which the doctor pricked your finger and then placed a capillary tube on the blood. The blood climbed up into the tube. This climbing up in the tube is called capillary action. There are forces of attraction between the water and the glass. These forces of attraction between two different materials are called adhesive forces. Forces between molecules of the same material are called cohesive forces. In the water-glass example, what occurs is that the adhesive forces between the water and the glass are strong enough that some water molecules move up the surface of the glass a little. The surface tension of the water keeps the surface of the water intact, so the whole water surface gets pulled up, lifting the water level in the tube. This continues until the attractive forces pulling the water up are balanced by the force of gravity pulling down on the water column.

Adhesive and cohesive forces also determine if a liquid will have a surface that curves up or down in a container. This curved surface is called a meniscus. If the adhesive forces between the liquid and the container are greater than the cohesive forces between the molecules of the liquid, the meniscus will have a downward curve. This downward curve is caused by the fact that some of the liquid moves up the surface of the container. This is the case for water in a glass container. If, however, the cohesive forces between the particles of the liquid are greater than the adhesive forces between the liquid and the container, the meniscus will curve upward. This is the case for mercury in glass.

exaggerated view of   exaggerated view of
water in a glass tube   mercury in a glass tube

Example 13-8: Comparing acetone (CH_3COCH_3) and 1-propanol ($CH_3CH_2CH_2OH$),

　　　　a. Which would be expected to have the greater surface tension?
These molecules have similar molar masses so their London dispersion forces should be similar. As discussed earlier, acetone is polar. Its molecules have dipole-dipole interactions. Because there is no hydrogen bonded to F, O, or N, acetone cannot hydrogen bond to itself. 1-propanol is polar, because of the –OH group attached to the hydrocarbon chain. Its molecules can engage in dipole-dipole interactions. In addition, they can hydrogen bond to each other because of the –OH. The intermolecular forces between the molecules of 1-propanol are greater than those of acetone. Because 1-propanol has greater intermolecular forces, it should have the higher surface tension.

　　　　b. Which would be expected to have the greater viscosity?
These molecules have similar molar masses. Because it has greater intermolecular forces, 1-propanol is expected to have the greater viscosity.

Try Study Question 67 in Chapter 13 of your textbook now!

e) Use the Clausius-Clapeyron equation, which connects temperature, vapor pressure, and enthalpy of vaporization for liquids (Section 13.5).

The Clausius-Clapeyron equation relates the vapor pressure of a liquid to the absolute temperature and the standard molar enthalpy of vaporization of the liquid. The equation is

$$\ln P = -\left(\frac{\Delta H^{o}_{vap}}{RT}\right) + C$$

This is an equation of a line, y = mx + b. In this case, $y = \ln P$, $x = 1/T$, $m = -\Delta H^{o}_{vap}/R$, and b = C. This indicates that if we plot ln P on the y-axis of a graph and 1/T on the x-axis of a graph, then the slope of the graph will be equal to $-\Delta H^{o}_{vap}/R$. To obtain the value of ΔH^{o}_{vap}, we change the sign of the slope and multiply it by R. Because we are dealing with energies in units of joules, we need to use 8.3145 J/(mole K) as the value for R. Keep in mind that the x-axis is not T but 1/T and also keep in mind that you must convert the temperatures to kelvin before taking their reciprocals.

Try Study Question 21 in Chapter 13 of your textbook now!

If you are only given two values of P and T, there is a two point form of the Clausius-Clapeyron equation that you can use:

$$\ln P_2 - \ln P_1 = \frac{\Delta H^{o}_{vap}}{R}\left[\frac{1}{T_1} - \frac{1}{T_2}\right]$$

Notice that the 1 and 2 switch places on the left and right sides of the equation. This is due to the minus sign that was present on the right side of the original equation. The results from this two point equation are, of course, not as good as from using multiple points and the graphical method.

• Understand cubic unit cells.

a) Describe the three types of cubic unit cells: primitive or simple cubic (sc), body-centered cubic (bcc), and face-centered cubic (fcc) (Section 13.6).

Most solids have a regular repeating pattern of atoms, ions, or molecules within their structure. This solid regular repeating pattern is called the crystal lattice. The smallest repeating unit of the crystal lattice is called a unit cell. There are several geometric shapes on which unit cells are based; we shall focus on those based on a cubic structure.

There are three types of cubic unit cells: simple cubic, body-centered cubic, and face-centered cubic. In the simple cubic unit cell, there is an atom at each of the eight corners of the cube. A body-centered unit cell has eight corner atoms and one atom in the middle of the cube. A face-centered unit cell has the eight corner atoms and one atom in the middle of each of the six sides (faces) of the cube. The face-centered cubic unit cell is sometimes also referred to as cubic close-packing. See Figure 13.27 for drawings of these unit cells.

How much of an atom belongs to a particular unit cell? An atom at the corner of a unit cell will not only be at the corner of the unit cell under consideration but will also be at the corner of other unit cells that adjoin that unit cell. Only a portion of the atom belongs fully to the unit cell under consideration. A corner atom in a cubic unit cell is shared by eight unit cells, thus only 1/8 of the atom belongs to a single unit cell. The fractions for different locations of atoms in a unit cell are shown in the following table:

Location of Atom in Unit Cell	# of Unit Cells to Which This Atom Belongs	Fraction of the Atom Belonging to One Unit Cell
Corner	8	1/8
Body	1	1
Face	2	1/2
Edge	4	1/4

Example 13-9: Determine how many atoms belong to each of the cubic unit cells.

Simple Cubic:
There are eight corner atoms in this type of cell. The factor for a corner atom is 1/8.

$$8 \text{ corner atoms x } \frac{1}{8} = 1 \text{ atom}$$

A simple cubic unit cell thus contains the equivalent of only 1 atom.

Body-Centered Cubic:
There are eight corner atoms and one atom in the body.

$$8 \text{ corner atoms x } \frac{1}{8} = 1 \text{ atom}$$

$$1 \text{ body atom x } 1 = 1 \text{ atom}$$

There are two atoms in a body-centered cubic unit cell.

Face-Centered Cubic:
There are eight corner atoms and six atoms in the faces.

$$8 \text{ corner atoms x } \frac{1}{8} = 1 \text{ atom}$$

$$6 \text{ face atoms x } \frac{1}{2} = 3 \text{ atoms}$$

There are 4 atoms in a face-centered cubic unit cell.

Type of Unit Cell	Number of Atoms Contained in a Single Unit Cell
Simple Cubic	1
Body-Centered Cubic	2
Face-Centered Cubic	4

b) Relate atom size and unit cell dimensions.

This objective is best illustrated by working some examples.

Example 13-10: The metal polonium has a simple cubic crystal structure. If one edge of the cube has a length of 334 pm, what is the radius of an atom of polonium?

The first step in this problem is to draw a picture of the unit cell. For the simple cubic unit cell, we can draw a cross-section of one side.

The next step is to decide what the edge corresponds to in this figure. We do not measure from the outer border of each atom to the outer border of the next. Instead, we go from the nucleus of one atom to the nucleus of the next.

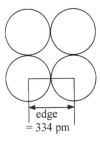

From this figure, we can see that for a simple cubic unit cell the edge corresponds to 2 atomic radii. The radius of one atom is therefore

$$\text{Radius} = \frac{\text{edge}}{2} = \frac{334 \text{ pm}}{2} = 167 \text{ pm}$$

Example 13-11: The density of copper is 8.95 g/cm³. Copper has a face-centered cubic crystal structure. What is the radius of a copper atom?

This is a much more challenging problem than the last one. Not only are we not given the edge length directly, but also the unit cell is more complicated.

Our goal in the first part of the problem is to determine the edge length of the unit cell, so that we will be at the same starting point as in the last problem. We are given the density. IF we knew the mass of the unit cell, we could calculate its volume. This volume is equal to the edge length cubed: volume of a cube = length x width x height = (length)³. The length of one edge is the cube root of the volume. BUT how can we figure out the mass of a unit cell? We know the molar mass. This has units of g/mole. Using Avogadro's number, we can calculate the mass per atom. There are four atoms in a face-centered cubic unit cell, so we can calculate the mass per unit cell.

Let's start down this path. We will calculate the mass per unit cell:

$$\frac{63.55 \text{ g Cu}}{\text{mole Cu}} \times \frac{1 \text{ mole Cu}}{6.022 \times 10^{23} \text{ atoms Cu}} = \frac{1.055 \times 10^{-22} \text{ g}}{\text{atom Cu}}$$

$$\frac{1.055 \times 10^{-22} \text{ g Cu}}{\text{atom}} \times \frac{4 \text{ atoms}}{\text{unit cell}} = \frac{4.221 \times 10^{-22} \text{ g Cu}}{\text{unit cell}}$$

Now we can use the density to calculate the volume of the unit cell.

$$\frac{4.221 \times 10^{-22} \text{ g Cu}}{\text{unit cell}} \times \frac{\text{cm}^3 \text{ Cu}}{8.95 \text{ g Cu}} = \frac{4.71 \times 10^{-23} \text{ cm}^3}{\text{unit cell}}$$

From the volume of the unit cell, we can calculate the length of one edge.

$$\text{Edge} = (\text{volume})^{1/3} = \left(4.71 \times 10^{-23} \text{ cm}^3\right)^{1/3} = 3.61 \times 10^{-8} \text{ cm}$$

$$3.61 \times 10^{-8} \text{ cm} \times \frac{1 \text{ m}}{100 \text{ cm}} \times \frac{10^{12} \text{ pm}}{1 \text{ m}} = 361 \text{ pm}$$

Now we have the length of one edge. We need to relate this to the structure of the unit cell. A picture again helps. Here is a sketch of one side of the unit cell.

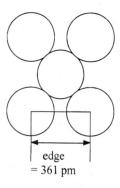

edge
= 361 pm

In this case, you can see that the edge does not correspond to two radii because there is open space between the atoms along the edge. We need to find a place where we have only atoms in a row. This is the diagonal. This diagonal corresponds to four radii. We can determine the diagonal length by setting up a right triangle and using the Pythagorean theorem.

diagonal = 4r   edge
                  = 361 pm

edge
= 361 pm

$$(edge)^2 + (edge)^2 = (diagonal)^2$$

$$(361 \text{ pm})^2 + (361 \text{ pm})^2 = (diagonal)^2$$

$$2.61 \times 10^5 \text{pm}^2 = (diagonal)^2$$

$$diagonal = 511 \text{ pm}$$

Finally, we can divide this diagonal length by 4 to obtain the radius of a single atom.

$$radius = \frac{diagonal}{4} = \frac{511 \text{ pm}}{4} = 128 \text{ pm}$$

Try Study Questions 49 and 55 in Chapter 13 of your textbook now!

- ## Relate unit cells for ionic compounds to formulas (Section 13.7).

  The unit cells for many ionic compounds consist of one type of ion placed in one of the unit cells we have been studying with the other element's ions placed in "holes" within the lattice. If an ion is in an octahedral hole, then it will have six ions of the other element around it in an octahedral geometry. If an ion is in a tetrahedral hole, then it will have four of the other type of ion surrounding it in a tetrahedral geometry.

We can determine the formula of an ionic compound by looking at its unit cell. We use the relationships in the earlier table showing the fraction of an atom belonging to a unit cell.

Example 13-12: Look at Figure 13.32b in your textbook showing the unit cell for sodium chloride. Based on this unit cell, what is the empirical formula of sodium chloride?

First, let's count up the sodium ions in the unit cell. There are twelve sodium ions along the edges and one sodium ion in the body.

$$12 \text{ edge Na}^+ \times \frac{1}{4} = 3 \text{ Na}^+$$

$$1 \text{ body Na}^+ \times 1 = 1 \text{ Na}^+$$

There are four Na^+ in this unit cell.

There are eight chloride ions at the corners and six in the faces.

$$8 \text{ corner Cl}^- \times \frac{1}{8} = 1 \text{ Cl}^-$$

$$6 \text{ face Cl}^- \times \frac{1}{2} = 3 \text{ Cl}^-$$

There are four Cl^- in this unit cell.

The yields a ratio of 4 Na^+/4 Cl^-. This can be reduced to 1 Na^+/ 1 Cl^-. The formula is NaCl.

Try Study Questions 25 and 27 in Chapter 13 of your textbook now!

• **Describe the properties of solids**

a) Characterize different types of solids: metallic (e.g., copper), ionic (e.g., NaCl and CaF$_2$), molecular (e.g., water and I$_2$), network (e.g., diamond), and amorphous (e.g., glass and many synthetic polymers) (Sections 13.6-13.8 and Table 13.6).

Table 13.6 in your textbook lists each of these types of solids and provides examples of the solids, the types of structural units making up the solids, the forces holding together the various types of solids, and typical properties of each type of solid. This table is a nice summary of the key information about each type of solid.

Metallic and ionic solids are made up of atoms or ions that are locked into place in crystal lattices. As Table 13.6 indicates, these are not the only types of solids.

Molecular Solids. The basic units of a molecular solid are, not surprisingly, molecules. An example of a molecular solid is ice. In a molecular solid, the molecules are held in place by means of the intermolecular forces we have been studying: dipole-dipole, hydrogen bonding, and induced dipole/induced dipole. The molecules tend to align with each other in the solid so as to maximize the intermolecular forces of attraction.

Network Solids. Network solids are composed of a three-dimensional array of covalently bonded atoms. The forces holding together a network solid are covalent bonds. Most network solids have high melting points and boiling points. This is due to the fact that covalent bonds make up the lattice, and it takes much energy to break these. The key examples of network solids are graphite, diamond, silicon, and silicates.

Diamond consists of a network in which each carbon atom is bonded to four others in a tetrahedral arrangement, forming a three-dimensional structure. In graphite, each carbon atom is bonded to only three other carbon atoms in a trigonal planar arrangement. These form flat sheets of hexagonal rings of carbon atoms. The attractions between different sheets are not covalent bonds but only weak attractions. The bonding within a sheet is very strong, but the sheets can easily slide past one another. Silicates, certain types of structures composed of silicon and oxygen, are also network solids. Silicates are present in many rocks.

Amorphous Solids. Amorphous solids do not have a regular structure at the particulate level. The molecules are locked into place, but they do not form a regular pattern such as a crystal lattice. Because they do not have a regular structure at the particulate level, amorphous materials do not have a sharp melting point but instead start melting at some temperature and continue to melt over a range of temperatures, until at some higher temperature, all of the material has melted. Glass and some polymers are examples of some amorphous solids.

b) Define the enthalpy of fusion and use it in a calculation (Section 13.9).

The process of a solid converting into a liquid is called melting (this probably doesn't come as a big surprise). The enthalpy change accompanying the transformation of 1 mole of a solid material under standard conditions into a liquid is called the standard molar enthalpy of fusion ($\Delta H^{\circ}_{fusion}$). It takes energy to convert a solid into a liquid, so melting is endothermic.

$$\text{Solid} \xrightarrow{\text{melting}} \text{Liquid} \quad \Delta H = \Delta H^{\circ}_{fusion}$$

In general, materials with low melting points have low enthalpies of fusion, and high melting points are associated with high enthalpies of fusion. This makes sense because if it takes more energy to melt a material, we will need to go to higher temperatures to have the particles have this higher energy.

Nonpolar molecular solids tend to have lower melting points than do polar molecular solids of a similar size and molar mass. This is not surprising because of the greater intermolecular forces holding together polar molecules. Within a series of related molecules, the melting points increase as the size and molar mass of the molecules increase. In general, molecular materials have lower melting points than do ionic materials.

When we freeze something, we convert it from a liquid into a solid. The standard molar enthalpy of crystallization (ΔH°_{cryst}) of a material is the enthalpy change to freeze 1 mole of the material under standard conditions. Because melting and freezing are reverse processes, $\Delta H^{\circ}_{fusion}$ and ΔH°_{cryst} have equal magnitudes but opposite signs. Freezing is exothermic.

$$\text{Liquid} \xrightarrow{\text{freezing}} \text{Solid} \quad \Delta H = \Delta H^{\circ}_{cryst} = -\Delta H^{\circ}_{fusion}$$

Example 13-13: How much heat is required to melt 15.0 g of tungsten. $\Delta H^{\circ}_{fusion}$ of tungsten is 35.2 kJ/mole.

We are given the mass of tungsten and its standard molar enthalpy of fusion. $\Delta H^{\circ}_{fusion}$ will allow us to convert from moles to kJ, but we were given the mass of tungsten. We will need to convert from grams of tungsten to moles of tungsten using the molar mass of tungsten.

$$15.0 \text{ g W} \times \frac{1 \text{ mole W}}{183.8 \text{ g W}} = 0.0816 \text{ moles W}$$

$$0.0816 \text{ moles W} \times \frac{35.2 \text{ kJ}}{\text{mole W}} = 2.87 \text{ kJ}$$

Try Study Question 31 in Chapter 13 of your textbook now!

• **Understand the nature of phase diagrams.**

Identify the different points (triple point, normal boiling point, freezing point) and regions (solid, liquid, vapor) of a phase diagram, and use the diagram to evaluate the vapor pressure of a liquid and the relative densities of a liquid and a solid (Section 13.10).

A typical phase diagram is shown below.

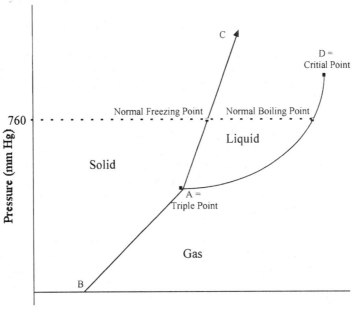

A phase diagram has regions for the solid, liquid, and gas phases. The solid lines are the conditions of pressure and temperature under which two phases are in equilibrium, where a phase change is occurring. Crossing line segment AD corresponds to boiling in one direction and condensation in the other. Line segment AC corresponds to melting in one direction and freezing in the other. Line segment BA corresponds to sublimation in one direction (and deposition in the other).

Point A is the single point where all three states of matter are in equilibrium. It is called the triple point. Point D is the point where the liquid region ends. It is the critical point.

To determine the melting or boiling point at a given pressure, we find the pressure of interest on the pressure axis, go over to the line segment of interest (AC for melting or AD for boiling), go down to the temperature axis, and read the corresponding temperature.

The normal melting point is the temperature at which melting will occur under 1 atm (760 mm Hg) pressure. To determine it, we simply read across from 760 mm Hg on the pressure axis until we get to line segment AC. Likewise, the normal boiling point can be determined by identifying the point on line segment AD that corresponds to 760 mm Hg pressure.

Line segment AD is also simply the vapor pressure *vs.* temperature curve that we discussed earlier. To determine the vapor pressure of a material at a given temperature, we go from the temperature axis up to line segment AD and then read the pressure from the pressure axis.

We can determine if the solid form of a material is more or less dense than the liquid from the slope of line segment AC. If this line segment has a positive slope, the solid is more dense than the liquid. If it has a negative slope, the liquid is more dense than the solid form.

Example 13-14: Use the following phase diagram for an imaginary substance to answer the following questions.

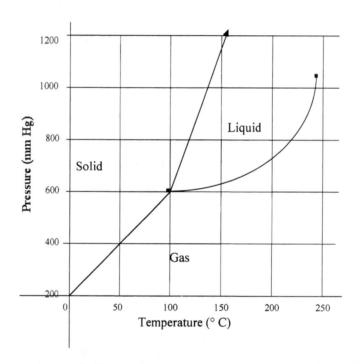

a. In what state is this material at 125°C and a pressure of 300 mm Hg?
We estimate 125°C on the temperature axis and 300 atm on the pressure axis. The two lines from these points intersect in the gas region of the diagram.

b. Is the solid form or the liquid form of this material more dense?
The slope of the line connecting the solid and liquid regions is positive. This means that the solid form is more dense. You can also tell this by drawing a vertical line that passes through this line segment. If, on increasing pressure, you pass from the liquid to the solid region, the solid is more dense. If, however, you pass from the solid to the liquid region, the liquid is more dense.

c. What change of state occurs if one goes from a temperature of 150°C to a temperature of 250°C while holding the pressure constant at 700 mm Hg?
Find these two points on the graph. The first is in the liquid region. The second is in the gas region. In order to get from one to the other, the liquid must have boiled.

Try Study Question 33 in Chapter 13 of your textbook now!

Other Notes

A solid converting directly into a gas without passing through the liquid state is called sublimation. The standard enthalpy change for 1 mole of a material to undergo sublimation is called the standard molar enthalpy of sublimation ($\Delta H^{\circ}_{sublimation}$). Sublimation is an endothermic process.

$$\text{Solid} \xrightarrow{\text{sublimation}} \text{Gas} \quad \Delta H = \Delta H^{\circ}_{sublimation}$$

CHAPTER 14: Solutions and Their Behavior

Chapter Overview

In this chapter, we examine solutions in more detail. A solution is a homogeneous mixture in which the components are evenly distributed in each other down to an atomic or molecular scale. The solute in a solution is the material present in the smaller amount; it is the dissolved material. The solvent is the material present in the larger amount; it is the medium in which the solute is dissolved. Although there can be solutions in any state of matter, we shall focus on those in which the solvent is a liquid. We first examine some more concentration units: molality, mole fraction, weight percent, and parts per million. You will learn how to use these concentration units and how to convert between them.

In a saturated solution, the concentration of solute is equal to the solubility. In an unsaturated solution, the concentration of solute is less than the solubility. A supersaturated solution is an unstable solution in which the concentration of solute is temporarily above the solubility of the material. Liquids that mix to an appreciable extent to form a solution are said to be miscible. Those that do not mix to form a solution are immiscible. Like dissolves like is a general guideline that helps us to predict which materials will be soluble in which others. Associated with the solution process is an energy change called the heat of solution. Heats of solution may be either endothermic or exothermic. Gases are more soluble at higher pressures and lower temperatures. Henry's law relates the solubility of a gas to its partial pressure. Most solids are more soluble at higher temperatures, but some are less soluble. We can use Le Chatelier's Principle to predict the pressure and temperature effects on solubility. Le Chatelier's Principle states that a change in any of the factors determining an equilibrium will cause the system to adjust so as to reduce or minimize the effect of the change.

Colligative properties are properties of the solvent that are different in a solution and which depend only on the number of solute particles per solvent molecule and not on the identity of the solute. Those we examine are vapor pressure lowering, boiling point elevation, freezing point depression, and osmotic pressure. An ionic material has a greater effect on colligative properties than a molecular solute at the same concentration. The factor by which this is true is the van't Hoff factor. As we go to lower solute concentrations, it approaches the number of ions per formula unit of the ionic compound.

Colloids are mixtures with properties between those of a solution and a suspension. The dispersed phase is dispersed throughout the dispersing medium. The particles are larger than in a solution but do not settle. We examine various types of colloids and also discuss how soaps and detergents work.

Key Terms

In this chapter, you will need to learn and be able to use the following terms:

Aerosol: a colloid in which either a liquid or solid is dispersed in a gas.

Colligative property: a property of a solvent that is different in the presence of a solute but that varies only according to the number of solute particles present per molecule of solvent, not on the identity of the solute.

Colloid (colloidal dispersion): an intermediate state between a solution and a suspension. The dispersed phase is dispersed throughout the dispersing medium, but the particles are larger than in a solution. Unlike a suspension, the particles do not settle out.

Emulsifying agent: a material that allows two otherwise insoluble liquids to form an emulsion (a colloid formed when the dispersing medium and the dispersed state are both liquids).

Emulsion: a colloid formed when the dispersing medium and the dispersed state are both liquids.

Enthalpy of hydration: for an ionic compound, this is the enthalpy change that corresponds to taking the separated ions in the gas phase and transforming them to the hydrated ions in solution.

Entropy: a measure of the disorder of a system; the more disordered a system is, the higher the entropy.

Foam: a colloid in which a gas is dispersed in either a liquid or solid.

Gel: a colloid in which a liquid is dispersed in a solid.

Henry's Law: the solubility of a gas in a liquid is proportional to the partial pressure of the gas.
$$S_g = k_H P_g$$

Hydrophilic: water-loving; materials that are strongly attracted to water are said to be hydrophilic.

Hydrophobic: water-fearing; materials that are not strongly attracted to water or that are insoluble in water are said to be hydrophobic.

Hypertonic solution: a solution of higher solute concentration than a cell.

Hypotonic solution: a solution of lower solute concentration than a cell.

Ideal solution: a solution that obeys Raoult's law; solutions approach ideality as the solute concentration is decreased and as the strength of the solute-solvent interactions more closely approximate those of the solvent-solvent interactions.

Immiscible: a term used to describe two liquids that are not soluble in each other to an appreciable extent.

Isotonic solution: a solution containing an equal solute concentration as a cell.

Le Chatelier's Principle: a change in any of the factors determining an equilibrium will cause the system to adjust so as to reduce or minimize the effect of the change.

Miscible: a term used to describe two liquids that are soluble in each other to an appreciable extent.

Molality: moles of solute per kilogram of solvent.

Mole fraction: the number of moles of one component of a mixture divided by the total number of moles of all components in the mixture. The sum of all of the mole fractions in a mixture is 1.

Osmosis: the movement of solvent molecules through a semipermeable membrane from a region of lower solute concentration into a region of higher solute concentration.

Osmotic pressure: during the process of osmosis, the pressure exerted by the solution of higher concentration when equilibrium has been established between solvent flowing into the solution due to osmosis and flowing out of the solution due to the pressure of the solution.

Parts per million: the number of grams of solute per million grams of solution; it may be calculated by taking the mass of solute, dividing by the mass of the solution, and then multiplying by 1,000,000. It also corresponds to mg of solute per kilogram of solution, which (for a dilute solution) is also approximately equal to the mg of solute per liter of solution.

Raoult's law: the vapor pressure exerted by the vapor of the solvent in a solution is equal to the mole fraction of the solvent in the solution multiplied by the vapor pressure of the pure solvent:
$$P_{solvent} = X_{solvent} P^o_{solvent}$$

Reverse osmosis: a process in which a pressure sufficient to overcome the osmotic pressure is applied to the solution of higher solute concentration such that solvent will flow in the opposite direction as would occur in the process of osmosis.

Saturated solution: a solution in which the concentration of the solute equals its solubility.

Semipermeable membrane: a membrane that allows some species to pass through but not others.

Sol: a colloid in which a solid is dispersed in a liquid.

Solid sol: a colloid in which a solid is dispersed in another solid.

Solubility: the concentration of solute in a solution in which there is an equilibrium between dissolved solute and undissolved solute.

Solute: in a solution, the component present in the smaller amount; it can be considered to be the dissolved material in the solution.

Solution: a homogeneous mixture of two or more substances in a single phase.

Solvent: in a solution, the component present in the largest amount; it can be considered to be the medium in which the other component in the solution is dissolved.

Standard heat of solution: the enthalpy change that occurs when sufficient solute is dissolved to form a solution with a 1 molal concentration.

Supersaturated solution: an unstable solution that temporarily contains a concentration of solute greater than the solute's solubility.

Surfactant (surface-active agent): a substance that affects the properties of surfaces.

Suspension: in a suspension, the particles of a material are temporarily dispersed in a solvent as large aggregates but these particles settle out.

Tyndall effect: the scattering of light that occurs when a beam of light is passed through a colloid.

Unsaturated solution: a solution in which the concentration of the solute is less than its solubility.

Van't Hoff factor: the ratio of the experimentally measured freezing point depression of a solution to the value calculated assuming that the solute is a molecular solute.

Weight percent: the number of grams of solute per hundred grams of solution; it may be calculated by taking the mass of solute, dividing by the mass of the solution and then multiplying by 100.

Chapter Goals

By the end of this chapter you should be able to:

- ## Learn additional methods of expressing solution concentration.

    ### a) Define the terms *solution, solvent, solute, and colligative properties* (Section 14.1).

    This chapter deals with solutions. A solution is a homogeneous mixture of two or more substances in a single phase. The component present in the largest amount in a solution is called the solvent. It is the medium in which the other materials are dissolved. The dissolved material is called the solute. It is the material present in the smaller amount in the solution. You are probably most familiar with solutions in which a liquid is the solvent, but it is possible to have gaseous and solid solutions as well.

    This chapter focuses on solutions in which the solvent is a liquid. We shall study in some detail properties of the solvent that are different in the presence of a solute but which depend only on the number of solute particles per solvent molecule and not on the identity of the solute. These properties are called colligative properties. We shall consider four colligative properties: vapor pressure depression, boiling point elevation, freezing point depression, and osmotic pressure.

b) Use the following concentration units: molality, mole fraction, weight percent, and parts per million (Section 14.1).

Up to this point, we have primarily used only one concentration unit, molarity.

$$\text{Molarity (M)} = \frac{\text{amount of solute (moles)}}{\text{volume of solution (L)}}$$

There are many other concentration units. In learning each concentration unit, you should pay attention to which quantities refer to the solute, which to the solvent, and which to the solution. The molality of a solution is defined to be the number of moles of solute per kilogram of solvent:

$$\text{Molality } (m) = \frac{\text{amount of solute (moles)}}{\text{mass of solvent (kg)}}$$

The mole fraction of any component in a solution can be calculated as

$$\text{Mole fraction of A } (X_A) = \frac{\text{moles of A}}{\text{total moles in solution}}$$

The weight percent of a material in a solution can be calculated as

$$\text{Weight \% of A} = \frac{\text{mass of A}}{\text{total mass of solution}} \times 100$$

Percent represents parts per hundred. For more dilute solutions, we sometimes use the concentration unit parts per million:

$$\text{Concentration in Parts Per Million} = \frac{\text{mass of A}}{\text{total mass of solution}} \times 1,000,000$$

This can also be calculated as follows:

$$\text{Concentration in Parts Per Million} = \frac{\text{mg of solute}}{\text{kg of solution}} \approx \frac{\text{mg of solute}}{\text{L of solution}}$$

<u>Example 14-1:</u> Calculate the molality of a solution prepared by taking 10.0 g of NaCl and dissolving it in 500. g of water.

One of the most important things to do in solving problems involving concentrations is to keep straight which quantities deal with the solute, which with the solvent, and which with the solution (or combination of solute and solvent). The solute in this case is the NaCl, and the water is the solvent. In solving many concentration problems, it is suggested that you set up a table with the columns: solute, solvent, and solution. The usual rows that you will need are mass, molar mass, and moles. Sometimes, you may also need to include rows for volume and density. For this problem, we construct the following table:

	Solute	Solvent	Solution
Mass	10.0 g	500. g	
Molar Mass			
Moles			

We are asked for the molality. Write out the definition for this unit.

$$\text{Molality } (m) = \frac{\text{moles of solute}}{\text{kg of solvent}}$$

We are given the mass of NaCl. We need to convert this to moles. To do this, we need the molar mass of NaCl, which is 58.44 g/mole.

$$10.0 \text{ g NaCl} \times \frac{1 \text{ mole NaCl}}{58.44 \text{ g NaCl}} = 0.171 \text{ moles NaCl}$$

The updated table is

	Solute	Solvent	Solution
Mass	10.0 g	500. g	
Molar Mass	58.44 g/mole		
Moles	0.171 moles		

We are given the mass of the water in g. We need for it to be in kg.

$$500. \text{ g H}_2\text{O} \times \frac{1 \text{ kg}}{1000 \text{ g}} = 0.500 \text{ kg}$$

	Solute	Solvent	Solution
Mass	10.0 g	500. g = 0.500 kg	
Molar Mass	58.44 g/mole		
Moles	0.171 moles		

We now can calculate the molality:

$$\text{Molality} = \frac{\text{moles of solute}}{\text{kg of solvent}} = \frac{0.171 \text{ moles NaCl}}{0.500 \text{ kg H}_2\text{O}} = 0.342 \, m \text{ NaCl}$$

Example 14-2: Calculate the mole fraction of NaCl in an aqueous 5.00% NaCl solution.

In solving this type of problem in which we convert from one concentration unit to another, we will again use our table of solute, solvent, and solution. How should we fill it in for a 5.00% solution? We will pick a sample size; it does not really matter what sample size we pick, but based on what percent means, the easiest sample size to pick is 100 g of solution because in that mass of solution, there will be 5.00 g of solute.

	Solute	Solvent	Solution
Mass	5.00 g		100 g
Molar Mass			
Moles			

The definition of mole fraction of NaCl is

$$X_{\text{NaCl}} = \frac{\text{moles of NaCl}}{\text{total moles in solution}}$$

$$= \frac{\text{moles of NaCl}}{\text{moles of NaCl} + \text{moles H}_2\text{O}}$$

We need to calculate the moles of NaCl and the moles of solvent. We have the mass of NaCl. We can convert this to moles using the molar mass of NaCl, 58.44 g/mole.

$$5.00 \text{ g NaCl} \times \frac{1 \text{ mole NaCl}}{58.44 \text{ g NaCl}} = 0.0856 \text{ moles NaCl}$$

	Solute	Solvent	Solution
Mass	5.00 g		100 g
Molar Mass	58.44 g/mole		
Moles	0.0856 moles		

In order to calculate the number of moles of water, we need the mass of the water. We know the mass of the NaCl and the mass of the solution, which consists of NaCl and water. To obtain the mass of the water, we subtract the mass of the NaCl from the mass of the solution.

mass solvent = mass solution – mass solute = 100 g – 5.00 g = 95.00 g

	Solute	Solvent	Solution
Mass	5.00 g	95.00 g	100 g
Molar Mass	58.44 g/mole		
Moles	0.0856 moles		

We just need to obtain the moles of H_2O, using the molar mass of H_2O (18.02 g/mole).

$$95.00 \text{ g } H_2O \times \frac{1 \text{ mole } H_2O}{18.02 \text{ g } H_2O} = 5.272 \text{ moles } H_2O$$

	Solute	Solvent	Solution
Mass	5.00 g	95.00 g	100 g
Molar Mass	58.44 g/mole	18.02	
Moles	0.0856 moles	5.272 moles	

We can now calculate the mole fraction of NaCl

$$X_{NaCl} = \frac{\text{moles of NaCl}}{\text{moles of NaCl + moles } H_2O} = \frac{0.0856 \text{ moles}}{0.0856 \text{ moles} + 5.272 \text{ moles}} = 0.0160$$

As we saw, the best sample size to assume when given a percent was 100 g of solution. That way, we could simply use the value of the % as the number of grams of solute. Each concentration unit has its own best sample size to use.

If the given concentration unit is ...	The numerical value corresponds to ...	When the sample size is ...
Weight percent	Mass of solute	100 g of solution
Molality	Moles of solute	1 kg of solvent
Mole Fraction of Solute	Moles of solute	Moles solute + moles solvent = 1 mole
Molarity	Moles of solute	1 L of solution

Example 14-3: Calculate the weight percent of a solution of sodium hypochlorite that has a concentration of 0.75 M and a density of 1.06 g/cm^3.

For problems that involve molarity, we need to add two more rows to the table, ones for density and volume. The sample size to use when given molarity is 1 L of solution, and the value of the molarity then corresponds to the number of moles of solute. The density given is the density of the solution.

	Solute	Solvent	Solution
Mass			
Molar Mass			
Moles	0.75 moles		
Density			1.06 g/cm^3
Volume			1 L

In this case, we are asked for the weight percent.

$$\text{Weight \% of NaClO} = \frac{\text{mass of NaClO}}{\text{total mass of solution}} \times 100$$

We have moles of NaClO. We can calculate the mass using the molar mass, 74.44 g/mole.

$$0.75 \text{ moles NaClO} \times \frac{74.44 \text{ g NaClO}}{1 \text{ mole NaClO}} = 56 \text{ g NaClO}$$

	Solute	Solvent	Solution
Mass	56 g		
Molar Mass	74.44 g/mole		
Moles	0.75 moles		
Density			1.06 g/cm^3
Volume			1 L

We now need to calculate the mass of the solution. We can do this using the volume and density of the solution.

$$1000 \text{ mL} \times \frac{1.06 \text{ g sol' n}}{\text{mL sol' n}} = 1060 \text{ g sol' n}$$

	Solute	Solvent	Solution
Mass	56 g		1060 g
Molar Mass	74.44 g/mole		
Moles	0.75 moles		
Density			1.06 g/cm^3
Volume			1 L = 1000 mL

We can now calculate the weight percent.

$$\text{Weight \% of NaClO} = \frac{\text{mass of NaClO}}{\text{total mass of solution}} \times 100 = \frac{56 \text{ g NaClO}}{1060 \text{ g sol' n}} \times 100 = 5.3\%$$

Try Study Questions 3, 5, 9, and 11 in Chapter 14 of your textbook now!

c) Understand the distinctions between saturated, unsaturated, and supersaturated solutions (Section 14.2).

For most solutes, if we continue adding solute to a solvent, we eventually reach a point where it appears that no more solute will dissolve. Any more solute appears to remain undissolved, often at the bottom of the container. In actuality, quite a bit is still going on. Solute is dissolving. As fast as solute dissolves, however, other solute particles come out of solution to form undissolved solute. The rates at which these two reverse processes are occurring are equal. A dynamic equilibrium has been established. The concentration of solute in a solution in which there is an equilibrium with undissolved solute is the solubility of the solute. A solution containing this concentration of solute has the maximum amount of solute it can hold at that temperature.

A solution in which the concentration of the solute equals the solubility of the solute is called a saturated solution. It cannot hold any more solute at that temperature. If more solute is added, it will not dissolve. (Actually, it will not appear to dissolve; remember that dissolution continues, but reprecipitation occurs at the same rate; a better phrasing would be that there is no net dissolving of the solute.) A sign that a solution is saturated is if there is undissolved solute present after sufficient time and mixing.

A solution in which the concentration of the solute is less than the solubility of the solute is called an unsaturated solution. If more solute is added to an unsaturated solution, it will dissolve (until the concentration of the solute equals the solubility).

A supersaturated solution is an unstable solution that temporarily contains a concentration of solute greater than the solubility of the solute. If we are working at a constant temperature, we cannot form a supersaturated solution, because once we reach the point of having a saturated solution, we cannot get any more solute to dissolve. The way that we form a supersaturated solution is to play around with the temperature. The solubility of many solutes increases if we go to higher temperatures. To make a supersaturated solution, what we do is to make the solution at a temperature where the solute is more soluble. We then carefully bring the solution back to the temperature at which we desire to have the supersaturated solution. Sometimes, we can get the solution back to this temperature without having the excess solute crystallize. The solution now contains a greater concentration than it should. This is an unstable situation. It usually does not take much to cause this excess solute to come out of solution. Sometimes, all we have to do is add a single crystal of the solute to the supersaturated solution. This crystal serves as a template on which the excess solute crystallizes. What we will end up with after the addition of the seed crystal is the excess solute in the undissolved state and the concentration of solute in the remaining solution equal to the solubility, a saturated solution.

Example 14-4: You come across a bottle containing a solution of NaCl. You are told that it has been sitting there for weeks. It looks like the following:

Is the solution unsaturated, saturated, or supersaturated?

There is undissolved solute present. There has been sufficient time for equilibrium to set up. The solution contains as much solute as it can hold. It cannot be a supersaturated solution because there is excess solute present. This would have been enough to cause the excess solute in a supersaturated solution to crystallize. This solution must be a saturated solution.

---

d) Define and illustrate the terms *miscible* and *immiscible* (Section 14.2).

Two liquids that are soluble in each other to an appreciable extent are said to be miscible. Water and ethanol are miscible in all proportions. No matter how much water and how much ethanol we combine, they will form a solution. Two liquids that are not soluble in each other are said to be immiscible. When two immiscible liquids are combined, they do not form a solution. Instead, layers will form. An example of two immiscible liquids is oil and water. When we mix these, we do not get a solution. Two layers form.

A rough rule for determining which liquids will be miscible is the "like dissolves like" rule we discussed in the last chapter. Thus, polar liquids tend to be miscible in other polar liquids. Nonpolar liquids tend to be miscible in other nonpolar liquids.

Example 14-5: Predict whether water and hexane ($CH_3CH_2CH_2CH_2CH_2CH_3$) should be miscible with each other.

Water is a polar compound. Hexane is a hydrocarbon and is nonpolar. We predict that the two liquids will not be miscible.

Try Study Question 13 in Chapter 14 of your textbook now!

• **Understand the solution process.**

a) Describe the process of dissolving a solute in a solvent, including the energy changes that may occur (Section 14.2).

In dissolving a solute in a solvent, the solvent molecules surround the solute molecules, and the solute molecules move out into the solvent.

Why does a solution form? There are two main factors that determine whether any process is spontaneous or not. One of these factors is the enthalpy change of the process. We have already learned that a negative enthalpy change is favorable. The other factor is the entropy change. Entropy is a measure of the disorder of a system. The higher the disorder, the higher the entropy is. A positive entropy change occurs when a system becomes more disordered. A positive entropy change contributes toward a process being spontaneous. Whether a process is spontaneous or not depends upon the interplay of these two factors: the enthalpy change and the entropy change. Sometimes they work in the same direction, both causing a process to be spontaneous or not spontaneous. Sometimes they work in opposite directions, where one favors a process occurring and the other favoring the process not occurring.

Let us consider an ionic material dissolving in water. In this case, the entropy change usually favors solution formation. The solution is more disordered than the two separated pure materials. If the enthalpy change is negative, also favoring the process being spontaneous, then formation of the solution is spontaneous. If, however, the enthalpy change is positive, it depends on the relative sizes of the enthalpy and entropy terms whether or not the process ends up spontaneous. If the enthalpy factor dominates, the formation of the solution will not be favored. If the entropy factor dominates, the formation of the solution will be favored.

b) Understand the relationship of lattice energy and enthalpy of hydration to the enthalpy of solution for an ionic solute (Section 14.2).

We are now going to concentrate on the enthalpy factor. What goes into determining whether it will be positive or negative? We can break this enthalpy factor down into two processes:

1) Energy must be supplied to separate the ions from each other. The differently charged ions in the solid are held together by ion-ion attractions, and it takes energy to separate them. This enthalpy change corresponds to $-\Delta H_{lattice}$ and has a positive value. This process works against favoring solution formation.

$$MX\ (s) \rightarrow M^{n+}\ (g) + X^{n-}\ (g)$$

2) Energy is given off when the ions are transferred into water and become hydrated. This corresponds to setting up ion-dipole forces between the ions and the water. This enthalpy change is $\Delta H_{hydration}$ and has a negative value.

$$M^{n+}\ (g) + X^{n-}\ (g) \rightarrow M^{n+}\ (aq) + X^{n-}\ (aq)$$

The overall process of forming the solution is the sum of these two processes:

$$MX\ (s) \rightarrow M^{n+}\ (aq) + X^{n-}\ (aq)$$

The enthalpy change accompanying this overall process corresponds to the sum of the two enthalpies ($\Delta H_{solution} = -\Delta H_{lattice} + \Delta H_{hydration}$) and is called the heat of solution. If we get back more energy from the hydration of the ions than we put in to pull the ions apart, then the heat of solution will be negative. If, however, we had to put in more energy to pull the ions apart than we get back in the process of hydrating them, then the heat of solution will be positive.

Even though we know that these two factors are involved, it is very difficult to predict ahead of time whether a particular salt will have a positive or negative heat of solution because both the unfavorable pulling the ions apart and the favorable hydration of the ions are increased by an ion being smaller or more highly charged.

The standard enthalpy (heat) of solution ($\Delta H_{solution}^{o}$) for a solute is the heat associated with forming a 1 molal solution.

Example 14-6: Using the data in Table 14.1 in your textbook, calculate the heat of solution for sodium hydroxide.

The equation for this process is

$$NaOH\ (s) \rightarrow NaOH\ (aq)$$

Table 14.1 gives us ΔH_{f}^{o} for solid sodium hydroxide and for aqueous sodium hydroxide.

$$\Delta H_{rxn}^{o} = \Delta H_{solution}^{o} = \Sigma\left[\Delta H_{f}^{o}(products)\right] - \Sigma\left[\Delta H_{f}^{o}(reactants)\right]$$

$$\Delta H_{solution}^{o} = \left[1\ mole\ \Delta H_{f}^{o}(NaOH\ (aq))\right] - \left[1\ mole\ \Delta H_{f}^{o}(NaOH\ (s))\right]$$

$$= \left[1\ mole\left(-469.2\ \frac{kJ}{mole}\right)\right] - \left[1\ mole\left(-425.9\ \frac{kJ}{mole}\right)\right]$$

$$= -43.3\ kJ$$

The heat of solution of sodium hydroxide is –44.5 kJ/mole NaOH. This is a large value for a heat of solution.

Try Study Questions 15 and 19 in Chapter 14 of your textbook now!

c) Describe the effect of pressure and temperature on the solubility of a solute (Section 14.3).

Changing the pressure affects the solubility of a gas in a liquid. In thinking about a solution of a gas in a liquid, keep in mind what you already know about such solutions from observing soda. A bottle of soda is packaged under conditions where the partial pressure of the carbon dioxide is very high. This causes more carbon dioxide to dissolve. When you open the bottle, the partial pressure of the carbon dioxide drops sharply because the carbon dioxide can escape into the room. When this occurs, more carbon dioxide comes out of solution. From these observations, you can put together the relationship between pressure and solubility of a gas. The solubility of a gas in a liquid is directly proportional to the partial pressure of the gas. If the partial pressure is raised, then more gas will dissolve. If the partial pressure is lowered, some of the gas will come out of solution. This relationship is summarized by Henry's law, the topic of the next objective.

You also probably already know the relationship between temperature and solubility of a gas from your observations of soda. If you heat up a bottle of soda, the soda will go flat. Gases are less soluble at higher temperatures. The reason behind this has to do with the fact that gases have an exothermic heat of solution. Usually, if a material has a negative heat of solution, the material will be less soluble at higher temperatures.

Many solids have positive heats of solution. Most of these are more soluble at higher temperatures. Some solids have negative heats of solution. These are usually less soluble at higher temperatures. There are some exceptions to these general rules relating the heat of solution and the temperature dependence of the solubility, but they do well for a first guess.

d) Use Henry's law to calculate the solubility of a gas in a solvent (Section 14.3).

Henry's law relates the partial pressure of a gas and its solubility:
$$S_g = k_H P_g$$
S_g is the solubility of the gas at a particular temperature, k_H is the Henry's law constant (which varies with temperature), and P_g is the partial pressure of the gas (not the total pressure). This equation lines up with the trend for solubility and partial pressure that we discussed qualitatively in the last section: gases are more soluble at higher partial pressures.

Example 14-7: What concentration of nitrogen should be present in a glass of water at room temperature? Assume a temperature of 25°C, a total pressure of 1.0 atm, and a mole fraction of nitrogen in air of 0.78.

We are asked for the solubility of a gas at a particular pressure. We can use Henry's law.
$$S_g = k_H P_g$$

We can find k_H for nitrogen at 25°C in Table 14.2 in your textbook, 8.42×10^{-7} M/mm Hg. The P_g in the equation is the partial pressure of the gas of interest (nitrogen), not the total pressure. From our work with Dalton's law in the gas laws chapter, we know that
$$P_{N_2} = X_{N_2} P_{total} = (0.78)(1.0 \text{ atm}) = 0.78 \text{ atm}$$

The Henry's law constant is per mm Hg, so we must convert atm to mm Hg:
$$0.78 \text{ atm} \times \frac{760 \text{ mm Hg}}{1 \text{ atm}} = 590 \text{ mm Hg}$$

We can now solve the Henry's law problem:
$$S_{N_2} = k_H P_{N_2} = \left(8.42 \times 10^{-7} \frac{M}{\text{mm Hg}}\right)(590 \text{ mm Hg}) = 5.0 \times 10^{-4} \text{ M}$$

Try Study Question 21 in Chapter 14 of your textbook now!

e. Apply Le Chatelier's principle to the change in solubility of gases with pressure and temperature changes (Section 14.3).

Le Chatelier's principle deals with how a system initially at equilibrium responds when something is done that causes it to no longer be at equilibrium. The system will respond by trying to go to a new equilibrium situation. It does so by shifting in a direction that reduces the effect of the change. Thus, Le Chatelier's principle can be stated as follows: a change in any of the factors determining an equilibrium will cause the system to adjust so as to reduce or minimize the effect of the change. We can see how this principle allows us to predict the pressure and temperature effects on the solubility of gases and also the overall trends in the solubility of solids.

We can write the equation for the formation of a solution of a gas as follows:
Gas + Liquid Solvent <=> Solution

Let us now imagine that we increase the partial pressure of the gas. The change is that we now have a greater pressure than we should have for the system to be at equilibrium. What can we do to minimize the effect of this change? Reduce the pressure of the gas some. In order to do this, more of the gas will dissolve in the solvent to obtain a greater concentration of the gas in the solution. The solubility of a gas is thus greater when the partial pressure of the gas is higher.

Conversely, if we initially have a system at equilibrium and decrease the partial pressure of the gas, we will then have a smaller pressure of the gas than we should have. How can the system compensate to reduce the effect of this change? By having gas come out of the solution to cause the partial pressure to come back up some (it won't go all the way back up). At a lower partial pressure, the solubility of a gas goes down.

Let's now see how we can use Le Chatelier's principle to predict solubility effects when the temperature changes. In order to do this, we look at whether the heat of solution is endothermic or exothermic and then write heat into the equation. If the heat of solution is endothermic, then heat will be a reactant. If the heat of solution is exothermic, then heat will be a product.

Endothermic (Positive) $\Delta H_{sol'n}$ Heat + Solute + Solvent <=> Solution
Exothermic (Negative) $\Delta H_{sol'n}$ Solute + Solvent <=> Solution + Heat

If we increase the temperature, then we have more heat. The system will respond by shifting in the direction that uses up heat. Almost all gases have a negative $\Delta H_{sol'n}$
Solute + Solvent <=> Solution + Heat
When the reaction proceeds to the right, we get more heat. When it proceeds to the left, we use up heat. To reduce the stress of increasing the heat, the reaction will proceed in the direction that uses up heat: to the left. Solute will come out of solution. Gases and some solids have negative values for $\Delta H_{sol'n}$. These materials are less soluble at higher temperatures.

What if $\Delta H_{sol'n}$ is positive?
Heat + Solute + Solvent <=> Solution
The heat is on the opposite side of the equation as in the last case we looked at. In this case, proceeding to the right uses up heat, and proceeding to the left produces heat. Increasing the temperature increases the heat. The system still reacts by shifting in the direction that uses up heat, but that direction is to the right in this case. The solute will be more soluble at higher temperatures. This is the case for many (but not all) solid solutes.

Let's summarize these results:

Solute	Change	Effect on Solubility
Gas	Increase Partial Pressure	Increase Solubility
	Decrease Partial Pressure	Decrease Solubility
Material with Endothermic $\Delta H_{sol'n}$ (most solids)	Increase Temperature	Increase Solubility
	Decrease Temperature	Decrease Solubility
Material with Exothermic $\Delta H_{sol'n}$ (gases and some solids)	Increase Temperature	Decrease Solubility
	Decrease Temperature	Increase Solubility

Example 14-8: Sodium acetate is more soluble at higher temperatures. Predict whether its heat of solution is positive or negative.

The two possibilities for the equilibrium are
Endothermic   Heat +   Solute + Solvent <=> Solution
Exothermic   Solute + Solvent <=> Solution + Heat

We will consider increasing the temperature – increasing the heat – in both cases. For the first case, increasing the heat will cause the equilibrium to shift to the right to use up some of the heat. This corresponds to more solute dissolving. This matches up with the case for sodium acetate. $\Delta H_{sol'n}$ is predicted to be endothermic, negative.

We can double check that it is not the second case. In that case, increasing the heat would cause the reaction to shift to the left, causing solute to come out of solution. This is not what is observed for sodium acetate.

Example 14-9: Use Le Chatelier's Principle to explain the fact that lowering the temperature increases the solubility of carbon dioxide gas.

Carbon dioxide has a negative heat of solution; it is exothermic. The equation is
carbon dioxide + Solvent <=> Solution + Heat
If we lower the temperature, this amounts to taking away heat. The system will respond by shifting in the direction that produces more heat. In this case, it will shift to the right. More carbon dioxide will go into solution. The gas is more soluble.

• **Understand and use the colligative properties of solutions.**

a) **Calculate the mole fraction of a solute or solvent ($X_{solvent}$) and the effect of a solute on solvent vapor pressure ($P_{solvent}$) using Raoult's law (Section 14.4).**

We have already discussed mole fractions. Up to this point, we have largely considered the mole fraction of solute in a solution:
$$X_{solute} = \frac{\text{moles of solute}}{\text{total moles in solution}}$$

The mole fraction of the solvent is
$$X_{solvent} = \frac{\text{moles of solvent}}{\text{total moles in solution}}$$

Because all of the mole fractions must add up to equal 1, if we have a solution consisting of one solute and one solvent, then if we know the mole fraction of the solute, we can calculate the mole fraction of the solvent by using the equation
$$X_{solvent} = 1 - X_{solute}$$

The first colligative property we shall study is dependent upon $X_{solvent}$. In the last chapter, we learned about the vapor pressure of a liquid. This is the pressure of the vapor state of a substance when the liquid and the vapor are in equilibrium. If we have a solution of a nonvolatile solute in a solvent, the vapor pressure exerted by the solvent is less than the vapor pressure of the pure solvent would be at that temperature. It turns out that the vapor pressure exerted by the solvent in the solution is dependent upon how much of the solution actually is the solvent. Raoult's law states that

$$P_{solvent} = X_{solvent}P^o_{solvent}$$

where $P_{solvent}$ is the vapor pressure of the solvent in the solution, $X_{solvent}$ is the mole fraction of the solvent in the solution, and $P^o_{solvent}$ is the vapor pressure of the pure solvent. Notice that the mole fraction used in this equation is the mole fraction of the solvent, not the mole fraction of the solute. If we have pure solvent, then $X_{solvent}$ is 1, and the vapor pressure predicted is that of the pure solvent. If 97% of the molecules in a solution are solvent molecules, the mole fraction of solvent is 0.97, and the vapor pressure is 0.97 times that of the pure solvent.

Solutions that obey Raoult's law are called ideal solutions. Just as there is no such thing as a truly ideal gas, there is no such thing as a truly ideal solution, but solutions approach ideality as their concentrations get lower and as the solute-solvent interactions more closely match the solvent-solvent interactions.

Example 14-10: Suppose 12.0 g of sucrose ($C_{12}H_{22}O_{11}$) (about 1 tablespoon of table sugar) is dissolved in 250. g of water at 90°C. What is the vapor pressure of water over the solution?

This problem asks us to calculate the vapor pressure of the solvent over the solution. This is a Raoult's law problem.

$$P_{solvent} = X_{solvent}P^o_{solvent}$$

We can look up the vapor pressure of the pure solvent in a table. Appendix G in your textbook lists this as 525.8 mm Hg at 90°C. The only thing we need is the mole fraction of the solvent:

$$X_{solvent} = \frac{\text{moles of solvent}}{\text{total moles in solution}}$$

To calculate this, we need the moles of solvent and the total moles of solution (moles solute + moles solution). The molar mass of sucrose is 342.3 g/mole and that of water is 18.02 g/mole.

$$\text{moles } C_{12}H_{22}O_{11} = 12.0 \text{ g } C_{12}H_{22}O_{11} \times \frac{1 \text{ mole } C_{12}H_{22}O_{11}}{342.3 \text{ g } C_{12}H_{22}O_{11}} = 0.0351 \text{ moles } C_{12}H_{22}O_{11}$$

$$\text{moles } H_2O = 250. \text{ g } H_2O \times \frac{1 \text{ mole } H_2O}{18.02 \text{ g } H_2O} = 13.9 \text{ moles } H_2O$$

We can now calculate the mole fraction of solvent.

$$X_{solvent} = \frac{\text{moles of solvent}}{\text{moles solute} + \text{moles solvent}} = \frac{13.9 \text{ moles}}{0.0351 \text{ moles} + 13.9 \text{ moles}} = 0.997$$

We then use Raoult's law to calculate the vapor pressure over the solution.

$$P_{solvent} = X_{solvent}P^o_{solvent} = (0.997)(525.8 \text{ mm Hg}) = 524 \text{ mm Hg}$$

Try Study Question 25 in Chapter 14 of your textbook now!

b) Calculate the boiling point elevation or freezing point depression caused by a solute in a solvent (Section 14.4).

With the last objective we learned that the vapor pressure exerted by a solvent is lower for the solvent when it is present in a solution than when it is by itself. The boiling point of a liquid is the temperature at which its vapor pressure is equal to the external pressure. In order to get the vapor pressure to equal the external pressure, we will need to go to a higher temperature if we have a solution than if we have the pure solvent. The boiling point of the solvent in the solution is higher than for the pure solvent. The amount that it is higher is given by the equation:

$$\Delta T_{bp} = K_{bp} m_{solute}$$

In this equation, ΔT_{bp} is the amount that the boiling point is higher than for the pure solvent, K_{bp} is the molal boiling point elevation constant (each solvent has its own K_{bp}), and m_{solute} is the molality of the solution.

<u>Example 14-11:</u> A solution of glycerol ($C_3H_8O_3$) in water was prepared by dissolving glycerol in 500. g of water. This solution has a boiling point of 100.42°C at 760 mm Hg. What mass of glycerol was dissolved to make this solution?

This is a boiling elevation problem. The equation for boiling point elevation is

$$\Delta T_{bp} = K_{bp} m_{solute}$$

The normal boiling point of water is defined to be 100°C. The change in boiling point is therefore 100.42°C – 100°C = 0.42°C. We can obtain K_{bp} from Table 14.3 in your textbook. It is 0.5121 °C/m.

$$\Delta T_{bp} = K_{bp} m_{solute}$$

$$0.42°C = \left(0.5121 \frac{°C}{m}\right) m_{solute}$$

$$m_{solute} = 0.82\, m$$

Molality is defined to be moles of solute per kilogram of solvent. We know the mass of water in g. We can convert this to kg and then use the molality to figure out how many moles of glycerol are present.

$$500.\ g\ H_2O \times \frac{1\ kg}{1000\ g} = 0.500\ kg\ H_2O$$

$$0.500\ kg\ H_2O \times \frac{0.82\ moles\ glycerol}{1\ kg\ H2O} = 0.41\ moles\ glycerol$$

We can then convert from moles of glycerol to the mass of glycerol using the molar mass of glycerol (92.09 g/mole).

$$0.41\ moles\ glycerol \times \frac{92.09\ g\ glycerol}{1\ mole\ glycerol} = 38\ g\ glycerol$$

Try Study Question 33 in Chapter 14 of your textbook now!

The freezing point of a solvent is lowered in a solution. Keep in mind that the boiling point is elevated; the freezing point is lowered. The liquid region is thus extended in both directions. The equation for freezing point depression is very similar to that for boiling point elevation:

$$\Delta T_{fp} = K_{fp} m_{solute}$$

In this equation, ΔT_{fp} is the amount that the freezing point goes down, K_{fp} is the molal freezing point depression constant, and m_{solute} is the molality of the solution.

Example 14-12: A solution is prepared by adding 0.500 g of caffeine ($C_8H_{10}O_2N_4$) to 100.0 g of benzene (C_6H_6). Calculate the freezing point of benzene in this solution. The normal freezing point of pure benzene is 5.50°C, and K_{fp} for benzene is –5.12°C/m.

We desire to determine the freezing point of benzene in the solution. We know an equation to calculate the freezing point depression:

$$\Delta T_{fp} = K_{fp} m_{solute}$$

We will solve for ΔT_{fp}. We are given K_{fp}, the mass of caffeine, and the mass of benzene. To use this equation, we need to calculate the molality of the solution. Molality is moles of solute per kg of solvent. We can calculate the moles of caffeine using the mass of caffeine and its molar mass (194.2 g/mole). We can also convert from g of benzene to kg of benzene.

$$\text{moles caffeine} = 0.500 \text{ g } C_8H_{10}O_2N_4 \times \frac{1 \text{ mole } C_8H_{10}O_2N_4}{194.2 \text{ g } C_8H_{10}O_2N_4} = 0.00257 \text{ moles } C_8H_{10}O_2N_4$$

$$100.0 \text{ g } C_6H_6 \times \frac{1 \text{ kg}}{1000 \text{ g}} = 0.1000 \text{ kg } C_6H_6$$

We can now calculate the molality.

$$m_{solute} = \frac{\text{moles } C_8H_{10}O_2N_4}{\text{kg } C_6H_6} = \frac{0.00257 \text{ moles } C_8H_{10}O_2N_4}{0.1000 \text{ kg } C_6H_6} = 0.0257 \, m$$

We then use the freezing point depression equation to calculate ΔT_{fp}.

$$\Delta T_{fp} = K_{fp} m_{solute} = \left(-5.12 \, \frac{°C}{m}\right)(0.0257 \, m) = -0.132°C$$

The question does not ask us to stop here, but to calculate the actual freezing point of benzene in the solution. It will be 0.132°C lower than for pure benzene.

$$T_{fp} = 5.50°C - 0.132°C = 5.37°C$$

Try Study Question 37 in Chapter 14 of your textbook now!

c) Use colligative properties to determine the molar mass of a solute (Section 14.4).

The molar mass of a material can be obtained by taking the mass of the material and dividing by the number of moles of the material.

$$\text{Molar Mass} = \frac{\text{mass}}{\text{moles}}$$

We can use colligative properties to determine how many moles of a material are present in a sample. If we also know the mass of the sample, we can calculate the molar mass.

Example 14-13: 15 g of an unknown molecular material was dissolved in 450. g of water. The resulting solution freezes at –0.34°C. What is the molar mass of the unknown material?

We are asked for the molar mass of an unknown material. We know the mass, 15 g. In order to calculate the molar mass, we will need to determine the number of moles of the material. We can do this using the freezing point depression. The normal freezing point of water is 0°C. This solution freezes at –0.34°C. The freezing point depression is

$$\Delta T_{fp} = -0.34°C - 0°C = -0.34°C$$

$$\Delta T_{fp} = K_{fp} m_{solute}$$

$$-0.34°C = -1.86 \frac{°C}{m} m_{solute}$$

$$m_{solute} = 0.18 \ m$$

Molality is moles of solute per kilogram of solvent. We know the mass of solvent used, so we can calculate the number of moles of solute present.

$$450. \ g \ H_2O \times \frac{1 \ kg}{1000 \ g} = 0.450 \ kg \ H_2O$$

$$0.450 \ kg \ H_2O \times \frac{0.18 \ moles \ solute}{1 \ kg \ H_2O} = 0.082 \ moles \ solute$$

We can now calculate the molar mass.

$$Molar \ Mass = \frac{mass}{moles} = \frac{15 \ g}{0.082 \ moles} = 1.8 \times 10^2 \ \frac{g}{mole}$$

Try Study Questions 39 and 45 in Chapter 14 of your textbook now!

d) Characterize the effect of ionic solutes on colligative properties (Section 14.4).

Colligative properties depend not on what is dissolved but only on the number of particles of solute per solvent particle. Because ionic compounds separate into their separate ions in aqueous solutions, they produce more particles in solution than do molecular materials. Consider sodium chloride. Its dissociation can be represented as

$$NaCl(s) \rightarrow Na^+ \ (aq) + Cl^- \ (aq)$$

According to this equation, 1 mole of sodium chloride produces 1 mole of sodium ions and 1 mole of chloride ions, a total of two moles of particles. The effect of sodium chloride on colligative properties is thus approximately twice that of a molecular compound that produces only one mole of particles per mole dissolved.

The dissociation of magnesium nitrate can be represented as

$$Mg(NO_3)_2 \ (s) \rightarrow Mg^{2+} \ (aq) + 2NO_3^- \ (aq)$$

For each mole of magnesium nitrate that dissolves, we get a total of three solute particles (one magnesium ion and two nitrate ions) in solution. The colligative properties caused by magnesium nitrate are approximately three times that caused by a molecular compound that produces only one mole of particles per mole dissolved.

We can make predictions like these for any ionic compound by figuring out how many particles we get in solution per formula unit that dissolves. We can introduce a correction term, i, into the equations for each of our colligative properties to account for this. The revised equation for freezing point depression is

$$\Delta T_{fp} = K_{fp} m_{solute} i$$

The correction factor i is the van't Hoff factor. Its predicted value is the same as the number of particles produced per formula unit that dissolves: 2 for NaCl, 3 for $Mg(NO_3)_2$, etc.

You can see that we can solve this equation for i.

$$i = \frac{\Delta T_{fp}}{K_{fp} m_{solute}}$$

If we use values for ΔT_{fp} actually measured in the laboratory, we do not get a value of 2 for sodium chloride or of 3 for magnesium nitrate. The actual van't Hoff factors obtained experimentally are always less than the number of ions produced. Your textbook has a table that shows the actual i values for NaCl ranging from 1.83 up to 1.94 for solutions ranging in concentration from 0.349 *m* down to 0.0120 *m*. The reason these come out lower than predicted is that the ions do not function as completely independent particles in the solutions; we get some ion pairing. Because of this, we never get the full effect predicted for the number of ions expected. As we go to more dilute solutions, the van't Hoff factors actually obtained get closer to the expected values.

e) Use the van't Hoff factor, i, in calculations involving colligative properties (Section 14.4).

<u>Example 14-14:</u> Predict the freezing point depression expected for a 0.0711 *m* aqueous solution of sodium sulfate. If this solution actually freezes at –0.320°C, what is the actual value of the van't Hoff factor for sodium sulfate at this concentration? K_{fp} for water is –1.86°C/*m*.

First, we will calculate the predicted freezing point depression assuming that the sodium sulfate dissociates completely and that the particles act as completely independent particles. The formula for sodium sulfate is Na_2SO_4. Its dissociation can be represented as

$$Na_2SO_4 \text{ (s)} \rightarrow 2Na^+ \text{ (aq)} + SO_4^{2-} \text{ (aq)}$$

We get three ions for each formula unit of Na_2SO_4, so the limiting value of i is 3.

$$\Delta T_{fp} = K_{fp}m_{solute}i = \left(-1.86\frac{°C}{m}\right)(0.0711\ m)(3) = -0.397°C$$

The normal freezing point of water is 0°C. If we lower it by 0.397°C, then the freezing point of the solvent in the solution is –0.397°C.

For the second part, we are told that the solution actually freezes at –0.320°C. We can use this to calculate the true value for i at this concentration.

$$i = \frac{\Delta T_{fp}}{K_{fp}m_{solute}} = \frac{-0.320°C}{\left(-1.86\frac{°C}{m}\right)(0.0711\ m)} = 2.42$$

Instead of the predicted value of 3, the van't Hoff factor is only 2.42.

Try Study Questions 47 and 49 in Chapter 14 of your textbook now!

f) Calculate the osmotic pressure (Π) for solutions and use the equation defining osmotic pressure to determine the molar mass of a solute (Section 14.4).

Osmosis is the movement of solvent molecules through a semipermeable membrane from a region of lower solute concentration to a region of higher solute concentration. A semipermeable membrane is a membrane that allows some species to pass through but not others. Look at Figure 14.16 in your textbook. Inside the bag is a solution of higher solute concentration. Outside the bag is a solution of lower solute concentration. Separating these two solutions is the bag, which is a semipermeable membrane. The system will try to equalize the concentrations on both sides of the membrane. The problem is that solute particles cannot move across the membrane; only solvent molecules can. What will occur is that solvent molecules will move into the bag diluting the solution inside the bag. As this occurs, the liquid level in the tube will go up. There will continue to be a net flow of solvent into the bag until the force exerted downward by the liquid in the tube is equal to the force of the water coming into the bag. At this point, the rate of flow of solvent into the bag will equal the rate of flow of solvent out of the bag; a dynamic equilibrium will have been established. The pressure exerted by the column of liquid in the tube when the system is at equilibrium is called the osmotic pressure.

Osmotic pressure is a colligative property. The higher the concentration of the solute, the higher the osmotic pressure will be. Its equation is

$$\Pi = cRT$$

In this equation, Π is the osmotic pressure, c is the concentration expressed in molarity (M), R is the gas constant, and T is the absolute temperature.

Example 15-15: What is the osmotic pressure of a 0.1 M solution of sucrose at 25°C?

The equation for osmotic pressure is

$$\Pi = cRT$$

We are given the concentration. We must convert the temperature to kelvin; 25°C corresponds to 298 K. We use our usual value of R, 0.082057 L atm/(mole K).

$$\Pi = cRT = \left(0.100 \ \frac{mole}{L}\right)\left(0.082057 \ \frac{L \ atm}{mole \ K}\right)(298 \ K) = 2.45 \ atm$$

Example 15-16: 100 mg of a protein are dissolved in enough water to make 10.0 mL of a solution. If this solution has an osmotic pressure of 13.3 mm Hg at 25.0°C, what is the molar mass of the protein?

The molar mass is mass/moles. We know the mass. We need to obtain moles. We can do this using the osmotic pressure. In order to use our value of R, we will need the osmotic pressure in atm and the temperature in K.

$$13 \ mm \ Hg \ x \ \frac{1 \ atm}{760 \ mm \ Hg} = 0.017 \ atm$$

$$(25.0°C + 273.15°C)\frac{1 \ K}{1°C} = 298.15 \ K$$

We can use the osmotic pressure equation to solve for the concentration of the solution.
$$\Pi = cRT$$

$$c = \frac{\Pi}{RT} = \frac{0.017 \ atm}{\left(0.082057 \ \frac{L \ atm}{mol \ K}\right)(298.15 \ K)} = 7.0 \ x \ 10^{-4} \ M$$

The volume of the solution is 10.0 mL. We will convert this to L and find out how many moles of the solution are in this volume of solution.

$$10.0 \ mL \ sol'n \ x \ \frac{1 \ L}{1000 \ mL} \ x \ \frac{7.0 \ x \ 10^{-4} \ moles \ protein}{L} = 7.0 \ x \ 10^{-6} \ moles \ protein$$

We now have the mass and the number of moles. We can calculate the molar mass.

$$100 \ mg \ x \ \frac{1 \ g}{1000 \ mg} = 0.100 \ g$$

$$Molar \ Mass \ of \ Protein = \frac{mass}{moles} = \frac{0.100 \ g}{7.0 \ x \ 10^{-6} \ moles} = 1.4 \ x \ 10^4 \ g/mole$$

Try Study Question 53 in Chapter 14 of your textbook now!

• **Describe colloids and their applications.**

a) Recognize the difference among a homogeneous solution, a suspension, and a colloid (or colloidal dispersion) (Section 14.5).

A solution is a homogeneous mixture. The solute particles are present in the solution as separated ions or small molecules. There will be no settling of the solute. A solution also does not exhibit the Tyndall effect. This means that it does not scatter visible light.

In a suspension, particles of one material are temporarily dispersed in the solvent, but they stay together as aggregates, and they will eventually settle out.

A colloid is intermediate between a solution and a suspension. The particles of a colloid are larger than ions and small molecules. They are either larger molecules or aggregates of ions or molecules. On the other hand, these particles do not settle out; they will remain dispersed in the colloid. A colloid exhibits the Tyndall effect; it scatters visible light shined through it. There are several classifications of colloids based upon the physical state of the dispersing phase (corresponds to the solvent) and the dispersed phase (corresponds to the solute).

Type of Colloid	Dispersing Medium	Dispersed Phase
Aerosol	Gas	Liquid OR Solid
Foam	Liquid OR Solid	Gas
Emulsion	Liquid	Liquid
Gel	Solid	Liquid
Sol	Liquid	Solid
Solid sol	Solid	Solid

b) Recognize hydrophobic and hydrophilic colloids (Section 14.5).

Colloids in which the dispersing medium is water can be classified as being either hydrophobic or hydrophilic. In a hydrophobic colloid, the particles of the dispersed phase are attracted to water only by weak attractive forces. Examples of hydrophobic colloids are dispersions of metals and nearly insoluble salts. In a hydrophilic colloid, there are strong attractive forces between the particles of the dispersed phase and water. The particles in hydrophilic colloids often have –OH and/or –NH_2 groups.

c) Describe the action of a surfactant (Section 14.5).

A surfactant is a material that affects the properties of surfaces. Soaps and detergents are examples of surfactants. We use soaps (and detergents) to remove nonpolar substances such as oils from our clothes and bodies. These nonpolar materials are not very soluble in water, so rinsing with water alone does not remove them. The reason that soaps (and detergents) can remove these materials has to do with the fact that the two ends of a soap (or detergent) molecule are different. One end is ionic; this part is hydrophilic. Sometimes it is referred to as a hydrophilic head. The other end is a long hydrocarbon; this part is hydrophobic. It is sometimes referred to as a hydrophobic tail.

The nonpolar regions on the soap or detergent associate with the nonpolar material (oil, for example) we are trying to remove from the surface we are trying to clean. The ionic region of the soap or detergent points out to the water, where it has favorable interactions. Eventually enough soap molecules associate with the oil particle to pull the particle off the surface to which it was attached. The oil surrounded by the soap goes into the water and can now be rinsed away. Your textbook shows this in Figure 14.23.

Other Notes

1. We learned that the vapor pressure exerted by the vapor of the solvent over a solution is given by Raoult's law.

$$P_{solvent} = X_{solvent} P^o_{solvent}$$

The mole fraction in this equation is the mole fraction of the solvent in the solution.

The amount that this is lower than the vapor pressure of the pure solvent is related to the mole fraction of the solute:

$$\Delta P_{solvent} = X_{solute} P^o_{solvent}$$

Notice that $P_{solvent} + \Delta P_{solvent} = P^o_{solvent}$

2. A cell membrane is a semipermeable membrane. Osmosis can occur through it. A solution that has a lower solute concentration than the cell is said to be hypotonic. If a cell is placed in a hypotonic solution, water flows from the solution into the cell. The cell will burst.

A solution that has a higher solute concentration than the cell is said to be hypertonic. If a cell is placed in a hypertonic solution, water flows from the cell into the solution. The cell shrivels.

A solution that has the same solute concentration as the cell is said to be isotonic.

3. Reverse osmosis can be used to purify water. In reverse osmosis, we still have two solutions of different concentration separated by a semipermeable membrane. In the process of osmosis, solvent would flow from the solution of lower concentration to the solution of higher concentration. In reverse osmosis, we apply a pressure to the solution of higher concentration that is greater than the osmotic pressure. The result of this is that solvent flows in the reverse direction: from the solution of higher concentration to the solution of lower concentration.

CHAPTER 15: Principles of Reactivity: Chemical Kinetics

Chapter Overview

Chemical kinetics deals with the study of chemical reaction rates. The rate of a reaction is the change in concentration of a material divided by the change in time. Reaction rates are always positive, so if a material is a reactant, then we must change the sign of Δ concentration. The rate of appearance or disappearance of different materials in a reaction can be related using the stoichiometric coefficients in the balanced chemical equation. The average rate over a time period can be calculated by finding the slope of the line that connects the two points corresponding to the beginning and ending times of this time interval on a concentration *vs.* time graph. The instantaneous rate at a given time can be found by determining the slope of a line tangent to this curve at that time. Among the factors affecting reaction rates are concentration, temperature, and the presence of a catalyst. A catalyst is a material that speeds up a chemical reaction but that is not itself permanently changed in the reaction. For a solid reactant, the surface area of the solid also affects the rate. The rate law shows the relationship between rate and concentration. For the reaction, $aA + bB \rightarrow$ products, the rate law has the form

$$\text{Rate} = k[A]^m[B]^n$$

In this equation, k is the rate constant, and m and n are the orders of the reactants. The orders must be determined experimentally. The sum of the orders gives the total order. You will learn how to determine a rate law by using the method of initial rates. Often, we wish to know the direct relationship between concentration and time. Integrated rate laws give us this information. We will study the equations for the integrated rate laws of reactions in which there is one reactant and the reaction order is zero-, first-, or second- order. You will need to be able to use these equations to determine any of the variables involved. Each equation can be written in a form that yields a linear graph. One way to determine the order of a reaction is to plot each of the graphs and see which yields a line. The half-life is the time it takes for the reactant concentration to fall to half its initial value. We will study only the first order half-life. Chemists have devised a theory on the submicroscopic level called collision theory to explain reaction rates. In this theory, in order for a reaction to occur, the reactant molecules must collide with each other. In addition, these collisions must have sufficient energy and be in the proper orientation for the reaction to occur. There is an energy barrier that molecules must surmount for the reaction to occur. The energy required to surmount this barrier is called the activation energy. The higher the activation energy, the slower the reaction will be. At the peak of this barrier, the chemical species that exists is called the transition state. The equation that describes the relationship between the rate constant and temperature is the Arrhenius equation. You will learn how to use various forms of this equation to determine the activation energy of a chemical reaction. A catalyst functions by providing an alternative pathway that has a lower activation energy. The individual steps that occur in a chemical reaction make up the reaction mechanism. These steps are called elementary steps. In the rate law for an elementary step, the orders for the reactants can be determined from the coefficients of those species in the equation for the elementary step. In reaction mechanisms, we often propose forming reaction intermediates, species that are formed in one step of the reaction and consumed in another. A reaction can proceed no faster than the slowest step in the mechanism. Often, this step is slow enough that it determines the overall rate. You will learn how to determine the rate law predicted by some mechanisms, given the identity of the slowest step.

Key Terms

In this chapter, you will need to learn and be able to use the following terms:

Activation energy (E_a): the minimum energy that must be added to the reactants for a reaction to occur; on a reaction coordinate diagram, it corresponds to the energy required to go from the reactants up to the highest point in energy on the diagram.

Arrhenius equation: an equation that relates the rate constant, temperature, and activation energy:

$$k = Ae^{\frac{-E_a}{RT}}$$

Average rate: the rate over a period of time; this is calculated by taking two points on the concentration vs. time graph and dividing the change in concentration by the change in time.

Bimolecular: a term used to describe an elementary step involving two molecules (or ions, atoms, or free radicals).

Catalyst: a material that speeds up a reaction without being permanently changed in the reaction.

Chemical kinetics: the study of the rates of chemical reactions.

Collision theory: a theory of chemical reaction rates that proposes that reactant molecules must collide with each other in order to react and that the molecules must collide with the proper orientation and with sufficient energy.

Elementary step: an individual step in a reaction mechanism.

Enzyme: a biological catalyst.

Half-life ($t_{1/2}$): the time required for the concentration of a reactant to decrease to half its initial value.

Heterogeneous catalyst: a catalyst that is present in a different phase from the reacting substances.

Homogeneous catalyst: a catalyst that is present in the same phase as the reacting substances.

Initial rate: the instantaneous reaction rate at the start of a reaction.

Instantaneous rate: the rate at a particular time; this is calculated by calculating the slope of the line tangent to the concentration *vs.* time graph at this time.

Integrated rate law: an equation that relates concentration and time; this equation is derived from the rate law using integral calculus.

Intermediate: a material that is produced in one step of a mechanism but is consumed in a later step.

Molecularity: the number of reactant molecules (or ions, atoms, or free radicals) involved in an elementary step.

Order: the exponent to which a reactant's concentration is raised in the rate law.

Rate constant: the proportionality constant in the rate equation.

Rate equation (rate law): the mathematical relationship between reactant concentrations and the reaction rate.

Reaction coordinate diagram: a diagram on which the y-axis is energy (or enthalpy) and the x-axis is a measure of a reaction's progress.

Reaction mechanism: the sequence of bond-making and bond-breaking steps that occurs during the conversion of reactants to products during a chemical reaction.

Reaction rate: the change in concentration of a substance divided by the change in time; if the material is a reactant, we must change the sign of this calculation.

Termolecular: a term used to describe an elementary step involving three molecules (or ions, atoms, or free radicals); termolecular steps are very rare.

Total order: the sum of all the exponents in a rate law.

Transition state: the chemical species present at the maximum in energy in a reaction coordinate diagram; it consists of atoms in some intermediate state of bond breaking and bond formation.

Unimolecular: a term used to describe an elementary step involving only one molecule (or ion, atom, or free radical).

Chapter Goals

By the end of this chapter you should be able to:

- ## Understand rates of reaction and the conditions affecting rates.

  ### a) Explain the concept of reaction rate (Section 15.1).

  In this chapter we explore the realm of chemical kinetics, the study of the rates of chemical reactions. The rate gives us an idea of how fast a reaction is occurring; the greater the rate, the faster the reaction is occurring. The rate of a chemical reaction is defined to be the change in concentration of a substance per unit of time.

  This last statement needs a little modification. Reaction rates are always positive. If we are measuring the appearance of a product, the final concentration will be larger than the initial concentration, so Δ concentration will be positive. So far, so good. For a product in a chemical reaction, the equation is what we said above.

  $$\text{Rate of appearance of a product} \ = \ \frac{\Delta \text{ concentration of product}}{\Delta \text{ time}}$$

  On the other hand, if we are measuring the disappearance of a reactant, the final concentration will be smaller than the initial concentration, so Δ concentration will be negative. In order to come out with a positive rate, we will need to change the sign of Δ concentration.

  $$\text{Rate of disappearance of a reactant} \ = \ - \ \frac{\Delta \text{ concentration of reactant}}{\Delta \text{ time}}$$

  Sometimes we like to compare the rates for the appearance or disappearance of the different materials involved in a particular chemical reaction. Consider the reaction
  $$2SO_2 \ (g) + O_2 \ (g) \rightarrow 2SO_3 \ (g)$$

  Because two SO_2 molecules must disappear for each oxygen molecule that disappears, the rate of disappearance of SO_2 must be twice that of the rate of disappearance of O_2. Another way of saying the same thing is that the rate of disappearance of oxygen is half the rate of disappearance of SO_2. Likewise, the rate of appearance of SO_3 must be twice the rate at which oxygen disappears and equal to the rate at which SO_2 appears. We can see that if we divide each rate by its stoichiometric coefficient, we will obtain the same number for all reactants and products involved in a chemical reaction. In this case,

  $$-\frac{1}{2}\left(\frac{\Delta[SO_2]}{\Delta t}\right) \ = \ -\frac{\Delta[O_2]}{\Delta t} \ = \ \frac{1}{2}\left(\frac{\Delta[SO_3]}{\Delta t}\right)$$

  In this equation, we are using the square brackets to stand for molar concentrations.

Example 15-1: Give the relative rates of disappearance of reactants and formation of products for the following reaction:

$$H_2 \ (g) + Cl_2 \ (g) \rightarrow 2HCl \ (g)$$

First, keep in mind that we need a negative sign in front of any reactant's expression to make sure that the rate will be positive. Secondly, we can obtain the relative rates by dividing each rate by the stoichiometric coefficient that goes with each substance.

$$-\frac{\Delta[H_2]}{\Delta t} = -\frac{\Delta[Cl_2]}{\Delta t} = \frac{1}{2}\frac{\Delta[HCl]}{\Delta t}$$

Try Study Question 1 in Chapter 15 of your textbook now!

b) Derive the average and instantaneous rates of a reaction from experimental information (Section 15.1).

The slope of the line y = mx + b is calculated as

$$\text{slope} = m = \frac{\Delta y}{\Delta x}$$

This has the same form as our equation for the rate of a chemical reaction. If we plot concentration on the y-axis and time on the x-axis, then the slope of this plot will be

$$\text{slope} = \text{rate} = \frac{\Delta \text{ concentration}}{\Delta \text{ time}}$$

The problem with this is that the plot of concentration *vs.* time is usually not a straight line. It is a curve. Unlike a line, the slope of this curve changes with time. To determine the average rate over a period of time, we mark the points of interest on the curve, draw a line connecting them, and determine the slope of this line.

To get better and better estimates of the rate at a particular point in time, what we could do would be to pick points that are closer and closer to the time of interest. To obtain the instantaneous rate at that point in time, we take this to the extreme. We draw a line that is tangent to the curve at the time of interest. A tangent line is a line that intersects the curve at one point only and has the same slope as the curve does at that point. So, we draw a tangent line and then determine the slope of this tangent line. This gives us the instantaneous rate.

Example 15-2: The following is data for the decomposition of dinitrogen monoxide on a gold surface at 900°C according to the following chemical equation:

$$N_2O \ (g) \rightarrow N_2 \ (g) + 1/2 \ O_2 \ (g)$$

Time (min)	$[N_2O]$ (mole/L)
15	0.0835
30	0.0680
80	0.0350
120	0.0220

a. Calculate the average rate of disappearance of N_2O from 20 to 110 minutes.
The first step will be to plot the data with concentration on the y-axis and time on the x-axis and to draw the best-fit curve for the data. After that, we mark the points at 20 minutes and at 110 minutes and connect them with a line. The following figure shows this.

The slope of this line gives us the average rate over that entire time period. The concentration at 20 min is 0.077 M and that at 110 min is 0.024 M.

$$\text{Average Rate} = -\frac{\Delta[N_2O]}{\Delta t} = -\frac{0.024\text{ M} - 0.076\text{ M}}{110\text{ min} - 20\text{ min}} = 0.00058\ \frac{M}{\min}$$

b. Calculate the instantaneous rate at 100 min.

We start off with the same plot of concentration *vs.* time. This time, however, we draw a tangent line to the curve at a time of 100 minutes. We then pick two points on this tangent line and determine the slope of the tangent line. This is the instantaneous rate.

Two points on the tangent line are (60 min, 0.044 M) and (115 min, 0.022 M). The slope of the line connecting these is

$$\text{Instantaneous Rate} = -\frac{0.022\text{ M} - 0.044\text{ M}}{115\text{ min} - 60\text{ min}} = 0.0040\ \frac{M}{\min}$$

You can see that the instantaneous rate at this point is different from the average rate over the whole time period.

Try Study Question 5 in Chapter 15 of your textbook now!

c. Describe factors that affect reaction rate (i.e., reactant concentrations, temperature, presence of a catalyst, and the state of the reactants (Section 15.2).

There are various factors that affect the rate of a chemical reaction. We shall examine reactant concentration, temperature, the presence of a catalyst, and the state of the reactants.

Reactant Concentration: In most cases, increasing the concentration of a reactant will increase the rate of a chemical reaction.

Temperature: Increasing the temperature of a reaction increases the reaction rate. We shall see that the reason for this is that at higher temperatures more molecules of the reactants possess enough energy to overcome the activation energy barrier for the reaction.

Presence of a Catalyst: A catalyst is a substance that accelerates a chemical reaction without being permanently changed in the reaction. In order for a material to be a catalyst, it must be present at the beginning of the reaction and also at the end of the reaction.

State of the Reactants: Atoms and molecules in the gas phase and in solution are mobile, so reactions are often carried out using mixtures of gases or in solution. The presence of a solid in a chemical reaction adds another factor that can affect reaction rates. The only place on a solid where reaction can occur is on the surface. The more surface exposed, the greater the reaction rate will be. One way to increase surface area is to break a solid up into small pieces. Before breaking up a large piece of a solid, the only surfaces available are those on the outside of the one big piece. Much of the solid is in the interior. If we break up the solid, each little piece has surfaces that are exposed to the other reactants. Parts of the solid that were once on the interior are now on the surface of the little pieces. One way to remember the surface area effect is to think of dissolving sugar in water. Granulated sugar dissolves more quickly than a big lump. This is because the granulated sugar is made up of smaller particles, each with their own surfaces. The surface area in the granulated sugar is larger.

- **Derive the rate equation, rate constant, and reaction order from experimental data.**

  a) Define the various parts of a rate equation (the rate constant and order of reaction) and understand their significance (Section 15.3).

  We have stated that the concentration of the reactants often affects the rate of a reaction. For the generic chemical equation

  $$aA + bB \rightarrow xX$$

  we can represent this concentration dependence mathematically as follows

  $$\text{Rate} \propto [A]^m[B]^n$$

  Just as we learned in the chapter dealing with the gas laws, we can turn a proportionality into an equality by introducing a proportionality constant. We will call this the rate constant, k.

  $$\text{Rate} = k[A]^m[B]^n$$

  This equation is called the rate equation or rate law. In the rate law, we have the concentration of each reactant raised to some power. In some cases, other species, such as a catalyst, may appear in the rate law as well.

The power to which a concentration is raised is called the order with respect to that reactant. Often, the orders are natural numbers (1, 2, …), but they can be zero, fractions, and negative numbers. The total reaction order is the sum of the exponents of the concentration terms.

The orders in a rate law must be determined by means of *experiments*. There is no way to know ahead of time what the orders will be.

Example 15-3: State the order for each concentration term in the following rate law. What is the over reaction order?

$$\text{Rate} = k[A][B]^2$$

The exponent for A is 1, and the exponent for B is 2. The reaction is said to be first order with respect to A and second order with respect to B.

To obtain the total order, we add together the exponents: $1 + 2 = 3$. The reaction is third order overall.

Try Study Question 7 in Chapter 15 of your textbook now!

One issue that arises with the rate constant is what its units should be. The units of the rate constant are different depending upon what order the reaction is overall. The units for a rate are always a concentration divided by time. The rate constant's units must be such that when all of the concentrations and orders are taken into account, we end up with only one concentration term and with a time unit in the denominator.

Example 15-4: Determine the correct units for the rate constant in each of the following cases. Assume that the time unit in each case is seconds.

    a. A first order reaction.

The generic form of a first order reaction is

$$\text{Rate} = k[A]$$

The units for the rate itself will be $\dfrac{\text{mole}}{L \bullet s}$. (Remember that molarity is moles/L.) The concentration of A provides the mole/L part of this unit. The only thing left is to introduce the time unit in the denominator. A first order rate constant has units of

$$\frac{1}{s} = s^{-1}$$
.

    b. A second order reaction.

The generic form of a second order reaction is

$$\text{Rate} = k[A]^2 \text{ or Rate} = k[A][B]$$

It doesn't matter which we choose because the units will work out the same.

The units for the rate itself will still be $\dfrac{\text{mole}}{L \bullet s}$. This time, the concentration terms will give units of $\dfrac{\text{mole}^2}{L^2}$. We need one of those moles to cancel and one of the liters to cancel. We also need to introduce the time unit in the denominator. Our units for the rate constant need to be $\dfrac{L}{\text{mole} \bullet s} = L\,\text{mole}^{-1}\,s^{-1}$.

b) Derive a rate equation from experimental information (Section 15.3).

To determine the rate law, we often use the method of initial rates. In order to determine the order for one of the reactants, we run two trials. The easiest type of experiment to analyze is one in which we start out with the same concentration of all of the reactants except for the one we are examining. The thing that we change is the concentration of this one reactant. Based on the effect that this change has on the initial rate, we can determine the order for that reactant. The following table summarizes some of the typical results.

If we change the concentration of the reactant by a factor of	And the rate increases by a factor of	Then the order for this reactant is
2	2	1
2	4	2
2	8	3
3	3	1
3	9	2
etc.		

Do not try to memorize this table! There is nothing magical about it. It is all based on what you already know about exponents: $2^1 = 2$, $2^2 = 4$, $2^3 = 8$, $3^1 = 3$, etc. Once we have determined the order for one reactant, we perform the same type of analysis where we vary the concentration of another reactant. We will see this in action in the next example.

Once we know the orders, all we need to do to determine the value of the rate constant is to substitute one trial's concentration values and rate into the rate law and solve for k.

Example 15-5: Determine the rate law for the following reaction
$$2A + 3B \rightarrow 2C$$
given the following concentrations and initial rates for a fictitious reaction.

Trial	[A] (mole/L)	[B] (mole/L)	Initial Rate (mole/(L·h))
1	0.0010	0.0030	0.0020
2	0.0020	0.0030	0.0040
3	0.0010	0.0090	0.018

In comparing Trials 1 and 2, we can see that we kept the concentration of B constant but doubled the concentration of A. We will divide the one with the larger rate by the one with the smaller rate.

$$\frac{\text{Rate 2}}{\text{Rate 1}} = \frac{k[A]_2^m[B]_2^n}{k[A]_1^m[B]_1^n}$$

$$\frac{0.0040}{0.0020} = \frac{k[0.0020]^m[0.0030]^n}{k[0.0010]^m[0.0030]^n}$$

You can see that the terms dealing with B cancel because we used the same concentrations of B in the two trials and that k will cancel as well. We are left with the following:

$$\frac{0.0040}{0.0020} = \frac{[0.0020]^m}{[0.0010]^m}$$

$$2.0 = (2.0)^m$$

To what power must 2 be raised to get 2 as the answer? $2^1 = 2$, so m must be equal to 1.

Let's now go after the exponent for B. The trials we will look at for this will be 1 and 3. In comparing these, we see that the concentration of A was kept constant but that the concentration of B was changed. We will divide rate 3 by rate 1. On the right side, we will use the corresponding rate law expressions.

$$\frac{\text{Rate 3}}{\text{Rate 1}} = \frac{k[A]_3^m[B]_3^n}{k[A]_1^m[B]_1^n}$$

Because A was kept constant, its terms will cancel, as will k. We can jump to the following line in the solution:

$$\frac{0.018}{0.0020} = \left(\frac{0.0090}{0.0030}\right)^n$$

$$9.0 = (3.0)^n$$

To what power must 3 be raised to get 9 as the answer? $3^2 = 9$ so n must be equal to 2.

We now know that the general form of the rate law is
$$\text{Rate} = k[A][B]^2$$

All that remains is to determine the value of k. For this, we can select any of the trials. Let's use the data from trial 2.

$$0.0040\ \frac{\text{mole}}{L \cdot s} = k\left(0.0020\ \frac{\text{mole}}{L}\right)\left(0.0030\ \frac{\text{mole}}{L}\right)^2$$

$$k = 2.2 \times 10^5\ \frac{L^2}{\text{mole}^2\ s} = 2.2 \times 10^5\ L^2\ \text{mole}^{-2}\ s^{-1}$$

The full rate law is Rate = $(2.2 \times 10^5\ L^2\ \text{mole}^{-2}\ s^{-1})\ [A]\ [B]^2$.

Try Study Questions 11 and 15 in Chapter 15 of your textbook now!

• **Use integrated rate laws.**

a) Describe and use the relationships between reactant concentration and time for zero-order, first-order, and second-order reactions (Section 15.4 and Table 15.1).

The rate laws we have studied involve the relationship between the rate and concentration. While this type of equation is useful for many applications, it does not give us a convenient way to relate time and concentration. We often wish to know what concentration of a reactant is present after some time period or we wish to know the amount of time that it will take for a reactant concentration to fall to a particular value. Using integral calculus, we can derive equations from our rate equations that allow us to carry out such calculations. We will concentrate on the equations we end up with, not on the calculus that is used to derive them. These equations are called integrated rate laws. We will look at only the simplest cases: reactions in which one reactant (R) decomposes to give products:
R → Products

We will consider only those cases where this reaction is first order, second order, or zero order. A different equation is obtained in each case.

First Order

There are two different forms of the equation that you can use.

$$\ln \frac{[R]_t}{[R]_0} = -kt$$

and

$$\ln[R]_t = -kt + \ln[R]_0$$

You're not going to go wrong with either equation for a first order process, but some problems can be solved more quickly or more easily using one or the other.

Second Order

$$\frac{1}{[R]_t} = kt + \frac{1}{[R]_0}$$

Zero Order

$$[R]_t = -kt + [R]_0$$

Example 15-6: A first order reaction has a rate constant of 0.203 day^{-1}. Originally the reactant makes up 89% of the dissolved solute in a solution. Calculate the amount of time it takes for the level of the reactant to fall to 45%.

We are told this is a first order reaction. One of the equations for a first order process is

$$\ln \frac{[R]_t}{[R]_0} = -kt$$

In this case, $[R]_t$ = 45%, $[R]_0$ = 89%, and k = 0.203 day^{-1}.

$$\ln \frac{[45\%]}{[89\%]} = -(0.203 \text{ day}^{-1})\, t$$

$$-0.68 = -(0.203 \text{ day}^{-1})\, t$$

$$t = 3.4 \text{ days}$$

Example 15-7: A first order reaction has a rate constant of 0.00510 min^{-1}. If we begin with a 0.10 M concentration of the reactant, how much of the reactant will remain after 3.0 hours?

We are told this is a first order reaction. One of the equations for a first order process is

$$\ln \frac{[R]_t}{[R]_0} = -kt$$

In this case, we are looking for $[R]_t$. We know that $[R]_0$ = 0.10 M, k = 0.00510 min^{-1}, and t = 3.0 hours. The time unit for the rate constant and the time do not match up. We begin by converting the time to minutes. After that, we can substitute into the first order equation.

$$3.0 \text{ hours} \times \frac{60 \text{ minutes}}{1 \text{ hour}} = 180 \text{ minutes}$$

$$\ln \frac{[R]_t}{[0.10 \text{ M}]} = -(0.00510 \text{ min}^{-1})(180 \text{ min})$$

$$\ln \frac{[R]_t}{[0.10 \text{ M}]} = -0.92$$

We need to get rid of the natural log. Ln and e are inverses of one another, so we will use both sides of the equation as exponents of e.

$$e^{\left(\ln\frac{[R]_t}{[0.10\ M]}\right)} = e^{-0.92}$$

As we said, ln and e undo each other, so the left side of the equation becomes the ratio of the two concentration terms:

$$\frac{[R]_t}{[0.10\ M]} = e^{-0.92}$$

$$\frac{[R]_t}{[0.10\ M]} = 0.40$$

$$[R]_t = 0.040\ M$$

Example 15-8: The rate constant for a second order process is 15 L mole^{-1} min^{-1}. If the initial concentration of the reactant is 0.10 M, what concentration remains after 20. min?

This is a second order process, so we use the second order equation:

$$\frac{1}{[R]_t} = kt + \frac{1}{[R]_0}$$

We are to solve for $[R]_t$, and we know k = 15 L mole^{-1} min^{-1}, t = 20. min, and $[R]_0$ = 0.10 M.

$$\frac{1}{[R]_t} = \left(15\ \frac{L}{mole\ min}\right)(20.\ min) + \frac{1}{[0.10\ M]}$$

$$\frac{1}{[R]_t} = 3.0 \times 10^2\ \frac{L}{mole} + 10\ \frac{L}{mole}$$

$$\frac{1}{[R]_t} = 310\ \frac{L}{mole}$$

We then take the reciprocal of both sides to obtain $[R]_t$.

$$[R]_t = 0.0032\ M$$

Example 15-9: The rate constant for a zero order process is 0.0030 mole L^{-1} s^{-1}. How long will it take for the initial concentration of the reactant to fall from 0.10 M to 0.075 M?

This is a zero order process, so we use the zero order equation:

$$[R]_t = -kt + [R]_0$$

We are asked to solve for time. We are given that $[R]_t$ = 0.075 M, $[R]_0$ = 0.10, and k = 0.0030 mole L^{-1} s^{-1}.

$$0.075\ M = -\left(0.0030\ \frac{mole}{L\ s}\right)t + 0.10\ M$$

$$-0.025\ M = -\left(0.0030\ \frac{mole}{L\ s}\right)t$$

$$t = 8.3\ s$$

Try Study Questions 17, 21, and 23 in Chapter 15 of your textbook now!

b) Apply graphical methods for determining reaction order and the rate constant from experimental data (Section 15.4 and Table 15.1).

Each of the reaction types we have studied (first order, second order, and zero order) has its own particular integrated rate law. Each of these can be arranged into the form of an equation of a line: y = mx + b. The following table summarizes this.

Order	Equation	y-axis	x-axis	Slope	y-intercept
First	$\ln[R]_t = -kt + \ln[R]_0$	$\ln[R]_t$	time	$-k$	$\ln[R]_0$
Second	$\dfrac{1}{[R]_t} = kt + \dfrac{1}{[R]_0}$	$\dfrac{1}{[R]_t}$	time	k	$\dfrac{1}{[R]_0}$
Zero	$[R]_t = -kt + [R]_0$	$[R]_t$	time	$-k$	$[R]_0$

There are two major uses of this type of graph. One is to use it to determine the value of k. We plot the correct type of graph for the particular reaction with which we are working and determine the slope of the line. It will either be equal to the rate constant or the negative of the rate constant. The other major use is to help us determine the order of the reaction. If we know that a reaction is one of these orders but do not know which one it is, we can construct all three graphs. The one that gives the best line is the one that corresponds to the correct order. A word of caution is in order. If you are doing this in the laboratory, make sure you collect data points for a sufficient period of time. At the very start of the reaction, all three graphs usually look pretty good. If you go out far enough, the deviations from linearity become more apparent.

Example 15-10: The following data were collected for the decomposition of a particular compound. Does this reaction obey first, second, or zero order kinetics? Also, determine the rate constant for the reaction.

Time (days)	$[R]_t$ (M)
0.00	0.101
3.00	0.0486
4.00	0.0376
6.11	0.0221

What we will do is to plot the data as if the reaction were each of the orders. Whichever graph produces a line will indicate which order the reaction really is. First, let's construct a data table that contains all of the information we need for the different graphs.

Time (days)	$[R]_t$ (M)	$\ln[R]_t$	$1/[R]_t$
0.00	0.101	−2.293	9.90
3.00	0.0486	−3.024	20.6
4.00	0.0376	−3.281	26.6
6.11	0.0221	−3.812	45.2

Here are the three plots:

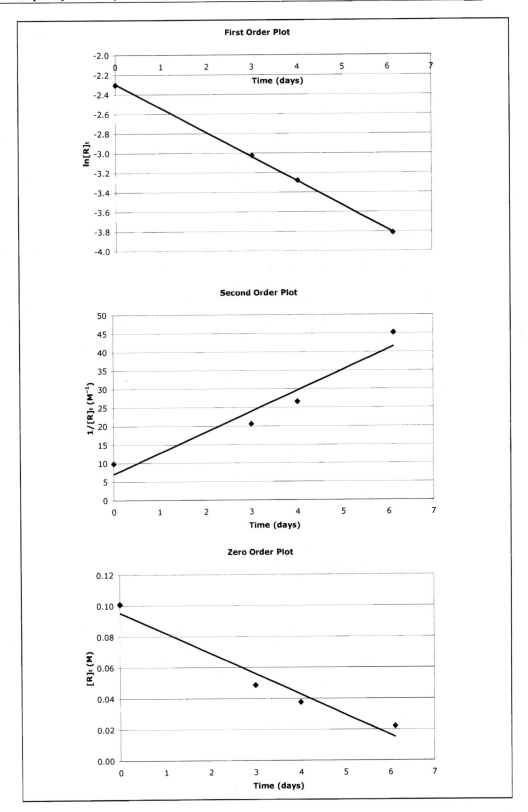

We can tell just by looking at these graphs that the second order plot and the zero order plot are not linear. The best-fit lines drawn in these graphs do not pass through the data points very well at all. The first order plot, however, is nicely linear. The best-fit line passes through the data points very well. The data is not inconsistent with the reaction being first order. The slope of the best-fit line in the first order plot can be calculated by taking two points on the line and calculating the slope or by having a computer program calculate the slope. It is -0.249. This corresponds to $-k$, so the value of the rate constant is 0.249 day^{-1}.

Try Study Question 33 in Chapter 15 of your textbook now!

c) Use the concept of half-life ($t_{1/2}$), especially for first-order reactions (Section 15.4).

The half-life of a reaction is the time required for the concentration of a reactant to decrease to one-half its initial value. The larger the half-life, the slower the reaction is. Each reaction type we have studied has its own equation to calculate the half-life. We will only worry about the equation for the half-life of a first order reaction.

$$t_{1/2} \;=\; \frac{\ln 2}{k} \;=\; \frac{0.693}{k}$$

For a first order reaction, the half-life is independent of the starting concentration of the reactant; this is not true for second order and zero order reactions.

After one half-life, the concentration of the reactant will be half its initial value. After two half-lives, it will be half of that amount, which corresponds to $1/2(1/2) = (1/2)^2 = 1/4$ of the original concentration. After three half-lives, it will be half of that amount, which corresponds to $1/2(1/2)(1/2) = (1/2)^3 = 1/8$ of the original concentration. In general, after n half-lives, $(1/2)^n$ of the original concentration remains.

Example 15-11: What is the half-life of a first-order reaction with $k = 0.015 \text{ min}^{-1}$?

We are told that this is a first order reaction. The equation for the half-life of a first order reaction is

$$t_{1/2} \;=\; \frac{0.693}{k}$$

We simply substitute into this equation.

$$t_{1/2} \;=\; \frac{0.693}{0.015 \text{ min}^{-1}} \;=\; 46 \text{ min}$$

If we had not been given the value of k, we would have had to determine the value of k using one of the ways that we have already learned.

Example 15-12: What fraction of the original concentration of the reactant in a first order reaction remains after 7 half-lives?

The concentration remaining after n half-lives is $(1/2)^n$. After seven half-lives, this corresponds to $(1/2)^7 = 1/128$. This corresponds to only 0.78% of the original concentration.

Try Study Questions 25 and 31 in Chapter 15 of your textbook now!

- **Understand the collision theory of reaction rates and the role of activation energy.**

  ### a) Describe the collision theory of reaction rates (Section 15.5).
  According to the collision theory of chemical reaction rates, in order for a chemical reaction to occur between molecules

  > 1. The reacting molecules must collide with one another.
  >
  > 2. The reacting molecules must collide with sufficient energy.
  >
  > 3. The reacting molecules must collide in an orientation that can lead to the proper rearrangement of the atoms.

  ### b) Relate activation energy (E_a) to the rate and thermodynamics of a reaction (Section 15.5).

  One of the tenets of collision theory is that the reacting molecules must collide with sufficient energy. Recall from our work with kinetic molecular theory, that at a given temperature there is a distribution of speeds and thus of the kinetic energies of the molecules in a gas sample. At a given temperature where reaction occurs, there is a fraction of the molecules that have enough energy to react. If we increase the temperature, we increase the fraction of molecules possessing this minimum energy. Because more molecules possess this energy, the reaction rate increases.

  This minimum energy required for reaction is called the activation energy (E_a) of the reaction. It can be pictured as an energy hill that the reaction must cross. On the diagram below it corresponds to the energy required to go from the reactants up to the highest point in energy on the diagram. This type of diagram is called a reaction coordinate diagram. It plots energy (or enthalpy) on the y-axis *vs.* some measure of the progress of the reaction.

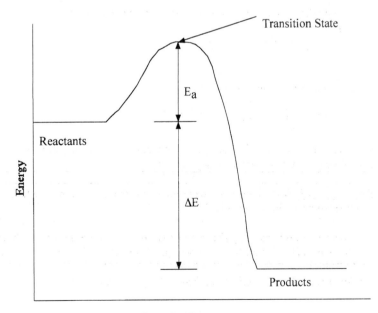

Reaction Progress

Why is there a minimum energy that molecules need in order to react with each other? We learned previously that breaking chemical bonds requires energy and that making chemical bonds releases energy. As a chemical reaction occurs, we get both bond breakage and bond formation. In the initial part of a chemical reaction, we need to add energy to start breaking the bonds in the reactants. The potential energy of the system goes up. Only after the reaction has proceeded part of the way do we start to get back enough energy from forming the new bonds to bring the potential energy back down. There is thus a peak in the energy coordinate diagram. The energy that it takes to go from the reactant up to this peak is the activation energy. The chemical species that exists at this point in the reaction is called the transition state. In the transition state, the original bonds in the reactants will be partially broken and the new bonds in the products will be partially formed.

The size of the activation energy is related to the rate of the reaction. The larger the activation energy, the slower the reaction will be. The smaller the activation energy, the faster the reaction will be. The reason has to do with the fact that more molecules will have sufficient energy to make it over the hill for a reaction that has a lower activation energy.

The difference in energy between the reactants and products is the change in energy for the reaction. This is marked as ΔE in the diagram. ΔE is related to the thermodynamics of the reaction. If the products are downhill from the reactants, then the reaction is exothermic. If the products are uphill from the reactants, then the reaction is endothermic. The previous diagram is an example of an exothermic reaction. The following diagram shows an example of an endothermic reaction.

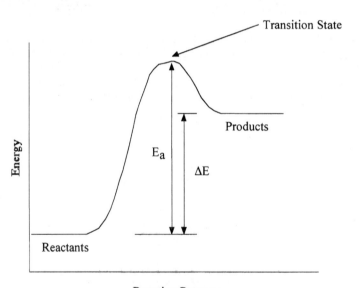

A reaction coordinate diagram thus has two distinct regions. The part dealing with the activation energy has to do with chemical kinetics – the rate of the reaction. The part dealing with ΔE (or ΔH) has to do with the thermodynamics of the reaction.

c) Use collision theory to describe the effect of reactant concentration on reaction rate (Section 15.5).

The rate of a chemical reaction is dependent upon the number of collisions between reactant molecules. If we increase the concentration of a reactant, there will be an increased number of collisions. Thus, increasing the concentration of a reactant will increase the rate of the reaction. (A more complete way of stating this using some information from later objectives dealing with mechanisms would be that increasing the concentration of a reactant in a given elementary step increases the rate of that elementary step. Because that elementary step may involve reaction intermediates or may not be the rate-determining step, we do not always see a first order relationship between a given overall reactant and the overall reaction rate.)

d) Understand the effect of molecular orientation on reaction rate (Section 15.5).

The third tenet of collision theory is that the molecules must collide with the proper orientation. The molecules must collide in such a way that the atoms that must form bonds with each other can do so. In the following diagram, assume that each type of circle (white, grey, and hatch-marked) represents a different kind of atom. We start out with the hatch-marked atom by itself and with the grey and white atoms linked together in a molecule. The reaction that occurs is that the white atom gets pulled off the grey atom and ends up attached to the hatch-marked atom. The top figure shows an example of an effective orientation for reaction because the hatch-marked atom and the white atom bump into each other. In this orientation we can break the bond between the grey and white atoms and form a bond between the white and hatch-marked atoms. The bottom figure shows an example of an orientation that does not lead to reaction. The white atom and the hatch-marked atoms do not come into contact so there is no way to form a bond between them. This would be an ineffective collision.

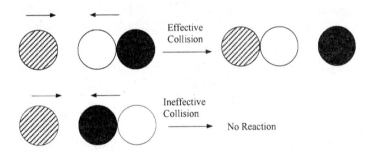

Try Study Question 97 in Chapter 15 of your textbook now!

e) Describe the effect of temperature on reaction rate using the collision theory of reaction rates and the Arrhenius equation (Equation 15.7 and Section 15.5).

Collision theory predicts that increasing the temperature should increase the rate of a chemical reaction. This is because at a higher temperature the molecules have a greater kinetic energy. More collisions will have sufficient energy to overcome the activation energy barrier and have a successful reaction.

The fact that the rates of chemical reactions increase with higher temperature is evidenced by the fact that the rate constant is a constant only at a given temperature. If we change the temperature, the rate constant will change; it will increase with increasing temperature. The Arrhenius equation is the mathematical relationship between the rate constant, temperature and activation energy:

$$k = Ae^{\frac{-E_a}{RT}}$$

In this equation, k is the rate constant, E_a is the activation energy, T is the temperature (in kelvins), R is the gas constant (use 8.3145×10^{-3} kJ/(mole K)), and A is a constant called the frequency factor. The value of A will be different for different reactions. It is related to the number of collisions and to the fraction of collisions that have the proper orientation. The full term of $e^{\frac{-Ea}{RT}}$ is interpreted as the fraction of the molecules that have the minimum energy required for reaction. The Arrhenius equation therefore predicts that the rate constant will be a function of the number of collisions having the proper orientation and the fraction of molecules possessing enough energy for reaction. The Arrhenius equation correctly predicts that the rate constant will increase with increasing temperature.

f) Use Equations 15.5, 15.6, and 15.7 to calculate the activation energy from experimental data (Section 15.5).

In the lab, we can do experiments to determine the rate constant for a reaction and we can also measure temperature. We can use these data to determine the activation energy for a chemical reaction. An analysis using the actual Arrhenius equation is not very useful because this equation leads to a curved graph. While modern computers can interpret such graphs, we often try to make the analysis more straightforward by using a rearranged form of this equation that will yield a linear graph. This rearranged form is

$$\ln k = -\frac{E_a}{RT} + \ln A$$

This has the form $y = mx + b$, the equation of a line. To obtain the linear graph, we plot $\ln k$ on the y-axis and $1/T$ on the x-axis. The slope will be $-\frac{E_a}{R}$ and the y-intercept will be $\ln A$. In order to determine the activation energy, we take the slope and multiply it by $-R$.

The most common mistakes that people make in solving a problem involving this equation are to forget to 1) take the natural logs of the rate constants, 2) convert the temperatures to kelvins, 3) take the reciprocals of the temperatures, 4) multiply the slope obtained by $-R$, and 5) use the correct value of R (8.3145×10^{-3} kJ/(mole K)).

Example 15-13: Imagine that temperature and concentration data were collected for a particular reaction. From the concentration data, the experimenters derived values of the rate constant for this reaction. From the rate constant and temperature data, determine the activation energy for this reaction.

Temperature (°C)	k (L mole^{-1} s^{-1})
25	3.8×10^{-8}
50	1.4×10^{-6}
100	6.9×10^{-4}
200	3.9

We are given rate constant and temperature data and asked for the activation energy. To solve this problem, we will construct an Arrhenius plot and determine the slope of the line, which is related to the activation energy. For the Arrhenius plot, we need to calculate $\ln k$ and $1/T$ where T is in kelvins. We will construct a data table that contains this information.

Temperature (°C)	k (L mole^{-1} s^{-1})	T (K)	1/T (K^{-1})	ln k
25	3.8×10^{-8}	298	0.00336	−17.09
50	1.4×10^{-6}	323	0.00310	−13.48
100	6.9×10^{-4}	373	0.00268	−7.28
200	3.9	473	0.00211	1.36

We next plot ln k on the y-axis and 1/T on the x-axis and draw the best-fit line.

Arrhenius Plot for Example 15-12

$y = -14907x + 32.793$
$R^2 = 0.9997$

We need to calculate the slope. To do this, we select two points on the line and determine $\Delta y/\Delta x$. The slope of the best-fit line is -1.5×10^4 K. This is equal to $-E_a/R$. To determine E_a, we multiply the slope by $-R$.

$$E_a = -R \cdot \text{slope} = - \left(8.3145 \times 10^{-3} \ \frac{kJ}{(\text{mole K})}\right)\left(-1.5 \times 10^4 \ K\right)$$

$$= 1.2 \times 10^2 \ \frac{kJ}{\text{mole}}$$

Try Study Question 77 in Chapter 15 of your textbook now!

If only two corresponding rate constants and temperatures are available, there is a two-point form of the Arrhenius equation that can be used.

$$\ln\left(\frac{k_2}{k_1}\right) = - \frac{E_a}{R}\left(\frac{1}{T_2} - \frac{1}{T_1}\right)$$

Example 15-14: For a particular reaction, it was found that the rate constant at 25°C was 0.0409 min^{-1} and that the rate constant at 45°C was 0.157 min^{-1}. What is the activation energy for this reaction?

We will use the two-point form of the Arrhenius equation. We first need to obtain the temperatures in kelvins. 25°C corresponds to 298 K, and 45°C corresponds to 318 K.

$$\ln\left(\frac{k_2}{k_1}\right) = - \frac{E_a}{R}\left(\frac{1}{T_2} - \frac{1}{T_1}\right)$$

$$\ln\left(\frac{0.157 \ \text{min}^{-1}}{0.0409 \ \text{min}^{-1}}\right) = - \frac{E_a}{8.3145 \times 10^{-3} \ \dfrac{kJ}{\text{mole K}}}\left(\frac{1}{318 \ K} - \frac{1}{298 \ K}\right)$$

$$\ln(3.84) = - \frac{E_a}{8.3145 \times 10^{-3} \ \dfrac{kJ}{\text{mole K}}}\left(-0.000211 \ K^{-1}\right)$$

$$1.345 = 0.0254 \, E_a$$
$$E_a = 53.0 \, \frac{kJ}{mole}$$

Try Study Question 41 in Chapter 15 of your textbook now!

• **Relate reaction mechanisms and rate laws.**

a) Describe the functioning of a catalyst and its effect on the activation energy and mechanism of a reaction (Section 15.5).

A catalyst is a material that speeds up a chemical reaction without itself being permanently changed in the reaction. This does not mean that the catalyst is not involved in the reaction. How does a catalyst speed up a reaction? When the catalyst is present, there is a different reaction pathway available to get from reactants to products; this different pathway has a lower activation energy than does the uncatalyzed reaction. Because the activation energy required is lower, the reaction is faster when the catalyst is present. A catalyst will be present at the beginning of the reaction and will be formed again in a later step in the reaction. It thus does not undergo a permanent change in the reaction. In a reaction mechanism, we can identify a catalyst by looking for a material that starts out on the left side of one of the steps but which is then formed again on the right side of one of the later steps.

A homogeneous catalyst is present in the same phase as the reacting substances. A heterogeneous catalyst is present in a different phase than the reacting substances.

b) Understand reaction coordinate diagrams (Section 15.5).

A drawing of a reaction coordinate diagram follows. One the y-axis is plotted energy (or enthalpy) and on the x-axis is plotted some measure of the progress of the reaction. The energy at first goes up to a maximum. The energy required to get to this maximum is the activation energy (E_a). The chemical species that exists at this point in the reaction is called the transition state. The size of the activation energy is related to the rate of the reaction. The larger the activation energy, the slower the reaction will be. The smaller the activation energy, the faster the reaction will be.

The difference in energy between the reactants and products is the change in energy for the reaction. This is marked as ΔE in the diagram. ΔE is related to the thermodynamics of the reaction. If the products are downhill from the reactants, then the reaction is exothermic. If the products are uphill from the reactants, then the reaction is endothermic. The diagram shown is an example of a reaction coordinate for an exothermic reaction.

The diagram shown is for a one-step reaction with no reaction intermediates. If there were a reaction intermediate, we would see that the energy would go up, then go down somewhat, then go up again to another transition state, and finally come down to the energy state of the final products. The reaction intermediate is present at the valley (the local minimum) between two humps on a reaction coordinate diagram. The overall activation energy for a multistep reaction is the energy difference between the reactants and the highest energy transition state on the diagram.

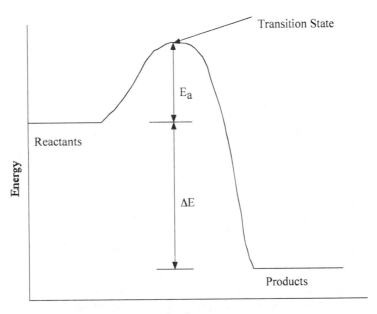

Reaction Progress

c) Understand the concept of a reaction mechanism (the sequence of bond-making and bond-breaking steps that occurs during the conversion of reactants to products) and the relation of the mechanism to the overall, stoichiometric equation for a reaction (Section 15.6).

The reaction mechanism for a chemical reaction is the series of bond-breaking and bond-making steps that occurs in the transformation of reactants to products. All of the steps in a reaction mechanism must add up to give the overall balanced equation for the reaction.

We can never have proof that we have guessed the reaction mechanism correctly; all we can do is make our best guess by proposing a mechanism that does not conflict with the known data. If other data come along that show our mechanism to be false, we must change our proposed mechanism so that it agrees with the data. Studies of reaction rates are one of the best ways to obtain evidence about reaction mechanisms.

d) Describe the elementary steps of a mechanism and give their molecularity (Section 15.6).

The individual steps in a reaction mechanism are called elementary steps. The number of reactant molecules (or ions, atoms, or free radicals) that react in an elementary step is called the molecularity of that step. If there is only one reactant molecule in an elementary step, it is a unimolecular step. If there are two reactant molecules in an elementary step, it is a bimolecular step. In a bimolecular step, the reactant molecules may be different species or the same species. It is very rare to have an elementary step that involves three reactant molecules; when that does happen it is called a termolecular step.

When we first began studying rates of chemical reactions, we said that we could not determine the orders of the reactants in the overall rate law just by looking at the chemical equation. This is absolutely true. These must be determined experimentally! Things are different, however, for an elementary step. In the rate law for a given elementary step, the order for each reactant in the rate law for that step is given by the stoichiometric coefficient for that reactant in the chemical equation for that step. The total order for an elementary step's rate will be the same as its molecularity. Keep in mind that these things are true only for elementary steps, not for an overall reaction.

<u>Example 15-15:</u> Identify the molecularity of the following elementary step in a chemical reaction. What is the rate law that will go along with this step?

$$Br_2 \ (g) + NO \ (g) \rightarrow Br_2NO \ (g)$$

There are two species on the reactant side of this step: one molecule of Br_2 and one molecule of NO. This is a bimolecular step.

Because this is an elementary step, the orders of the reactants are given by the stoichiometric coefficients. This step will be first order in Br_2 and first order in NO.

Rate = k $[Br_2]$ $[NO]$

Try Study Question 47 in Chapter 15 of your textbook now!

e) Define the rate-determining step in a mechanism and identify any reaction intermediates (Section 15.6).

A reaction can proceed no faster than the rate of the slowest elementary step in the process. The rate of an overall reaction is limited by the combined rates of all elementary steps up through the slowest step in the mechanism. Often, the overall reaction rate and the rate of the slowest step are nearly the same. If this is the case, the slowest step is called the rate-determining step.

A reaction intermediate is a material that is produced in one step of a reaction mechanism but is consumed in a later step. Just like a catalyst, it will not appear in the overall balanced equation for the reaction. Be careful to distinguish between a reaction intermediate and a transition state. A transition state is a maximum on the reaction coordinate diagram. A reaction intermediate is a local minimum on the reaction coordinate diagram.

A typical mechanism problem in general chemistry is to determine the rate law predicted by a given mechanism. The key thing to keep in mind is that the overall rate law is dependent upon the rate law for the slowest step and that the rate law should involve reactants, products, and/or catalysts, no reaction intermediates.

<u>Example 15-16:</u> Determine the rate law predicted by the following reaction mechanism:

The first thing to do is to identify the rate-determining step. This is the slowest step, step 1 in this case. The rate law for this step is

Rate = $k_1[(CH_3)_3I]$

In this case, the predicted rate law is already in terms of just reactants, so this is the rate law for this reaction. (Also note that the cation produced in Step 1 is an intermediate; it is produced in Step 1 but consumed in Step 2.)

Example 15-17: This is based on Exercise 15.13 in your textbook. One possible mechanism for the decomposition of nitryl chloride, NO_2Cl, is

$$\text{Step 1} \qquad NO_2Cl\ (g) \underset{k_{-1}}{\overset{k_1}{\rightleftharpoons}} NO_2\ (g) + Cl\ (g) \qquad \text{Fast, Equilibrium}$$

$$\text{Step 2} \qquad NO_2Cl\ (g) + Cl\ (g) \xrightarrow{k_2} NO_2\ (g) + Cl_2\ (g) \qquad \text{Slow}$$

What is the overall rate law predicted by this mechanism? Identify any intermediates.

The slow step in this reaction is the second step. The rate law predicted for this elementary step is
$$\text{Rate} = k_2 [NO_2Cl][Cl]$$

The problem with this rate law is that it includes [Cl], which is an intermediate. What we need to do is to determine the value of [Cl] in terms of reactants and/or products.

To do this, we will take advantage of the fact that Cl is also a part of step 1 and that step 1 is an equilibrium. We know that in an equilibrium the rate of the forward reaction is equal to the rate of the reverse reaction. For the first step, therefore,
$$k_1 [NO_2Cl] = k_{-1} [NO_2][Cl]$$
where k_1 is the rate constant of the forward direction of step 1 and k_{-1} is the rate constant of the reverse direction.

We now solve this equation for [Cl]
$$[Cl] = \frac{k_1}{k_{-1}} \frac{[NO_2Cl]}{[NO_2]}$$

We can substitute this expression into our previous expression for the rate-determining step.
$$\text{Rate} = k_2 [NO_2Cl][Cl]$$
$$= k_2 [NO_2Cl] \left(\frac{k_1}{k_{-1}} \frac{[NO_2Cl]}{[NO_2]} \right)$$

All of the constants can be combined into one overall constant.
$$\text{Rate} = k_{obs} \frac{[NO_2Cl]^2}{[NO_2]}$$

To determine if there are any intermediates present in the mechanism, we look for a material produced in one step and used up in a later step. We can see that we produce Cl in Step 1 and that it is used up in Step 2. The Cl is an intermediate.

Try Study Question 49 and 51 in Chapter 15 of your textbook now!

CHAPTER 16: Principles of Reactivity: Chemical Equilibria

Chapter Overview

Many chemical reactions are reversible. With this chapter, we begin our study of chemical equilibrium in earnest. Chemical equilibria are dynamic equilibria: both the forward and reverse reactions still occur, but they do so at equal rates. For a given chemical reaction

$$aA + bB <=> cC + dD$$

we can calculate a quantity called the reaction quotient, Q

$$Q = \frac{[C]^c[D]^d}{[A]^a[B]^b}$$

where the quantities in brackets are the concentrations of the reactants and products in units of moles/L. As the reaction proceeds, Q eventually levels off at a particular value for each reaction. This value is called the equilibrium constant. In the equilibrium constant expression, we do not include concentrations for solids, pure liquids, and solvents. An equilibrium constant based on concentrations, such as we have been discussing, is sometimes called K_c. For gases, we can define another type of equilibrium constant in which we use the partial pressures of the gases instead of the concentrations. This type of K is called K_p. K_c and K_p can be related with the expression $K_p = K_c(RT)^{\Delta n}$ If K >> 1, a reaction is product-favored. If K << 1, it is reactant favored. We can use the reaction quotient to predict whether a reaction will be spontaneous in going from reactants to products or vice versa. If Q < K, then the reaction will proceed from reactants to products. If Q = K, then the reaction is at equilibrium. If Q > K, then the reaction will proceed from products to reactants.

You will learn how to determine an equilibrium constant given information about reactants and products. You will also learn how to calculate equilibrium concentrations of materials given the value of K and the initial concentrations. You will learn how to determine the K for a given chemical equation given information about the equilibrium constants for other related chemical reactions. If we multiply a chemical equation by a factor, n, then the equilibrium constant for the new equation will be $K_{new} = (K_{original})^x$. If we reverse a chemical equation, we must take the reciprocal of the equilibrium constant. If we add two equations, we multiply their equilibrium constants.

We can predict the direction a chemical equilibrium will shift when the equilibrium conditions of the system are changed. Le Chatelier's Principle states that a change in any of the factors determining an equilibrium will cause the system to adjust so as to reduce or minimize the effect of the change. You will learn how to interpret Le Chatelier's Principle for cases involving changes of concentration (or partial pressure) of a reactant or product, changing the volume of a system involving gases, or changing the temperature of the system.

Key Terms

In this chapter, you will need to learn and be able to use the following terms:

Dynamic equilibrium: a situation in which two reverse processes are occurring at the same rate so that there is no net change in the concentrations of reactants or products.

Equilibrium constant expression: a mathematical expression that relates the concentrations of the reactants and products at equilibrium. The equilibrium concentrations of the products appear in the numerator and the equilibrium concentrations of the reactants appear in the denominator. Each of these equilibrium concentrations is raised to the power of its stoichiometric coefficient in the balanced chemical equation for the reaction.

Equilibrium constant: the value of the equilibrium constant expression for a given reaction. For a given reaction at a given temperature, there will be only one value for the equilibrium constant, no matter how the equilibrium was achieved.

Le Chatelier's Principle: a change in any of the factors determining an equilibrium will cause the system to adjust so as to reduce or minimize the effect of the change.

Product-favored: a reaction that at equilibrium has more products than reactants.

Reactant-favored: a reaction that at equilibrium has more reactants than products.

Reaction quotient: a mathematical expression that relates the concentrations of the reactants and products in a chemical reaction. The concentrations of the products appear in the numerator and the concentrations of the reactants appear in the denominator. Each of these concentrations is raised to the power of its stoichiometric coefficient in the balanced chemical equation for the reaction.

Chapter Goals

By the end of this chapter you should be able to:

- **Understand the nature and characteristics of chemical equilibria.**

  Chemical reactions are reversible and equilibria are dynamic (Section 16.1).

  In previous chapters, we have examined some physical processes in which an equilibrium state was achieved: the equilibrium between a liquid and its vapor in a closed container and also the equilibrium between undissolved solute and a solution in a saturated solution. Let's review some key points of one of these: the liquid-vapor case. If we place a liquid in a closed container, then initially we see that the liquid evaporates. The rate of evaporation is faster than the rate of condensation. As we get more vapor molecules in the air above the liquid, the odds of a vapor molecule colliding with the surface of the liquid and condensing increases. The rate of condensation increases; simultaneously, the rate of evaporation slows down. Eventually, we get to a state where the rate of evaporation equals the rate of condensation. At this point, we have a state of dynamic equilibrium. We have two reverse processes occurring at the same rate. There is no further net change in the composition of the liquid and vapor states, but both processes are still occurring. We represent this as follows where the double arrows indicate that we have a process that is reversible.

  Liquid <=> Vapor

  In this chapter, we begin our detailed study of chemical equilibrium. Chemical equilibria have many of the same features as the physical equilibria we have already studied. Chemical reactions are reversible, they can occur in both the forward and reverse directions. So far, we have usually considered the forward direction of chemical reactions. For example,

$$N_2 \text{ (g)} + 3H_2 \text{ (g)} \rightarrow 2NH_3 \text{ (g)}$$

On the other hand, the reverse reaction can also occur.

$$2NH_3 \text{ (g)} \rightarrow N_2 \text{ (g)} + 3H_2 \text{ (g)}$$

We can summarize these two equations by using the double arrow that indicates that we have a reversible process.

$$N_2 \text{ (g)} + 3H_2 \text{ (g)} <=> 2NH_3 \text{ (g)}$$

In the case of a chemical equilibrium, we have two reverse processes occurring, the forward reaction and the reverse reaction, in a closed system (a system that does not exchange matter with the surroundings, not necessarily a closed container). Equilibrium is established when the rate of the forward reaction is equal to the rate of the reverse reaction. If we start out with only reactants present, the forward reaction proceeds more quickly at first. As the concentration of the products increases, the odds of them colliding and reacting in the reverse reaction increases. As the reaction proceeds, the reverse reaction speeds up and the forward reaction slows down until they are occurring at the same rate. When this occurs, the system is at equilibrium. Chemical equilibria, just like the other equilibria we have studied, are dynamic equilibria. The forward and reverse reactions do not stop. Both are occurring, but they are occurring at the same rate so that the overall concentrations of reactants and products do not change once equilibrium has been established.

- **Understand the significance of the equilibrium constant, K, and reaction quotient, Q.**

  a) Write the reaction quotient, Q, for a chemical reaction (Section 16.2). When the system is at equilibrium, the reaction quotient is called the equilibrium constant expression and has a constant value called the equilibrium constant, which is symbolized by K (Section 16.2).

  For a given reversible chemical reaction,

  $$aA + bB <=> cC + dD$$

  we can calculate a quantity called the reaction quotient, Q,

  $$Q = \frac{[C]^c [D]^d}{[A]^a [B]^b}$$

  where [A], [B], [C], and [D] are equal to the molar concentrations of these species present and where the exponents a, b, c, and d are equal to the stoichiometric coefficients in the balanced chemical equation. Notice that the products go in the numerator of this quotient, and the reactants go in the denominator.

  The concentrations that we use in the reaction quotient are the concentrations that we happen to have in the reaction. As such, Q may have any value. As the reaction proceeds toward the equilibrium state, the value of Q will change as the concentrations of the reactants and products change. When the system arrives at equilibrium, the concentrations of the reactants and products stop changing, and the value of the reaction quotient stops changing. The value of the reaction quotient at equilibrium is called the equilibrium constant. The definition of the equilibrium constant looks just like the reaction quotient, but the difference is that the concentrations are the concentrations present *at equilibrium.*

  $$K = \frac{[C]^c [D]^d}{[A]^a [B]^b}$$

The equilibrium constant is constant in more senses than just that the value of the reaction quotient stops changing. For a given chemical reaction at a given temperature, it does not matter what concentrations we start with, the value of the equilibrium constant will be the same. This does not mean that every possible combination of initial reactant and product concentrations will give the same equilibrium reactant and product concentrations. It means that all possible combinations of initial and product concentrations will yield equilibrium concentrations whose ratio in the proper format will be the same number. If we are at equilibrium for a given reaction at a given temperature, the value of the equilibrium constant will be the same.

Example 16-1: Write the equilibrium constant expression for the following reaction:
$$N_2 \ (g) + 3 \ H_2 \ (g) \Longleftrightarrow 2NH_3 \ (g)$$

K has the product in the numerator raised to the power of its stoichiometric coefficient. It has the reactants in the denominator each raised to the power of its stoichiometric coefficient.

$$K = \frac{\left[NH_3 \right]^2}{\left[N_2 \right] \left[H_2 \right]^3}$$

---

b) Recognize that the concentrations of solids, pure liquids, and solvents (e.g., water) are not included in the equilibrium constant expression (Equation 16.1, Section 16.2).

The concentrations of solids, pure liquids, and solvents are not included in the equilibrium constant expression. The reason for this has to do with the fact that these concentrations do not change during the course of a reaction. For example, the concentration of a solid, the number of moles per liter is a constant, which can be obtained from the density and the molar mass. The density does not change during the course of a chemical reaction, so the concentration does not change, only the amount of the solid of that density that is present.

Example 16-2: Write the equilibrium constant expression for the following reaction:
$$Cu(OH)_2 \ (s) + 2HCl \ (aq) \rightarrow 2H_2O \ (l) + CuCl_2 \ (aq)$$

In the equilibrium constant expression, we will not include the $Cu(OH)_2$ because it is a solid or the H_2O because it is the solvent for this reaction. The equilibrium constant expression is

$$K = \frac{\left[CuCl_2 \right]}{\left[HCl \right]^2}$$

Try Study Question 1 in Chapter 16 of your textbook now!

c) Recognize that a large value of K (K >> 1) means the reaction is product-favored, and the product concentrations are greater than the reactant concentrations at equilibrium. A small value of K (K << 1) indicates a reactant-favored reaction in which the product concentrations are smaller than the reactant concentrations at equilibrium (Section 16.2).

One of the major issues we shall examine at various points in upcoming chapters is whether a reaction is product- or reactant-favored, whether when all is said and done, we have more products or reactants present at equilibrium.

For the chemical reaction,
$$aA + bB \Longleftrightarrow cC + dD$$
the equilibrium constant expression is

$$K = \frac{[C]^c [D]^d}{[A]^a [B]^b}$$

If K is much larger than 1, then the numerator is larger than the denominator by a significant extent. Because the products are in the numerator, this means that there must be a greater concentration of products at equilibrium than of reactants. The reaction is product-favored.

If K is much smaller than 1, then the denominator is larger than the numerator by a significant extent. Because the reactants are in the denominator, this means that there must be a greater concentration of reactants than of products. The reaction is reactant-favored.

When K is close to 1, it is not immediately apparent whether the reactant concentrations or the product concentrations are greater. It depends on the number of reactants and products and on the reaction stoichiometry.

If K is equal to 1, it is the very rare condition that the numerator and denominator have exactly the same value: $[C]^c [D]^d = [A]^a [B]^b$

We can summarize the key findings in the following table:

K	Reactant or Product Favored?
K >> 1	Product-Favored
K = 1	$[C]^c [D]^d = [A]^a [B]^b$
K << 1	Reactant Favored

Example 16-3: Is the following reaction product- or reactant-favored?
$$O_3 \ (g) + NO \ (g) <=> O_2 \ (g) + NO_2 \ (g) \qquad K = 6.0 \times 10^{34}$$

We look to see if K is much greater or much less than 1. It is much greater. If K >> 1, then the reaction is product-favored.

Try Study Question 65 in Chapter 16 of your textbook now!

d) Appreciate the fact that equilibrium concentrations may be expressed in terms of reactant and product concentrations (in moles per liter), and K is then sometimes designated as K_c. Alternatively, concentrations of gases may be represented by partial pressures, and K for such cases is designated K_p (Section 16.2).

So far, we have calculated all of our equilibrium constants by using the concentrations of the reactants and products (in moles per liter). Sometimes, we explicitly indicate that we have used concentrations in calculating an equilibrium constant by writing a subscript c, K_c. For a generic chemical reaction

$$aA + bB <=> cC + dD$$

in which A, B, C, and D are either solutes or are gases (remember that we do not include solids, pure liquids, or solvents in the equilibrium constant expressions) we define K_c as

$$K_c = \frac{[C]^c [D]^d}{[A]^a [B]^b}$$

This is the same definition we have been using for K up to this point.

When we have gases involved in a chemical reaction, instead of using the concentrations, we can also use the partial pressures of the gases because the partial pressure and the concentration in moles per liter (n/V) are proportional to (but not equal to) each other. According to the ideal gas law, they are related by the factor RT. If we use the partial pressures of the gases in an equilibrium constant, we refer to the equilibrium constant as K_p.

For our generic chemical equation, if all of the species are gases, the equilibrium constant written in terms of partial pressures is

$$K_p = \frac{P_C{}^c P_D{}^d}{P_A{}^a P_B{}^b}$$

where P_A refers to the partial pressure of gas A, P_B refers to the partial pressure of gas B, etc.

It can be shown that K_c and K_p can be related by the following equation:

$$K_p = K_c (RT)^{\Delta n}$$

In this equation, Δn = total moles of gaseous products – total moles of gaseous reactants. One consequence of this is that if we have the same number of moles of gaseous products as gaseous reactants, then K_p will equal K_c. Otherwise, K_p will not equal K_c.

Example 16-4: The following chemical reaction is carried out at 25°C:
$$2NO_2 (g) <=> N_2O_4 (g)$$
At equilibrium, it is found that the partial pressure of NO_2 is 0.30 atm and that the partial pressure of N_2O_4 is 0.60 atm.

a. Determine the value of K_p.
First, we write out the equilibrium constant expression for Kp and then we substitute these partial pressures into the expression.

$$K_p = \frac{P_{N_2O_4}}{P_{NO_2}{}^2} = \frac{0.60}{(0.30)^2} = 6.7$$

b. What is the value of K_c?
We know that
$$K_p = K_c (RT)^{\Delta n}$$

For this reaction, we have 1 mole of gaseous products and 2 moles of gaseous reactants, so $\Delta n = 1 - 2 = -1$.

$$K_p = K_c (RT)^{-1} = \frac{K_c}{RT}$$

We are asked for Kc, so we must multiply both sides of the equation by RT.
$$K_p RT = K_c$$
$$K_c = 6.7 \left(0.082057 \frac{L\ atom}{mole\ K}\right)(298\ K) = 1.6 \times 10^2$$

Try Study Question 67a in Chapter 16 of your textbook now!

- **Understand how to use K in quantitative studies of chemical equilibria.**

a) Use the reaction quotient (Q) to decide whether a reaction is at equilibrium (Q = K), or if there will be a net conversion of reactants to products (Q < K) or products to reactants (Q > K) to attain equilibrium (Section 16.2).

Another issue that we shall revisit at various points in the rest of the course is whether a particular mixture of reactants and products is at equilibrium or not. If it is not at equilibrium, then we want to be able to answer the question as to which direction the reaction will proceed: to make more products or to make more reactants. (This is a separate issue from the earlier one we discussed as to whether *at equilibrium*, the reaction is product-favored or reactant-favored. For that issue, we found that the relationship of K to the number 1 gave us a way to answer the question.)

There are a variety of ways to ask this new question about whether the reaction is at equilibrium or not and if not, in which direction the reaction will proceed to get to equilibrium. Some examples of phrasings that can be used are 1) whether the reaction will result in a net conversion of reactants to products or of products to reactants, 2) whether the reaction is spontaneous in the direction written or in the opposite direction, or 3) whether there will be a shift to the right (production of products) or to the left (production of reactants). All of these different ways of asking the question really relate back to the same basic idea. How do we tell if a reaction is at equilibrium and if it is not at equilibrium, how do we tell in which direction the reaction will proceed in order to get to equilibrium?

We have already learned that at equilibrium, the value of the reaction quotient will be equal to the equilibrium constant. With that statement, we have a way to tell if the reaction is at equilibrium or not at equilibrium. If Q = K, then the reaction is at equilibrium. If Q ≠ K, then the reaction is not at equilibrium.

That answers part of our question, but we would like a way to tell in which direction the reaction will proceed if it is not at equilibrium.

Think back to our initial definition of Q for the generic reaction aA + bB <=> cC + dD.

$$Q = \frac{[C]^c[D]^d}{[A]^a[B]^b}$$

In this equation, the concentrations of the products are in the numerator, and the concentrations of the reactants are in the denominator. If Q is greater than K (Q > K), Q must get smaller in order to reach the value of K. The way for Q to get smaller is for the top of the fraction in the definition of Q to get smaller and the bottom to get bigger. The concentrations of the products must come down, and the concentrations of the reactants must go up. The reaction will proceed from products to reactants, to the left.

On the other hand, if Q is less than K (Q < K), then Q must get bigger to reach the value of K. The top of the fraction in the definition of Q must get bigger, and the bottom must get smaller. The concentrations of the products must go up, and the concentrations of the reactants must come down. The reaction will proceed from reactants to products, to the right.

We can summarize this in the following table:

Q *vs.* K	Outcome (1, 2, and 3 are different ways of saying the same thing.)
Q > K	1) Reaction proceeds from products to reactants. 2) Reaction proceeds to the left. 3) Reaction is not spontaneous as written; it is spontaneous in the opposite direction.
Q = K	The reaction is at equilibrium.
Q < K	1) Reaction proceeds from reactants to products. 2) Reaction proceeds to the right. 3) Reaction is spontaneous as written.

Example 16-5: At 700°C, K_c for the reaction C *(s)* + CO_2 *(g)* <=> 2CO *(g)* is equal to 0.025. If a 1.0 L flask contains 4.0×10^{-3} moles of CO and 6.0×10^{-3} moles of CO_2, and 5.0×10^{-2} moles of C, is the system at equilibrium? If not, in which direction will the reaction proceed in order to get to equilibrium?

In this problem, we are given the number of moles of each species and the volume of the flask. We are also given the equilibrium constant. We are asked to determine if the reaction is at equilibrium. We will do this by calculating Q and comparing it to K.

To calculate Q, we need the concentrations of CO and CO_2. Since the volume of the flask is 1.0 L, the concentrations of CO and CO_2 are the same as the number of moles. We do not need to worry about the C since it is a solid and will not appear in our expressions for Q and K.

$$Q = \frac{[CO]^2}{[CO_2]} = \frac{[4.0 \times 10^{-3}]^2}{[6.0 \times 10^{-3}]} = 0.0027$$

K = 0.025 and Q = 0.0027. Q $\neq$ K, so the system is not at equilibrium. Furthermore, Q < K. In order to get to equilibrium, Q will need to increase. To do this, the concentrations of the products will need to increase, and the concentrations of the reactants will need to decrease. The reaction will proceed to the right, from reactants to products.

Try Study Question 5 in Chapter 16 of your textbook now!

b) Calculate an equilibrium constant given the reactant and product concentrations at equilibrium (Section 16.3).

We can use the reactant and product concentrations present at equilibrium to calculate an equilibrium constant. There are two basic types of problems you will run into. In the first type, you are given the concentrations of all of the reactants and products at equilibrium. To solve this type of problem, you just substitute these values into the equilibrium constant expression.

Example 16-6: For the following equilibrium at 25°C, N_2O_4 *(g)* <=> 2NO_2 *(g)*, it was found that the equilibrium concentration of N_2O_4 was 0.0027 moles/L and that of NO_2 was 0.0040 moles/L. Calculate the value of the equilibrium constant.

We are given the concentrations of the reactants and products at equilibrium and are asked to determine the value of K. We substitute these concentrations into the equilibrium constant expression.

$$K = \frac{[NO_2]^2}{[N_2O_4]} = \frac{(0.0040)^2}{0.0027} = 5.9 \times 10^{-3}$$

Try Study Question 7 in Chapter 16 of your textbook now!

The more frequent type of problem you will run across will be one in which you will be given information so that you can calculate the initial (not the equilibrium) concentrations of the materials and also given information about one of the equilibrium concentrations. From this information, you will need to determine the equilibrium concentrations of all of the reactants and products (using the balanced chemical equation to help you). Once you have all of the equilibrium concentrations, you can calculate the equilibrium constant.

In solving this type of problem (and many others as well), it is useful to set up an ICE table underneath the chemical equation. ICE stands for initial, change, and equilibrium.

Example 16-7: A mixture consisting of 0.400 moles of nitrogen gas and 0.400 moles of hydrogen gas was introduced into a 2.00 L flask at 450°C. The following equilibrium set up:

$$N_2 \ (g) + 3H_2 \ (g) <=> 2NH_3 \ (g)$$

At equilibrium, it was determined that the concentration of ammonia in the flask was 0.013 moles/L. Determine the equilibrium constant for this reaction.

The equilibrium constant expression for this reaction is

$$K = \frac{[NH_3]^2}{[N_2][H_2]^3}$$

The concentrations in this expression are equilibrium concentrations. We are given the equilibrium concentration of ammonia, but the initial amounts of nitrogen and hydrogen, not the equilibrium concentrations. We need to determine their equilibrium concentrations.

The first step is to set up the basic form of our ICE table.

	$N_2 \ (g)$	+	$3H_2 \ (g)$	<=>	$2NH_3 \ (g)$
Initial					
Change					
Equilibrium					

We will first determine the initial concentrations. We have 0.400 moles of nitrogen and hydrogen in a 2.0 liter flask.

$$\text{Initial Concentrations of } N_2 \text{ and } H_2 = \frac{0.400 \text{ moles}}{2.00 \text{ L}} = 0.200 \ \frac{\text{moles}}{\text{L}}$$

The initial concentrations of nitrogen and hydrogen are each 0.200 moles/L. At the beginning of the reaction, there was no ammonia, so its initial concentration was zero moles/L. We fill these into the table.

	$N_2 \ (g)$	+	$3H_2 \ (g)$	<=>	$2NH_3 \ (g)$
Initial	0.200 moles/L		0.200 moles/L		0 moles/L
Change					
Equilibrium					

As the reaction proceeds, some of the nitrogen will react, some of the hydrogen will react, and some ammonia will be formed. The balanced chemical equation tells us the relative amounts by which this will be true. If we let x represent the change in concentration of the nitrogen, then 3x will be the change in concentration of the hydrogen because the balanced equation indicates that for every 1 nitrogen that reacts, 3 hydrogen molecules react. The change in the concentration of ammonia will be 2x because there will be two ammonia molecules produced for each nitrogen molecule that reacts. In filling these into the table, we use a minus sign to indicate that the concentration is going down, and a plus to indicate that the concentration is going up.

	$N_2 \ (g)$	+	$3H_2 \ (g)$	<=>	$2NH_3 \ (g)$
Initial	0.200 moles/L		0.200 moles/L		0 moles/L
Change	$-x$		$-3x$		$+2x$
Equilibrium					

At equilibrium, therefore, the concentration of nitrogen will be 0.200 −x, the concentration of hydrogen will be 0.200 − 3x, and the concentration of ammonia will be 2x.

	N_2 (g)	+	$3H_2$ (g)	$<=>$	$2NH_3$ (g)
Initial	0.200 moles/L		0.200 moles/L		0 moles/L
Change	−x		−3x		+2x
Equilibrium	0.200 − x		0.200 − 3x		2x

We are given that the concentration of ammonia at equilibrium is 0.013 moles/L so 2x = 0.013 moles/L.

	N_2 (g)	+	$3H_2$ (g)	$<=>$	$2NH_3$ (g)
Initial	0.200 moles/L		0.200 moles/L		0 moles/L
Change	−x		−3x		+2x
Equilibrium	0.200 − x		0.200 − 3x		2x = 0.013 moles/L

We can now solve this equation for x.

$$2x = 0.013 \ \frac{moles}{L}$$

$$x = 0.0065 \ \frac{moles}{L}$$

Now that we know the value of x, we can figure out the equilibrium concentrations of each of the other species.

$$[N_2] = 0.200 - x = 0.200 - 0.0065 = 0.194 \ \frac{moles}{L}$$

$$[H_2] = 0.200 - 3x = 0.200 - 3(0.0065) = 0.180 \ M$$

Once we have all the equilibrium concentrations for each of the species, we can calculate the equilibrium constant.

$$K = \frac{[NH_3]^2}{[N_2][H_2]^3} = \frac{[0.013]^2}{[0.194][0.180]^3} = 0.15$$

Try Study Questions 11 and 33 in Chapter 16 of your textbook now!

Students sometimes wonder what to let x equal. So long as you use the proper relationships in the balanced equation, it does not really matter, but some choices are easier than others. If there is a species in the equation with a coefficient of 1, it is easiest to let x equal the change in its concentration. That way, you can use the coefficients as we did in this example.

c) Use equilibrium constants to calculate the concentration (or pressure) of a reactant or a product at equilibrium (Section 16.4).

This type of problem is the reverse of the last problem. In this type of problem, we are given the value of K and are asked to calculate the equilibrium concentrations. Once again, we shall find that the ICE table helps. These problems are largely all set up similarly; what differs is the mathematics we need to use to get to the answer. While there are a wide variety of mathematical cases you might run into, you should make sure you can solve at least the three following types.

Type 1: A problem involving a perfect square.

Example 16-8: At 986°C, $K_c = 1.6$ for the reaction

$$H_2 \ (g) + CO_2 \ (g) <=> H_2O \ (g) + CO \ (g)$$

If we introduce 0.100 moles each of CO_2 and hydrogen into a 1.0 L container at 986°C, what will be the equilibrium concentrations of each of the reactants and products?

Because we are using a 1.0 L flask, the initial concentrations of H_2 and CO_2 will each be 0.100 moles/L. The initial concentrations of H_2O and CO will be zero. As the system goes to equilibrium, some of the H_2 and CO2 will react to form H_2O and CO. We will let x = the change in concentration of hydrogen. Our completed ICE table follows:

	$H_2 \ (g)$	+	$CO_2 \ (g)$	<=>	$H_2O \ (g)$	+	CO (g)
Initial	0.100 moles/L		0.100 moles/L		0		0
Change	$-x$		$-x$		$+x$		$+x$
Equilibrium	$0.100 - x$		$0.100 - x$		x		x

The equilibrium constant expression for this reaction is

$$K = \frac{[H_2O][CO]}{[H_2][CO_2]}$$

We substitute the expressions from our ICE table into this expression.

$$1.6 = \frac{[x][x]}{[0.100 - x][0.100 - x]}$$

$$1.6 = \frac{x^2}{(0.100 - x)^2}$$

On the right side, we have something squared over something squared. We can simplify this equation by taking the square root of both sides of the equation.

$$\pm 1.3 = \frac{x}{0.100 - x}$$

If we use the positive root, we obtain

$1.3 \ (0.100 - x) = x$

$0.13 - 1.3 \ x = x$

$0.13 = 2.3 \ x$

$x = 0.056$

If we use the negative root, we obtain

$-1.3(0.100 - x) = x$

$-0.13 + 1.3 \ x = x$

$-0.13 = -0.26 \ x$

$x = 0.48$

The result from the negative root, while mathematically possible, is not possible for this real-life situation because it would lead to negative concentrations for the equilibrium concentrations of hydrogen and carbon dioxide. The correct root is x = 0.056 moles/L.

Be careful at this point. Sometimes students stop here because they have solved for x. The question did not ask this. It asked for the equilibrium concentrations.

$[H_2] = [CO_2] = 0.100 - x = 0.100 - 0.056 = 0.044$ moles/L

$[H_2O] = [CO] = x = 0.056$ moles/L

It is always good at the end of one of these problems to make sure that these concentrations do give the correct equilibrium constant (within round-off error). In this case, we obtain

$$K = \frac{[H_2O][CO]}{[H_2][CO_2]} = \frac{[0.056][0.056]}{[0.044][0.044]} = 1.6$$

This is the correct value for K.

Type 2: A problem where we must use the quadratic formula.

Example 16-9: At 25°C, K_p for the following reaction is 6.8.

$$2NO_2\ (g) <=> N_2O_4\ (g)$$

If we introduce NO_2 into a flask to a partial pressure of 1.0 atm, what will be the equilibrium partial pressures of NO_2 and N_2O_4?

We start this problem off, in much the same way as the previous problem, by filling in our ICE table. In this case, we will let x equal the change in the concentration of N_2O_4. This will lead to easier values in the table because the stoichiometric coefficient of N_2O_4 is 1.

	2NO₂ *(g)*	<=>	N₂O₄ *(g)*
Initial	1.0 atm		0
Change	− 2x		+ x
Equilibrium	1.0 − 2x		x

We write the equilibrium constant expression for K_p and substitute in the values from our ICE table.

$$K_p = \frac{P_{N_2O_4}}{\left(P_{NO_2}\right)^2}$$

$$6.8 = \frac{x}{(1.0 - x)^2}$$

This is not a perfect square. It also does not fit the criteria we shall discuss a little later for making a simplifying assumption. We must solve this algebra problem as it is.

$$6.8 = \frac{x}{1.0 - 4.0\,x + 4x^2}$$

$$6.8\left(1.0 - 4.0\,x + 4x^2\right) = x$$

$$6.8 - 27.2\,x + 27.2\,x^2 = x$$

$$27.2\,x^2 - 28.2\,x + 6.8 = 0$$

This is an equation of the form $ax^2 + bx + c = 0$. We can solve this using the quadratic formula.

$$x = \frac{-b \pm \sqrt{b^2 - 4\,a\,c}}{2\,a}$$

$$x = \frac{28.2 \pm \sqrt{(-28.2)^2 - 4\,(27.2)\,(6.8)}}{2\,(27.2)}$$

$$x = 0.66\ \ OR\ \ x = 0.38$$

The first root is not physically possible because it would lead to a negative concentration for the equilibrium partial pressure of NO_2. The root that we want is x = 0.38 atm. We now can determine the equilibrium partial pressures.

$$P_{N_2O_4} = x = 0.38 \text{ atm}$$

$$P_{NO_2} = 1.0 - 2x = 1.0 - 2(0.38) = 0.24 \text{ atm}$$

Checking that these work, we obtain

$$K_p = \frac{0.38}{(0.24)^2} = 6.6$$

The K we were given was 6.8. While this does not match exactly, the difference can be attributed to round-off error.

Sometimes, we can make a simplifying assumption that eliminates the need to use the quadratic formula. This is the case when we have an expression of the form $([A]_0 - x)$ in an equilibrium constant expression where $[A]_0$ represents the initial concentration of a reactant. The simplifying assumption is that $([A]_0 - x) \approx [A]_0$. In other words, when we subtract x from the initial concentration, the value of the concentration that results is not significantly different from the initial concentration. When can we make this assumption? We can make it when two things are true: 1) K << 1 and 2) 100 • K << $[A]_0$. In general, we cannot make this assumption when K is about or greater than 1.

Type 3: A problem where we can make the simplifying assumption that x is small in comparison to an initial concentration.

Example 16-10: The dissociation of iodine gas to atomic iodine
$$I_2 \ (g) <=> 2I \ (g)$$
at 500 K has an equilibrium constant of 5.6 x 10^{-12}. If 0.050 moles of iodine gas (I_2) is introduced into a 2.0 L container at 500 K, what will be the equilibrium concentrations of iodine gas and atomic iodine.

The initial concentration of iodine gas is
$$[I_2]_0 = \frac{0.050 \text{ moles}}{2.0 \text{ L}} = 0.025 \frac{\text{mole}}{\text{L}}$$

We set up our ICE table, letting x equal the concentration of iodine that converts into I.

	I_2 *(g)*	<=>	2I *(g)*
Initial	0.025 moles/L		0
Change	– x		+ 2x
Equilibrium	0.025 – x		2x

We set up our equilibrium expression.
$$K = \frac{[I]^2}{[I_2]}$$

$$5.6 \times 10^{-12} = \frac{(2x)^2}{(0.025 - x)}$$

Before jumping into expanding this equation and solving it with the quadratic formula, let's check to see if we can use our simplifying assumption. Is K less than 1? Yes, by quite a bit.

Is $100 \cdot K$ much less than the initial concentration? $100 \cdot 5.6 \times 10^{-12} = 5.6 \times 10^{-10}$ This is a lot smaller than 0.025. We can use the simplifying assumption here.

Assume $x \ll 0.025$ so that $0.025 - x \approx 0.025$

$$5.6 \times 10^{-12} = \frac{(2x)^2}{(0.025)}$$

This is a much simpler equation to solve for x.

$$1.4 \times 10^{-13} = 4x^2$$

$$x^2 = 3.5 \times 10^{-14}$$

$$x = \pm 1.9 \times 10^{-7}$$

The negative root is not possible, so $x = 1.9 \times 10^{-7}$ moles/L. We can now calculate the final concentrations. We can see that the concentration of I_2 will still be 0.025 moles/L to all of the significant figures we can report. The concentration of I is $2x = 2(1.9 \times 10^{-7}) = 3.8 \times 10^{-7}$ moles/L.

Try Study Questions 13, 15, and 17 in Chapter 16 of your textbook now!

d) Know how K changes as different stoichiometric coefficients are used in a balanced equation, if the equation is reversed, or if several equations are added to give a new net equation (Section 16.5).

When the stoichiometric coefficients of a balanced equation are multiplied by some number, the equilibrium constant for the new equation will be equal to the equilibrium constant of the original equation raised to an exponent that is equal to the number by which we multiplied the stoichiometric coefficients.

When we reverse the equation for a chemical reaction, the equilibrium constant of the new equation will be the reciprocal of the equilibrium constant for the original equation.

If we add two or more chemical equations, the equilibrium constant for the equation that results can be obtained by multiplying together the equilibrium constants of the original equations.

Action to Original Equation(s)	New K
Multiply coefficients by n	$K_{new} = \left(K_{original}\right)^n$
Reverse	$K_{new} = \dfrac{1}{K_{original}}$
Add reaction 1, 2, 3, ...	$K_{new} = K_1 \cdot K_2 \cdot K_3 \ldots$

Example 16-11: The equilibrium constant for the reaction

$H_2 (g) + I_2 (g) \Longleftrightarrow 2HI (g)$

is 55.64 at 425°C. What is the equilibrium constant for the following reaction at the same temperature?

$4HI (g) \Longleftrightarrow 2H_2 (g) + 2I_2 (g)$

In this case, we multiplied the original equation by 2 and also reversed it. We will need to raise the original equilibrium constant to the second power and take its reciprocal.

$$K_{new} = \frac{1}{\left(K_{original}\right)^2} = \frac{1}{(55.64)^2} = 3.230 \times 10^{-4}$$

The most common type of problem in which we use this information is when we wish to piece together equations whose K's we know in order to determine the equilibrium constant for a reaction whose K we do not know. (This is similar to a Hess's Law problem to determine ΔH instead of K.)

Example 16-12: Given the following chemical equations and equilibrium constants:

Ag^+ *(aq)* + Cl^- *(aq)* <=> AgCl *(s)* $K_1 = 5.6 \times 10^9$
Ag^+ *(aq)* + $2NH_3$ *(aq)* <=> $Ag(NH_3)_2^+$ *(aq)* $K_2 = 1.6 \times 10^7$

Calculate the equilibrium constant for the reaction

AgCl *(s)* + $2NH_3$ *(aq)* <=> $Ag(NH_3)_2^+$ *(aq)* + Cl^- *(aq)*

We need to find a way to piece together the reactions to add up to the desired equations. AgCl (s) is a reactant in the desired equation, so we need to reverse equation 1. This also places Cl^- on the correct side. When we reverse a chemical equation, we take the reciprocal of its equilibrium constant.

AgCl *(s)* <=> Ag^+ *(aq)* + Cl^- *(aq)* $K_1^{'} = \frac{1}{5.6 \times 10^9} = 1.8 \times 10^{-10}$

The second equation is already written in the proper direction, so we do not need to change it or its K. In this case, we do not need to multiply any equations by a coefficient, but if we did, we would do that next. Finally, we add up the equations to make sure that they give the proper chemical equation. Since we are adding chemical equations, we then multiply their equilibrium constants to obtain the overall equilibrium constant.

AgCl *(s)* <=> Ag^+ *(aq)* + Cl^- *(aq)* $K_1^{'} = \frac{1}{5.6 \times 10^9} = 1.8 \times 10^{-10}$

$\underline{Ag^+\ (aq) + 2NH_3\ (aq) <=> Ag(NH_3)_2^+\ (aq)}$ $\underline{K_2 = 1.6 \times 10^7}$

AgCl *(s)* + $2NH_3$ *(aq)* <=> $Ag(NH_3)_2^+$ *(aq)* + Cl^- *(aq)*

$$K_{net} = K_1^{'} \cdot K_2 = \left(1.8 \times 10^{-10}\right) \cdot \left(1.6 \times 10^7\right) = 2.9 \times 10^{-3}$$

Try Study Questions 19 and 23 in Chapter 16 of your textbook now!

e) Know how to predict, using Le Chatelier's principle, the effect of a disturbance on a chemical equilibrium – a change in temperature, a change in concentrations, or a change in volume or pressure for a reaction involving gases (Section 16.6 and Table 16.2).

Le Chatelier's principle deals with how a system initially at equilibrium responds when something is done that causes it to no longer be at equilibrium. The system will respond by trying to go to a new equilibrium situation. It does so by shifting in a direction that reduces the effect of the change. Let us examine the direction the equilibrium will shift in order to react to particular changes.

Changing the concentration or partial pressure of a reactant or product.
Increasing reactant concentration (or partial pressure). The stress is that there is too much of that reactant for the other concentrations of reactants and products. The system responds by using up some of the reactant and producing more products. This is a shift to the right.

Decreasing reactant concentration (or partial pressure). The stress is that there is not enough of that reactant for the other concentrations of reactants and products present. The system responds by producing more reactant, using up some products. This is a shift to the left.

Increasing product concentration (or partial pressure). The stress is that there is too much of the product for the other concentrations of reactants and products. The system responds by using up some products and producing more reactants. This is a shift to the left.

Decreasing product concentration (or partial pressure). The stress is that there is not enough product for the concentrations of reactants and products. The system responds by producing more product and using up reactant. This is a shift to the right.

Note that adding more of an insoluble solid or insoluble liquid does not change the position of the equilibrium. This is because these have a fixed concentration related to the temperature and density. Remember that they are not included in the equilibrium constant expressions.

All of these changes could have been predicted using the reaction quotient, Q. If we increase the concentration of a product, then Q will be too large because we have increased the concentration of something in the numerator of Q. In order to get back to K, Q must decrease. This means a shift from products to reactants, just like we predict with Le Chatelier's Principle.

Changing the Total Pressure
Changing the pressure can affect equilibria that involve gases.

If the pressure is changed by adding an inert gas, this will not affect the equilibrium. We do not affect the partial pressures at all, so Q does not change from equaling K.

On the other hand, if we change the pressure by changing the volume of the container, the partial pressures will be affected, and the equilibrium may shift. If the volume is decreased, which results in an increase in pressure, the equilibrium will shift to whichever side has fewer gas molecules. If the volume is increased, which results in a decrease in pressure, the equilibrium will shift to whichever side has more gas molecules. If there are equal numbers of gas molecules on both sides of the equation, then changing the volume will not affect the position of the equilibrium.

Changing the Temperature
We examined this issue in Chapter 13. Let's review this. We can treat heat as if it is a reactant or product depending upon whether the reaction is endothermic or exothermic. If the reaction is endothermic as written, then heat will be a reactant. If it is exothermic as written, then heat will be a product.

Endothermic (Positive) ΔH	Heat + Reactants <=> Products
Exothermic (Negative) ΔH	Reactants <=> Products + Heat

If we increase the temperature, then we have more heat. The system will respond by shifting in the direction that uses up heat. If the reaction is endothermic as written, then it will shift to the right. If it is exothermic as written, it will shift to the left.

If we decrease the temperature, then we have less heat. The system will respond by shifting in the direction that produces heat. If the reaction is endothermic as written, then it will shift to the left. If the reaction is exothermic as written, then it will shift to the right.

Changing the temperature is the only change that we have studied that actually changes the value of the equilibrium constant. Equilibrium constants have different values at different temperatures. If we increase the temperature and the value of K goes up, then the higher temperature results in a shift to more products and less reactants. Shifting to the right must use up heat. The reaction must be endothermic in the direction written. On the other hand, if the equilibrium constant goes down at higher temperatures, then the higher temperature results in less products and more reactants. Shifting to the left must use up heat. The reaction must be exothermic in the direction written.

<u>Example 16-13:</u> Predict how the following system will respond to the following changes.

$$2NO_2 \ (g) <=> N_2O_4 \ (g) \qquad\qquad \Delta H \ (25°C) = -57 \ kJ$$

a. Introduce more NO_2 into the flask.

This increases the concentration of NO_2, a reactant. This will result in a shift to the right. More products will be produced. We could also view the stress as increasing the partial pressure of the NO_2, which results in the same conclusion, a shift to the right.

b. Decrease the partial pressure of N_2O_4.

This decrease in the partial pressure of N_2O_4, a product, will result in a shift to the right, producing more products to reduce this stress.

c. Transfer the reaction mixture to a smaller flask.

This will decrease the volume. A decrease in volume increases the pressure. The system will shift in the direction that has fewer gas molecules. The left side has two moles of gases. The right side has one. The right side has fewer gas molecules. This change will result in a shift to the right.

d. Increase the temperature.

The ΔH value indicates that this is an exothermic reaction going from left to right. Heat is a product.

$$2NO_2 \ (g) <=> N_2O_4 \ (g) + Heat$$

The stress of increasing the temperature results in too much heat. The system will use up heat to reduce this. The direction that uses up heat is going from right to left. The system will shift to the left.

Try Study Question 25 in Chapter 16 of your textbook now!

CHAPTER 17: Principles of Reactivity: The Chemistry of Acids and Bases

Chapter Overview

In this chapter, we reexamine the topic of acids and bases. We introduce two new definitions of acids and bases. According to the Brønsted-Lowry definition, an acid is a proton donor, and a base is a proton acceptor. A Lewis acid is an electron pair acceptor, and a Lewis base is an electron pair donor. For most of the chapter, we use the Brønsted-Lowry definition. A conjugate acid-base pair consists of two species that differ by the presence of one H^+. An acid-base reaction involves two conjugate acid-base pairs. The equilibrium will always lie on the side with the weaker acid and the weaker base.

Water molecules react with each other to a small extent to form H_3O^+ and OH^-. The equilibrium constant for this process is called K_w. In any aqueous solution, $[H_3O^+][OH^-] = K_w$. The pH of a solution is equal to $-\log[H_3O^+]$. At 25°C, acidic solutions have a pH < 7, and basic solutions have a pH > 7. The pOH equals $-\log[OH^-]$. At 25°C, pH + pOH = 14.00.

The equilibrium constant for the ionization of an acid in water is called K_a and that for the ionization of a base in water is called K_b. The stronger an acid is, the weaker its conjugate base. $K_a \cdot K_b = K_w$. We can determine the value of K_a or K_b from pH or pOH data. Similarly, we can calculate the equilibrium concentrations of the different species present in an ionization equilibrium by using the K_a (or K_b) and the initial concentration of the acid or base. From this, we can calculate the pH of the solution. You will also learn how to calculate the pH of a solution of a polyprotic acid for which the ionization constants of the different ionization steps differ by a large amount.

Salt solutions can be acidic, basic, or neutral. The acidity or basicity of salt solutions is caused by reactions of their ions with water to produce hydronium or hydroxide ions. You will learn how to predict whether a solution will be acidic, basic, or neutral and how to calculate the pH of many salt solutions. In a similar way, the pH of a solution after equal molar amounts of an acid and a base have been added together may be acidic, basic, or neutral due to reaction of the product ions with water.

Finally, you will learn about some of the factors of molecular structure that influence the strength of an acid or base.

Key Terms

In this chapter, you will need to learn and be able to use the following terms:

Acid ionization constant (K_a): the equilibrium constant for the ionization of an acid in water. The general form for a generic acid HA is

$$K_a = \frac{[H_3O^+][A^-]}{[HA]}$$

Acid-base adduct: the product from a Lewis acid-base reaction. In the acid-base adduct, the lone pair donated by the Lewis base is shared with an atom in the Lewis acid.

Amphiprotic: a substance that can behave either as a Brønsted acid or as a Brønsted base.

Amphoteric: a substance that can behave as either an acid or a base.

Arrhenius acid: a material that produces H^+ ions in aqueous solutions.

Arrhenius base: a material that produces OH^- ions in aqueous solutions.

Autoionization: a process in which molecules of the same material react with each other to produce ions. The autoionization of water produces hydronium and hydroxide ions.

Base ionization constant (K_b): the equilibrium constant for the ionization of a base in water. The general form for a generic base B is

$$K_b = \frac{\left[BH^+\right]\left[OH^-\right]}{[B]}$$

Brønsted acid: a material that can donate a proton to another substance.

Brønsted base: a material that can accept a proton from another substance.

Complex ion (coordination complex): a metal ion that has Lewis bases joined to it by coordinate covalent bonds.

Conjugate acid-base pair: two species that differ from each other by the presence of one H^+.

Inductive effect: the attraction of electrons from adjacent bonds by an electronegative atom.

Ionization constant for water (K_w): in an aqueous solution, $K_w = [H_3O^+][OH^-]$. At 25°C, the value of this constant is 1.0×10^{-14}.

Lewis acid: a material that accepts an electron pair in a chemical reaction.

Lewis base: a material that donates an electron pair in a chemical reaction.

Monoprotic acid: a material that is capable of donating one proton per formula unit of the acid.

Monoprotic base: a material capable of accepting only one proton per formula unit of the base.

pH: the negative logarithm of the hydronium ion concentration ($pH = -\log[H_3O^+]$).

pOH: the negative logarithm of the hydroxide ion concentration ($pOH = -\log[OH^-]$).

Polyprotic acid: a material capable of donating more than one proton per formula unit of the acid.

Polyprotic base: a material capable of accepting more than one proton per formula unit of the base.

Chapter Goals

By the end of this chapter you should be able to:

- **Use the Brønsted-Lowry and Lewis theories of acids and bases.**

   a) Define and use the Brønsted concept of acids and bases (Section 17.2).

   In Chapter 5, we dealt with the Arrhenius definition of acids and bases. According to this definition, an acid produces hydrogen ions (H^+), and a base produces hydroxide ions (OH^-) in aqueous solutions. H^+ is a proton. It cannot exist independently in aqueous solutions; it combines with water to produce the hydronium ion (H_3O^+) and higher hydrates. Because of this, a modified Arrhenius definition is that an acid produces H_3O^+ in aqueous solutions.

Other chemists have proposed different definitions of acids and bases that allow us to understand not only the behavior of these compounds but also other compounds and reactions in terms of acid/base behavior. We examine these more general definitions in this chapter. Which definition to use depends on the given reaction we are examining. Different processes will be easier to explain using one or another of the definitions.

According to the Brønsted-Lowry definition, an acid is a material that can donate a proton (H^+) to another substance. A Brønsted base is a material that can accept a proton from another substance. This definition of acids and bases is based on how molecules behave in chemical reactions. We must see how the substance behaves, whether it is a proton acceptor or proton donor in chemical reactions. It also frees us from having to concentrate solely on aqueous solutions and allows us to consider reactions as acid/base reactions that would not have been classified as such by the Arrhenius definition.

Example 17-1: Consider the reaction between hydrogen chloride gas and ammonia gas. Is this an acid/base reaction according to the Brønsted-Lowry definition of acids and bases? If so, identify which material is the acid and which is the base.

The chemical reaction we are considering is

$$HCl \ (g) + NH_3 \ (g) \rightarrow NH_4Cl \ (s)$$

In this reaction, the H^+ is transferred from the HCl to the NH_3, forming NH_4^+ and Cl^-. We can consider this to be an acid/base reaction according to the Brønsted-Lowry definition. (Note that we would not have defined it as an acid/base reaction according to the Arrhenius definition because it is not occurring in aqueous solution.) The H^+ is transferred from the HCl to the NH_3. The HCl is the proton donor; it started out with the H^+ and gave it to something else. It is the Brønsted acid. The NH_3 ends up with one more H^+ than it started with. It is the proton acceptor, the Brønsted base.

---

b) Recognize common monoprotic and polyprotic acids and bases and write balanced equations for their ionization in water (Section 17.2).

A monoprotic acid, according to the Brønsted definition, is a material that is capable of donating one proton per formula unit of the acid. Polyprotic acids are capable of donating more than one proton per formula unit of the acid. A monoprotic base is a material that is capable of accepting only one proton per formula unit of the base. Polyprotic bases are materials that are capable of accepting more than one proton per formula unit of the base.

Table 5.2 in your textbook lists some common molecular acids and bases. You should be familiar with this table already. There are many other acids than are listed on this table, but this is a start in recognizing the identities of acids and bases. In addition to molecular acids and bases, some ions can also act as acids and bases. In order for something to function as an acid, it must have at least one hydrogen atom as part of its formula. In addition, this hydrogen atom must be ionizable. In order for something to act as a base, it must have at least one lone pair of electrons where the hydrogen ion can attach itself.

You should learn how to write the equations for the ionization of acids and bases in water. For an acid, we show the acid reacting with H_2O on the left side of the equation. On the right side, we show that H^+ has been transferred from the acid to the H_2O (forming H_3O^+ and a species with one less hydrogen than the acid). The charge of the species left over from the acid is 1 less than the original acid. Charge must be balanced in the equation – the sum of the charges on the left side of the equation must equal the sum of the charges on the right side.

$$HCl \ (aq) + H_2O \ (l) \rightarrow H_3O^+ \ (aq) + Cl^- \ (aq)$$
$$CH_3COOH \ (aq) + H_2O \ (l) <=> H_3O^+ \ (aq) + CH_3COO^- \ (aq)$$
$$H_3PO_4 \ (aq) + H_2O \ (l) <=> H_3O^+ \ (aq) + H_2PO_4^- \ (aq)$$
$$H_2PO_4^- \ (aq) + H_2O \ (l) <=> H_3O^+ \ (aq) + HPO_4^{2-} \ (aq)$$

For the ionization of bases in water, we show the base reacting with water on the left side of the equation. H^+ transfers from the water to the base. On the right, we have OH^- and a species with one more hydrogen than the base. The charge of the species that gained the H^+ is 1 more than the original base.

$$NH_3 \ (aq) + H_2O \ (l) <=> NH_4^+ \ (aq) + OH^- \ (aq)$$
$$PO_4^{3-} \ (aq) + H_2O \ (l) <=> HPO_4^{2-} \ (aq) + OH^- \ (aq)$$

Example 17-2: Write chemical equations showing the ionization of the following materials in water.

    a. nitrous acid

Nitrous acid is not on our list of strong acids. It is a weak acid. On the left side of the equation, we write the formulas of the acid (HNO_2) and water (H_2O). On the right side, we write the formula for hydronium ion (H_3O^+) and the formula for what remains of the acid once the hydrogen ion has been transferred (NO_2^-). The charge is 1– because we started with a neutral acid and removed a positive charge from it, leaving a charge of 1–.
$$HNO_2 \ (aq) + H_2O \ (l) <=> H_3O^+ \ (aq) + NO_2^- \ (aq)$$

    b. oxalate ion

The oxalate ion ($C_2O_4^{2-}$) cannot be an acid because it does not have an ionizable H^+. It is a base. It has lone pairs that can accept H^+ ions. On the left side of the equation, we write the formula of the ion ($C_2O_4^{2-}$) and water (H_2O). H^+ transfers from the water to the oxalate. What is left from the water is OH^-. The base has now gained H^+. Its formula is $HC_2O_4^-$.
$$C_2O_4^{2-} \ (aq) + H_2O \ (l) <=> HC_2O_4^- \ (aq) + OH^- \ (aq)$$

---

c) Appreciate when a substance can be amphiprotic (Section 17.2).

A material is amphiprotic if it can behave either as a Brønsted acid or as a Brønsted base. In order to be amphiprotic, a material must have a hydrogen ion that is ionizable and also a lone pair that can accept a hydrogen ion.

One of the most important amphiprotic materials is water.

In the following equation, water acts as a base, accepting H^+ from HNO_3.
$$HNO_3 \ (aq) + H_2O \ (l) \rightarrow H_3O^+ \ (aq) + NO_3^- \ (aq)$$

On the other hand, it acts as an acid in the following equation, donating H^+ to NH_3.
$$NH_3 \ (aq) + H_2O \ (l) <=> NH_4^+ \ (aq) + OH^- \ (aq)$$

Another situation where we run into amphiprotic behavior is with ions that are derived from polyprotic acids (or bases) in which the acid or base has lost (or gained) some of the H^+ that it can but not all. Carbonic acid is a polyprotic acid (H_2CO_3). In the hydrogen carbonate ion, H_2CO_3 has lost one of the protons, but not both. It is amphiprotic. The hydrogen carbonate ion is shown acting as an acid in the first equation below and as a base in the second.
$$HCO_3^- \ (aq) + H_2O \ (l) <=> H_3O^+ \ (aq) + CO_3^{2-} \ (aq)$$
$$HCO_3^- \ (aq) + H_2O \ (l) <=> H_2CO_3 \ (aq) + OH^- \ (aq)$$

<u>Example 17-3:</u> The different ionization states of phosphoric acid are H_3PO_4, $H_2PO_4^-$, HPO_4^{2-}, and PO_4^{3-}. Which of these can only be an acid, only a base, or amphiprotic?

Phosphoric acid (H_3PO_4) is the fully protonated form. As such, under normal conditions, it will only act as an acid.

$$H_3PO_4\ (aq) + H_2O\ (l) <=> H_3O^+\ (aq) + H_2PO_4^-\ (aq)$$

The phosphate ion is the fully deprotonated form. As such, it does not have a hydrogen attached, so it cannot act as an acid. It acts only as a base.

$$PO_4^{3-}\ (aq) + H_2O\ (l) <=> HPO_4^{2-}\ (aq) + OH^-\ (aq)$$

Both $H_2PO_4^-$ and HPO_4^{2-} are in some intermediate state of protonation. They can go in either direction, losing another proton or gaining a proton. They are amphiprotic.

$$H_2PO_4^-\ (aq) + H_2O\ (l) <=> H_3O^+\ (aq) + HPO_4^{2-}\ (aq) \quad \text{acid}$$
$$H_2PO_4^-\ (aq) + H_2O\ (l) <=> H_3PO_4\ (aq) + OH^-\ (aq) \quad \text{base}$$

$$HPO_4^{2-}\ (aq) + H_2O\ (l) <=> H_3O^+\ (aq) + PO_4^{3-}\ (aq) \quad \text{acid}$$
$$HPO_4^{2-}\ (aq) + H_2O\ (l) <=> H_2PO_4^-\ (aq) + OH^-\ (aq) \quad \text{base}$$

Later in the chapter, we shall see how to assess which property predominates.

---

d) Recognize the Brønsted acid and base in a reaction and identify the conjugate partner of each (Section 17.2).

We have already talked about how to recognize Brønsted acids and bases in reaction. The acid is the material that starts out with the hydrogen ion and transfers it to the base. The base is the substance that starts out without the hydrogen ion and ends up with it.

For example, in the following reaction,

$$HF\ (aq) + H_2O\ (l) <=> H_3O^+\ (aq) + F^-\ (aq)$$

in the forward direction, HF is the Brønsted acid. It starts out with the H^+ and transfers it to the base. H_2O is the Brønsted base. It ends up with the H^+ that gets transferred.

This is a reversible reaction. The reverse reaction is also an acid-base reaction. In this case, the H_3O^+ is the acid, and F^- is the base. Putting this all together,

$HF\ (aq)$	+	$H_2O\ (l)$	<=>	$H_3O^+\ (aq)$	+	$F^-\ (aq)$
acid		base		acid		base

A conjugate acid-base pair consists of two species that differ from each other by the presence of one hydrogen ion. Thus, HF and F^- comprise a conjugate acid-base pair. In this pair, HF is the acid, and F^- is the base. There is a second conjugate acid-base pair in the above equation: H_2O and H_3O^+. Once again, these two species differ from each other by the presence of one hydrogen ion. We can see that the acid-base reaction above involves two conjugate acid-base pairs. This will be true of all acid-base reactions. Every reaction between a Brønsted acid and Brønsted base involves H^+ transfer and has two conjugate acid-base pairs.

<u>Example 17-4:</u> Identify the requested conjugates.

a. What is the conjugate acid of NH_3?

We are asked for the conjugate acid of NH_3. This will have one more H^+ than NH_3. The conjugate acid is NH_4^+. Notice that in going from the conjugate base to the conjugate acid, the charge will always go up by 1.

b. What is the conjugate base of CH_3COOH?
The conjugate base will have one less H^+ than CH_3COOH. The conjugate base is CH_3COO^-. Notice that the charge went down due to the loss of H^+.

c. What is the conjugate acid for CO_3^{2-}?
To get the conjugate acid, we add one H^+. Doing so increases the charge by 1. The conjugate is HCO_3^-. Notice that we add only one proton. We do *not* go all the way back to H_2CO_3, but instead stop after one proton has been added, HCO_3^-.

Example 17-5: In the following acid-base reaction, identify the Brønsted acid and base on the left and their conjugate partners on the right.
$$OH^-\ (aq) + HCl\ (aq) <=> H_2O\ (l) + Cl^-\ (aq)$$

In this reaction, the HCl starts out with the H^+ and transfers it to the OH^-. The acid on the left is HCl; its conjugate base is Cl^-. The base on the left is OH^-; its conjugate acid is H_2O.

Try Study Questions 1and 7 in Chapter 17 of your textbook now!

e) Understand the concept of water autoionization and its role in Brønsted acid-base chemistry. Use the water ionization constant, K_w (Section 17.3).

We have seen that water is amphiprotic; it can act either as an acid or as a base. In fact, two molecules of water can react with each other, with one taking on the role of an acid and one taking on the role of a base.
$$2H_2O\ (l) <=> H_3O^+\ (aq) + OH^-\ (aq)$$

This process is called autoionization because molecules of the same material react with each other to produce ions.

The equilibrium constant for the autoionization of water is given by
$$K_w\ =\ [H_3O^+]\ [OH^-]$$

This equilibrium constant is designated K_w to indicate that it is the equilibrium constant for the ionization of water. The value of this equilibrium constant at 25°C is 1.0×10^{-14}.
$$K_w\ =\ [H_3O^+]\ [OH^-]\ =\ 1.0 \times 10^{-14}$$

In pure water, the concentration of hydronium ions will be equal to the concentration of hydroxide ions. At 25°C in pure water, $[H_3O^+] = [OH^-] = 1.0 \times 10^{-7}$ M. In a solution that is said to be neutral, this will also be true. The following table summarizes the relationships between $[H_3O^+]$ and $[OH^-]$ in acidic, neutral, and basic solutions.

Type of Solution	Relationship between $[H_3O^+]$ and $[OH^-]$	$[H_3O^+]$ at 25°C	$[OH^-]$ at 25°C
Acidic	$[H_3O^+] > [OH^-]$	$[H_3O^+] > 1.0 \times 10^{-7}$ M	$[OH^-] < 1.0 \times 10^{-7}$ M
Neutral	$[H_3O^+] = [OH^-]$	$[H_3O^+] = 1.0 \times 10^{-7}$ M	$[OH^-] = 1.0 \times 10^{-7}$ M
Basic	$[H_3O^+] < [OH^-]$	$[H_3O^+] < 1.0 \times 10^{-7}$ M	$[OH^-] > 1.0 \times 10^{-7}$ M

Example 17-6: What is the concentration of hydroxide ions in an aqueous solution that has a concentration of $[H_3O^+] = 3.7 \times 10^{-3}$ M at 25°C? Is this solution acidic or basic?

This is an aqueous solution, so the fundamental relationship between hydronium and hydroxide is given by K_w.

$$K_w = [H_3O^+][OH^-] = 1.0 \times 10^{-14}$$

$$[OH^-] = \frac{1.0 \times 10^{-14}}{[H_3O^+]} = \frac{1.0 \times 10^{-14}}{3.7 \times 10^{-3}} = 2.7 \times 10^{-12} \text{ M}$$

The concentration of H_3O^+ is greater than that of OH^-, indicating that the solution is acidic. Because the temperature is 25°C, we could also have noted that the concentration of hydronium is greater than 1.0×10^{-7} M, once again indicating an acidic solution.

f) Use the pH concept (Section 17.3).

In Chapter 5, the concept of pH was introduced.

$$pH = -\log[H_3O^+]$$

The equation above differs from the one we gave in Chapter 5 in that we now show it using the more realistic hydrated form of H^+, H_3O^+. pH is a logarithmic scale; changing the pH by 1 implies a tenfold change in the hydronium ion concentration. For a solution at 25°C, the following relationships between the pH and acidity or basicity of a solution hold.

Solution	pH Value
Acidic	< 7.00
Neutral	= 7.00
Basic	> 7.00

Example 17-7: What is the pH of a solution in which $[H_3O^+] = 4.2 \times 10^{-9}$ M?

To solve this problem, we use the equation for pH.

$$pH = -\log[H_3O^+] = -\log\left[4.2 \times 10^{-9}\right] = 8.38$$

Just as there is the quantity pH that is related to the concentration of hydronium ions, there is a quantity called the pOH that is related to the concentration of hydroxide ions.

$$pOH = -\log[OH^-]$$

Because the concentrations of hydronium and hydroxide are related to each other, there is also a relationship between the pH and the pOH. At 25°C, this relationship is

$$pH + pOH = 14.00$$

Example 17-8: Calculate the pOH of a solution in which the hydroxide ion concentration is 6.5×10^{-4} M.

For this problem, we use the definition of pOH.

$$pOH = -\log[OH^-] = -\log\left[6.5 \times 10^{-4}\right] = 3.19$$

Example 17-9: What is the pH in the solution in Example 17-8?

We know the pOH, so we can calculate the pH using the fact that the sum of the pH and pOH is 14.00.

$$pH + pOH = 14.00$$
$$pH = 14.00 - pOH = 14.00 - 3.19 = 10.81$$

Try Study Question 9 in Chapter 17 of your textbook now!

g) Identify common strong acids and bases (Tables 5.2 and 17.3).

You should learn the strong acids and bases listed in Table 5.2 in your textbook. Some of these are also listed in Table 17.3. The strong acids are the acids for which the K_a in Table 17.3 is reported as being "large". The strong bases are the bases for which the K_b is reported as being "large".

h) Recognize some common weak acids and understand that they can be neutral molecules (such as acetic acid), cations (such as NH_4^+ or hydrated metal ions such as $Fe(H_2O)_6^{2+}$, or anions (such as HCO_3^-) (Table 17.3).

Table 5.2 and Table 17.3 list some weak acids. In Table 17.3, these are the materials with a K_a less than 1. The general rule of thumb to follow is to memorize the strong acids. If we have an acid that is not one of these strong acids, then it probably is a weak acid.

Molecular Weak Acids
There are many more compounds that are weak acids than there are that are strong acids. An example of one of these is acetic acid. It ionizes some in water, but at equilibrium there are many more molecules than there are ions. The equation for the ionization of acetic acid is
$$CH_3COOH \ (aq) + H_2O \ (l) <=> H_3O^+ \ (aq) + CH_3COO^- \ (aq)$$

Cationic Weak Acids
There are two major types of cationic weak acids. The first is the conjugate acid of a weak base. NH_3 is a weak base. Its cation is NH_4^+. To write a chemical equation showing its acidity, we write the formula for the ion and water on one side. On the other side, we will form the conjugate base of the ion. This conjugate base is formed by the ion NH_4^+ transferring a proton to water, forming H_3O^+.
$$NH_4^+ \ (aq) + H_2O \ (l) <=> NH_3 \ (g) + H_3O^+ \ (aq)$$

The other type of cationic weak acid is a hydrated metal ion (a metal ion surround by a certain number of water molecules. Sometimes these react with another water molecule removing a proton from one of the water molecules surrounding the metal ion.
$$Fe(H_2O)_6^{2+} \ (aq) + H_2O \ (l) <=> Fe(H_2O)_5(OH)^+ \ (aq) + H_3O^+ \ (aq)$$

Ions Derived from a Polyprotic Acid
An ion derived from a polyatomic acid that has not been fully deprotonated can also be a weak acid. Phosphoric acid (H_3PO_4) is a polyprotic acid. The fully deprotonated form is PO_4^{3-}. The intermediate states ($H_2PO_4^-$ and HPO_4^{2-} have some acidic properties.
$$H_2PO_4^- \ (aq) + H_2O \ (l) <=> HPO_4^{2-} \ (aq) + H_3O^+ \ (aq)$$

- ## Apply the principles of chemical equilibrium to acids and bases in aqueous solution.

a) Write equilibrium constant expressions for weak acids and bases (Section 17.4).

The chemical equation for a generic acid (HA) ionizing in water is

$$HA \; (aq) + H_2O \; (l) <=> H_3O^+ \; (aq) + A^- \; (aq)$$

There is an equilibrium between the molecular and ionic forms. We can write an equilibrium constant for this reaction.

$$K_a = \frac{\left[H_3O^+\right]\left[A^-\right]}{[HA]}$$

Remember that we do not include our solvent H_2O in the equilibrium constant expression. This equilibrium constant is called an ionization constant and is designated as K_a. The larger the value of K_a, the stronger the acid is. A strong acid (an acid whose ionization is product-favored) has $K_a > 1$. A weak acid (an acid whose ionization is reactant-favored) has $K_a < 1$.

The equation for a Brønsted base ionizing in water is

$$B \; (aq) + H_2O \; (l) <=> BH^+ \; (aq) + OH^- \; (aq)$$

We can write an ionization constant for this reaction, which we shall call K_b.

$$K_b = \frac{\left[BH^+\right]\left[OH^-\right]}{[B]}$$

The larger the value of K_b is, the stronger the base. Strong bases have $K_b > 1$, and weak bases have $K_b < 1$.

b) Calculate pK$_a$ from K$_a$ (or K$_a$ from pK$_a$) and understand how pK$_a$ is correlated with acid strength (Section 17.4).

We can calculate a quantity called the pK_a. The "p" indicates to take the negative logarithm. The smaller the pK_a, the stronger the acid is.

$$pK_a = - \log K_a$$

To obtain a K_a from a pK_a, we raise 10 to the $-pK_a$ power.

$$K_a = 10^{-pK_a}$$

In a similar fashion, we can define a quantity called the pK_b for a base. The smaller the pK_b, the stronger the base is.

$$pK_b = - \log K_b$$

We can calculate K_b from a pK_b using the equation

$$K_b = 10^{-pK_b}$$

Example 17-10: The K_a for propanoic acid is 1.3×10^{-5}. What is the pK_a of this acid?

We are given a K_a and asked for a pK_a. pK_a is the negative logarithm of the K_a.

$$pK_a = - \log K_a = - \log(1.3 \times 10^{-5}) = 4.89$$

Example 17-11: Calculate the K_a for an acid that has a pK_a of 9.14.

In this case, we are given the pK_a and asked for K_a.
$$K_a = 10^{-pK_a} = 10^{-9.14} = 7.2 \times 10^{-10}$$

Try Study Questions 23, 25, and 27 in Chapter 17 of your textbook now!

c) Understand the relationship between K_a for a weak acid and K_b for its conjugate base (Section 17.4).

The greater the strength of an acid, the greater is its tendency to give up its hydrogen and transfer it as H^+ to something else. The greater this tendency, the smaller is the tendency that the species formed, the conjugate base, will grab H^+ back. In more scientific language, the greater the acid strength of a material, the smaller is the base strength of its conjugate base. The greater the K_a of the acid, the smaller is the K_b of the conjugate base. Likewise the greater the K_b of a base, the smaller is the K_a of its conjugate acid.

Beyond just this ordering of strengths, there is a mathematical relationship between K_a and K_b of a conjugate acid-base pair.
$$K_a \cdot K_b = K_w$$

Using pK_a and pK_b, this equation becomes
$$pK_a + pK_b = pK_w$$

At 25°C, these relationships become
$$K_a \cdot K_b = 1.0 \times 10^{-14}$$
$$pK_a + pK_b = 14.00$$

Example 17-12: The K_a for hydrocyanic acid, HCN, is 4.0×10^{-10} at 25°C. What is the conjugate base of this acid and what is its K_b?

The acid is hydrocyanic acid, HCN. Its conjugate base will have one less hydrogen and have a charge that has been reduced by 1. The conjugate base is the cyanide ion, CN^-.

We are given the K_a at 25°C. The relationship between K_a and K_b of a conjugate pair is
$$K_a \cdot K_b = K_w$$
$$K_b = \frac{K_w}{K_a}$$

At 25°C,
$$K_b = \frac{K_w}{K_a} = \frac{1.0 \times 10^{-14}}{4.0 \times 10^{-10}} = 2.5 \times 10^{-5}$$

Try Study Questions 15 and 29 in Chapter 17 of your textbook now!

d) Write equations for acid-base reactions and decide whether they are product- or reactant favored (Section 17.5 and 17.6 and Table 17.5).

We have seen that the reaction of an acid with a base involves two conjugate acid-base pairs. Consider the reaction of hydrofluoric acid and sodium hydroxide:
$$HF \ (aq) + NaOH \ (aq) \iff NaF \ (aq) + H_2O \ (l)$$

In this reaction, the sodium ions are spectator ions. HF and F^- are one conjugate pair where HF is the conjugate acid and F^- is the conjugate base. The other conjugate pair is OH^- and H_2O where hydroxide is the conjugate base and H_2O is the conjugate acid.

We can use acid and base strengths (which we can determine by looking at K_a and K_b) to determine whether a given acid-base reaction is product- or reactant-favored. We will see at various points in this course that more reactive things react to form less reactive things. At equilibrium, therefore, we will end up with more of the less reactive things. Putting this in terms of acid-base reactions, we will see that at equilibrium, we will have more of the materials that are the weaker conjugate acids and bases. This is best seen with examples.

Example 17-13: Will the reaction between HF and NaOH be reactant- or product- favored?

The chemical reaction is the following:
$$HF\ (aq) + NaOH\ (aq) <=> NaF\ (aq) + H_2O\ (l)$$

We must identify the two conjugate pairs. HF and F^- is one, and OH^- and H_2O is the second. Each conjugate pair has one acid and one base. We must compare the relative strengths of the two acids and the relative strengths of the two bases.

The two acids are HF and H_2O. Which is stronger? Table 17.3 in your textbook lists K_a for HF as 7.2×10^{-4}, and K_a for H_2O as 1.0×10^{-14}. HF has the larger K_a so it is the stronger acid.

The two bases are OH^- and F^-. Which is stronger? K_b for OH^- is 1.0, and K_b for F^- is 1.4×10^{-11}. OH^- is the stronger base. Let us label these.

$HF\ (aq)$	+	$NaOH\ (aq)$	$<=>$	$NaF\ (aq)$	+	$H_2O\ (l)$
stronger		stronger		weaker		weaker
acid		base		base		acid

At equilibrium, we will have less of the stronger materials present and more of the weaker materials (the stronger ones will have reacted more). The equilibrium will lie on the right side in this equation. This is a product-favored reaction.

You might have noticed that once we had decided that HF was a stronger acid than H_2O, we could have predicted how the base results would have come out. The stronger acid will have the weaker conjugate base, so F^- must be a weaker base than OH^-.

Example 17-14: Will the reaction between H_2S and F^- be reactant- or product-favored?

The chemical reaction is
$$H_2S\ (aq) + F^-\ (aq) \rightarrow HF\ (aq) + HS^-\ (aq)$$

One conjugate pair is H_2S (acid) and HS^- (base). The other is F^- (base) and HF (acid).

The two acids are H_2S and HF. Which is stronger? Table 17.3 gives the K_a of H_2S as 1.0×10^{-7} and that of HF as 7.2×10^{-4}. HF is the stronger acid. The weaker acid will have the stronger conjugate base, so HS^- must be a stronger base than F^-. The weaker acid is H_2S, and the weaker base is F^-. These are on the reactant side, so this is a reactant-favored reaction.

Try Study Question 35 in Chapter 17 of your textbook now!

e) Calculate the equilibrium constant for a weak acid (K_a) or a weak base (K_b) from experimental information (such as pH, [H_3O^+], or [OH^-]) (Section 17.7 and Example 17.4).

We can determine the value of K_a and K_b from pH measurements because these allow us to know the concentration of hydronium in a solution.

Example 17-15: The pH of a 0.100 M solution of propanoic acid ($CH_3CH_2CO_2H$) is 2.94. What is the value of K_a for propanoic acid?

We will first write out the chemical equation for the equilibrium for this weak acid.
$$CH_3CH_2CO_2H \ (aq) + H_2O \ (l) \Longleftrightarrow H_3O^+ \ (aq) + CH_3CH_2CO_2^- \ (aq)$$

Next, we begin our ICE table. The initial concentration of propanoic acid is 0.100 M. The initial concentration of H_3O^+ is 10^{-7} M, but we can assume that this will be overwhelmed by the hydronium that will form due to the presence of the acid. The initial concentration of propanoate ion is 0 M.

	$CH_3CH_2CO_2H$	+	H_2O	$\Longleftrightarrow$	H_3O^+	+	$CH_3CH_2CO_2^-$
Initial	0.100 M				0 M		0 M
Change							
Equilibrium							

As the system approaches equilibrium, some of the $CH_3CH_2CO_2H$ will react, and we will form H_3O^+ and $CH_3CH_2CO_2^-$. Because of the reaction stoichiometry, we know that for each $CH_3CH_2CO_2H$ that reacts, we will get one H_3O^+ and one $CH_3CH_2CO_2^-$.

	$CH_3CH_2CO_2H$	+	H_2O	$\Longleftrightarrow$	H_3O^+	+	$CH_3CH_2CO_2^-$
Initial	0.100 M				0 M		0 M
Change	$-x$				$+x$		$+x$
Equilibrium	$0.100 - x$				x		x

We are given the pH of the solution. This will allow us to calculate the concentration of hydronium ions in the solution.
$$\left[H_3O^+\right] = 10^{-pH} = 10^{-2.94} = 1.1 \times 10^{-3} \text{ M}$$

This corresponds to the value of x in our table. Knowing this, we can determine the concentrations of all the relevant species.
$$\left[CH_3CH_2CO_2^-\right] = \left[H_3O^+\right] = x = 1.1 \times 10^{-3} \text{ M}$$
$$\left[CH_3CH_2CO_2H\right] = 0.100 - x = 0.100 - 1.1 \times 10^{-3} \text{ M} = 0.099 \text{ M}$$

	$CH_3CH_2CO_2H$	+	H_2O	$\Longleftrightarrow$	H_3O^+	+	$CH_3CH_2CO_2^-$
Initial	0.100 M				0 M		0 M
Change	$-x$				$+x$		$+x$
Equilibrium	$0.100 - x =$ 0.099 M				$x =$ 1.1×10^{-3} M		$x =$ 1.1×10^{-3} M

Now that we know the equilibrium concentrations of the different species, we can calculate the equilibrium constant.
$$K_a = \frac{\left[H_3O^+\right]\left[CH_3CH_2CO_2^-\right]}{\left[CH_3CH_2CO_2H\right]} = \frac{\left[1.1 \times 10^{-3}\right]\left[1.1 \times 10^{-3}\right]}{[0.099]} = 1.2 \times 10^{-5}$$

If we are trying to calculate a K_b, we can use the pH to determine the pOH (pH + pOH = 14.00). Once we have the pOH, we can determine the concentration of hydroxide ion.

Try Study Question 41 and 43 in Chapter 17 of your textbook now!

f) Use the equilibrium constant and other information to calculate the pH of a solution of a weak acid or weak base (Section 17.7 and Examples 17.5 and 17.6).

This problem is like those we studied in the last chapter in which we used an equilibrium constant to determine the equilibrium concentrations of the reactants and products. Often, we will be able to use the simplifying assumption that the equilibrium concentration of the acid will about equal the initial concentration of the acid: $[A]_0 - x \approx [A]_0$. We can make this assumption if K_a is less than one (we have a weak acid) and $K_a \cdot 100 < [A]_0$.

<u>Example 17-16:</u> What is the pH of 0.050 M CH_3COOH? $K_a(CH_3COOH) = 1.8 \times 10^{-5}$.

The only things we know are the initial concentration of acetic acid and its K_a. We start by filling in our ICE table.

	CH_3COOH	+	H_2O	$\Longleftrightarrow$	H_3O^+	+	CH_3COO^-
Initial	0.050 M				0 M		0 M
Change	$-x$				$+x$		$+x$
Equilibrium	$0.050 - x$				x		x

We substitute the expressions for the equilibrium concentrations into the equation for K_a.

$$K_a = \frac{\left[H_3O^+\right]\left[CH_3COO^-\right]}{\left[CH_3COOH\right]}$$

$$1.8 \times 10^{-5} = \frac{x \cdot x}{0.050 - x}$$

Let's see if we can avoid using the quadratic equation.
$$100 \cdot K_a = 100 \cdot 1.8 \times 10^{-5} = 1.8 \times 10^{-3}$$

This is less than the initial concentration of 0.050 M, so we can assume $0.050 - x \approx 0.050$.

$$1.8 \times 10^{-5} = \frac{x^2}{0.050}$$
$$9.0 \times 10^{-7} = x^2$$
$$x = \pm 9.5 \times 10^{-4}$$

The negative root is not possible because x is a concentration. The concentration of H_3O^+ is 9.5×10^{-4} M.
$$pH = -\log\left[H_3O^+\right] = -\log(9.5 \times 10^{-4}) = 3.02$$

Sometimes, even when we cannot use our simplifying assumption, we can still avoid using the quadratic equation by using something called the method of successive approximations. What we do in this method is to make our simplifying assumption and solve the problem. Then we take this answer and use it as our x in the expression for the concentration of the acid and solve the problem again to get a new x. We keep repeating this until the answer stops changing out to the number of significant figures we want. This method is very easy to use with a spreadsheet program on a computer.

Example 17-17: What is the pH of a 0.010 M solution of HF? K_a for HF is 7.2×10^{-4}.

We fill in our ICE table.

	HF	+	H_2O	$\Longleftrightarrow$	H_3O^+	+	F^-
Initial	0.010 M				0 M		0 M
Change	$-x$				$+x$		$+x$
Equilibrium	$0.010 - x$				x		x

We set up our equilibrium constant expression.

$$K_a = \frac{\left[H_3O^+\right]\left[F^-\right]}{[HF]}$$

$$7.2 \times 10^{-4} = \frac{x^2}{0.010 - x}$$

We check to see if we can use our simplifying assumption.

$$100 \cdot K_a = 100 \cdot 7.2 \times 10^{-4} = 0.072$$

This is not smaller than our initial concentration. We cannot use the simplifying assumption. We could solve the quadratic equation, but let's use the method of successive approximations. We first use the simplifying assumption and solve for x.

$$7.2 \times 10^{-4} = \frac{x^2}{0.010}$$

$$7.2 \times 10^{-4} = \frac{x^2}{0.010}$$

$$7.2 \times 10^{-6} = x^2$$

$$x = \pm 2.7 \times 10^{-3}$$

We reject the negative root as being impossible. $x = 2.7 \times 10^{-3}$ M. We know that this is not the correct answer because we know that we cannot use the simplifying assumption. We now use this value as the value of x in the denominator and solve for x again.

$$7.2 \times 10^{-4} = \frac{x^2}{0.010 - x}$$

$$7.2 \times 10^{-4} = \frac{x^2}{0.010 - 2.7 \times 10^{-3}}$$

$$7.2 \times 10^{-4} = \frac{x^2}{0.0073}$$

$$5.3 \times 10^{-6} = x^2$$

$$x = \pm 2.3 \times 10^{-3}$$

We now use 2.3×10^{-3} M as the x in the denominator.

$$7.2 \times 10^{-4} = \frac{x^2}{0.010 - 2.3 \times 10^{-3}} \qquad x = \pm 2.4 \times 10^{-3}$$

We use 2.4×10^{-3} as the x in the denominator.

$$7.2 \times 10^{-4} = \frac{x^2}{0.050 - 2.4 \times 10^{-3}} \qquad x = \pm 2.3 \times 10^{-3}$$

The answer has now converged to 2.3×10^{-3} M. This is the value of x, which corresponds to the concentration of hydronium. We can now calculate the pH.

$$pH = -\log[H_3O^+] = -\log(2.3 \times 10^{-3}) = 2.64$$

Try Study Question 49, 51, and 55 in Chapter 17 of your textbook now!

g) Describe the acid-base properties of salts and calculate the pH of a solution of a salt of a weak acid or of a weak base (Section 17.7 and Example 17.7).

Salt solutions are not necessarily neutral in pH. They can be acidic, basic, or neutral depending on the specific ions that are present. These solutions can end up acidic or basic because some ions can react with water to form either hydronium or hydroxide ions. The following tables summarize the acid/base properties of various cations and anions.

Cations

Type of Ion	Example	Effect on pH	Example of Net Ionic Equation to Explain Acid/Base Behavior
Ions from Groups 1A and 2A	Na^+	No Effect	---
2+ (Other Than from Group 2A) and 3+ Metal Ions	Al^{3+}	Acidic	$[M(H_2O)_n]^{m+}$ *(aq)* + H_2O *(l)* <=> $[M(H_2O)_{n-1}(OH)]^{(m-1)+}$ *(aq)* + H_3O^+ *(aq)*
Cations Derived from a Weak Base	NH_4^+	Acidic	BH^+ *(aq)* + H_2O *(l)* <=> B *(aq)* + H_3O^+ *(aq)*

Anions

Type of Ion	Example	Effect on pH	Example of Net Ionic Equation to Explain Acid/Base Behavior
Anion Derived from a Strong Acid	Cl^-	No Effect	---
Anion Derived from a Weak Monoprotic Acid	CH_3COO^-	Basic	A^- *(aq)* + H_2O *(l)* <=> HA *(aq)* OH^- *(aq)*
Fully Deprotonated Anion from a Weak Polyprotic Acid	CO_3^{2-}	Basic	A^{n-} *(aq)* + H_2O *(l)* <=> $HA^{(n-1)-}$ *(aq)* + OH^- *(aq)*
Partially Deprotonated Anion from a Weak Polyprotic Acid	HCO_3^-	Acidic or Basic (Compare K_a and K_b of Ion)	HA^{n-} *(aq)* + H_2O *(l)* <=> $A^{(n+1)-}$ *(aq)* + H_3O^+ *(aq)* OR HA^{n-} *(aq)* + H_2O *(l)* <=> $H_2A^{(n-1)-}$ *(aq)* + OH^- *(aq)*

On the left of each net ionic equation, we write the ion and water. On the right, we write the conjugate acid or base and what is left from water when this conjugate acid or base forms.

Example 17-18: State if the following ions contribute toward an aqueous solution being acidic, basic, or neutral. If acidic or basic, write a net ionic equation to explain this behavior.

a. NO_3^-
Nitrate is an anion derived from the strong acid, nitric acid. It will have no effect on the pH.

b. PO_4^{3-}

Phosphate is derived from the polyprotic acid phosphoric acid, H_3PO_4. It is the fully deprotonated anion. It will tend to make the solution more basic.

$$PO_4^{3-} \text{ (aq)} + H_2O \text{ (l)} <=> HPO_4^{2-} \text{ (aq)} + OH^- \text{ (aq)}$$

The solution is basic because of the hydroxide ions formed.

c. HCO_3^-

This ion is derived from the weak diprotic acid carbonic acid, H_2CO_3. It is not the fully deprotonated form. It can act either as an acid or as a base.

$$HCO_3^- \text{ (aq)} + H_2O \text{ (l)} <=> CO_3^{2-} \text{ (aq)} + H_3O^+ \text{ (aq)} \text{ OR}$$
$$HCO_3^- \text{ (aq)} + H_2O \text{ (l)} <=> H_2CO_3 \text{ (aq)} + OH^- \text{ (aq)}$$

Which is it? Is it a better acid or a better base? To determine this, we look up its K_a and K_b values. K_a for HCO_3^- is 4.8×10^{-11}, K_b for HCO_3^- is 2.4×10^{-8}. K_b is larger, so it is a better base. It will make the solution basic, and the second of the net ionic equations is dominant.

In any salt, there will be both a cation and an anion. We must take into account both in determining if a solution of the salt will be acidic or basic.

Example 17-19: Will aqueous solutions of the following salts be acidic, basic, or neutral?

a. sodium chloride

The two ions are sodium and chloride ions. Na^+ is a Group 1A cation. It will not affect the pH. Cl^- is the anion derived from a strong acid, HCl. It also will not affect the pH. This solution should be neutral.

b. ammonium chloride

The ammonium ion (NH_4^+) is a cation derived from the weak base ammonia, NH_3. It will tend to make the solution acidic. As discussed for part a, Cl^- will not affect the pH. The overall effect is that the solution will be acidic due to the NH_4^+.

c. aluminum nitrate (the hydrated aluminum ion has 6 waters around the aluminum)

The aluminum ion is a 3+ ion. Its hydrated ion will be acidic. The nitrate ion is an anion derived from a strong acid and will have no effect on the pH. The overall effect is that the solution will be acidic due to the aluminum ion.

d. sodium carbonate

As discussed in part a, the sodium ion has no effect on the pH. Carbonate ion (CO_3^{2-}) is the fully deprotonated anion from the weak diprotic acid carbonic acid, H_2CO_3. This ion will cause the solution to be basic.

e. ammonium phosphate

As discussed in part b, the ammonium ion will tend to make a solution acidic. The phosphate ion is the fully deprotonated ion from the polyprotic acid, H_3PO_4. As such, it will tend to make the solution basic. These are competing against each other. It will come down to whether ammonium is a better acid or phosphate is a better base. We can determine this by comparing the K_a for NH_4^+ and the K_b for PO_4^{3-}. Whichever is larger will determine the overall effect of the salt. From Table 17.3 in your textbook, we can see that K_a for NH_4^+ is 5.6×10^{-10} and that the K_b for phosphate is 2.8×10^{-2}. The K_b of phosphate is bigger than the K_a of ammonium so the solution will be basic overall.

We already have the tools to calculate the expected pH of many salt solutions.

Example 17-20: What is the pH of a 0.10 M solution of NH_4Cl? The K_b of NH_3 is 1.8×10^{-5}.

The first step is to determine whether the solution should be acidic, basic, or neutral and the chemical equation that determines this. The NH_4^+ ion is acidic, and the Cl^- ion has no effect on the pH. This solution will be acidic. The equation that explains this is

$$NH_4^+ \ (aq) + H_2O \ (l) <=> NH_3 \ (aq) + H_3O^+ \ (aq)$$

This is the equation for the K_a of the ammonium ion. We are not given K_a for NH_4^+ but the K_b of its conjugate base, NH_3. We can calculate the K_a of NH_4^+.

$$K_a \cdot K_b = K_w$$

$$K_a = \frac{K_w}{K_b} = \frac{1.0 \times 10^{-14}}{1.8 \times 10^{-5}} = 5.6 \times 10^{-10}$$

Next, we can set up our ICE table.

	NH_4^+	+	H_2O	<=>	NH_3	+	H_3O^+
Initial	0.100 M				0 M		0 M
Change	− x				+ x		+ x
Equilibrium	0.100 − x				x		x

We can now substitute these expressions into K_a.

$$K_a = \frac{\left[NH_3\right]\left[H_3O^+\right]}{\left[NH_4^+\right]}$$

$$5.6 \times 10^{-10} = \frac{x \cdot x}{0.100 - x}$$

Let's check to see if we can use our simplifying assumption.

$$100 \cdot K_a = 100 \cdot 5.6 \times 10^{-10} = 5.6 \times 10^{-8}$$

This is much less than the initial concentration, so we can use the simplifying assumption.

$$5.6 \times 10^{-10} = \frac{x^2}{0.100}$$

$$x = \pm 7.5 \times 10^{-6}$$

We can reject the negative root.

$$x = [H_3O^+] = 7.5 \times 10^{-6} \ M$$

We can now calculate the pH.

$$pH = - \log\left[H_3O^+\right] = - \log(7.5 \times 10^{-6}) = 5.13$$

Try Study Questions 19, 21 and 59 in Chapter 17 of your textbook now!

h) Calculate the pH of a solution of a polyprotic acid or base (Section 17.8 and Example 17.9).

For a polyprotic acid, the K_a for each successive step of removing a proton gets smaller. Often, there is a large difference between the different ionization steps. Because of this, the pH of many polyprotic acids depends primarily on the hydronium generated in the first step. To calculate the concentration of the fully deprotonated ion, we must consider all the steps.

For polyprotic bases, the K_b for each successive step of adding a proton gets smaller. The pH of many polyprotic bases depends primarily on the OH^- produced in the first step. We must take into account all the steps to determine the concentration of the fully protonated species.

Example 17-21: Calculate the pH of a 0.100 M oxalic acid solution and the concentrations of oxalic acid, hydrogen oxalate, and oxalate present at equilibrium. K_{a1} for oxalic acid is 5.9 x 10^{-2} and K_{a2} = 6.4 x 10^{-5}.

For the first ionization, the chemical equation is
$$H_2C_2O_4 \ (aq) + H_2O \ (l) <=> HC_2O_4^- \ (aq) + H_3O^+ \ (aq)$$

We set up our ICE table:

	$H_2C_2O_4$	+	H_2O	<=>	$HC_2O_4^-$	+	H_3O^+
Initial	0.100 M				0 M		0 M
Change	$-x$				$+x$		$+x$
Equilibrium	$0.100-x$				x		x

We set up our K_{a1} expression and substitute the equilibrium expressions into it:
$$K_{a_1} = \frac{\left[H_3O^+\right]\left[HC_2O_4^-\right]}{\left[H_2C_2O_4\right]}$$
$$5.9 \times 10^{-2} = \frac{x \cdot x}{0.100 - x}$$

Let us check to see if we can use the simplifying assumption that $0.100 - x \approx 0.100$.
$$100 \cdot K_{a_1} = 100 \cdot 5.9 \times 10^{-2} = 5.9$$

This is not smaller than 0.100, so we cannot make the assumption. We can either use the quadratic formula or the method of successive approximations. Let us use the latter.

Initially, we make the guess that we can make our assumption (knowing that we cannot):
$$5.9 \times 10^{-2} = \frac{x \cdot x}{0.100}$$
$$x = \pm 0.077$$

We reject the negative root and use 0.077 M as x in the denominator in the next round.
$$5.9 \times 10^{-2} = \frac{x \cdot x}{0.100 - 0.077}$$
$$x = \pm 0.037$$

We reject the negative root and use 0.037 M as x in the denominator in the next round.
$$5.9 \times 10^{-2} = \frac{x \cdot x}{0.100 - 0.037}$$
$$x = \pm 0.061$$

We continue in this same manner.

$$5.9 \times 10^{-2} = \frac{x \cdot x}{0.100 - 0.061} \qquad x = \pm 0.047$$

$$5.9 \times 10^{-2} = \frac{x \cdot x}{0.100 - 0.047} \qquad x = \pm 0.056$$

$$5.9 \times 10^{-2} = \frac{x \cdot x}{0.100 - 0.056} \qquad x = \pm 0.051$$

$$5.9 \times 10^{-2} = \frac{x \cdot x}{0.100 - 0.051} \qquad x = \pm 0.054$$

$$5.9 \times 10^{-2} = \frac{x \cdot x}{0.100 - 0.054} \qquad x = \pm 0.052$$

$$5.9 \times 10^{-2} = \frac{x \cdot x}{0.100 - 0.052} \qquad x = \pm 0.053$$

$$5.9 \times 10^{-2} = \frac{x \cdot x}{0.100 - 0.053} \qquad x = \pm 0.053$$

The answer has now converged to two significant figures.

$$\left[H_3O^+\right] = \left[HC_2O_4^-\right] = x = 0.053 \text{ M}$$
$$\left[H_2C_2O_4\right] = 0.100 - x = 0.047 \text{ M}$$

Because there is a large difference between K_{a1} and K_{a2}, we can predict that the change in the concentration of H_3O^+ will be insignificant in the second ionization. We can therefore calculate the pH of the solution now.

$$pH = -\log\left[H_3O^+\right] = -\log(0.053) = 1.28$$

To determine the concentration of $C_2O_4^{2-}$, we need to consider the second ionization.

$$HC_2O_4^- \ (aq) + H_2O \ (l) <=> C_2O_4^{2-} \ (aq) + H_3O^+ \ (aq)$$

	$HC_2O_4^-$	+	H_2O	<=>	$C_2O_4^{2-}$	+	H_3O^+
Initial	0.053 M				0 M		0.053 M
Change	$-x$				$+x$		$+x$
Equilibrium	$0.053 - x$				x		$0.053 + x$

$$6.4 \times 10^{-5} = \frac{(0.053 + x) \cdot x}{0.053 - x}$$

Let us see if we can use our simplifying assumption that $0.053 - x \approx 0.053$

$$100 \cdot K_{a_2} = 100 \cdot 6.4 \times 10^{-5} = 6.4 \times 10^{-3}$$

This is less than 0.053 so we can make this assumption. This also means that we can make the assumption that $0.053 + x \approx 0.053$ for the concentration of hydronium ion.

$$6.4 \times 10^{-5} = \frac{(0.053) \cdot x}{0.053}$$
$$x = 6.4 \times 10^{-5}$$

When all the equilibria have been established, we know from our first calculation that $[H_2C_2O_4] = 0.047$ M, $[HC_2O_4^-] = 0.053$ M, and $[H_3O^+] = 0.053$ M. The concentrations of $HC_2O_4^-$ and H_3O^+ did not change appreciably in establishing the second equilibrium. We needed the second equilibrium to determine $[C_2O_4^{2-}]$, 6.4×10^{-5} M

Try Study Question 65 in Chapter 17 of your textbook now!

• Predict the outcome of reactions of acids and bases.

a) Recognize the type of acid-base reaction and describe its result (Section 17.6).

The following table summarizes different aspects of the reactions of various acids (HA) and bases (B). In the first column, SA = strong acid, SB = strong base, WA = weak acid, and WB = weak base.

Type	Net Ionic Equation	pH when equal molar amounts are mixed (pH at 25°C)	This pH depends on
SA + SB	H_3O^+ *(aq)* + OH^- *(aq)* <=> H_2O *(l)*	Neutral (pH = 7)	---
SA + WB	H_3O^+ *(aq)* + B *(aq)* <=> BH^+ *(aq)* + H_2O *(l)*	Acidic (pH < 7)	K_a of BH^+
WA + SB	HA *(aq)* + OH^- *(aq)* <=> A^- *(aq)* + H_2O *(l)*	Basic (pH > 7)	K_b of A^-
WA + WB	HA *(aq)* + B *(aq)* <=> BH^+ *(aq)* + A^- *(aq)*	Depends	K_a of BH^+ and K_b of A^-

---

Example 17-22: Write the net ionic equation for the reaction between acetic acid and sodium hydroxide. When equal molar amounts are mixed, will the solution be acidic, basic, or neutral?

Acetic acid is a weak acid. The predominant form in a solution of acetic acid is the molecular form. Sodium hydroxide is a strong base. It will be completely dissociated into sodium ions and hydroxide ions. The hydroxide ion is the active part.

$$CH_3COOH \ (aq) + OH^- \ (aq) <=> CH_3COO^- \ (aq) + H_2O \ (l)$$

The conjugate base of a weak acid has a significant basicity, so the acetate ion is basic. The solution present after mixing equal molar amounts of the acid and base will be basic due to the basicity of the acetate ion.

Let's take a closer look at the reaction of a weak acid and a weak base. In order to predict whether the reaction will produce an acidic or basic solution, we must know whether K_a of BH^+ or K_b of A^- is larger. If BH^+ is a stronger acid than A^- is a base, the solution will be acidic. If A^- is a stronger base than BH^+ is an acid, the solution will be basic.

Try Study Questions 37 and 63 in Chapter 17 of your textbook now!

b) Calculate the pH after an acid-base reaction (Section 17.7 and Example 17.8).

If we mix equal molar quantities of an acid and a base, we end up with a salt solution. These can be acidic, basic or neutral, as we have already discussed.

Example 17-23: What is the pH of the solution that results from the reaction of 100. mL of 0.100 M HCl with 100. mL of 0.100 M NaOH?

We are mixing equal volumes of equal concentrations of HCl and NaOH. HCl is a strong acid, and NaOH is a strong base. The resulting solution should be neutral with a pH of 7.

If one of the ions can react with water to form hydronium or hydroxide ions, then we need to solve a more complicated problem in order to determine the actual pH of the solution. We work this type of problem in two main parts: 1) a stoichiometry problem and 2) an equilibrium problem. As with other stoichiometry problems, we will solve the stoichiometry part using moles. We will set up a stoichiometry table modified from that we used in earlier chapters. The way that it is modified is that instead of just having one line for moles, we specify the initial number of moles, the change in moles, and the final number of moles. In some ways, it looks like one of our equilibrium tables except that we work in terms of moles instead of concentrations. The equilibrium part is worked as a normal equilibrium problem.

<u>Example 17-24</u>: What is the pH of the solution that results from the reaction of 100. mL of 0.100 M propanoic acid and 50.0 mL of 0.200 M sodium hydroxide? K_a for propanoic acid is 1.3×10^{-5}.

First, we need to solve the stoichiometry part. The acid-base reaction that occurs is.
$$CH_3CH_2COOH \ (aq) + NaOH \ (aq) \rightarrow NaOOCCH_2CH_3 \ (aq) + H_2O \ (l)$$

We next set up our modified stoichiometry table.

	CH_3CH_2COOH	+	$NaOH$	$\rightarrow$	$NaOOCCH_2CH_3$	+	H_2O
Measured	100. mL		50.0 mL				
Conversion Factor	0.100 moles/L		0.200 moles/L				
Initial Moles							
Change in Moles							
Final Moles							

We can calculate the number of moles of each of the reactants:
$$100. \ mL \ x \ \frac{1 \ L}{1000 \ mL} \ x \ \frac{0.100 \ moles \ CH_3CH_2COOH}{L}$$
$$= 0.0100 \ moles \ CH_3CH_2COOH$$
$$50.0 \ mL \ x \ \frac{1 \ L}{1000 \ mL} \ x \ \frac{0.200 \ moles \ NaOH}{L} = 0.0100 \ moles \ NaOH$$

Initially, we have no sodium propanoate present.

	CH_3CH_2COOH	+	$NaOH$	$\rightarrow$	$NaOOCCH_2CH_3$	+	H_2O
Measured	100. mL		50.0 mL				
Conversion Factor	0.100 moles/L		0.200 moles/L				
Initial Moles	0.0100 moles		0.0100 moles		0 moles		
Change in Moles							
Final Moles							

The two reactants are present in exactly the same amount. Because everything reacts in a 1:1 ratio, we can do the mole conversions without explicitly showing the calculations. The propanoic acid and the sodium hydroxide are present in exactly the right amounts, so they will completely react. In the process, we will form the same amount of sodium propanoate.

	CH_3CH_2COOH +	NaOH	→	$NaOOCCH_2CH_3$ +	H_2O
Measured	100. mL	50.0 mL			
Conversion Factor	0.100 moles/L	0.200 moles/L			
Initial Moles	0.0100 moles	0.0100 moles		0 moles	
Change in Moles	–0.0100 moles	–0.0100 moles		+0.0100 moles	
Final Moles	0 moles	0 moles		0.0100 moles	

At the end of this reaction, we have a solution of sodium propanoate. We have 0.0100 moles of sodium propanoate in a volume of 100. mL + 50.0 mL = 150. mL. The concentration of sodium propanoate is

$$\frac{0.0100 \text{ moles } NaOOCCH_2CH_3}{0.150 \text{ L}} = 0.0667 \text{ M}$$

Now we will solve an equilibrium problem. In essence the problem asks for the pH of a 0.0667 M sodium propanoate solution. This is like the problems we solved earlier for the pH of a salt solution.

We know that the sodium ion will have no effect on the pH. The propanoate ion is the anion from a monoprotic weak acid, so it makes the solution basic.

$$CH_3CH_2COO^- \text{ (aq)} + H_2O \text{ (l)} <=> CH_3CH_2COOH \text{ (aq)} + OH^- \text{ (aq)}$$

This is the chemical equation for the K_b of the propanoate ion. We were given the K_a of propanoic acid, not the K_b of propanoate. We can calculate the K_b, using the relationship between K_a and K_b for a conjugate pair.

$$K_b = \frac{K_w}{K_a} = \frac{1.0 \times 10^{-14}}{1.3 \times 10^{-5}} = 7.7 \times 10^{-10}$$

We now solve the equilibrium problem to determine the concentration of OH^-. We set up the ICE table. For this table, we use concentrations.

	$CH_3CH_2COO^-$ +	H_2O	<=>	CH_3CH_2COOH +	OH^-
Initial	0.0667 M			0 M	0 M
Change	– x			+ x	+ x
Equilibrium	0.0667 – x			x	x

$$K_b = \frac{\left[CH_3CH_2COOH\right]\left[OH^-\right]}{\left[CH_3CH_2COO^-\right]}$$

$$7.7 \times 10^{-10} = \frac{x^2}{0.0667 - x}$$

Let's check to see if we can use our simplifying assumption.

$$100 \cdot K_a = 100 \cdot 7.7 \times 10^{-10} = 7.7 \times 10^{-8}$$

This is much less than the initial concentration, so we can use the assumption.

$$7.7 \times 10^{-10} = \frac{x^2}{0.0667}$$

$$5.1 \times 10^{-11} = x^2$$
$$x = \pm 7.2 \times 10^{-6}$$

We can reject the negative root. $x = 7.2 \times 10^{-6}$ M. This corresponds to the concentration of OH^- ions. We can calculate the pOH from this.

$$pOH = -\log[OH^-] = -\log(7.2 \times 10^{-6}) = 5.14$$

We can then use the relationship that pH + pOH = 14.00 to calculate the pH.

$$pH = 14.00 - pOH = 14.00 - 5.14 = 8.86$$

Try Study Question 61 in Chapter 17 of your textbook now!

• Understand the influence of structure and bonding on acid-base properties.

a) Characterize a compound as a Lewis base (an electron-pair donor) or a Lewis acid (an electron-pair acceptor) (Section 17.9).

There is yet another definition of acids and bases, the Lewis definition. Just as going from the Arrhenius definition to the Brønsted-Lowry definition allowed us to expand the number of reactions that can be considered to be acid-base reactions, so too going to the Lewis definition expands this realm even further.

The Brønsted-Lowry definition focused on the transfer of H^+. The Lewis definition focuses on electrons. Electrons are negative, whereas protons are positive. According to the Brønsted-Lowry system, we called the base the proton acceptor. This will be the electron pair donor. The acid was the proton donor; we can look at this as the electron pair acceptor.

A Lewis acid is an electron-pair acceptor, and a Lewis base is an electron pair donor. In order for something to act as a Lewis base, it must have a lone pair of electrons.

Example 17-25: In the following reactions, identify the Lewis acids and the Lewis bases.

a. $H^+ + OH^- \rightarrow H_2O$

Let's look at the Lewis structures, paying attention to where the electrons go. In the following figures, a curved arrow represents the movement of an electron pair.

An electron pair that starts out on the OH^- ends up shared with what started out as H^+. OH^- is the electron pair donor, the Lewis base. H^+ is the electron pair acceptor, the Lewis acid.

b. $C(CH_3)^+ + Cl^- \rightarrow C(CH_3)_3Cl$

Let's look at the Lewis structures.

One of the lone pairs that starts out on the chloride ends up shared between the chlorine and what was the carbocation. The chloride ion is the Lewis base and the carbocation is the Lewis acid. Notice that this reaction would not have been classified as an acid-base reaction according to either the Arrhenius or the Brønsted-Lowry definitions.

c. $NH_3 + BF_3 \rightarrow NH_3BF_3$

The nitrogen in the ammonia starts out with a lone pair. The boron starts out with only six electrons around it. The lone pair in the ammonia ends up being shared between the N and the B. The ammonia is the Lewis base. The BF_3 is the Lewis acid.

Try Study Question 69 in Chapter 17 of your textbook now!

b) Appreciate the connection between the structure of a compound and its acidity or basicity (Section 17.10).

As one moves down a group on the periodic table, the binary acids tend to get stronger. For example, the acid strength of the hydrohalic acids goes in the order $HF < HCl < HBr < HI$. This is due to decreasing H–X bond strength as one moves down the group.

In a series of oxoacids with the same central atom, acid strength increases as the number of oxygen atoms increases. For example, H_2SO_4 is a stronger acid than H_2SO_3.

The presence of electronegative groups in another part of a molecule increases the acid strength. This is because of the inductive effect in which there is an attraction of electrons from adjacent bonds by an electronegative atom. This removes electron density from the bond to the hydrogen.

A material is also more acidic if the anion that is formed has resonance structures that allow for the delocalization of the negative charge over more than one atom.

Try Study Question 73 in Chapter 17 of your textbook now!

Other Notes

1. Calculating the pH of a Strong Acid or a Strong Base.
Calculating the pH of a solution that contains a strong acid or strong base is fairly straightforward because for each formula unit, we get ions in solution. For example, the strong acid HCl ionizes completely in water.

$$HCl \ (aq) + H_2O \ (l) \rightarrow H_3O^+ \ (aq) + Cl^- \ (aq)$$

The equilibrium constant for this reaction is large, indicating that this is very much a product-favored reaction. For each molecule of HCl we start with, we end up with a hydronium ion and a chloride ion.

The concentration of hydronium ion in the solution is therefore the same as the concentration of the acid we start with. In a 0.10 M solution of HCl, the concentration of H_3O^+ ion will be 0.10 M. In a 0.05 M solution of HCl, the H_3O^+ concentration will be 0.05 M.

For a strong base like NaOH, for each formula unit of NaOH we start with, we get one sodium ion and one hydroxide ion in solution. In a 1 M solution of NaOH, the OH^- concentration will be 1 M. For a base like $Sr(OH)_2$, for each formula unit of $Sr(OH)_2$ we start with, we get one Sr^{2+} ion and two OH^- ions (until we reach the solubility of $Sr(OH)_2$ in water). For a 0.01 M solution of $Sr(OH)_2$, the concentration of OH^- will be 0.02 M.

Example 17-26: Calculate the pH of each of the following solutions.

     a. 0.025 M HNO_3

HNO_3 is nitric acid, a strong acid. It ionizes completely in water to form H_3O^+ and NO_3^-.

$$HNO_3\ (aq) + H_2O\ (l) \rightarrow H_3O^+\ (aq) + NO_3^-\ (aq)$$

For each molecule of HNO_3 we start with, we end up with an ion of H_3O^+. The concentration of H_3O^+ is 0.025 M. (We can ignore the original 1.0×10^{-7} M concentration of H_3O^+ in the water because the 0.025 M from the HNO_3 will overwhelm its contribution.)

$$pH = -\log\left[H_3O^+\right] = -\log\,[0.025] = 1.60$$

     b. 0.0080 M $Sr(OH)_2$

The equation for the dissociation of $Sr(OH)_2$ in water is

$$Sr(OH)_2\ (s) \rightarrow Sr^{2+}\ (aq) + 2OH^-\ (aq)$$

For each formula unit of $Sr(OH)_2$ we start out with, we end up with two hydroxide ions in solution. The concentration of hydroxide in this solution will be 2(0.0080 M) = 0.016 M

$$pOH = -\log\left[OH^-\right] = -\log\,[0.016] = 1.80$$

We can use the relationship that pH + pOH = 14.00 to determine the pH.

$$pH = 14.00 - pOH = 14.00 - 1.80 = 12.20$$

Try Study Question 11 and 13 in Chapter 17 of your textbook now!

2. A complex ion is a metal ion that has Lewis bases joined to the cation by means of coordinate covalent bonds. The electrons in these bonds originally belonged solely to the Lewis bases. An example of a complex ion is $[Cu(NH_3)_4]^{2+}$.

CHAPTER 18: Principles of Reactivity: Other Aspects of Aqueous Equilibria

Chapter Overview

In this chapter, we apply the principles of chemical equilibria to further examples of acid/base phenomena and also to solubility equilibria. Buffers resist a change in pH when hydroxide or hydronium ions are added. They are usually solutions containing 1) a weak acid and a salt containing the conjugate base of the weak acid or 2) a weak base and a salt containing the conjugate acid of the weak base. A buffer can respond equally well to the addition of either acid or base if we have equal concentrations of the conjugate acid and base present; the pH of such a buffer will be equal to the pK_a of the acid form. You will learn how to calculate the pH of a buffer solution using the Henderson-Hasselbalch equation. You will also learn how to calculate the pH of a buffer after the addition of either strong acid or strong base. In order to do this, you will first solve a stoichiometry problem and then determine the new pH using the Henderson-Hasselbalch equation.

We next examine acid/base titrations in detail. You should become familiar with the shapes of titration curves for different types of acid/base titrations. You should be able to calculate the pH in a titration 1) before the addition of any titrant, 2) after the addition of some titrant but before the equivalence point, 3) at the equivalence point, and 4) after the equivalence point. These calculations usually involve a stoichiometry problem followed by a calculation to yield the pH. The pH at the equivalence point will be 7 for a strong acid/strong base titration, greater than 7 for a weak acid/strong base titration, and less than 7 for a strong acid/weak base titration. The pH at the half-equivalence point for a titration of a weak acid is equal to the pK_a of the weak acid. For a titration involving a weak base, it will be equal to the pK_a of the conjugate acid of the weak base. In titrations involving weak polyprotic acids, the protons are removed stepwise. Acid-base indicators are often weak acids or bases themselves where the acid form is one color and the base form is another.

The solubility equilibrium for a sparingly soluble ionic compound is expressed mathematically by the solubility product constant, K_{sp}. You will learn how to estimate the value of K_{sp} from solubility data and vice versa. We can use Q and K_{sp} to determine if a precipitation reaction will occur. A slightly soluble salt is less soluble in a solution containing another source of one of the ions. Salts containing anions from weak acids are more soluble in an acidic solution than in water. Salts are more soluble if we add a ligand that can bind to the cation to form a complex ion. We can also use solubilities to determine procedures to separate different ions in solution from each other by selective precipitation.

Key Terms

In this chapter, you will need to learn and be able to use the following terms:

Acid-base indicator: a material that is one color below a particular pH and another color above it; typically acid-base indicators are themselves weak acids or bases.

Buffer capacity: a measure of how much hydronium or hydroxide can be added to a buffer solution before the buffer can no longer control the pH.

Buffer solution: a solution that resists a change in pH when hydroxide or hydronium ions are added.

Common ion effect: 1) Adding the conjugate base (or conjugate acid) to a solution of an acid (or base) increases the concentration of the acid (or base). 2) Adding a common ion to a saturated solution of a salt will lower the salt solubility.

Equivalence point: the point in a titration at which one reactant has been exactly consumed by addition of the other reactant.

Formation constant: the equilibrium constant for the formation of a complex ion.

Henderson-Hasselbalch equation: an equation that allows us to calculate the pH of a buffer:

$$pH = pK_a + \log\frac{\left[\text{conjugate base}\right]}{\left[\text{acid}\right]}$$

Solubility product constant: the equilibrium constant for the dissolution of a sparingly soluble salt. For the salt A_xB_y, it has the form: $K_{sp} = [A^{y+}]^x\,[B^{x-}]^y$

Titrant: the substance added during a titration.

Chapter Goals

By the end of this chapter you should be able to:

- **Understand the common ion effect.**

  Predict the effect of the addition of a "common ion" on the pH of the solution of a weak acid or base (Section 18.1).

  Consider the equilibrium for the ionization of a weak acid, such as acetic acid, in water.
  $$CH_3COOH\ (aq) + H_2O\ (l) <=> H_3O^+\ (aq) + CH_3COO^-\ (aq)$$

  Now consider adding acetate ions, perhaps by adding some sodium acetate. Le Chatelier's Principle predicts that the equilibrium will shift to the left in response to the stress of having too great a concentration of acetate. When equilibrium is reestablished, we will have more acetic acid than we had in the initial equilibrium, less hydronium ions, and more acetate ions (though less than we had when we first stressed the system). The key is that we have more acetic acid and less hydronium ions. By adding the acetate, the equilibrium shifted to the less ionized species. Because the concentration of hydronium ions is less, the pH will be higher. The common ion effect is the limiting of acid ionization caused by addition of the conjugate base. In our example, there was less ionization of the acid when we added its conjugate base, acetate ion.

  The common ion effect applies to bases too. Consider the equilibrium in a solution of ammonia.
  $$NH_3\ (aq) + H_2O\ (l) <=> NH_4^+\ (aq) + OH^-\ (aq)$$

  If the concentration of NH_4^+ is increased, the equilibrium shifts to the left, toward the unionized species. This reduces the concentration of OH- ions. The pH will be lower.

  Try Study Question 1 in Chapter 18 of your textbook now!

- **Understand the control of pH in aqueous solutions with buffers (Section 18.2).**

a) Describe the functioning of buffer solutions.

A buffer solution resists a change in pH when hydronium or hydroxide ions are added. How does a buffer do this? A buffer contains species that react with added hydronium and hydroxide ions. Something that reacts with hydronium ions is a base, and something that reacts with hydronium ions is an acid. A buffer must therefore contain both an acid and a base. We also need for the acid and base in the buffer not to react with each other. There are two main types of buffer solutions 1) those containing a weak acid and the salt containing the conjugate base of the weak acid and 2) those containing a weak base and a salt containing the conjugate acid of the weak base.

How does this work? Imagine that we have a buffer consisting of a weak acid (HA) and a salt containing its conjugate base (A^-). Now imagine that we add some H_3O^+ to this buffer. With which component of the buffer will H_3O^+ react? With the base, A^-.

$$H_3O^+ \ (aq) + A^- \ (aq) \rightarrow HA \ (aq) + H_2O \ (l)$$

At the end of this reaction, the H_3O^+ has reacted; it is no longer present as H_3O^+. It therefore does not affect the pH. There is only a very small change in pH that occurs due to a shift in the equilibrium for the ionization of HA in water.

Now let us imagine that we again have our original buffer solution. We add some OH^-. With which component of the buffer will OH^- react? With the acid, HA.

$$OH^- \ (aq) + HA \ (aq) \rightarrow A^- \ (aq) + H_2O \ (l)$$

At the end of this reaction, the OH^- has reacted. It does not affect the pH. Once again, there is only a very small change in pH that occurs due to a shift in the equilibrium for the ionization of HA in water.

Our buffer can thus resist changes in pH because it contains components that can react with either added H_3O^+ or OH^-.

Similar reactions occur in a buffer made from a weak base (B) and a salt containing the conjugate acid (BH^+) of the weak base. Added hydronium will react with the base.

$$H_3O^+ \ (aq) + B \ (aq) \rightarrow BH^+ \ (aq) + H_2O \ (l)$$

Added hydroxide will react with the conjugate acid.

$$OH^- \ (aq) + BH^+ \ (aq) \rightarrow B \ (aq) + H_2O \ (l)$$

Just as was true for our other buffer system, the hydronium and hydroxide are removed from the solution, and the only change in pH is due to the shifts that occur as the equilibrium between B and BH^+ reestablishes.

In each of these cases, note that we cannot add an infinite amount of the acid or base and get only minimal change in the pH. Eventually, we will reach the point where all of one of the components of the buffer has reacted. Once that has occurred, the buffer can no longer control the pH.

Try Study Question 99 in Chapter 18 of your textbook now!

b) Use the Henderson-Hasselbalch equation (Equation 18.2) to calculate the pH of a buffer solution of given composition.

The pH of a buffer solution can be calculated using the Henderson-Hasselbalch equation:

$$pH = pK_a + \log\frac{[\text{conjugate base}]}{[\text{acid}]}$$

This equation is simply a rearranged form of the expression for K_a. Technically, the concentrations that should be used in this equation are the equilibrium concentrations of the conjugate base and acid. If we are in the region where a buffer works, however, we can use the initial concentrations of the conjugate base and acid because the changes in their concentrations are insignificant compared to their initial concentrations. This makes this equation much easier to use.

Example 18-1: What is the pH of an acetic acid/sodium acetate buffer that has an acetic acid concentration of 0.100 M and a sodium acetate concentration of 0.080 M? The K_a for acetic acid is 1.8×10^{-5}.

The important components in this buffer are acetic acid and acetate ion. We will determine the pH of this buffer using the Henderson-Hasselbalch equation:

$$pH = pK_a + \log\frac{[\text{conjugate base}]}{[\text{acid}]}$$

The K_a is 1.8×10^{-5}, from which we can calculate the pK_a. The concentration of acetic acid is 0.100 M and that of acetate is 0.080 M.

$$pH = pK_a + \log\frac{[\text{conjugate base}]}{[\text{acid}]} = -\log\left(1.8 \times 10^{-5}\right) + \log\left(\frac{0.080}{0.100}\right) = 4.65$$

The pH is less than the pKa because there is more acid than base in the buffer.

There are a couple of things to note about the Henderson-Hasselbalch equation. One is that the pKa is a pivot for the pH. If $[HA] = [A^-]$, then the pH will be equal to the pKa. If we have more acid present than the conjugate base, $[HA] > [A^-]$, then the pH will be less than the pKa. If we have more of the conjugate base present than acid, $[HA] < [A^-]$, then the pH will be greater than the pKa.

We can use this equation if we have a buffer that consists of a weak acid and the salt containing the conjugate base of the weak acid, such as we have already seen. We can also use this equation if we have a weak base with a salt containing the conjugate acid of the weak base. In this case, we use the pKa of the conjugate acid (BH^+), [B] as the concentration of the conjugate base, and $[BH^+]$ as the concentration of the acid.

Example 18-2: What is the pH of an ammonia/ammonium chloride buffer in which the concentration of ammonia is 0.50 M and the concentration of ammonium ion is 0.40 M? K_b of ammonia is 1.8×10^{-5}.

This is a buffer problem. We will use the Henderson-Hasselbalch equation to solve it.

$$pH = pK_a + \log\frac{[\text{conjugate base}]}{[\text{acid}]}$$

For this equation, we need K_a of the acid. We are given K_b for the base. We can calculate K_a.

$$K_a \cdot K_b = K_w$$

$$K_a = \frac{K_w}{K_b} = \frac{1.0 \times 10^{-14}}{1.8 \times 10^{-5}} = 5.6 \times 10^{-10}$$

The initial concentration of the acid form is 0.40 M and that of the base form is 0.50 M. We substitute all of this into the Henderson-Hasselbalch equation.

$$pH = pK_a + \log\frac{[\text{conjugate base}]}{[\text{acid}]} = -\log(5.6 \times 10^{-10}) + \log\left(\frac{0.50}{0.40}\right) = 9.35$$

Try Study Questions 7 and 13 in Chapter 18 of your textbook now!

c) Describe how a buffer solution of a given pH can be prepared.

There are a few considerations that should be taken into account in preparing a buffer solution:

1. In selecting a buffer system to use, the pKa of the acid should be around the pH that is desired for the buffer. Slight deviations from this pH can be made by adjusting the ratio of the concentrations of the acid and its conjugate base.

2. The concentration of the buffer should be large enough to counteract the amount of acid or base that might be added. In scientific terms, we want to make sure that the amount of acid or base that might be added will not exceed the buffer capacity. Often, buffers are prepared with concentrations of the acid and conjugate base in the range of 0.1 – 1.0 M. No matter what the buffer is, if we add sufficient acid or base, we will eventually exceed the buffer capacity.

3. Because the acid and the conjugate base are present in the same solution, we can use either the concentrations or the number of moles of these species in the Henderson-Hasselbalch equation. The reason for this is that the volume terms will be the same for both species and will cancel in the ratio. Sometimes, using moles rather than concentrations will save a step in a calculation.

Example 18-3: Describe how you would prepare 1.0 L of a buffer that has a pH of 10.50. It is desired to have the concentration of the basic component of the buffer be 0.10 M. Use the table in Example 18.4 in your textbook to help you select the appropriate species.

The first step in this problem is to determine which materials to use in this buffer. When we look at Example 18.4 in your textbook, there are three different buffer systems listed. We wish to use the one with a pKa around the value of the pH we desire. HCO_3^- has a pKa of 10.32. This is close to the pH value so we will set up a HCO_3^-/CO_3^{2-} buffer system.

Next we need to find the ratio of conjugate base to acid that we should use. For this, we will use the Henderson-Hasselbalch equation.

$$pH = pK_a + \log\frac{[\text{conjugate base}]}{[\text{acid}]}$$

$$10.50 = 10.32 + \log \frac{[\text{conjugate base}]}{[\text{acid}]}$$

$$\log \frac{[\text{conjugate base}]}{[\text{acid}]} = 0.18$$

$$\frac{[\text{conjugate base}]}{[\text{acid}]} = 10^{0.18}$$

$$\frac{[\text{conjugate base}]}{[\text{acid}]} = 1.5$$

This tells us that we need to have 1.5 times the amount of base as we have acid. We are told that we want to have the concentration of the base, CO_3^{2-} be 0.10 M. The concentration of HCO_3^- must therefore be

$$\frac{0.10 \text{ M}}{[\text{acid}]} = 1.5$$

$$[\text{acid}] = \frac{0.10 \text{ M}}{1.5} = 0.067 \text{ M}$$

We, therefore, wish to prepare a 1.0 L solution with $[CO_3^{2-}] = 0.10$ M and $[HCO_3^-] = 0.067$ M. We can weigh out amounts of sodium carbonate and sodium hydrogen carbonate to obtain these concentrations.

$$1.0 \text{ L} \times \frac{0.10 \text{ moles Na}_2CO_3}{\text{L sol' n}} \times \frac{106 \text{ g Na}_2CO_3}{\text{mole Na}_2CO_3} = 11 \text{ g Na}_2CO_3$$

$$1.0 \text{ L} \times \frac{0.067 \text{ moles NaHCO}_3}{\text{L sol' n}} \times \frac{84 \text{ g NaHCO}_3}{\text{mole NaHCO}_3} = 5.6 \text{ g NaHCO}_3$$

To make this buffer, we should weigh out 11 g of sodium carbonate and 5.6 g of sodium hydrogen carbonate. We should transfer these to a 1.0 L volumetric flask. We should dissolve these in water and then add sufficient water to bring the solution to 1.0 L.

There are several different ways to prepare a buffer. To prepare an acetic acid/sodium acetate buffer, for example, one could start with a solution containing the correct concentration of acetic acid and then add the appropriate mass of sodium acetate. One could also start with a more concentrated solution of acetic acid and add the correct amount of sodium hydroxide. The sodium hydroxide reacts with the acetic acid to form the acetate ion in the solution.

Try Study Questions 15 and 17 in Chapter 18 of your textbook now!

d) Calculate the pH of a buffer solution before and after adding acid or base.

Before we add acid or base to a buffer, we calculate the pH of a buffer solution using the Henderson-Hasselbalch equation.

Example 18-4: Calculate the pH of a buffer solution that is 0.10 M in acetic acid and 0.10 M in sodium acetate. K_a for acetic acid is 1.8×10^{-5}.

This is a buffer. We can determine its pH using the Henderson-Hasselbalch equation.

$$pH = pK_a + \log \frac{[\text{conjugate base}]}{[\text{acid}]}$$

To use this equation, we will need the pK_a of the acid, the concentration of the acid and the concentration of the conjugate base. The pK_a can be calculated from the K_a.

$$pK_a \ = \ - \log K_a \ = \ - \log(1.8 \times 10^{-5}) \ = \ 4.74$$

The concentration of acetic acid is 0.10 M. The concentration of the conjugate base, acetate ion, is also 0.10 M.

$$pH \ = \ 4.74 \ + \ \log \frac{0.10 \ M}{0.10 \ M} \ = \ 4.74$$

Notice that in this case the pH equals the pK_a. This will be true whenever we have equal concentrations of the acid and the conjugate base.

In order to determine the pH after adding acid or base to it, we need to solve a two-part problem such as the one we solved in Example 17-24 in the study guide. The first part of the problem is a stoichiometry problem in which we use a modified stoichiometry table to determine how the added acid or base changes the ratio of the buffer's conjugate acid and base. The second part is an equilibrium problem in which we determine the pH of a buffer having this new ratio. For this second part, we can use the Henderson-Hasselbalch equation.

<u>Example 18-5:</u> Suppose that we have 100. mL of the buffer from Example 18-5. Now, let us add 5.00 mL of a 1.00 M solution of NaOH. What should be the new pH of the solution?

The first part of this problem is a stoichiometry problem. We must first decide the chemical reaction that occurs when the NaOH solution is added to the acetic acid/sodium acetate buffer. NaOH is a base. It will react with the acid component of the buffer.

$$OH^- \ (aq) + CH_3COOH \ (aq) \rightarrow CH_3COO^- \ (aq) + H_2O \ (l)$$

Now we will set up our modified stoichiometry table. This stoichiometry part is solved in terms of moles. We are given the volumes and molarities of the sodium hydroxide, the acetic acid, and the sodium acetate.

	OH^-	+	CH_3COOH	$\rightarrow$	CH_3COO^-	+	H_2O
Measured	5.00 mL		100. mL		100. mL		
Conversion Factor	1.00 moles/L		0.100 moles/L		0.100 moles/L		
Initial Moles							
Change in Moles							
Final Moles							

From this information, we can calculate the number of moles of each in the reaction mixture.

$$5.00 \ mL \ NaOH \ \times \ \frac{1 \ L}{1000 \ mL} \ \times \ \frac{1.00 \ moles \ OH^-}{1 \ L} \ = \ 0.00500 \ moles \ OH^-$$

$$100. \ mL \ CH_3COOH \ \times \ \frac{1 \ L}{1000 \ mL} \ \times \ \frac{0.100 \ moles \ CH_3COOH}{1 \ L}$$
$$= \ 0.0100 \ moles \ CH_3COOH$$

$$100. \ mL \ CH_3COO^- \ \times \ \frac{1 \ L}{1000 \ mL} \ \times \ \frac{0.100 \ moles \ CH_3COO^-}{1 \ L}$$
$$= \ 0.0100 \ moles \ CH_3COO^-$$

We can enter this information in the table.

	OH⁻	+	CH_3COOH	→	CH_3COO^-	+	H_2O
Measured	5.00 mL		100. mL		100. mL		
Conversion Factor	1.00 moles/L		0.100 moles/L		0.100 moles/L		
Initial Moles	0.00500 moles		0.0100 moles		0.0100 moles		
Change in Moles							
Final Moles							

Because the stoichiometric ratio of hydroxide to acetic acid in this reaction is 1:1, we can look at the moles of the reactants and determine the limiting reactant. OH⁻ is the limiting reactant because it is present in the smaller amount. All of it will react. When it does this, the same amount of acetic acid will react and the same amount of acetate will be produced.

	OH⁻	+	CH_3COOH	→	CH_3COO^-	+	H_2O
Measured	5.00 mL		100. mL		100. mL		
Conversion Factor	1.00 moles/L		0.100 moles/L		0.100 moles/L		
Initial Moles	0.00500 moles		0.0100 moles		0.0100 moles		
Change in Moles	− 0.00500 moles		− 0.00500 moles		+ 0.00500 moles		
Final Moles							

We can now calculate the final number of moles.

	OH⁻	+	CH_3COOH	→	CH_3COO^-	+	H_2O
Measured	5.00 mL		100. mL		100. mL		
Conversion Factor	1.00 moles/L		0.100 moles/L		0.100 moles/L		
Initial Moles	0.00500 moles		0.0100 moles		0.0100 moles		
Change in Moles	− 0.00500 moles		− 0.00500 moles		+ 0.00500 moles		
Final Moles	0 moles		0.0050 moles		0.0150 moles		

This is the end of the stoichiometry part of the problem. We have determined that after the reaction, we will have 0.0050 moles of acetic acid and 0.0150 moles of acetate present. We now need to determine the pH of a new buffer that has those amounts of CH_3COOH and CH_3COO^-. To do this, we use the Henderson-Hasselbalch equation.

$$pH = pK_a + \log \frac{[\text{conjugate base}]}{[\text{acid}]}$$

$$pH = pK_a + \log \frac{[\text{conjugate base}]}{[\text{acid}]} = 4.74 + \log \frac{0.0150 \text{ moles}}{0.0050 \text{ moles}} = 5.22$$

Try Study Questions 19 and 21 in Chapter 18 of your textbook now!

e) Predict the pH of an acid-base reaction at its equivalence point (see also Sections 17.5 and 17.6).

This objective really fits in with the next goal dealing with acid-base titrations. In an acid-base titration, we incrementally add either an acid or base to the opposite, a base or an acid. The key point of the titration is what is called the equivalence point, the point at which one reactant has been completely consumed by addition of the other reactant. The pH at the equivalence point will depend upon whether the acid and base used were strong or weak.

Acid	Base	pH at Equivalence Point
Strong	Strong	= 7 (neutral)
Strong	Weak	< 7 (acidic)
Weak	Strong	> 7 (basic)

The results in this table should not come as a surprise. In the last chapter, we went over what the pH would be when equal molar quantities of the acid and base have been added. This is what we have at the equivalence point of a monoprotic titration. At the equivalence point, we essentially have just a salt solution where the cation of the salt came from the base and the anion from the acid. We learned in the last chapter that cations from strong bases and anions from strong acids do not affect the pH of the solution. The cations from weak bases make a solution acidic, and the anions from weak acids make a solution basic. This is because these ions react with water to form hydronium or hydroxide ions, respectively.

We did not include weak acid-weak base titrations in the above table. These titrations are generally not done because the equivalence point is sometimes difficult to see and cannot be judged accurately. Weak acid/weak base equivalence points may be acidic, basic, or neutral depending on the K_a and K_b values of the products formed.

Example 18-6: Will the equivalence point for the titration of ammonia with hydrochloric acid be acidic, basic, or neutral?

Ammonia is a weak base. Hydrochloric acid is a strong acid. The equivalence point for a strong acid-weak base titration is acidic due to the reaction of the ammonium ion with water.

• **Evaluate the pH in the course of acid-base titrations (Section 18.3).**

a) Calculate the pH at the equivalence point in the reaction of a strong acid with a weak base, or in the reaction of a strong base with a weak acid.

In a strong acid-weak base or strong base-weak acid titration, we can actually calculate the pH of the solution at the equivalence point. We do this using the knowledge we have from the last chapter about the pH of salt solutions.

Example 18-7: Suppose that 25.00 mL of 0.10 M NH_3 is titrated with 0.20 M HCl. What is the pH after 12.50 mL of the HCl solution has been added?

First of all, we have a chemical reaction going on. Ammonia and hydrochloric acid react.
$$NH_3 \ (aq) + HCl \ (aq) \rightarrow NH_4Cl \ (aq)$$

We set up our modified stoichiometry table.

NH₃	+	HCl	→	NH₄Cl
Measured	25.00 mL		12.50 mL		
Conversion Factor	0.10 moles/L		0.20 moles/L		
Initial Moles					
Change in Moles					
Final Moles					

We calculate the initial number of moles present of the NH_3 and the HCl.

$$25.00\ mL \times \frac{1\ L}{1000\ mL} \times \frac{0.10\ moles\ NH_3}{1\ L} = 0.0025\ moles\ NH_3$$

$$12.50\ mL \times \frac{1\ L}{1000\ mL} \times \frac{0.20\ moles\ HCl}{1\ L} = 0.0025\ moles\ HCl$$

NH₃	+	HCl	→	NH₄Cl
Measured	25.00 mL		12.50 mL		
Conversion Factor	0.10 moles/L		0.20 moles/L		
Initial Moles	0.0025 moles		0.0025 moles		0 moles
Change in Moles					
Final Moles					

We have stoichiometrically equivalent amounts of NH_3 and HCl. This is the equivalence point. They each react completely. In the process, we will form 0.0025 moles of NH_4Cl.

NH₃	+	HCl	→	NH₄Cl
Measured	25.00 mL		12.50 mL		
Conversion Factor	0.10 moles/L		0.20 moles/L		
Initial Moles	0.0025 moles		0.0025 moles		0 moles
Change in Moles	− 0.0025 moles		− 0.0025 moles		+ 0.0025 moles
Final Moles	0 moles		0 moles		0.0025 moles

At the end of the reaction, we have a solution of ammonium chloride in water. We have 0.0025 moles of NH_4Cl in 37.50 mL of solution. Now we will work an equilibrium problem to figure out the pH of this solution. The chloride ions will not affect the pH. Ammonium is the conjugate acid of a weak base. It will make the solution be acidic.

$$NH_4^+\ (aq) + H_2O\ (l) <=> NH_3\ (aq) + H_3O^+\ (aq)$$

The initial concentration of NH_4^+ is

$$\left[NH_4^+\right] = \frac{0.0025\ moles\ NH_4^+}{0.03750\ L} = 6.7 \times 10^{-2}\ M$$

	NH₄⁺	+	H₂O	<=>	NH₃	+	H₃O⁺
Initial	6.7×10^{-2} M				0		0
Change	− x				+ x		+ x
Equilibrium	$6.7 \times 10^{-2} - x$				x		x

The chemical equation is the equation for the K_a of NH_4^+. This is 5.6×10^{-10}.

$$K_a = \frac{[NH_3][H_3O^+]}{[NH_4^+]}$$

$$5.6 \times 10^{-10} = \frac{x \cdot x}{6.7 \times 10^{-2} - x}$$

We can assume that x is much less than 6.7×10^{-2} because $100 \cdot K_a$ is much less than the initial concentration.

$$5.6 \times 10^{-10} = \frac{x^2}{6.7 \times 10^{-2}}$$

$$x = \pm 6.1 \times 10^{-6}$$

We can reject the negative root, so $x = 6.1 \times 10^{-6}$ M. This is the concentration of H_3O^+. From this, we can calculate the pH.

$$pH = -\log[H_3O^+] = -\log(6.1 \times 10^{-6}) = 5.21$$

Try Study Question 23 in Chapter 18 of your textbook now!

b) Understand the differences between the titration curves for a strong acid-strong base titration versus cases in which one of the substances is weak.

Figure 18.4 in your textbook shows the titration curve for the titration of 0.100 M HCl with 0.100 M NaOH. Figure 18.5 shows the curve for the titration of 0.100 M CH_3COOH with 0.100 M NaOH. The acid is strong in one case and weak in the other.

The weak acid titration starts out at a higher pH. The weak acid titration has a buffer zone around the pK_a of the weak acid, where the pH largely levels off. The pH jump at the equivalence point is smaller for the weak acid. The equivalence point is in the basic region for the weak acid titration instead of being neutral. Following the equivalence points, the two graphs are identical.

In comparing a weak base-strong acid titration with a strong base-strong acid titration, similar issues are found. For a weak base-strong acid titration, the initial pH is lower, there is a buffer zone around the pK_a of the conjugate acid of the weak base, the pH drop is smaller, and the equivalence point is acidic instead of neutral.

In all of the titration curves, the equivalence point is the mid-point in the vertical portion of the pH versus volume curve.

In the case of a weak acid-strong base titration curve, if we take the volume at the equivalence point, divide by 2, and then read off the pH at this half-neutralization point, the pH will be equal to the pK_a of the acid. In the case of a weak base-strong acid titration, the pH at the half neutralization point is equal to the pK_a of the conjugate acid of the weak base.

Try Study Question 27 in Chapter 18 of your textbook now!

c) Describe how an indicator functions in an acid-base titration.

An acid-base indicator is a material that is one color below a particular pH and another color above that pH. Acid-base indicators are usually weak acids and bases themselves. The conjugate acid is one color and the conjugate base is another. At the pK_a of the indicator, we have half of the indicator present in its acidic form and half in the basic form. The acid form predominates below the indicator's pK_a, and the base form predominates above that pH. In an acid-base titration, we detect the end point of a titration by using an acid-base indicator that changes color in the vicinity of the pH of the equivalence point. In other words, the pK_a of the indicator should be around the value of the equivalence point expected. It does not need to be exactly the pH of the equivalence point because the titration curves rise so sharply around the equivalence point, but it should be within about two pH units.

Example 18-8: Would phenolphthalein be a good choice to use as an indicator in a titration of NH_3 with HCl?

This is a weak base-strong acid titration. The pH at the equivalence point will be in the acidic region. Figure 18.10 in your textbook shows that phenolphthalein has a color change that occurs in the basic region. This would not be a good choice. A better choice would be something like methyl red, which has a slightly acidic end point.

- **Apply chemical equilibrium concepts to the solubility of ionic compounds.**

a) Write the equilibrium constant expression – the solubility product constant, K_{sp} – for any insoluble salt (Section 18.4).

Given a solubility equilibrium for a sparingly soluble salt,
$$A_xB_y \ (s) <=> xA^{y+} \ (aq) + yB^{x-} \ (aq)$$
we can write an equilibrium constant expression, called the solubility product constant, K_{sp}
$$K_{sp} = [A^{y+}]^x [B^{x-}]^y$$

Example 18-9: Write the solubility product constant expressions for the following salts.

   a. $CaCO_3$
The first step is to write the chemical equation for the dissociation of the compound in water.
$$CaCO_3 \ (s) <=> Ca^{2+} \ (aq) + CO_3^{2-} \ (aq)$$
Once we have the chemical equation, we write the equilibrium constant expression in the usual manner; this is the solubility product constant expression.
$$K_{sp} = [Ca^{2+}] [CO_3^{2-}]$$

   b. $Ca_3(PO_4)_2$
The dissolution equilibrium is
$$Ca_3(PO_4)_2 \ (s) <=> 3Ca^{2+} \ (aq) + 2PO_4^{3-} \ (aq)$$
The solubility product constant is
$$K_{sp} = [Ca^{2+}]^3 [PO_4^{3-}]^2$$

Try Study Question 37 in Chapter 18 of your textbook now!

b) Calculate K_{sp} values from experimental data (Section 18.4).

We can use experimental data about the solubility of a salt to determine its K_{sp}. The key step is to recognize how the concentration given for the salt relates to the concentrations of the ions that actually end up in solution.

Example 18-10: Assume that the molar solubility of manganese(II) hydroxide in water is approximately 1.1×10^{-4} moles/liter of solution. What is the K_{sp} for magnesium hydroxide?

First, we will write the balanced equation for the dissolution process.
$$Mn(OH)_2 \ (s) <=> Mn^{2+} \ (aq) + 2OH^- \ (aq)$$

There are initially no Mn^{2+} ions or OH^- ions in solution (actually there is 10^{-7} M OH^-, but we shall assume this is negligible compared to the hydroxide produced by our salt). Some of the $Mn(OH)_2$ dissolves. We will let x represent the nominal concentration of $Mn(OH)_2$ that dissolves. In reality, we do not have the compound but dissociated ions in solution. For each formula unit of $Mn(OH)_2$ that dissolves, we get one Mn^{2+} ion and two OH^- ions.

	$Mn(OH)_2 \ (s)$	<=>	$Mn^{2+} \ (aq)$	+	$2OH^- \ (aq)$
Initial			0		0
Change			+ x		+ 2x
Equilibrium			x		2x

We are told that the solubility of magnesium fluoride is 1.1×10^{-4} M. This is our x.

	$Mn(OH)_2 \ (s)$	<=>	$Mn^{2+} \ (aq)$	+	$2OH^- \ (aq)$
Initial			0		0
Change			+ x		+ 2x
Equilibrium			x $= 1.1 \times 10^{-4}$ M		2x $= 2(1.1 \times 10^{-4}$ M) $= 2.2 \times 10^{-4}$ M

We can now determine K_{sp}.
$$K_{sp} = \left[Mn^{2+}\right]\left[OH^-\right]^2 = \left(1.1 \times 10^{-4}\right)\left(2.2 \times 10^{-4}\right)^2 = 5.3 \times 10^{-12}$$

Try Study Questions 39 and 43 in Chapter 18 of your textbook now!

c) Estimate the solubility of a salt from the value of K_{sp} (Section 18.4).

We can also go in the other direction. Given a $K_{sp,}$ we can estimate the solubility.

Example 18-11: Estimate the solubility of Ag_2SO_4 in water based on its K_{sp}, 1.2×10^{-5}.

The equation for the dissolution of Ag_2SO_4 is
$$Ag_2SO_4 \ (s) <=> 2Ag^+ \ (aq) + SO_4^{2-} \ (aq)$$

We set up our ICE table. Once again, we let x equal the nominal concentration of the dissolved salt. For each formula unit of silver sulfate that dissolves, we get two silver ions and one sulfate ion. At equilibrium, we will have a silver ion concentration of 2x M and a sulfate ion concentration of x M.

	$Ag_2SO_4 \ (s)$	<=>	$2Ag^+ \ (aq)$	+	$SO_4^{2-} \ (aq)$
Initial			0		0
Change			+ 2x		+ x
Equilibrium			2x		x

We substitute these values into K_{sp}.

$$K_{sp} = \left[Ag^+\right]^2 \left[SO_4^{2-}\right]$$
$$1.2 \times 10^{-5} = (2x)^2 (x)$$
$$1.2 \times 10^{-5} = 4x^3$$
$$3.0 \times 10^{-6} = x^3$$

We now take the cube root of both sides of the equation to determine the value of x.
$$x = 1.4 \times 10^{-2}$$

The solubility of silver sulfate is thus 1.4×10^{-2} M.

These calculations only give an estimate of the solubility. The solubility of some salts is dependent upon more factors than just those that determine the value of K_{sp}. Sometimes the solubility can be very different from that predicted from these calculations.

Try Study Questions 45 and 47 in Chapter 18 of your textbook now!

d) Calculate the solubility of a salt in the presence of a common ion (Section 18.4).

Let us consider the solubility equilibrium for Ag_2SO_4.
$$Ag_2SO_4 \ (s) <=> 2Ag^+ \ (aq) + SO_4^{2-} \ (aq)$$

What should happen if we increase the concentration of the sulfate ion? Le Chatelier's Principle predicts that the equilibrium should shift to the left. The solubility of the salt will be less in the presence of the added sulfate. This is another type of common ion effect. Adding a common ion to a saturated solution of a salt lowers the solubility of the salt.

Example 18-12: Estimate the solubility of Ag_2SO_4 in a solution containing 0.10 M Na_2SO_4.

The key difference between this example and the last one is that the concentration of sulfate does not start out as 0 M. We start out with 0.10 M SO_4^{2-}.

	Ag_2SO_4 (s)	<=>	$2Ag^+$ (aq)	+	SO_4^{2-} (aq)
Initial			0		0.100 M
Change			+ 2x		+ x
Equilibrium			2x		0.100 + x

$$K_{sp} = \left[Ag^+\right]^2 \left[SO_4^{2-}\right]$$
$$1.2 \times 10^{-5} = (2x)^2 (0.100 + x)$$

We can make the assumption that x is much smaller than 0.100 M. ($100 \cdot K_{sp} < 0.100$).
$$1.2 \times 10^{-5} = (2x)^2 (0.100)$$
$$1.2 \times 10^{-4} = 4x^2$$
$$x = 5.5 \times 10^{-3}$$

The solubility of silver sulfate in this solution is therefore 5.5×10^{-3} M. This is less than what we found in the last example, in line with the prediction from the common ion effect.

Try Study Question 55 in Chapter 18 of your textbook now!

e) Understand the effect of basic anions on the solubility of a salt (Section 18.4).

Consider the solubility equilibrium that goes along with K_{sp} for a salt, MA.
$$MA\ (s) <=> M^{n+}\ (aq) + A^{n-}\ (aq)$$

When A^{n-} is the conjugate base of a weak acid (i.e., acetate, carbonate, hydroxide, phosphate, sulfide), there are some extra considerations because these conjugate bases have a significant basicity. (These considerations do not apply to salts containing the conjugate bases of strong acids.)

1) Any salt containing an anion that is the conjugate base of a weak acid will dissolve in water to a greater extent than given by K_{sp}.

The reason for this is that the conjugate base of a weak acid reacts with water as follows:
$$A^-\ (aq) + H_2O\ (l) <=> HA\ (aq) + A^-\ (aq)$$

In effect, this reaction removes some A^- from the solution. According to Le Chatelier's Principle, this will shift the equilibrium to the right, allowing more salt to dissolve.

2) Insoluble salts in which the anion is the conjugate base of a weak acid dissolve in strong acids.

The reason for this is that these anions have a significant basicity and react with the acid.
$$A^-\ (aq) + H_3O^+\ (aq) <=> HA\ (aq) + H_2O\ (l)$$

This reaction will be even more product favored than the reaction above with water because the strong acid is a stronger acid than water. Once again, the effect is to remove A^- from the solution, shifting the solubility equilibrium to the right.

Example 18-13: Will the following salts be more soluble in an acidic solution or not?

a. calcium phosphate
A salt is more soluble in an acidic solution if the salt contains the conjugate base of a weak acid. Phosphate is the conjugate base of the weak acid, HPO_4^-. It will have a significant basicity and react with the acid. This will shift the equilibrium so that more calcium phosphate will dissolve.

b. silver chloride
The chloride ion is the conjugate base of a strong acid. It does not have a significant basicity. This salt will not be more soluble in the acidic solution than in water.

Try Study Question 57 in Chapter 18 of your textbook now!

f) Decide whether a precipitate will form when the ion concentrations are known (Section 18.5).

To determine if a precipitate will form when we mix together particular solutions, we calculate Q, the reaction quotient. In the case of solubility equilibria, we find the following relationships between Q and K_{sp}.

If $Q < K_{sp}$, the solution is unsaturated. No precipitation will occur.
If $Q = K_{sp}$, the solution is saturated.
If $Q > K_{sp}$, precipitation will occur.

Example 18-14: 100. mL of a 0.100 M $AgNO_3$ solution is combined with 150. mL of a 0.0500 M NaCl solution. Will a precipitate form?

First of all, we must identify what the precipitate could be. The possible products from this reaction are sodium nitrate and silver chloride. Sodium nitrate is very soluble. Silver chloride is predicted to be insoluble, but have we really exceeded its solubility?

$$AgCl \ (s) <=> Ag^+ \ (s) + Cl^- \ (aq) \qquad K_{sp} = 1.8 \times 10^{-10}$$

We must calculate the concentrations of the ions in the final solution and compare Q and K_{sp}. The final solution has a volume of 250. mL. We determine the concentration of each ion by means of a dilution calculation.

Concentration of Ag^+

$$c_c V_c = c_d V_d$$

$$c_d = \frac{c_c V_c}{V_d}$$

$$= \frac{(0.100 \ M) \ (100. \ mL)}{250. \ mL}$$

$$= 0.0400 \ M$$

Concentration of Cl^-

$$c_c V_c = c_d V_d$$

$$c_d = \frac{c_c V_c}{V_d}$$

$$= \frac{(0.0500 \ M) \ (150. \ mL)}{250. \ mL}$$

$$= 0.0300 \ M$$

We now calculate Q

$$Q = \left[Ag^+ \right]\left[Cl^- \right] = (0.0400) \ (0.0300) = 1.20 \times 10^{-3}$$

This is larger than K_{sp} so the concentration of the ions is too large for the system to be at equilibrium. Precipitation should occur.

Try Study Question 59 in Chapter 18 of your textbook now!

g) Calculate the ion concentrations that are required to begin the precipitation of an insoluble salt (Section 18.5).

Precipitation occurs when Q is greater than K_{sp}. To determine the concentration of an ion needed to begin the precipitation of a salt, we determine what concentration of that ion we need in order for Q to equal K. Any concentration greater than this will cause precipitation.

Example 18-15: A solution contains 0.0080 M $Ca(NO_3)_2$. What concentration of fluoride ions is required to begin precipitating CaF_2? The K_{sp} of CaF_2 is 5.3×10^{-11}.

First, we write the chemical equation for the K_{sp} of CaF_2 and the K_{sp} expression.

$$CaF_2 \ (s) <=> Ca^{2+} \ (aq) + 2F^- \ (aq) \qquad K_{sp} = [Ca^{2+}] \ [F^-]^2$$

We know that the concentration of Ca^{2+} in the solution is 0.0080 M. We substitute this into the equation and solve for the fluoride ion concentration.

$$5.3 \times 10^{-11} = 0.0080 \ x^2$$

$$x = \pm 8.1 \times 10^{-5}$$

We reject the negative root because x is a concentration. Q = K when the concentration of F^- equals 8.1×10^{-5} M. Once the concentration of F^- exceeds 8.1×10^{-5} M, precipitation begins.

Try Study Question 61 in Chapter 18 of your textbook now!

h) Understand that the formation of a complex ion can increase the solubility of an insoluble salt (Section 18.6).

Recall that a complex ion is a metal ion that has Lewis bases joined to it by coordinate covalent bonds. The groups joined to the metal ion are called ligands. In water, a metal ion is present as a complex ion of the cation surrounded by water molecules. If we add a stronger ligand to the solution, the water molecules are displaced and a new complex ion is formed. The equilibrium constant that goes along with this process is called a formation constant, $K_{formation}$ (K_f). For example, nickel ions form a complex ion with cyanide ions:

$$Ni^{2+} (aq) + 4CN^- (aq) <=> [Ni(CN)_4]^{2-} (aq)$$

$$K_f = \frac{\left[Ni(CN)_4^{2-} \right]}{\left[Ni^{2+} \right]\left[CN^- \right]^4}$$

To determine the charge of a complex ion, we add together the charges of all of the species that went into forming it. In this case, we obtain $+2 + 4(-1) = -2$

Salts are more soluble when we add to the solution a ligand that can form a complex ion with the metal ion. Why? Let's say we have a sparingly soluble salt, MX.

$$MX (s) <=> M^{n+} (aq) + X^{n-} (aq)$$

If we add to the solution a ligand that forms a complex ion with the metal ion that has a large enough formation constant, the complex ion will form, effectively removing M^{n+} from the solution (the metal ion will now be present not as M^{n+} (aq) but as the complex ion). According to Le Chatelier's Principle, the solubility equilibrium should shift to the right to regenerate some M^{n+} (aq). The net effect is that some more of the original salt dissolves.

<u>Example 18-16:</u> What is the value of the equilibrium constant K_{net} for dissolving $Ni(OH)_2$ in a solution containing the cyanide ion?

We can view this process as involving two steps. The first is the solubility equilibrium of $Ni(OH)_2$ in water. We can find the equilibrium constant for this (K_{sp}) in Appendix J.

$$Ni(OH)_2 (s) <=> Ni^{2+} (aq) + 2OH^- (aq) \qquad K_{sp} = 5.5 \times 10^{-16}$$

The second process is the formation of the complex ion between Ni^{2+} and CN^-. The equilibrium constant for this process K_f can be found in Appendix K.

$$Ni^{2+} (aq) + 4CN^- (aq) <=> [Ni(CN)_4]^{2-} (aq) \qquad K_f = 1.0 \times 10^{31}$$

To obtain the overall K, we need to add together these two chemical equations. When we add together equations, we multiply their equilibrium constants.

$Ni(OH)_2 (s) <=> Ni^{2+} (aq) + 2OH^- (aq)$	$K_{sp} = 5.5 \times 10^{-16}$
$Ni^{2+} (aq) + 4CN^- (aq) <=> [Ni(CN)_4]^{2-} (aq)$	$K_f = 1.0 \times 10^{31}$
$Ni(OH)_2 (s) + 4CN^- (aq) <=> [Ni(CN)_4]^{2-} (aq) + 2OH^- (aq)$	$K_{net} = K_{sp} \cdot K_f = 5.5 \times 10^{15}$

The equilibrium constant for the overall process is much greater than one. The overall process is thus product-favored. $Ni(OH)_2$, while not very soluble at all in water can be dissolved in a cyanide solution.

Try Study Question 65 in Chapter 18 of your textbook now!

i) Use K_{sp} values to devise a method of separating ions in solution from one another (Section 18.7).

In separating ions in solution from one another, the best-case scenario is to find an ion that will form a precipitate with one of the ions but not with the other(s).

Example 18-17: A solution is known to contain Ba^{2+} ions and Cu^{2+} ions. Devise a way to separate these ions by precipitating one as an insoluble salt and leaving the other in solution.

In order to separate these ions, we must determine an anion that we can add to the solution that will precipitate one of the ions but not the other. The major tools to do this will be the solubility product constants in Appendix J and the solubility rules we learned back in Chapter 5. There may sometimes be more than one way to achieve the desired outcome.

Using our solubility guidelines, we know that $BaSO_4$ in not very soluble but that $CuSO_4$ is. What we can do is to add to our solution a source of sulfate ions, perhaps Na_2SO_4. The Ba^{2+} should precipitate as $BaSO_4$, whereas the Cu^{2+} should be left in solution. In selecting the salt to add, the sodium salt was selected because sodium salts are soluble and should not interact with the anions already present in the solution to form an undesired precipitate.

Try Study Question 67 in Chapter 18 of your textbook now!

Other Notes

1. Calculations Involving the Titration of a Strong Base with a Strong Acid

Example 18-18: Suppose we transfer 25.00 mL of 0.100 M NaOH to an Erlenmeyer flask. We titrate this with 0.200 M HCl.

a. Calculate the initial pH.
Before we add any acid, this is a solution of the strong base sodium hydroxide. The nominal concentration of sodium hydroxide is 0.100 M. For each formula unit of NaOH, we will have one OH^- ion in the solution, so the concentration of hydroxide ion is 0.100 M.

$$pOH = -\log\left[OH^-\right] = -\log(0.100) = 1.00$$
$$pH = 14.00 - pOH = 14.00 - 1.00 = 13.00$$

b. What is the pH after 5.00 mL of the acid has been added?
The chemical reaction that occurs between hydrochloric acid and sodium hydroxide is
$$HCl\ (aq) + NaOH\ (aq) \rightarrow NaCl\ (aq) + H_2O\ (l)$$

We first work a stoichiometry problem using our modified stoichiometry table.

	HCl	+	NaOH	$\rightarrow$	NaCl	+	H_2O
Measured	5.00 mL		25.00 mL				
Conversion Factor	0.200 mole/L		0.100 moles/L				
Initial Moles							
Change in Moles							
Final Moles							

We calculate the initial number of moles of each.
$$5.00\text{ mL x } \frac{1\text{ L}}{1000\text{ mL}} \text{ x } \frac{0.200\text{ moles HCl}}{1\text{ L}} = 0.00100\text{ moles HCl}$$
$$25.00\text{ mL x } \frac{1\text{ L}}{1000\text{ mL}} \text{ x } \frac{0.100\text{ moles NaOH}}{1\text{ L}} = 0.00250\text{ moles NaOH}$$

	HCl	+	NaOH	→	NaCl	+	H₂O
Measured	5.00 mL		25.00 mL				
Conversion Factor	0.200 mole/L		0.100 moles/L				
Initial Moles	0.00100 moles		0.00250 moles		0 moles		
Change in Moles							
Final Moles							

Because this is a 1:1 reaction, we can tell that hydronium will be the limiting reactant.

	HCl	+	NaOH	→	NaCl	+	H₂O
Measured	5.00 mL		25.00 mL				
Conversion Factor	0.200 mole/L		0.100 moles/L				
Initial Moles	0.00100 moles		0.00250 moles		0 moles		
Change in Moles	− 0.00100 moles		− 0.00100 moles		+ 0.00100 moles		
Final Moles	0 moles		0.00150 moles		0.00100 moles		

The NaCl formed will not affect the pH. The remaining NaOH will. We have 0.00150 moles of NaOH in our solution. The volume is now 5.00 mL + 25.00 mL = 30.00 mL = 0.03000 L. We can calculate the concentration of hydroxide, the pOH, and finally the pH.

$$\left[OH^-\right] = \frac{0.00150 \text{ moles OH}^-}{0.03000 \text{ L}} = 0.0500 \text{ M}$$

$$pOH = -\log\left[OH^-\right] = -\log(0.0500) = 1.301$$

$$pH = 14.00 - pOH = 14.00 - 1.30 = 12.70$$

c. What is the pH at the equivalence point?

This titration involves a strong acid and a strong base. The pH at the equivalence point should be 7.00.

d. What is the pH after 17.00 mL of the acid has been added?

We again set up the modified stoichiometry table and calculate the number of initial number of moles of HCl and NaOH.

	HCl	+	NaOH	→	NaCl	+	H₂O
Measured	17.00 mL		25.00 mL				
Conversion Factor	0.200 mole/L		0.100 moles/L				
Initial Moles	0.00340 moles		0.00250 moles		0 moles		
Change in Moles							
Final Moles							

This time, the NaOH is the limiting reactant.

	HCl	+	NaOH	→	NaCl	+	H₂O
Measured	17.00 mL		25.00 mL				
Conversion Factor	0.200 mole/L		0.100 moles/L				
Initial Moles	0.00340 moles		0.00250 moles		0 moles		
Change in Moles	− 0.00250 moles		− 0.00250 moles		+ 0.00250 moles		
Final Moles	0.00090 moles		0 moles		0.00250 moles		

The NaCl still does not affect the pH but now we have leftover HCl. We can calculate the concentration of hydronium ions using the number of moles of HCl and the total volume of the solution. We then calculate the pH.

$$\left[H_3O^+\right] = \frac{0.00090 \text{ moles } H_3O^+}{(0.01700 \text{ L} + 0.02500 \text{ L})} = 0.021 \text{ M}$$

$$pH = -\log\left[H_3O^+\right] = -\log(0.021) = 1.67$$

2. Calculations Involving the Titration of a Weak Acid with a Strong Base

Example 18-19: Suppose we transfer 25.00 mL of 0.200 M acetic acid to an Erlenmeyer flask. We titrate this with 0.100 M NaOH. The K_a of acetic acid is 1.8×10^{-5}.

a. Calculate the initial pH.
We have a solution of a weak acid. We calculate its pH using its K_a.

	CH₃COOH	+	H₂O	<=>	CH₃COO⁻	+	H₃O⁺
Initial	0.200 M				0		0
Change	− x				+ x		+ x
Equilibrium	0.200 − x				x		x

$$K_a = \frac{\left[H_3O^+\right]\left[CH_3COO^-\right]}{\left[CH_3COOH\right]}$$

$$1.8 \times 10^{-5} = \frac{x^2}{0.200 - x}$$

We can use the assumption that x is much less than 0.200 because $100 \cdot K_a$ is less than 0.200.

$$1.8 \times 10^{-5} = \frac{x^2}{0.200}$$

$$x = \pm 1.9 \times 10^{-3}$$

We reject the negative root because x represents a concentration.

$$\left[H_3O^+\right] = x = 1.9 \times 10^{-3} \text{ M}$$

$$pH = -\log\left[H_3O^+\right] = -\log\left(1.9 \times 10^{-3}\right) = 2.72$$

b. Calculate the pH after 10.00 mL of the 0.100 M NaOH has been added.
This type of problem involves two steps: a stoichiometry problem and an equilibrium problem. The reaction that occurs between acetic acid and sodium hydroxide is

$$CH_3COOH \ (aq) + NaOH \ (aq) \rightarrow CH_3COONa \ (aq) + H_2O \ (l)$$

We fill in our modified stoichiometry table.

	CH_3COOH	+	NaOH	$\rightarrow$	CH_3COONa	+	H_2O
Measured	25.00 mL		10.00 mL				
Conversion Factor	0.200 moles/L		0.100 moles/L				
Initial Moles							
Change in Moles							
Final Moles							

We can determine the number of moles of acetic acid and of sodium hydroxide.

$$25.00 \ mL \ x \ \frac{1 \ L}{1000 \ mL} \ x \ \frac{0.200 \ moles \ CH_3COOH}{1 \ L} = 0.00500 \ moles \ CH_3COOH$$

$$10.00 \ mL \ x \ \frac{1 \ L}{1000 \ mL} \ x \ \frac{0.100 \ moles \ NaOH}{1 \ L} = 0.00100 \ moles \ NaOH$$

We substitute these into our table as the initial number of moles. Initially, there is no sodium acetate present. The sodium hydroxide is the limiting reactant. It will all react. When it does so, the same number of moles of acetic acid will react, and the same number of moles of sodium acetate will be produced.

	CH_3COOH	+	NaOH	$\rightarrow$	CH_3COONa	+	H_2O
Measured	25.00 mL		10.00 mL				
Conversion Factor	0.200 moles/L		0.100 moles/L				
Initial Moles	0.00500 moles		0.00100 moles		0 moles		
Change in Moles	− 0.00100 moles		− 0.00100 moles		+ 0.00100 moles		
Final Moles	0.00400 moles		0 moles		0.00100 moles		

We now have a solution containing significant amounts of both acetic acid and sodium acetate. We are in the buffer region of the titration curve. We can determine its pH using the Henderson-Hasselbalch equation. Because the volume will be the same for both the acetate and the acetic acid, we can use moles in the equation.

$$pH = pK_a + log \ \frac{\left[conjugate \ base \right]}{\left[acid \right]} = - \ log \left(1.8 \ x \ 10^{-5} \right) + log \left(\frac{0.00100}{0.00400} \right)$$

$$= 4.14$$

c. What will be the pH at the half-equivalence point?
At the half-equivalence point, the pH will be equal to the pK_a. This is because the concentrations of acetic acid and sodium acetate will be equal, and the Henderson-Hasselbalch equation becomes $pH = pK_a$.

$$pH = pKa = - \ log \left(1.8 \ x \ 10^{-5} \right) = 4.74$$

e. Calculate the pH at the equivalence point.

At the equivalence point, our stoichiometry table is

	CH₃COOH	+	NaOH	→	CH₃COONa	+	H₂O
Measured	25.00 mL		50.0 mL				
Conversion Factor	0.200 moles/L		0.100 moles/L				
Initial Moles	0.00500 moles		0.00500 moles		0 moles		
Change in Moles	− 0.00500 moles		− 0.00500 moles		+ 0.00500 moles		
Final Moles	0 moles		0 moles		0.00500 moles		

At this point, we have no acetic acid or hydroxide left. This is a solution of sodium acetate. The sodium ions do not affect the pH. The acetate ions do. They react with water to produce a basic solution according to the equation:

$$CH_3COO^- \ (aq) + H_2O \ (l) <=> CH_3COOH \ (aq) + OH^- \ (aq)$$

This is the chemical equation that goes along with K_b of acetate ion. We can calculate K_b from the K_a.

$$K_b = \frac{K_w}{K_a} = \frac{1.0 \times 10^{-14}}{1.8 \times 10^{-5}} = 5.6 \times 10^{-10}$$

To solve the equilibrium problem, we need the initial concentration of acetate. We know the number of moles. The final volume is 25.00 mL + 50.0 mL = 75.0 mL = 0.075 L

$$\left[CH_3COO^- \right] = \frac{0.00500 \text{ moles}}{0.0750 \text{ L}} = 0.0667 \text{ M}$$

We now set up our equilibrium table.

	CH₃COO⁻	+	H₂O	<=>	CH₃COOH	+	OH⁻
Initial	0.0667 M				0		0
Change	− x				+ x		+ x
Equilibrium	0.0667 − x				x		x

$$K_b = \frac{\left[CH_3COOH \right]\left[OH^- \right]}{\left[CH_3COO^- \right]}$$

$$5.6 \times 10^{-10} = \frac{x \cdot x}{0.0667 - x}$$

We assume that x is small in comparison to 0.0667 because 100 • K_b is less than 0.0667 M.

$$5.6 \times 10^{-10} = \frac{x \cdot x}{0.0667}$$

$$x = \pm 6.1 \times 10^{-6}$$

We can reject the negative root.

$$\left[OH^- \right] = x = 6.1 \times 10^{-6} \text{ M}$$

$$pOH = - \log\left[OH^- \right] = - \log\left(6.1 \times 10^{-6} \right) = 5.21$$

$$pH = 14.00 - 5.21 = 8.79$$

f. Calculate the pH after 75.0 mL of the sodium hydroxide solution has been added. Once again, we set up our stoichiometry table.

	CH₃COOH	+	NaOH	→	CH₃COONa	+	H₂O
Measured	25.00 mL		75.0 mL				
Conversion Factor	0.200 moles/L		0.100 moles/L				
Initial Moles	0.00500 moles		0.00750 moles		0 moles		
Change in Moles							
Final Moles							

Now, the acetic acid is the limiting reactant.

	CH₃COOH	+	NaOH	→	CH₃COONa	+	H₂O
Measured	25.00 mL		75.0 mL				
Conversion Factor	0.200 moles/L		0.100 moles/L				
Initial Moles	0.00500 moles		0.00750 moles		0 moles		
Change in Moles	− 0.00500 moles		− 0.00500 moles		+ 0.00500 moles		
Final Moles	0 moles		0.00250 moles		0.00500 moles		

After this reaction, we have sodium hydroxide present and sodium acetate present. Both contribute to the solution being basic. The hydroxide in the solution from the sodium hydroxide overwhelms the little bit of hydroxide from the sodium acetate reaction with water, so we just need to worry about the hydroxide from the sodium hydroxide. This will be true once we get just a little beyond the equivalence point.

We have 0.00250 moles of sodium hydroxide in a volume of 100.0 mL = 0.100 L. The concentration of sodium hydroxide is

$$[NaOH] = [OH^-] = \frac{0.00250 \text{ moles}}{0.100 \text{ L}} = 0.0250 \text{ M}$$

$$pOH = -\log[OH^-] = -\log(0.0250) = 1.60$$

$$pH = 14.00 - 1.60 = 12.40$$

Try Study Question 29 in Chapter 18 of your textbook now!

CHAPTER 19: Principles of Reactivity: Entropy and Free Energy

Chapter Overview

In this chapter, we return to the area of thermodynamics. In Chapter 6, you learned about the first law of thermodynamics and about the state function enthalpy (H). In this chapter, you will learn about two more state functions, entropy (S) and the Gibbs free energy (G). Entropy is a measure of the disorder of a system. It is related to the dispersal of energy and matter and has its foundations in probability. The zero point of entropy is established by the third law of thermodynamics, which states that the entropy of a perfect crystal at zero kelvin is zero. The second law of thermodynamics states that the entropy of the universe increases in a spontaneous change. A spontaneous change is one that occurs without outside intervention. Thus, in order to determine if a process is spontaneous, we could determine the entropy changes of the system and of the surroundings and add them together. If their sum is greater than zero, then the process is spontaneous. If we have standard conditions, we can calculate the standard entropy change for the system in a chemical reaction by using the equation:

$$\Delta S^o_{reaction} = \Sigma S^\circ(products) - \Sigma S^\circ(reactants)$$

The standard entropy change of the surroundings can be calculated as

$$\Delta S^o_{surroundings} = \frac{-\Delta H^o_{system}}{T}$$

The Gibbs free energy change (ΔG) is another measure of spontaneity.

$$\Delta G = \Delta H - T\Delta S$$

If ΔG is less than zero, then the process is spontaneous. If ΔG is greater than zero, then it is not spontaneous. If ΔG is equal to zero, then the system is at equilibrium. You will learn different methods to calculate ΔG°, the Gibbs free energy change under standard conditions:

$$\Delta G^\circ = \Delta H^\circ - T\Delta S^\circ$$
$$\Delta G^\circ = \Sigma \Delta G^o_f(products) - \Sigma \Delta G^o_f(reactants)$$

You will also learn which combinations of ΔH and ΔS lead to spontaneous reactions. A negative ΔH (exothermic process) and a positive ΔS (going to greater disorder) contribute to a reaction being spontaneous. If these two driving forces act in opposite directions for a given process, the direction predicted by ΔS will win at higher temperatures. You will learn that there is a relationship between the value of ΔG° and the equilibrium constant:

$$\Delta G^\circ = -RT \ln K$$

Because of this relationship, there is a relationship between whether a reaction is reactant-favored or product-favored at equilibrium and whether it is spontaneous or not under standard conditions. You will also learn how to predict whether a reaction will be spontaneous or not based upon the relationship between Q and K or between ΔG and zero.

Key Terms

In this chapter, you will need to learn and be able to use the following terms:

Entropy (S): a measure of the disorder of a system.

Gibbs free energy change (ΔG): the maximum energy available to do useful work in a process. ΔG is a measure of spontaneity; if ΔG is less than zero for a process, then it is a spontaneous process.

Product-favored process: a process that, when equilibrium is established, has a greater concentration of products than reactants.

Reactant-favored process: a process that, when equilibrium is established, has a greater concentration of reactants than products.

Second law of thermodynamics: the entropy of the universe increases in a spontaneous change.

Spontaneous change: a change that occurs without outside intervention

Standard free energy of formation (ΔG_f^o): the free energy change when forming one mole of a compound from its component elements with all of the reactants and products in their standard states.

Third law of thermodynamics: the entropy of a perfect crystal at zero kelvin is zero.

Chapter Goals

By the end of this chapter you should be able to:

- **Understand the concept of entropy and its relationship to spontaneity.**

    a) Understand that entropy is a measure of matter and energy dispersal or disorder (Section 19.3).

    Entropy (S) is a measure of the disorder of a system. The change in entropy (ΔS) that occurs in a process is related to dispersal of energy. When entropy increases (positive ΔS), there is a dispersal (spreading out) of energy. If we start out with energy concentrated in a portion of the possible energy states, the direction that increases entropy will be that in which the energy gets spread out over the possible states.

    It is rarely obvious how to calculate the different energy levels of a system and how to discern the distribution of the total energy among them. Fortunately, we can often tell whether a system is going to higher or lower entropy by looking at whether *matter* ends up getting more dispersed. For example, the production of a gas from a liquid solution leads to a greater spreading out and disorder of the matter. The entropy of the system increases in this process. Once again, this tendency is related to probability. Consider tossing ten coins in the air. There is only one state with the coins landing all heads up and one state with all of them heads down. On the other hand, there are many states in which they are disordered with some heads up and some heads down. There are more disordered states, so it is much more probable that the coins will end up in one of these more disordered states.

    b) Recognize that entropy can be determined experimentally as the heat change of a reversible process (A Closer Look, Section 19.4).

    The entropy change ΔS for a process can be calculated as the heat change for the process to occur by a reversible pathway divided by the absolute temperature:

$$\Delta S = \frac{q_{rev}}{T}$$

A reversible pathway is a pathway in which heat is added or removed in extremely small increments so that we can return to our starting point without altering the surroundings any from what they were at the beginning. Real processes that are spontaneous do not occur by means of a reversible pathway. To determine an entropy change by using this equation, we would need to identify a reversible pathway by which the process could occur.

From this equation, we can tell that the units for entropy are joules per kelvin, J/K.

c) Know how to calculate entropy changes from tables of entropy values for compounds (Section 19.4).

The entropies of a number of substances have been measured and tabulated in listings such as Appendix L in your textbook. These values are what are called standard entropies ($S°$). This means that the conditions used were standard conditions (1 bar pressure). For solutes, the concentration will be 1 molal. For Appendix L, the temperature chosen was 25°C, but this is not part of the definition of standard conditions.

Entropy is a state function. To calculate $\Delta S°$ for the system in a chemical reaction, all we need to worry about is the final and the initial states.

$$\Delta S^o_{reaction} = \Delta S^o_{system} = \Sigma S°(\text{products}) - \Sigma S°(\text{reactants})$$

This equation is similar to one you learned in Chapter 6 for calculating $\Delta H°$ from ΔH^o_f data.

Example 19-1: Use $S°$ values to calculate the entropy change, $\Delta S°$, for the following process and comment on the sign of the change.

$$2Mg \text{ (s)} + O_2 \text{ (g)} \rightarrow 2MgO \text{ (s)}$$

We are asked to calculate $\Delta S°$ from $S°$ values. We can find $S°$ values in Appendix L. The equation that we shall use is

$$\Delta S^o_{reaction} = \Delta S^o_{system} = \Sigma S°(\text{products}) - \Sigma S°(\text{reactants})$$

To use this equation, we must take into account the values of $S°$ and the number of moles of each substance in the balanced chemical equation. Whereas ΔH^o_f for an element in its standard state is equal to zero, this is not true for $S°$ of an element.

$$\Delta S^o_{reaction} = 2 \text{ moles } S°(MgO_{(s)}) - \left[2 \text{ moles } S°(Mg_{(s)}) + 1 \text{ mole } S°(O_{2(g)}) \right]$$

$$= 2 \text{ moles} \left(26.85 \ \frac{J}{\text{mole K}} \right)$$

$$- \left[2 \text{ moles} \left(32.67 \ \frac{J}{\text{mole K}} \right) + 1 \text{ mole} \left(205.07 \ \frac{J}{\text{mole K}} \right) \right]$$

$$= -216.71 \ \frac{J}{K}$$

In this case, the entropy change for the reaction is negative. This indicates that this reaction involves the system going to a state of less disorder, more order.

Try Study Questions 5 and 7 in Chapter 19 of your textbook now!

d) Identify common processes that are entropy-favored (Section 19.4)

A substance becomes increasingly disordered in going from a solid to a liquid to a gas. ΔS will be positive in going from a solid to a liquid or from a liquid to a gas. We say that these phase changes are entropy-favored.

The entropy of any substance increases with temperature. This is because adding heat corresponds to a positive q in the equation q_{rev}/T, so ΔS ends up positive.

The entropy of a gas increases with an increase in volume.

Reactions that increase the number of moles of gases in the system are accompanied by an increase in entropy.

Mixing things together often leads to an increase in entropy and thus are entropy-favored. For example, mixing together two gases results in an increase of disorder and is entropy-favored. Forming many solutions also leads to an increase in disorder. The solute and solvent start out as pure, ordered substances. In the solution they are mixed up with each other and are more disordered - have a higher entropy. (There are some cases, however, when forming a solution does not lead to an increase in disorder or entropy due to the ordering of the water that occurs around some solute particles.)

Example 19-2: Which substance has the higher entropy in each of the following pairs?

      a. liquid water or water vapor
The entropy of a gas is higher than a liquid's. Water vapor should have the higher entropy.

      b. a solution of sodium chloride or pure sodium chloride
We would predict that the solution should have a greater entropy because the matter is more dispersed in the solution.

Example 19-3: Would the following process be expected to be entropy-favored or not?
    $2KClO_3 \ (s) \rightarrow 2KCl \ (s) + 3O_2 \ (g)$

This reaction involves the production of a gas from a solid. The gas has a higher entropy than the solid, so this process should be entropy-favored.

Try Study Question 1 in Chapter 19 of your textbook now!

- **Predict whether a process will be spontaneous.**

a) Use entropy and enthalpy changes to predict whether a reaction will be spontaneous (Section 19.5 and Table 19.2).

A spontaneous process is one that will occur without outside intervention under the conditions (including the concentrations of reactants and products) that happen to be present. This is a different issue from that of whether a reaction is product-favored or reactant-favored. A reaction is product-favored if, when equilibrium has been established, there is a greater concentration of products than reactants. A reaction is reactant-favored if, when equilibrium has been established, there is a greater concentration of reactants than products. We have already learned that the relationship between the equilibrium constant and the number 1 indicates whether a reaction is reactant-favored or product-favored. We have also learned that we can determine whether a reaction will be spontaneous by the relationship between the reaction quotient and the equilibrium constant. An example contrasting these terms is the following. Silver chloride is a sparingly soluble salt. The equation for its dissolution is

$$AgCl\ (s) <=> Ag^+\ (aq) + Cl^-\ (aq) \qquad\qquad K_{sp} = 1.8 \times 10^{-10}$$

Because the value of K is less than 1, this is a reactant-favored process. There will be more reactants present at equilibrium than products. On the other hand, if we place some silver chloride in water, initially some silver chloride will dissolve until equilibrium is established. Silver chloride dissolving is spontaneous until equilibrium is achieved. When equilibrium is at last achieved, there will be more reactants present than products, as the equilibrium constant indicates.

In this chapter, we will examine these issues in more detail and determine other means to determine whether a reaction will be spontaneous or not and whether it will be reactant-favored or product-favored at equilibrium.

The second law of thermodynamics states that in a spontaneous process, the entropy of the universe increases. In other words, the change in entropy of the universe is positive. For a spontaneous change,

$$\Delta S_{universe} > 0$$

This can also be stated as

$$\Delta S_{system} + \Delta S_{surroundings} > 0$$

Notice that the second law does not prohibit the entropy of a system from decreasing, only the universe. The entropy of the system may decrease so long as there is an even greater increase in the entropy of the surroundings.

With the second law of thermodynamics, we have a measure of spontaneity. If we take into account all of the entropy changes in the system and the surroundings and if this value overall comes out greater than zero, then a process will be spontaneous. If it comes out less than zero, then the process will not be spontaneous in the forward direction, but will be spontaneous in the opposite direction.

In order to use this indicator of spontaneity, we must calculate the change in entropy of the universe. There will be two components: the entropy change for the system and the entropy change for the surroundings.

Often, we will use standard conditions in our problems, in part because we have tables that give entropies under standard conditions. If we have all of our reactants and products in their standard states, we can say that a process will be spontaneous *under standard conditions* if

$$\Delta S^o_{universe} > 0$$
$$\Delta S^o_{system} + \Delta S^o_{surroundings} > 0$$

To determine if a process is spontaneous under standard conditions, therefore, we must calculate the standard entropy change of the system and the standard entropy change of the surroundings. We have already discussed how to calculate the standard entropy change of the system for a chemical reaction:

$$\Delta S^o_{reaction} = \Delta S^o_{system} = \Sigma\ S°\big(products\big) - \Sigma\ S°(reactants)$$

We also need to calculate the entropy change for the surroundings. To do this, we shall use the relationship discussed earlier that

$$\Delta S = \frac{q_{rev}}{T}$$

For a chemical reaction, the heat for the surroundings is that which is transferred to the surroundings by the chemical reaction. For a reaction at constant pressure, this will be equal in magnitude to ΔH of the reaction, but it will be opposite in sign (heat leaving the system goes into the surroundings; what is exothermic for the system is endothermic to the surroundings. Thus, under standard conditions,

$$\Delta S^o_{surroundings} = \frac{-\Delta H^o_{system}}{T}$$

Now that we have a way to calculate ΔS° for the system and ΔS° for the surroundings, we simply need to add these together to get ΔS° of the universe.

$$\Delta S^o_{universe} = \Delta S^o_{system} + \Delta S^o_{surroundings}$$

$$= \Delta S^o_{system} - \frac{\Delta H^o_{system}}{T}$$

If this value is positive, the reaction will be spontaneous under standard conditions.

Example 19-4: Determine whether the following reaction is spontaneous or not under standard conditions and at 25.00°C by calculating ΔS° of the universe.

$$2H_2\ (g) + O_2\ (g) \rightarrow 2H_2O\ (l)$$

To determine ΔS° of the universe we will calculate ΔS° of the system and of the surroundings and then add them together to obtain ΔS° of the universe.

ΔS° of the system can be calculated using the S° values in Appendix L.

$$\Delta S^o_{system} = \Delta S^o_{reaction} = \Sigma\ S^\circ(products) - \Sigma\ S^\circ(reactants)$$

$$\Delta S^o_{system} = 2\ moles\ S^\circ\!\left(H_2O_{(l)}\right) - \left[2\ moles\ S^\circ\!\left(H_{2(g)}\right) - 1\ mole\ S^\circ\!\left(O_{2(g)}\right)\right]$$

$$= 2\ moles\left(69.95\ \frac{J}{mole\ K}\right)$$

$$- \left[2\ moles\left(130.7\ \frac{J}{mole\ K}\right) + 1\ mole\left(205.07\ \frac{J}{mole\ K}\right)\right]$$

$$= -\ 326.6\ \frac{J}{K}$$

The next step is to calculate ΔS° of the surroundings.

$$\Delta S^o_{surroundings} = \frac{-\Delta H^o_{system}}{T}$$

In order to use this equation, we need first to calculate ΔH° for the reaction (system). We can do this using standard molar enthalpies of formation.

$$\Delta H^o_{system} = \Sigma\ \Delta H^o_f(products) - \Sigma\ \Delta H^o_f(reactants)$$

$$\Delta H^\circ_{system} = 2 \text{ moles } \Delta H^\circ_f\left(H_2O_{(l)}\right) - \left[2 \text{ moles } \Delta H^\circ_f\left(H_{2(g)}\right) - 1 \text{ mole } \Delta H^\circ_f\left(O_{2(g)}\right)\right]$$

$$= 2 \text{ moles } \left(-285.83 \frac{kJ}{mole}\right)$$

$$- \left[2 \text{ moles }\left(0 \frac{kJ}{mole}\right) + 1 \text{ mole }\left(0 \frac{kJ}{mole}\right)\right]$$

$$= -571.66 \text{ kJ}$$

We are now ready to calculate ΔS° of the surroundings. T is 25.00°C = 298.15 K.

$$\Delta S^\circ_{surroundings} = \frac{-\Delta H^\circ_{system}}{T} = -\frac{-571.66 \text{ kJ}}{298.15 \text{ K}} = 1.9174 \frac{kJ}{K}$$

One thing to be careful about is that the units above for S° of the system were in joules, but the ΔH° value is in kJ. We need to make sure that we convert one or the other so that we are dealing with either J for both or kJ for both. Forgetting to do this is one of the most common mistakes that students make. In this case, we will convert ΔS° so that it is in J.

$$\Delta S^\circ_{system} = -326.6 \frac{J}{K} \times \frac{1 \text{ kJ}}{1000 \text{ J}} = -0.3266 \frac{kJ}{K}$$

Finally, we can calculate ΔS° of the universe.

$$\Delta S^\circ_{universe} = \Delta S^\circ_{system} + \Delta S^\circ_{surroundings}$$

$$\Delta S^\circ_{universe} = -0.3266 \frac{kJ}{K} + 1.9174 \frac{kJ}{K} = 1.5908 \frac{kJ}{K}$$

ΔS° for the universe is positive, therefore the reaction of hydrogen with oxygen to form liquid water is a spontaneous process.

Try Study Question 13 in Chapter 19 of your textbook now!

b) Recognize how temperature influences whether a reaction is spontaneous (Section 19.5).

As we have seen the entropy of the universe can be expressed as follows:

$$\Delta S^\circ_{universe} = \Delta S^\circ_{system} + \Delta S^\circ_{surroundings}$$

$$= \Delta S^\circ_{system} - \frac{\Delta H^\circ_{system}}{T}$$

ΔH of the system is divided by the temperature in the part of the equation dealing with the entropy change for the surroundings. This means that this part of the equation will have a smaller effect at higher temperatures. At higher temperatures, ΔS for the system becomes the dominant factor because the factor dealing with ΔH gets smaller.

- **Understand and use a new thermodynamic function, Gibbs free energy.**

a) Understand the connection between enthalpy and entropy changes and the Gibbs free energy change for a process (Section 19.6).

In the last objective, we found that the entropy change for the universe could be expressed as

$$\Delta S_{universe} = \Delta S_{system} - \frac{\Delta H_{system}}{T}$$

If we multiply both sides of this equation by $-T$, we obtain
$$-T\Delta S_{universe} = -T\Delta S_{system} + \Delta H_{system}$$
$$= \Delta H_{system} - T\Delta S_{system}$$

If we now define $-T\Delta S_{universe}$ as a new function, ΔG, we obtain
$$\Delta G_{system} = \Delta H_{system} - T\Delta S_{system}$$

In this equation, both ΔH and ΔS are the values for the system, so we usually drop the word "system" from the definition. Our equation becomes
$$\Delta G = \Delta H - T\Delta S$$

This new function, ΔG, is called the Gibbs free energy change. Under conditions of constant pressure, it is the measure of spontaneity most often used. Recall that a positive $\Delta S_{universe}$ indicated a spontaneous process. Because ΔG is equal to $-T\Delta S_{universe}$, a positive $\Delta S_{universe}$ implies a negative ΔG. A spontaneous process thus has a negative ΔG.

	$\Delta S_{universe}$	ΔG
Spontaneous	> 0	< 0
Equilibrium	$= 0$	$= 0$
Not Spontaneous	< 0	> 0

Often, we will deal with the reactants or products all being in their standard states. The Gibbs free energy change is then called $\Delta G°$, and the equation becomes
$$\Delta G° = \Delta H° - T\Delta S°$$

	$\Delta G°$
Spontaneous Under Standard Conditions	< 0
Equilibrium Under Standard Conditions	$= 0$
Not Spontaneous Under Standard Conditions	> 0

b) Calculate the change in free energy at standard conditions for a reaction from the enthalpy and entropy changes or from the standard free energy of formation of reactants and products (ΔG_f^o) (Section 19.6).

How can we calculate $\Delta G°$? One way is to use the equation
$$\Delta G° = \Delta H° - T\Delta S°$$

In this method, we calculate $\Delta H°$ and $\Delta S°$ and then combine them in the equation.

Example 19-5: Determine the value of $\Delta G°$ at 25°C for the following reaction by using the $\Delta H°$ and $\Delta S°$ values for the reaction.
$$2Hg\ (l) + O_2\ (g) \rightarrow 2HgO\ (s)$$

First, we will calculate $\Delta H°$ using standard molar enthalpies of formation from Appendix L.
$$\Delta H_{rxn} = \Sigma \Delta H_f^o(products) - \Sigma \Delta H_f^o(reactants)$$
$$= 2\ moles\ \Delta H_f^o(HgO_{(g)}) - \left[2\ moles\ \Delta H_f^o(Hg_{(l)}) - 1\ mole\ \Delta H_f^o(O_{2(g)})\right]$$

$$\Delta H^\circ_{rxn} = 2 \text{ moles} \left(-90.83 \ \frac{kJ}{mole}\right) - \left[2 \text{ moles} \left(0 \ \frac{kJ}{mole}\right) + 1 \text{ mole} \left(0 \ \frac{kJ}{mole}\right)\right]$$

$$= -181.66 \text{ kJ}$$

Next, we will calculate ΔS°, once again using Appendix L.

$$\Delta S^\circ_{system} = \Delta S^\circ_{reaction} = \Sigma \ S^\circ(\text{products}) - \Sigma \ S^\circ(\text{reactants})$$

$$\Delta S^\circ_{rxn} = 2 \text{ moles } S^\circ\left(HgO_{(l)}\right) - \left[2 \text{ moles } S^\circ\left(Hg_{(l)}\right) - 1 \text{ mole } S^\circ\left(O_{2(g)}\right)\right]$$

$$= 2 \text{ moles} \left(70.29 \ \frac{J}{mole \ K}\right)$$

$$- \left[2 \text{ moles} \left(76.02 \ \frac{J}{mole \ K}\right) + 1 \text{ mole} \left(205.07 \ \frac{J}{mole \ K}\right)\right]$$

$$= -216.53 \ \frac{J}{K}$$

ΔH is in units of kJ, whereas this entropy has units of J/K. We will convert from J/K to kJ/K.

$$-216.53 \ \frac{J}{K} \ \times \ \frac{1 \ kJ}{1000 \ J} = -0.21653 \ \frac{kJ}{K}$$

Now we use the equation

$$\Delta G^\circ = \Delta H^\circ - T\Delta S^\circ$$

$$\Delta G^\circ = \Delta H^\circ - T\Delta S^\circ = -181.66 \text{ kJ} - (298 \text{ K})\left(-0.21653 \ \frac{kJ}{mole}\right) = -117 \text{ kJ}$$

This reaction has a negative ΔG° so it is spontaneous under standard conditions.

Try Study Question 21 in Chapter 19 of your textbook now!

Another way to calculate ΔG° for a reaction is to use standard free energies of formation (ΔG°_f). ΔG°_f of a compound is the free energy change that occurs when one mole of the compound is formed from its component elements with products and reactants in their standard states. Standard free energies of formation are analogous to standard enthalpies of formation that we studied back in Chapter 6 except that we are dealing with free energy changes instead of enthalpy changes. The chemical reaction accompanying ΔG°_f for a compound is exactly the same as that for ΔH°_f. Just as was true for ΔH°_f, ΔG°_f to form an element in its standard state is zero. Values for ΔG°_f are listed in Appendix L.

The equation to calculate ΔG° using ΔG°_f values is

$$\Delta G^\circ = \Sigma \ \Delta G^\circ_f(\text{products}) - \Sigma \ \Delta G^\circ_f(\text{reactants})$$

Example 19-6: Using ΔG°_f values, calculate ΔG° for the reaction
$$2Na \ (s) + 2H_2O \ (l) \rightarrow 2NaOH \ (aq) + H_2 \ (g)$$

To solve this problem, we will use the equation
$$\Delta G^\circ = \Sigma \ \Delta G^\circ_f(\text{products}) - \Sigma \ \Delta G^\circ_f(\text{reactants})$$
and values of ΔG°_f from Appendix L.

$$\Delta G^\circ = \Sigma \, \Delta G_f^\circ \big(\text{products}\big) \; - \; \Sigma \, \Delta G_f^\circ \big(\text{reactants}\big)$$

$$\Delta G^\circ = 2 \text{ moles } \Delta G_f^\circ \big(\text{NaOH}_{(aq)}\big) + 1 \text{ mole } \Delta G_f^\circ \big(\text{H}_{2(l)}\big)$$

$$\qquad - \Big[2 \text{ moles } \Delta G_f^\circ \big(\text{Na}_{(s)}\big) + 2 \text{ moles } \Delta G_f^\circ \big(\text{H}_2\text{O}_{(l)}\big)\Big]$$

$$= 2 \text{ moles } \left(-418.09 \; \frac{\text{kJ}}{\text{mole}}\right) + 1 \text{ moles } \left(0 \; \frac{\text{kJ}}{\text{mole}}\right)$$

$$\qquad - \left[2 \text{ moles } \left(0 \; \frac{\text{kJ}}{\text{mole}}\right) + 2 \text{ moles } \left(-237.15 \; \frac{\text{kJ}}{\text{mole}}\right)\right]$$

$$= \; -361.88 \text{ kJ}$$

This reaction has a negative ΔG°, so it is spontaneous under standard conditions.

Try Study Questions 25 and 27 in Chapter 19 of your textbook now!

c) Know how free energy changes with temperature (Section 19.6).

What leads to a negative ΔG? There are two driving forces: the enthalpy change and the entropy change of the system. A negative ΔH (exothermic) will contribute to ΔG being negative. A positive ΔS (going to more disorder) will also contribute to ΔG being negative because of the minus sign before the entropy term in the equation. Let us examine different combinations of positive and negative ΔH and ΔS.

ΔH	ΔS	Spontaneous?
$\Delta H < 0$ Exothermic Favorable	$\Delta S > 0$ More disorder Favorable	ΔG will always be negative. The reaction will be spontaneous under all conditions.
$\Delta H > 0$ Endothermic Not Favorable	$\Delta S < 0$ More order Not Favorable	ΔG will always be positive. The reaction will not be spontaneous under any conditions.
$\Delta H < 0$ Exothermic Favorable	$\Delta S < 0$ More order Not Favorable	ΔG could be positive or negative depending upon the actual values of ΔH and ΔS. At lower temperatures, the reaction will be spontaneous, but at higher temperatures, ΔS will win, and the reaction will not be spontaneous.
$\Delta H > 0$ Endothermic Not Favorable	$\Delta S > 0$ More disorder Favorable	ΔG could be positive or negative depending upon the actual values of ΔH and ΔS. At higher temperatures, ΔS will win, and the reaction will be spontaneous.

Notice that in each of the cases where ΔH and ΔS are pulling the reaction in two different directions that the entropy term wins at higher temperatures. This is because the entropy term is multiplied by the temperature in the mathematical equation for ΔG.

Example 19-7: Consider the reaction of nitrogen dioxide to form dinitrogen tetraoxide.
$$2\text{NO}_2 \; (g) \rightarrow \text{N}_2\text{O}_4 \; (g)$$

a. Is this reaction spontaneous under standard conditions at 25°C?
We determine whether a reaction is spontaneous or not under standard conditions by calculating ΔG°. We could either use the equation

$$\Delta G° = \Delta H° - T\Delta S°$$

or use ΔG_f^o values. Looking ahead to a later part of the problem, we can see that it would be best to use the first equation because we will want to solve for a temperature later.

We calculate $\Delta H°$ and $\Delta S°$ for the reaction at 25°C.

$$\Delta H_{reaction}^o = \Sigma \Delta H_f^o(products) - \Sigma \Delta H_f^o(reactants)$$

$$\Delta H_{rxn}^o = 1 \text{ mole } \Delta H_f^o\left(N_2O_{4(g)}\right) - 2 \text{ moles } \Delta H_f^o\left(NO_{2(g)}\right)$$

$$= 1 \text{ mole } \left(9.08 \frac{kJ}{mole}\right) - 2 \text{ moles } \left(33.1 \frac{kJ}{mole}\right)$$

$$= -57.1 \text{ kJ}$$

$$\Delta S_{rxn}^o = \Sigma S°(products) - \Sigma S°(reactants)$$

$$\Delta S_{rxn}^o = 1 \text{ mole } S°\left(N_2O_{4(g)}\right) - 2 \text{ moles } \Delta S°\left(NO_{2(g)}\right)$$

$$= 1 \text{ mole } \left(304.38 \frac{J}{mole \ K}\right) - 2 \text{ moles } \left(240.04 \frac{J}{mole \ K}\right)$$

$$= -175.70 \frac{J}{K}$$

$$-175.70 \frac{J}{K} \times \frac{1 \text{ kJ}}{1000 \text{ J}} = -0.17570 \frac{kJ}{K}$$

$\Delta H°$ is favorable for the process to be spontaneous, but $\Delta S°$ is not. At 25°C, we find

$$\Delta G° = \Delta H° - T\Delta S° = -57.1 \text{ kJ} - (298 \text{ K})\left(-0.17570 \frac{kJ}{K}\right) = -4.7 \text{ kJ}$$

The reaction is spontaneous at 25°C under standard conditions.

b. Estimate the temperature beyond which this reaction is not spontaneous.
In this case, $\Delta H°$ favors reaction, whereas $\Delta S°$ does not. At higher temperatures, the entropy factor wins, and the reaction will not be spontaneous. To estimate the temperature at which this occurs, we will use the same values for $\Delta H°$, and $\Delta S°$ that we calculated at 25°C and determine the temperature at which $\Delta G°$ will equal zero. If we go above this temperature, then the reaction will not be spontaneous because $\Delta G°$ will then be positive.

$$\Delta G° = \Delta H° - T\Delta S°$$

$$0 = -57.1 \text{ kJ} - T\left(-0.17570 \frac{kJ}{K}\right)$$

$$57.1 \text{ kJ} = T\left(0.17570 \frac{kJ}{K}\right)$$

$$T = 325 \text{ K} = 52°C$$

This answer is only an estimate. We have made the assumption that $\Delta H°$ and $\Delta S°$ do not change when the temperature is changed. This is not correct, but it allows us to get an estimate of the temperature.

Try Study Questions 19 and 31 in Chapter 19 of your textbook now!

• **Understand the relationship of a free energy change for a reaction, its equilibrium constant, and whether the reaction is product- or reactant-favored.**

Describe the relationship between the free energy change and equilibrium constants and calculate K from ΔG^{o}_{rxn} (Section 19.7).

Whether ΔG is positive or negative tells us whether a reaction is spontaneous under whatever set of conditions that exist. If those conditions happen to be standard conditions, we refer to the ΔG for that special set of conditions as $\Delta G°$. We have already learned some methods to calculate $\Delta G°$. There is a relationship between $\Delta G°$ and ΔG. ΔG is equal to $\Delta G°$ plus a correction factor that involves our old friend the reaction quotient, Q.

$$\Delta G = \Delta G° + RT \ln Q$$

The temperature, as usual, is in kelvins. The value of R should be the one involving J/(mole K) or kJ/(mole K) because ΔG and $\Delta G°$ will be in J/mole or kJ/mole.

If we have a reaction at equilibrium, then ΔG is equal to zero, and Q is equal to K. When these are substituted into the above equation and the equation is rearranged, we obtain an equation that relates $\Delta G°$ and K:

$$\Delta G° = -RT \ln K$$

Example 19-8: At 1000 K, the equilibrium constant for
$$I_2 \ (g) <=> 2I \ (g)$$
is equal to 3.76×10^{-3}. What is the value of $\Delta G°$?

We are given the value of K and are looking for the value of $\Delta G°$.

$$\Delta G° = -RT \ln K$$

$$\Delta G° = -\left(8.3145 \times 10^{-3} \ \frac{kJ}{mole \ K}\right)(1000 \ K) \ln\left(3.76 \times 10^{-3}\right)$$

$$= 46.4 \ kJ/mole$$

Example 19-9: For the reaction,
$$C \ (graphite) + 2H_2 \ (g) <=> CH_4 \ (g)$$
$\Delta G°$ is equal to -50.8 kJ at 25°C. What is the value of the equilibrium constant?

This is the reverse type of problem. We are given $\Delta G°$ and are asked for K.

$$\Delta G° = -RT \ln K$$

We need to solve this equation for K.

$$-\frac{\Delta G°}{RT} = \ln K$$

$$K = e^{-\frac{\Delta G°}{RT}}$$

We now substitute our values into this equation.

$$K = e^{-\frac{(-50.8 \ kJ/mole)}{\left(8.3145 \times 10^{-3} \ \frac{kJ}{mole \ K}\right)(298 \ K)}} = 8.02 \times 10^{8}$$

Try Study Question 35 in Chapter 19 of your textbook now!

We have learned that the equilibrium constant, K, is related to whether a reaction will be reactant-favored or product favored at equilibrium. The fact that there is a mathematical relationship between $\Delta G°$ and K for a reaction implies that there is a relationship between whether a reaction is spontaneous under standard conditions and whether the reaction is reactant-favored or product-favored. Looking at the equation relating the two quantities

$$\Delta G° = -RT \ln K$$

we can see that a K greater than 1 will yield a $\Delta G°$ that is negative and that a K less than zero will yield a $\Delta G°$ greater than zero. For the reaction $aA + bB \iff cC + dD$, we can construct the following table.

K	$\Delta G°$	Reactant-favored or Product-favored *at Equilibrium*?	Spontaneous *Under Standard Conditions*?
$K \gg 1$	$\Delta G° < 0$	Product-favored	Spontaneous under standard conditions
$K = 1$	$\Delta G° = 0$	$[C]^c[D]^d = [A]^a[B]^b$ at equilibrium	At equilibrium under standard conditions
$K \ll 1$	$\Delta G° > 0$	Reactant-favored	Not spontaneous under standard conditions.

We can also relate our two measures of spontaneity, Q and ΔG. For these, we are not necessarily at equilibrium and we are not necessarily under standard conditions.

Q	ΔG	Spontaneous?
$Q < K$	$\Delta G < 0$	Spontaneous to the right.
$Q = K$	$\Delta G = 0$	At Equilibrium
$Q > K$	$\Delta G > 0$	Not spontaneous to the right; spontaneous to the left.

Example 19-10: Consider the dissolution of calcium fluoride in water.

$$CaF_2 \ (s) \iff Ca^{2+} \ (aq) + 2F^- \ (aq)$$

The equilibrium constant that describes this reaction is the solubility product constant of calcium fluoride, 5.3×10^{-11}.

      a. Is this process reactant-favored or product-favored at equilibrium?
$K \ll 1$, so this process is reactant-favored at equilibrium.

      b. Is this process spontaneous under standard conditions?
If $K < 1$, then $\Delta G°$ is positive. Dissolution is not spontaneous under standard conditions.

      c. Let us assume that we place some calcium fluoride in water. Will dissolution of any calcium fluoride occur?
When we place the calcium fluoride in water, we initially have no calcium ions or fluoride ions in the water. Q is thus equal to zero. This is less than the value of K. The reaction is spontaneous - dissolution will occur - until $Q = K$.

The process is thus spontaneous until equilibrium is achieved. Notice that even though the reaction is not spontaneous under standard conditions ($\Delta G° > 0$), the reaction is spontaneous ($\Delta G < 0$) under the conditions we have in this case, 0 M Ca^{2+} and 0 M F^-. Eventually, we will reach equilibrium, at which point Q will be equal to K and ΔG will be equal to zero.

Other Notes

The Third Law of Thermodynamics
The third law of thermodynamics provides a zero for the entropy scale. A perfect crystal at zero kelvin has an entropy of zero. Because there is a zero for entropy we can deal with actual entropy values, not just changes in value. (Recall that for enthalpy and free energy, we always dealt with changes in their values, not the actual values.)

The entropy of a substance can be obtained by determining the heat required to raise the temperature from zero K to the desired temperature by following a reversible pathway. Another consequence of the third law is that actual entropy values will be positive because we will always be working above absolute zero. (The actual entropies must be greater than zero; a *change* in entropy can be either positive or negative, depending on whether the entropy increased or decreased in the process.)

CHAPTER 20: Principles of Reactivity: Electron Transfer Reactions

Chapter Overview

In this chapter, we examine oxidation-reduction (redox) reactions in more detail. First, you will learn how to balance net ionic equations involving redox reactions carried out in acidic and basic solutions.

In a voltaic cell, we use a spontaneous redox reaction to do electrical work. The two half-reactions are separated, and the electrons are forced to travel through a wire to go from the material being oxidized to the material being reduced. A salt bridge allows ions to flow between the half-cells to complete the circuit and to maintain electrical neutrality. Reduction occurs at the cathode, and oxidation occurs at the anode. In a voltaic cell, the anode is negative, and the cathode is positive. The key features of various types of batteries are described, including LeClanché dry cells and alkaline batteries. These batteries are voltaic cells that cannot be recharged. In a rechargeable battery, the battery acts as a voltaic cell when it is being discharged and as an electrolytic cell when it is being recharged. The key features of two types of rechargeable batteries (lead storage batteries and Ni-cad batteries) are discussed. In a fuel cell, the reactants of a spontaneous redox reaction are continuously added to the reaction chamber from external reservoirs.

A table of half-cells and their standard potentials (measured *vs.* the standard hydrogen electrode) has been constructed. In the table, the half-reactions are written as reductions and are listed in order of their standard reduction potentials with the largest potential at the top. A half-reaction further up on the table will run as a reduction and force a half-reaction below it to run as an oxidation. The northwest-southeast rule states that the reducing agent always lies to the southeast of the oxidizing agent in a product-favored reaction. The potential of a cell ($E°$) can be calculated as

$$E^o_{cell} = E^o_{cathode} - E^o_{anode}$$

Reversing the direction of a chemical reaction changes the sign of its E. Multiplying an equation by a coefficient does not change the value of its E. The potential of a cell under nonstandard conditions can be calculated using the Nernst equation.

A positive E means a reaction will be spontaneous. E, Q, and ΔG are related. A positive $E°$ means that a reaction is spontaneous under standard conditions. This also means the reaction will be product-favored when equilibrium has been established. There are equations relating $E°$, K, and $\Delta G°$.

In electrolysis, we use electricity to carry out a chemical reaction. Electricity of sufficient voltage is used to overcome an unfavorable E. In an electrolytic cell, the cathode is still the electrode at which reduction occurs, and the anode is the electrode at which oxidation occurs. In an electrolytic cell, however, the cathode is negative, and the anode is positive. Typically, electrolysis is performed on molten salts and aqueous solutions of salts. In aqueous solutions, there is the added complication that half-reactions involving water can also occur. We can use the faraday to relate the amount of electricity used to the amounts of reactants and products in the chemical reaction.

Key Terms

In this chapter, you will need to learn and be able to use the following terms:

Ampere: a unit of current, 1 amp = 1 C/s.

Anode: the electrode at which oxidation occurs.

Cathode: the electrode at which reduction occurs.

Current: the amount of charge flowing per unit time.

Electrochemistry: the field of chemistry that considers chemical reactions that produce or are caused by electrical energy.

Electrode: a device such as a metal plate or wire for conducting electrons into and out of solutions in electrochemical cells.

Electrolysis: the use of electrical energy to produce a chemical change.

Electromotive force (emf): the difference in potential energy between two electrodes in an electrochemical cell (when no current flows).

Faraday constant (F): the charge carried by one mole of electrons (9.64853338×10^4 C/mole e$^-$).

Fuel cell: an electrochemical cell into which reactants are continuously added.

Galvanic cells: also called voltaic cells, these are electrochemical cells that use chemical reactions to produce an electric current.

Half-reactions: the two chemical equations into which the equation for an oxidation-reduction reaction can be divided, one representing the oxidation process and the other the reduction process.

Inert electrode: an electrode made of a material that conducts an electric current but that is not oxidized or reduced in the voltaic cell. They permit another material to participate in the cell reaction at that electrode.

Northwest-southeast rule: in a table of reduction potentials that lists the half-reactions in order with the greatest reduction potential at the top, the reducing agent always lies to the southeast of the oxidizing agent in a product-favored reaction.

Overvoltage: the voltage needed to make a reaction occur at a reasonable rate. Voltages higher than the minimum are usually used to speed up electrolytic reactions that otherwise take place only slowly.

Potential: the difference in potential energy measured by a voltmeter inserted in an electrochemical cell. A potential will have both a magnitude and a sign.

Primary battery: a battery that cannot be returned to its original state by recharging.

Reduction Potential: the potential that would be measured when a half-cell of interest is set up as the cathode in an electrochemical cell and the standard hydrogen electrode is set up as the anode.

Salt bridge: a device for maintaining the balance of ion charges in the compartments of an electrochemical cell.

Secondary battery (storage battery, rechargeable battery): a battery in which the redox reaction can be reversed and the battery can thus be recharged.

Volt: the potential difference needed to impart one joule of energy to an electric charge of one coulomb (1 V = 1 J/C).

Voltage: a value equal in magnitude to the potential but having no positive or negative sign.

Voltaic cells: also called galvanic cells, these are electrochemical cells that use chemical reactions to produce an electric current.

Chapter Goals

By the end of this chapter you should be able to:

• **Balance equations for oxidation-reduction reactions in acidic or basic solutions using the half-reaction approach (Section 20.1).**

In Chapter 5, we studied some basic features of oxidation-reduction (redox) reactions, reactions that involve the transfer (or apparent transfer) of electrons. You should review the sections of your textbook and study guide that deal with these. In particular, you should review the terms oxidation, reduction, oxidizing agent, reducing agent, and oxidation number. At that time, we also went over rules for determining oxidation numbers and how to recognize oxidation-reduction as reactions in which oxidation numbers change.

Balancing chemical equations for oxidation-reduction reactions can often be more challenging than for the other types of chemical reactions we have studied. Fortunately, there is a procedure that we can follow to balance such equations. Often, this author finds that students find this procedure challenging at first, but that once they have practiced the procedure sufficiently, they actually like balancing these equations because there is a set pattern to follow.

The key to balancing redox equations is that they must be balanced for both mass and charge. This is actually true for all chemical reactions, but for reactions that do not involve redox, the charge balance usually takes care of itself and we don't have to worry about it. For redox reactions, we must make sure that we take it into account.

The method that we shall go over for balancing redox reactions is called the half-reaction method. A redox reaction can be broken up into two half-reactions. A half-reaction describes one half of the process. One half-reaction describes the oxidation part, and the other half-reaction describes the reduction part. We also need to take into account whether a reaction is occurring in an acidic or basic solution.

Steps to Follow to Balance a Redox Reaction in Acidic Solutions:

1. Recognize the reaction as an oxidation-reduction reaction.

2. Separate the process into half-reactions.

3. Balance each half-reaction for mass (atoms).
    a. Balance all elements that are not hydrogen or oxygen first.
    b. If oxygen is still not balanced, add one water molecule for each oxygen that is needed to the side that needs more oxygen.
    c. If hydrogen is still not balanced, add one H^+ ion for each hydrogen that is needed to the side that needs more hydrogen.

4. Balance each half-reaction for charge.
    Add up the charge on both sides of the half-reaction. We need to make it so that the charge on both sides is equal. Because the half-reaction has already been balanced for mass, we cannot add atoms. To balance charge, we use electrons, which are negative. To the side that is more positive, add electrons so that the charge on that side is brought down to the same value as on the other side. Once this is done, our half-reactions are balanced.

5. Multiply each half-reaction by an appropriate factor.

In a redox process, the number of electrons gained in reduction must equal the number of electrons lost in oxidation. Multiply each half-reaction by the necessary factors so that the number of electrons gained in one is equal to the number of electrons lost in the other.

6. Add the half-reactions to produce the overall balanced equation.

7. Simplify by eliminating reactants and products that appear on both sides. (For example, if we have six of some species on the left side and four of that same species on the right side, four could be eliminated because they appeared on both sides of the equation. This would leave only two of that species on the left side.)

Example 20-1: Balance the following net ionic equation:
$$Zn\ (s) + Ag^+\ (aq) \rightarrow Zn^{2+}\ (aq) + Ag\ (s)$$

At first glance, you might think that this reaction is already balanced. It is balanced for mass, but it is not balanced for charge. The total charge on the left side is $0 + 1 = +1$, and that on the right side is $+2 + 0 = +2$. These are not the same value.

1. Redox?
This is a redox reaction. The charge of zinc changes from 0 to 2+ and that of silver changes from 1+ to 0.

2. Half-reactions.
We separate the reaction into two half-reactions. On one side of the half-reaction we will have one of the species that undergoes a change; on the other we will have what that species becomes in the products. The two half-reactions are
$$Zn\ (s) \rightarrow Zn^{2+}\ (aq)$$
$$Ag^+\ (aq) \rightarrow Ag\ (s)$$

3. Balance each half-reaction for mass.
For the top one, we have one zinc on the left and one on the right. It is balanced for mass already. For the bottom one, we have one silver on the left and one on the right. It is balanced for mass already.

4. Balance each half-reaction for charge.
Let's deal with the top half-reaction first. On the left, we have a zinc atom. Its charge is zero. On the right, we have a zinc ion; its charge is 2+.

	$Zn\ (s)$	$\rightarrow$	$Zn^{2+}\ (aq)$
Charge	0		+2

The only things we can add to balance charge are electrons, which are negative. They will always serve to bring the charge on a side down, so we add them to the side that starts out more positive. In this case, this is the right side. We want to bring the charge of this side down to the charge of the other side, zero. We add two electrons to the right side.

	$Zn\ (s)$	$\rightarrow$	$Zn^{2+}\ (aq)$	$+$	$2e^-$
Charge	0		+2		−2
Total Charge	0			0	

This half-reaction is now balanced for both mass and charge. Now, let's work on the silver half-reaction.

$$Ag^+ \ (aq) \quad \rightarrow \quad Ag \ (s)$$
Charge $+1$ 0

This time, the more positive side is the left side. We add one electron to the left side to bring its charge down to the value on the right side, zero.

$$1e^- \quad + \quad Ag^+ \ (aq) \quad \rightarrow \quad Ag \ (s)$$
Charge -1 $+1$ 0
Total Charge 0 0

This half-reaction is now balanced for both mass and charge.

5. Multiply by a factor to ensure that electrons lost = electrons gained.
Our two balanced half-reactions are

$$Zn \ (s) \quad \rightarrow \quad Zn^{2+} \ (aq) \quad + \quad 2e^-$$
$$1e^- \quad + \quad Ag^+ \ (aq) \quad \rightarrow \quad Ag \ (s)$$

In the top process, two electrons are lost. In the bottom process, one electron is gained. The number of electrons lost must equal the number of electrons gained. The bottom half-reaction must occur twice for every one time that the top half-reaction occurs. We multiply the bottom half-reaction by 2.

$$Zn \ (s) \quad \rightarrow \quad Zn^{2+} \ (aq) \quad + \quad 2e^-$$
$$2[1e^- \quad + \quad Ag^+ \ (aq) \quad \rightarrow \quad Ag \ (s)]$$

6. Add the two half-reactions. The electrons will cancel.

$$Zn \ (s) \quad \rightarrow \quad Zn^{2+} \ (aq) \quad + \quad 2e^-$$
$$2[1e^- \quad + \quad Ag^+ \ (aq) \quad \rightarrow \quad Ag \ (s)]$$
$$\overline{Zn \ (s) \quad + \quad 2Ag^+ \ (aq) \quad \rightarrow \quad Zn^{2+} \ (aq \quad + \quad 2Ag \ (s)}$$

7. In this case, there are no common terms to cancel. The net ionic equation is balanced.
$$Zn \ (s) + 2Ag^+ \ (aq) \rightarrow Zn^{2+} \ (aq) + 2Ag \ (s)$$

Example 20-2: Balance the net ionic equation for the following reaction in an acidic solution.
$$Cu \ (s) + NO_3^- \ (aq) \rightarrow Cu^{2+} \ (aq) + NO \ (g)$$

1. This is a redox reaction.

2. Half-reactions.
$$Cu \ (s) \rightarrow Cu^{2+} \ (aq)$$
$$NO_3^- \ (aq) \rightarrow NO \ (g)$$

3. Balance the half-reactions for mass.
The top half-reaction is already balanced for mass. There is one copper on each side.
$$Cu \ (s) \rightarrow Cu^{2+} \ (aq)$$

The bottom half-reaction is not balanced for mass. There is one nitrogen on each side of the equation, but there are three oxygens on the left and only one on the right. We need two more oxygens on the right. We balance oxygens by adding water molecules, so we add two water molecules to the right side.
$$NO_3^- \ (aq) \rightarrow NO \ (g) + 2H_2O \ (l)$$

The oxygens are now balanced, but now we have hydrogens that are not balanced. There are four hydrogens on the right side but no hydrogens on the left side. We balance hydrogen using H^+ ions. We add 4 H^+ to the left side.

$$4H^+ (aq) + NO_3^- (aq) \rightarrow NO (g) + 2H_2O (l)$$

This equation is now balanced for mass.

4. Balance the half-reactions for charge.

Let's do the copper half-reaction first.

$$Cu (s) \quad \rightarrow \quad Cu^{2+} (aq)$$
Charge   0   +2

We will add electrons to the more positive side and bring that charge on that side down to the charge on the other side. We add two electrons to the right side.

$$Cu (s) \quad \rightarrow \quad Cu^{2+} (aq) \quad + \quad 2e^-$$

Now, let's do the nitrate half-reaction.

$$4H^+ (aq) \quad + \quad NO_3^- (aq) \quad \rightarrow \quad NO (g) \quad + \quad 2H_2O (l)$$
Charge   4(+1) = +4   −1   0   0
                         +3   0

We add three electrons to the left side to bring its charge down to zero, the value on the right.

$$3e^- \quad + \quad 4H^+ (aq) \quad + \quad NO_3^- (aq) \quad \rightarrow \quad NO (g) \quad + \quad 2H_2O (l)$$

5. Multiply the half-reactions by the appropriate factor.

The copper half-reaction involves the transfer of two electrons. The nitrate half-reaction involves the transfer of three electrons. The number of electrons gained must equal the number lost. The number that two and three have in common is six. We will multiply the copper half-reaction by 3 so that 6 electrons will be lost. We will multiply the nitrate half-reaction by 2 so that 6 electrons will be gained.

$$3[Cu (s) \quad \rightarrow \quad Cu^{2+} (aq) \quad + \quad 2e^-]$$
$$2[4H^+ (aq) \quad + \quad NO_3^- (aq) \quad \rightarrow \quad NO (g) \quad + \quad 2H_2O (l)]$$

6. Add the half-reactions.

$$3[Cu (s) \quad \rightarrow \quad Cu^{2+} (aq) \quad + \quad 2e^-]$$
$$2[4H^+ (aq) \quad + \quad NO_3^- (aq) \quad \rightarrow \quad NO (g) \quad + \quad 2H_2O (l)]$$
$$3 Cu (s) \quad + \quad 8H^+ (aq) \quad + \quad 2NO_3^- (aq) \quad \rightarrow \quad 3Cu^{2+} (aq) \quad + \quad 2NO (g)$$
$$+ \quad 4H_2O (l)$$

7. Once again, this reaction does not require further simplification, so the answer is the equation we had at the end of step six.

Steps to Follow to Balance a Redox Reaction in Basic Solutions:

1. Balance the equation as if it were in acid.

2. In the final step, for each H^+ in the equation, add an OH^- on both sides of the equation. (So long as we do exactly the same thing on both sides of the equation, we do not mess up our mass or charge balance.)

3. On one side of the equation, there will be both H^+ and OH^-. Combine these to form H_2O.

4. Simplify the equation again.

Example 20-3: Balance the following net ionic equation in basic solution:
$$ClO^- \ (aq) + S^{2-} \ (aq) \rightarrow Cl^- \ (aq) + S \ (s)$$

1. This is a redox reaction in base. We will balance it first as if it were in acid.

2. The two half-reactions are
$$ClO^- \ (aq) \rightarrow Cl^- \ (aq)$$
$$S^{2-} \ (aq) \rightarrow S \ (s)$$

3. Balance the half-reactions for mass.

In the hypochlorite equation, the chlorines are balanced, but there is an extra oxygen on the left side. We balance oxygens with water, so we will add one water to the right side.
$$ClO^- \ (aq) \rightarrow Cl^- \ (aq) + H_2O \ (l)$$

This introduces two hydrogens on the right side. We balance hydrogens using H^+, so we add two H^+ to the left side.
$$2H^+ \ (aq) + ClO^- \ (aq) \rightarrow Cl^- \ (aq) + H_2O \ (l)$$

This half-reaction is balanced for mass. The S^{2-} half-reaction is already balanced for mass.
$$S^{2-} \ (aq) \rightarrow S \ (s)$$

4. Balance the half-reactions for charge.

Let's first deal with the hypochlorite equation:

	$2H^+ \ (aq)$	+	$ClO^- \ (aq)$	$\rightarrow$	$Cl^- \ (aq)$	+	$H_2O \ (l)$
Charge	$2(+1) = +2$		-1		-1		0
		$+1$				-1	

We add electrons to the more positive side. In this case we need to bring the charge on the left down from $+1$ to -1. This will require two electrons.
$$2e^- + 2H^+ \ (aq) + ClO^- \ (aq) \rightarrow Cl^- \ (aq) + H_2O \ (l)$$

Now, let's do the sulfide case.

	$S^{2-} \ (aq)$	$\rightarrow$	$S \ (s)$
Charge	-2		0

The more positive side is the right side. We need to bring its charge down from 0 to -2, so we will add two electrons to the right side.
$$S^{2-} \ (aq) \rightarrow S \ (s) + 2e^-$$

5. Multiply by a factor so that electrons lost = electrons gained.

In this case, two electrons are gained in the hypochlorite half-reaction and two electrons are lost in the sulfide half-reaction. No additional factor is needed.

6. Add the half-reactions.

$$2e^- \quad + \quad 2H^+ \ (aq) \quad + \quad ClO^- \ (aq) \quad \rightarrow \quad Cl^- \ (aq) \quad + \quad H_2O \ (l)$$
$$S^{2-} \ (aq) \quad \rightarrow \quad S \ (s) \quad + \quad 2e^-$$

$$2H^+ \ (aq) \quad + \quad ClO^- \ (aq) \quad + \quad S^{2-} \ (aq \quad \rightarrow \quad Cl^- \ (aq) \quad + \quad H_2O \ (l)$$
$$+ \quad S \ (s)$$

7. There is no simplification needed.

At this point, we would be finished if the reaction had been run in an acidic solution, but it was not. It was run in a basic solution. We will take this into account now. In our balanced equation above, there are $2H^+$ ions on the left side. We now add the same number of hydroxide ions to both sides of the equation. We add two OH^- to both sides.

$$2OH^- \ (aq) + \quad 2H^+ \ (aq) \quad + \quad ClO^- \ (aq) \quad + \quad S^{2-} \ (aq) \quad \rightarrow \quad Cl^- \ (aq) \quad + \quad H_2O \ (l)$$
$$+ \ S \ (s) \quad + \quad 2OH^- \ (aq)$$

On the left side, the H^+ and OH^- combine to form H_2O.
$$2H_2O \ (l) \quad + \quad ClO^- \ (aq) \quad + \quad S^{2-} \ (aq) \quad \rightarrow \quad Cl^- \ (aq) \quad + \quad H_2O \ (l)$$
$$+ \ S \ (s) \quad + \quad 2OH^- \ (aq)$$

There are two waters on the left and one on the right. We end up with one water on the left.
$$H_2O \ (l) \quad + \quad ClO^- \ (aq) \quad + \quad S^{2-} \ (aq) \quad \rightarrow \quad Cl^- \ (aq) \quad + \quad S \ (s)$$
$$+ \quad 2OH^- \ (aq)$$

Our equation is now fully balanced.

Try Study Questions 1, 3, and 5 in Chapter 20 of your textbook now!

- **Understand the principles underlying voltaic cells.**

  a) In a voltaic cell identify the half-reactions occurring at the anode and the cathode, the polarity of the electrodes, the direction of electron flow in the external circuit, and the direction of ion flow in the salt bridge (Section 20.2).

  So far, we have considered redox reactions in which we mix together all of the reactants in one container. There is electron transfer that occurs right there in the container between the material being oxidized and the material being reduced.

  Electricity is the flow of electrons. There was a flow of electrons between the material being oxidized and the material being reduced in our redox reactions, but we could not tap into it to do electrical work because the transfer occurred directly between the two materials. If we could force the electrons to go through a wire on their way from the material being oxidized to the material being reduced, then we could get electricity that we could use. This is what goes on in a voltaic (or galvanic) cell. In a voltaic cell, chemical reactions are used to produce an electric current.

  In designing a voltaic cell, we need to separate the two processes of oxidation and reduction from each other. In Figure 20.5 in your text, the two half-reactions are set up in different beakers. One beaker contains a copper strip and a solution of a salt containing Cu^{2+} ions, and the other beaker contains a silver strip and a solution of a salt containing Ag^+ ions.

  The electrode at which oxidation occurs is the anode. The electrode at which reduction occurs is the cathode. There is a mnemonic to help keep this straight: RED CAT AN OX, which stands for <u>red</u>uction occurs at the <u>cat</u>hode and the <u>an</u>ode involves <u>ox</u>idation. In a voltaic cell, the anode is assigned a negative sign, and the cathode is assigned a positive sign.

There also needs to be an electrical connection between the two half-reactions. In the figure, this is provided by a wire. It allows the electrons to flow from the copper half-reaction to the silver half-reaction, from the anode to the cathode.

A salt bridge completes the circuit and allows cations and anions to move between the two half-cells. This movement of ions allows the beakers to retain electrical neutrality. Negative electrons leave the anode compartment and go into the wire. Anions from the salt bridge must come into the anode compartment to maintain electrical neutrality. Anions therefore flow in the opposite direction that the electrons flow in the external circuit. Electrons flow from the anode to the cathode, so anions flow from the cathode compartment to the anode compartment. Cations flow in the opposite direction as anions, so they flow in the same direction as the electrons in the external circuit, from the anode to the cathode. The salt in the salt bridge should contain ions that will not react with the chemical reagents in the half-cells.

<u>Example 20-4:</u> If the following reaction is set up to run in a voltaic cell, which electrode will be the cathode and which will be the anode? Electrons will flow from which electrode to which electrode? Which will be the negative electrode and which will be the positive electrode? Describe the flow of ions that would occur in a potassium chloride salt bridge.

$$Zn\ (s) + Cu^{2+}\ (aq) \rightarrow Cu\ (s) + Zn^{2+}\ (aq)$$

In the zinc half-reaction, the oxidation number of zinc changes from zero to +2. The oxidation number has gone up, so the zinc is oxidized in the reaction. The electrode at which oxidation occurs is the anode, so the zinc electrode will be the anode.

In the copper half-reaction, the Cu^{2+} goes from an oxidation number of +2 to zero. Its oxidation number goes down; it is reduced. The electrode at which reduction occurs is the cathode, so the copper electrode is the cathode.

Electrons flow from the anode to the cathode. In this case, this means that they will go from the zinc electrode to the copper electrode.

The anode is assigned a negative sign, so the zinc electrode will be the negative electrode. The cathode is assigned a positive sign, so the copper electrode will be the positive electrode.

The electrons flow from the zinc electrode to the copper electrode. Anions flow in the opposite direction as the electrons; the anions will flow from the copper compartment to the zinc compartment. Cations will flow in the opposite directions as anions, so they will flow from the zinc compartment to the copper compartment.

Try Study Question 9 in Chapter 20 of your textbook now!

b) Appreciate the chemistry and advantages and disadvantages of dry cells, alkaline batteries, lead storage batteries, and Ni-cad batteries (Section 20.3).

For many years, the type of battery used in devices such as flashlights were LeChanché dry cells. Figure 20.10 in your textbook shows an example of one of these dry cells. The two half-reactions involved are

Cathode, reduction	$2NH_4^+\ (aq) + 2e^- \rightarrow 2NH_3\ (g) + H_2\ (g)$
Anode, oxidation	$Zn\ (s) \rightarrow Zn^{2+}\ (aq) + 2e^-$

The voltage produced by this reaction is 1.5 V. Notice that the cathode reaction produces two gases: NH_3 and H_2. To prevent these from building up in the cell, there were other materials present in the cell to react with the gases as they were formed:

$$Zn^{2+} (aq) + 2NH_3 (g) + 2Cl^- (aq) \rightarrow Zn(NH_3)_2Cl_2 (s)$$
$$2MnO_2 (s) + H_2 (g) \rightarrow Mn_2O_3 (s) + H_2O (l)$$

There are several problems associated with this type of cell. The first is that if current is drawn too quickly, the reactions that tie up the gases cannot do so quickly enough. When this occurs the voltage drops and there is also a risk that the battery could explode due to the gases building up in the battery. Another problem is that the zinc shell of the battery is in contact with the NH_4^+ ion, a weak acid. A reaction occurs which deteriorates the zinc shell. Eventually, the battery can leak acid.

Because of these problems, chemists came up with an alternative called the alkaline battery, which is much more widely used today. They are called alkaline batteries because of the involvement of OH^- ions in the chemistry.

Cathode, reduction:	$2MnO_2 (s) + H_2O (l) + 2e^- \rightarrow Mn_2O_3 (s) + 2OH^- (aq)$
Anode, oxidation	$Zn (s) + 2OH^- (aq) \rightarrow ZnO (s) + H_2O (l) + 2e^-$

The voltage produced by such a cell, 1.54 V, is almost the same as for the old LeClanché cells. These do not produce gases and do not leak acid. The voltage from an alkaline battery does not drop under high current loads like that of the LeClanché cells. Both the LeClanché dry cell and the alkaline cells are examples of primary batteries, electrochemical cells in which the redox reactions cannot be returned to their original state by recharging. When these batteries are dead, they are discarded.

Other types of batteries can be recharged. These are called secondary (or storage or rechargeable batteries). The two most common types of rechargeable batteries are lead storage batteries and Ni-cad batteries.

Automobile batteries are lead storage batteries. When it is being discharged to start your engine, the chemistry going on is

Cathode:	$PbO_2 (s) + 4H^+ (aq) + SO_4^{2-} (aq) + 2e^- \rightarrow PbSO_4 (s) + 2H_2O (l)$
Anode:	$Pb (s) + SO_4^{2-} (aq) \rightarrow PbSO_4 (s) + 2e^-$

After the car has started, the battery gets recharged. In this process, the solid $PbSO_4$ that formed during the discharge process gets converted back to the initial reactants.

Advantages of the lead storage battery are that it is fairly inexpensive, can produce a large initial current, and can be recharged numerous times. Disadvantages are that it is very heavy and that lead compounds are toxic making their disposal a complication.

Nickel-cadmium batteries (Ni-cad) batteries are used in many portable electronic devices such as cell phones. The chemistry involved when they are being discharged is

Cathode:	$NiO(OH) (s) + H_2O (l) + e^- \rightarrow Ni(OH)_2 (s) + OH^- (aq)$
Anode	$Cd (s) + 2OH^- (aq) \rightarrow Cd(OH)_2 (s) + 2e^-$

These batteries are lighter weight and also rechargeable. They produce a nearly constant voltage. Their disadvantages are that they are fairly expensive and their disposal is again an issue because their components are an environmental hazard due to the cadmium present.

Try Study Question 11 in Chapter 20 of your textbook now!

c) Understand how fuel cells work, and recognize the difference between batteries and fuel cells (Section 20.3).

In a typical battery, there is an enclosed container that contains a fixed quantity of reagents. If the battery is run long enough, equilibrium will be established, and the battery dies. A fuel cell differs from this in that fresh reactants are supplied to the fuel cell continuously. See Figure 20.12 in your textbook for a diagram of the most common type of fuel cell. In this fuel cell, the reactants are hydrogen and oxygen, and the eventual product is water. The reactant gases are contained in some external reservoir. The gases are pumped into the fuel cell, where the redox chemistry occurs. The products then leave the fuel cell.

Cathode: $O_2\ (g) + 2H_2O\ (l) + 4e^- \rightarrow 4OH^-\ (aq)$
Anode: $H_2\ (g) \rightarrow 2H^+\ (aq) + 2e^-$

• Understand how to use electrochemical potentials.

a) Understand the process by which standard reduction potentials are determined and identify standard conditions as applied to electrochemistry (Section 20.4).

Why is there a voltage in an electrochemical cell? There is a difference in the potential energy of electrons at the two different electrodes of an electrochemical cell. When the two electrodes are connected in an electrochemical cell, electrons move from the electrode where they have the higher potential energy to the electrode where they have the lower potential energy (from the anode to the cathode). The difference in potential energy per electrical charge between the two electrodes when no current flows is called the electromotive force (emf). The units for emf are volts. There is not a very convenient way to measure emf directly. Instead, we connect the two electrodes and allow current to flow. Using a voltmeter, we can measure what is called the potential (E) of the cell. Its value will be slightly less than the emf, but we usually don't worry about this difference and equate this with the emf.

Standard conditions for electrochemistry are defined to be any solids and liquids are present as pure solids and liquids, gaseous reactants or products are present at a partial pressure of 1.0 bar, and solutes in aqueous solutions have a concentration of 1.0 M.

A potential measured under standard conditions is designated as E^o_{cell}. Usually, 25°C is chosen as the temperature, but this is not part of the definition of standard conditions.

b) Describe the standard hydrogen electrode ($E° = 0.00$ V) and explain how it is used as the standard to determine the standard potentials of half-reactions (Section 20.4).

Is there a way to predict the potential for an electrochemical cell? If we knew the potential energy of each half-cell in an electrochemical cell, we could determine the difference between them. Unfortunately, we cannot measure the potentials of individual half-cells. In order to measure a potential, we must have a complete cell, not just a half-cell. In other words, we always have to measure a change in potential energy, not the actual potential energy of one half-cell. We get around this by measuring potentials relative to a reference electrode. We can then use these potentials to calculate the potential of any cell we want. Let's look at this mathematically. Let's say we want to predict the potential that would be measured when we connect half-cells A and B together. What we want to calculate is
$$E_{cell} = E_A - E_B$$

What is available for our calculation is the potential of a cell connecting A and some reference half-cell and the potential of a cell connecting B and the same reference cell. The potential for the cell connecting A and the reference half-cell can be represented as $E_A - E_{reference}$. (We don't know what E_A or $E_{reference}$ is, but we have measured their difference in the lab with a voltmeter.) The potential for the cell connecting B to the reference half-cell is $E_B - E_{reference}$. If we use these potentials, we can calculate the potential we desire.

$$\left(E_A - E_{reference}\right) - \left(E_B - E_{reference}\right) = E_A - E_B$$

It doesn't matter what half-cell is chosen to be the reference half-cell because its potential will be eliminated when the subtraction is done. The key is that we need to choose a reference and we must measure all of the potentials relative to it.

The reference that has been chosen is the standard hydrogen electrode. The standard hydrogen electrode consists of an inert platinum electrode over which hydrogen gas is bubbled at a pressure of 1 bar. This electrode is immersed in a solution with a concentration of H^+ at 1 M. The half-reaction (written as a reduction) for this electrode is

Reduction $2H^+ (aq) + 2e^- \rightarrow H_2 (g)$

This half-reaction is assigned an E° value of zero volts. Half-cells with a greater ability to proceed as reductions are assigned positive potentials relative to the standard hydrogen electrode. Half-cells with a smaller ability to proceed as reductions are assigned negative potentials relative to the standard hydrogen electrode.

c) Know how to use standard reduction potentials to determine cell voltages for cells under standard conditions (Equation 20.1).

In the last objective, we described how we could assign potentials to half-reactions, written as reductions, relative to the standard hydrogen electrode. If we have standard conditions, these are called standard reduction potentials (E°). Standard reduction potentials for a number of half-reactions are listed in Appendix M. The standard cell potential for an electrochemical cell can be calculated using the following equation:

$$E^o_{cell} = E^o_{cathode} - E^o_{anode}$$

A reaction that has a positive E^o_{cell} is spontaneous under standard conditions. A reaction that has E^o_{cell} equal to zero is at equilibrium under standard conditions. A reaction that has a negative E^o_{cell} is not spontaneous under spontaneous conditions. Such a reaction will be spontaneous in the opposite direction.

E°	Interpretation
> 0	Spontaneous Under Standard Conditions (Product-Favored Reaction at Equilibrium)
= 0	At Equilibrium Under Standard Conditions
< 0	Not Spontaneous Under Standard Conditions (Reactant-Favored Reaction at Equilibrium)

Notice that whereas a negative $\Delta G°$ implied a spontaneous reaction under standard conditions, a positive E° implies a spontaneous reaction under standard conditions.

Example 20-5: Calculate E° for the following reactions and indicate whether the reaction is spontaneous or not.

a. $Cu^{2+} (aq) + Zn (s) \rightarrow Cu (s) + Zn^{2+} (aq)$

In order to use the equation, we must identify cathode and the anode. In this reaction, the reduction part is the part involving copper. The copper is thus the cathode.

$Cu^{2+} (aq) + 2e^- \rightarrow Cu (s)$

The oxidation half-reaction involves zinc. The zinc is the anode.

$$Zn \ (s) \rightarrow Zn^{2+} \ (aq) + 2e^-$$

We need to locate the correct E°'s in Appendix M. We find that the E° for the copper reaction. This is $E^o_{cathode}$.

$$Cu^{2+} \ (aq) + 2e^- \rightarrow Cu \ (s) \qquad\qquad E° = 0.337 \ V$$

We do not find the oxidation reaction directly because it is an oxidation, not a reduction. The corresponding reduction that we find in the table is

$$Zn^{2+} \ (aq) + 2e^- \rightarrow Zn \ (s) \qquad\qquad E° = -0.763 \ V$$

This is E^o_{anode} .

Now, we use our equation:

$$E^o_{cell} \ = \ E^o_{cathode} \ - \ E^o_{anode} \ = \ 0.337 \ V \ - \ (-0.763 \ V) \ = \ 1.100 \ V$$

Because E° is positive, this reaction will be spontaneous under standard conditions.

b. $Cu^{2+} \ (aq) + 2Ag \ (s) \rightarrow 2Ag^+ \ (aq) + Cu \ (s)$

As written, the copper again is being shown as being reduced, so its E° will be $E^o_{cathode}$:

$$Cu^{2+} \ (aq) + 2e^- \rightarrow Cu \ (s) \qquad\qquad E° = 0.337 \ V$$

As written, the silver half-reaction is the oxidation process.

$$2Ag \ (s) \rightarrow 2Ag^+ \ (aq) + 2e^-$$

The corresponding equation in Appendix M is the reduction:

$$Ag^+ \ (aq) + e^- \rightarrow Ag \ (s) \qquad\qquad E° = 0.799 \ V$$

Its E° will be the value of E^o_{anode} .

Now, we can use our equation:

$$E^o_{cell} \ = \ E^o_{cathode} \ - \ E^o_{anode} \ = \ 0.337 \ V \ - \ 0.799 \ V \ = \ -0.462 \ V$$

Notice that we used the E° for the reaction with 1 silver even though our balanced equation had 2 silvers. Changing a stoichiometric coefficient does NOT change the value of E°.

The final E° for the reaction is negative. This means that this reaction is not spontaneous under standard conditions. Instead, the reverse reaction of silver ions with copper metal will be spontaneous. In other words, the copper is not the cathode, and the silver is not the anode. In reality, the silver is the cathode, and the copper is the anode for the spontaneous reaction.

Try Study Question 13 in Chapter 20 of your textbook now!

d) Know how to use a table of standard reduction potentials (Table 20.1 and Appendix M) to rank the strengths of oxidizing and reducing agents, to predict which substances can reduce or oxidize another species, and to predict whether redox reactions will be product-favored or reactant-favored (Sections 20.4 and 20.5).

Can we predict ahead of time, which half-reaction will be the cathode and which will be the anode in an electrochemical cell? When we set up an electrochemical cell, one of the half-reactions will proceed as an oxidation and one will proceed as a reduction. We can picture an electrochemical reaction as a tug of war for electrons. The half-reaction that is stronger as a reduction (the one with the greater reduction potential) will win the tug of war and actually get the electrons. It will proceed as a reduction. In other words, the one with the greater reduction potential will be the cathode in the cell.

The one with the lower reduction potential will lose the tug of war. It will not proceed as a reduction but will proceed in the opposite direction as an oxidation. It will be the anode.

Example 20-6: Write a balanced chemical equation for the reaction that is product-favored at equilibrium when an electrochemical cell is hooked up involving the following half-cells: 1) a lead electrode in a solution containing 1.0 M Pb^{2+} and 2) a zinc electrode in a solution containing 1.0 M Zn^{2+}.

We look at Appendix M for the half-reactions involving Pb and Pb^{2+} and Zn and Zn^{2+}, written as reductions. We find the following:

$$Pb^{2+} \; (aq) + 2e^- \rightarrow Pb \; (s) \qquad E° = -0.126 \text{ V}$$
$$Zn^{2+} \; (aq) + 2e^- \rightarrow Zn \; (s) \qquad E° = -0.763 \text{ V}$$

The lead half-reaction has the more positive E°. It will proceed as a reduction under standard conditions. The zinc half-reaction will proceed in the opposite direction, as an oxidation.

$$Pb^{2+} \; (aq) + 2e^- \rightarrow Pb \; (s)$$
$$\underline{Zn \; (s) \rightarrow Zn^{2+} \; (aq) + 2e^-}$$
$$Pb^{2+} \; (aq) + Zn \; (s) \rightarrow Pb \; (s) + Zn^{2+} \; (aq)$$

If something has a high reduction potential, it is a strong oxidizing agent. On our table, the oxidizing agents are listed on the left side, and their strength increases as we go up the table.

A reaction far down on the table does not proceed as a reduction with very many things. Most of the time, it proceeds in the reverse direction as an oxidation. The materials on the right sides of the reduction half-reactions are reducing agents. Strength as an oxidizing agent increases the further down on the table something is.

We can generalize things even further in predicting the direction of spontaneity. In a product-favored reaction, a substance on the left in the table (an oxidizing agent) will react with a substance lower than it on the right (a reducing agent). This is referred to as the northwest-southeast rule: the reducing agent always lies southeast of the oxidizing agent in a product-favored reaction.

Example 20-7: Are the following reactions product-favored or reactant-favored?

    a. $Sn^{2+} \; (aq) + Mn \; (s) \rightarrow Mn^{2+} \; (aq) + Sn \; (s)$
Looking at Appendix M, we find the half-reactions involving $Sn^{2+} \; (aq)$ and Sn (s) and involving $Mn^{2+} \; (aq)$ and Mn (s). Mn is southeast of Sn^{2+} (below it and to the right). This will be a product-favored reaction at equilibrium.

    b. $Sn^{2+} \; (aq) + Cu \; (s) \rightarrow Cu^{2+} \; (aq) + Sn \; (s)$
Looking at Appendix M, we find the half-reactions involving $Sn^{2+} \; (aq)$ and Sn (s) and involving $Cu^{2+} \; (aq)$ and Cu (s). Cu (s) is northeast, not southeast, of Sn^{2+} on the table. This is not the proper orientation for a product-favored reaction. This reaction will be reactant-favored at equilibrium.

Try Study Questions 17 and 19 in Chapter 20 of your textbook now!

e) Use the Nernst equation (Equations 20.2 and 20.3) to calculate the cell potential under nonstandard conditions (Section 20.6).

You might recall that in the last chapter we found that ΔG (the free energy change under whatever conditions exist) could be obtained by using ΔG° (the free energy change under standard conditions) and then adding a correction factor (RT ln Q) to take into account the fact that we were not at standard conditions. In a similar way, we can calculate E (the cell potential under whatever conditions exist) by taking E° (the cell potential under standard conditions) and adding a correction factor $\left(-\dfrac{RT}{nF} \ln Q \right)$. The equation that shows this is called the Nernst equation:

$$E = E° - \frac{RT}{nF} \ln Q$$

In this equation, R is the gas constant (8.314510 J/(mole K)), T is the temperature in kelvin, n is the number of moles of electrons transferred in the balanced chemical reaction, F is the Faraday constant (9.6485309 x 10^4 J/(V mole)), and Q is the reaction quotient.

At 25°C, this equation simplifies to

$$E = E° - \frac{0.0257}{n} \ln Q$$

While you should probably be familiar with the earlier form of the equation, the latter form is the one you will probably use most often.

A positive E for a process indicates that the process is spontaneous under the conditions that are present. A negative E indicates that the process is not spontaneous in the direction written (it will be spontaneous in the reverse direction). An E equal to zero indicates that the process is at equilibrium (a dead battery).

E	Interpretation
> 0	Spontaneous
= 0	At Equilibrium
< 0	Not Spontaneous

Example 20-8: A voltaic cell is set up at 25°C with the following half-cells: Ag^+ (0.0010 M)|Ag and Cu^{2+} (0.10 M)|Cu. What should be the cell potential that is observed?

We wish to determine the cell potential under nonstandard conditions (the solute concentrations are not 1 M). To do this, we will use the Nernst equation:

$$E = E° - \frac{0.0257}{n} \ln Q$$

In order to use this equation, we need the values of E°, n, and Q.

Let's first go after E°. We look up the half-reactions in Appendix M. We find:

Ag^+ *(aq)* + e^- → Ag *(s)*   E° = 0.7994 V
Cu^{2+} *(aq)* + $2e^-$ → Cu *(s)*   E° = 0.337 V

The silver reaction has the higher E°. It will be the one that goes as a reduction, so silver will be the cathode, and copper will be the anode.

$$E°_{cell} = E°_{cathode} - E°_{anode} = 0.7994 \text{ V} - 0.337 \text{ V} = 0.462 \text{ V}$$

To determine n, we must know the chemical reaction that occurs. We write the silver half-reaction as a reduction, just as it is. We must reverse the copper half-reaction.

$$Ag^+ \text{ (aq)} \quad + \quad e^- \quad \rightarrow \quad Ag \text{ (s)}$$
$$Cu \text{ (s)} \quad \rightarrow \quad Cu^{2+} \text{ (aq} \quad + \quad 2e^-$$

These half-reactions are already balanced for mass and charge. We need to adjust them so that the number of electrons gained equals the number of electrons lost. The top half-reaction needs to be multiplied by 2 so that two electrons will be gained in the silver half-reaction and lost in the zinc half-reaction. This means that n will be 2.

$$2[Ag^+ \text{ (aq)} \quad + \quad e^- \quad \rightarrow \quad Ag \text{ (s)]}$$
$$\underline{\qquad Cu \text{ (s} \quad \rightarrow \quad Cu^{2+} \text{ (aq} \quad + \quad 2e^- \qquad}$$
$$2Ag^+ \text{ (aq)} \quad + \quad Cu \text{ (s)} \quad \rightarrow \quad 2Ag \text{ (s)} \quad + \quad Cu^{2+} \text{ (aq)}$$

This balanced equation also leads us to the correct form for the reaction quotient. Remember that we do not include solids in these expressions.

$$Q = \frac{\left[Cu^{2+}\right]}{\left[Ag^+\right]^2}$$

We can now use the Nernst equation:

$$E = E° - \frac{0.0257}{n} \ln Q$$

$$= E° - \frac{0.0257}{2} \ln \frac{\left[Cu^{2+}\right]}{\left[Ag^+\right]^2}$$

$$= 0.462 \text{ V} - \frac{0.0257}{2} \ln \frac{[0.10]}{[0.0010]^2}$$

$$= 0.314 \text{ V}$$

Try Study Question 27 in Chapter 20 of your textbook now!

f) Explain how cell voltage relates to ion concentration and explain how this allows the determination of pH (Section 20.5) and other ion concentrations.

Because Q is involved in the Nernst equation, electrochemical potentials are dependent upon the concentrations of aqueous solutes in the electrochemical cell. Similarly, if we know the electrochemical potential of an electrochemical cell, we can relate this to the concentrations of the aqueous solutes in the cell.

Example 20-9: A copper-silver cell is set up as in Example 20-8. The copper(II) ion concentration again is 0.10 M, but we do not know the concentration of silver ions in the silver half-cell. Suppose that a potential of 0.422 V was measured. What is the concentration of silver ions in the silver half-cell?

We will again use the Nernst equation. The chemical equation is the same as that in Equation 20-8, so the form of Q will be the same.

$$E = E° - \frac{0.0257}{n} \ln Q$$

$$E = E° - \frac{0.0257}{2} \ln \frac{\left[Cu^{2+}\right]}{\left[Ag^+\right]^2}$$

We know the value of E and the $[Cu^{2+}]$. E° and n are the same as in Example 20-8.

$$0.422 \text{ V} = 0.462 \text{ V} - \frac{0.0257}{2} \ln \frac{[0.10]}{x^2}$$

$$-0.040 = -\frac{0.0257}{2} \ln \frac{[0.10]}{x^2}$$

$$3.1 = \ln \frac{[0.10]}{x^2}$$

We need to get x by itself, but it is part of the natural logarithm expression. Ln and e are inverses of each other, so we raise e to both sides of the equation.

$$e^{3.1} = e^{\ln \frac{[0.10]}{x^2}}$$

$$e^{3.1} = \frac{[0.10]}{x^2}$$

$$22\, x^2 = 0.10$$

$$x = 0.067 \text{ M}$$

The concentration of silver ions in this electrochemical cell is 0.067 M.

We take advantage of this relationship of E to ion concentration whenever we use a pH electrode to determine the pH of a solution. To have an electrochemical cell, we need to connect two half-cells. One half-cell is set up with a glass electrode and our solution of unknown pH. The glass electrode consists of a silver wire coated with AgCl in a solution of known concentration of HCl. This is encased in glass. At the bottom of the glass casing is a thin glass membrane. The glass electrode is immersed in the solution whose pH we wish to determine. The other half-cell is a reference half-cell. Often this is a calomel electrode (mercury-mercury(I)). Sometimes, the reference electrode is encased in the same device as the glass electrode in a so-called combination glass electrode. The voltage is measured when the glass electrode is immersed in the solution of unknown pH. The voltage measured across the glass membrane is related to the concentration of H^+ ions in the solution. This in turn determines the pH.

There are other ion selective electrodes in addition to pH electrodes. An ion selective electrode is an electrode that is sensitive to the concentration of a particular ion in solution. The only unknown in the electrochemical cell is the concentration of the ion of interest in the solution being examined. The voltage is measured, and this voltage is related to the concentration of the ion of interest.

Try Study Question 29 in Chapter 20 of your textbook now!

g) Use the relationships between cell voltage (E_{cell}) and free energy (ΔG) (Equations 20.5 and 20.6) and between $E°_{cell}$ and an equilibrium constant for the cell reaction (Equation 20.7) (Section 20.5).

Many of the recent chapters have dealt with developing ways to be able to answer two important questions of reactivity. One question is whether a reaction will be product-favored or reactant-favored when equilibrium is established. In the last chapter, we found that this question was also related to whether the process was spontaneous under standard conditions or not. The second question deals with whether a process will be spontaneous given the conditions actually present, whether they are standard conditions or some other conditions.

In previous chapters, we found that we were able to develop mathematical measures to determine the answers to these questions from two different theoretical foundations: 1) equilibrium and reaction quotient calculations involving determining concentrations of reactants and products and 2) thermodynamic calculations involving determining enthalpy, entropy, and free energy changes of the system. In the current chapter, we have learned yet another type of measurement that allows us to determine the answers to these questions: electrochemical cell potentials.

Let us first examine the question of whether a reaction is product-favored or reactant-favored at equilibrium. We have learned previously that K and $\Delta G°$ provide ways for determining this. In this chapter, we also learned that $E°$ is yet another method to determine this.

K	$\Delta G°$	$E°$	Reactant-favored or Product-favored *at Equilibrium?*	Spontaneous *Under Standard Conditions?*
$K \gg 1$	$\Delta G° < 0$	$E° > 1$	Product-favored	Spontaneous under standard conditions
$K = 1$	$\Delta G° = 0$	$E° = 0$	$[C]^c[D]^d = [A]^a[B]^b$ at equilibrium	At equilibrium under standard conditions
$K \ll 1$	$\Delta G° > 0$	$E° < 1$	Reactant-favored	Not spontaneous under standard conditions.

In the last chapter, we learned that there is a mathematical relationship between K and $\Delta G°$:
$$\Delta G° = -RT \ln K$$

In this chapter, a relationship between $\Delta G°$ and $E°$ is introduced:
$$\Delta G° = -nFE°$$

Since both of these expressions are equal to $\Delta G°$, they are also equal to each other.
$$-nFE° = -RT \ln K$$

We can solve this for ln K.
$$\ln K = \frac{nFE°}{RT}$$

At 25°C, this expression becomes,
$$\ln K = \frac{nE°}{0.0257}$$

We thus have three different measures of whether a reaction will be product-favored or reactant-favored at equilibrium: K, $\Delta G°$, and $E°$. You should know the implications of different values of these and also how to convert between them.

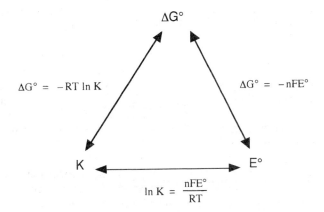

$$\Delta G° = -RT \ln K \qquad\qquad \Delta G° = -nFE°$$

$$\ln K = \frac{nFE°}{RT}$$

The second question deals with the issue of spontaneity under the set of conditions actually present in the reaction mixture. We have already learned two measures of this: ΔG and Q. In this chapter, we learned a third, E.

Q	ΔG	E	Spontaneous?
Q < K	ΔG < 0	E < 0	Spontaneous to the right.
Q = K	ΔG = 0	E = 0	At Equilibrium
Q > K	ΔG > 0	E > 0	Not spontaneous to the right; spontaneous to the left.

We can also interconvert between these:

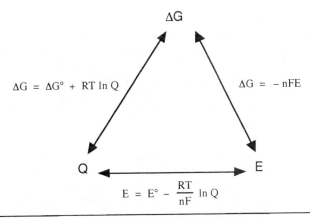

$$\Delta G = \Delta G° + RT \ln Q \qquad\qquad \Delta G = -nFE$$

$$E = E° - \frac{RT}{nF} \ln Q$$

Example 20-10: Calculate the equilibrium constant and ΔG° for the following reaction at 25°C:

$$Mg\ (s) + Cu^{2+}\ (aq) \iff Mg^{2+}\ (aq) + Cu\ (s)$$

The Cu^{2+} is reduced in this reaction so the copper reaction is the cathode half-reaction. The Mg gets oxidized, so its half-reaction is the process going on at the anode. We can calculate E° for the reaction using the values in Appendix M.

$$Cu^{2+}\ (aq) + 2e^- \rightarrow Cu\ (s) \qquad\qquad E° = 0.34\ V$$
$$Mg^{2+}\ (aq) + 2e^- \rightarrow Mg\ (s) \qquad\qquad E° = -2.37\ V$$

$$E^{\circ}_{cell} = E^{\circ}_{cathode} - E^{\circ}_{anode} = 0.34\,V - (-2.37\,V) = 2.71\,V$$

The equation that relates E° and K at 25°C is

$$\ln\,K = \frac{nE^{\circ}}{0.0257}$$

From the half-reactions, we can see that the number of electrons transferred is 2.

$$\ln\,K = \frac{2(2.71)}{0.0257} = 211$$

$$K = e^{211} = 3.90 \times 10^{91}$$

The equilibrium constant is thus very large, 3.90×10^{91}.

The equation that relates ΔG° and E° is

$$\Delta G^{\circ} = -nFE^{\circ}$$

$$\Delta G^{\circ} = -\left(2\,moles\,e^{-}\right)\left(96,500\,\frac{C}{mole\,e^{-}}\right)(2.71\,V)$$

$$= -5.23 \times 10^{5}\,J = -523\,kJ$$

ΔG° is a very large negative value, –523 kJ. Notice that all three measures predict that the reaction will be product-favored at equilibrium and spontaneous under standard conditions.

Try Study Questions 31 and 33 in Chapter 20 of your textbook now!

- ## Explore electrolysis, the use of electrical energy to produce chemical change.

a) Describe the chemical processes occurring in an electrolysis. Recognize the factors that determine which substances are oxidized and reduced at the electrodes (Section 20.7).

Up to this point, we have focused on the use of chemistry to produce electrical energy. In this section, we shift our focus to the opposite process: using electrical energy to produce a chemical change. The examples of this that we have seen so far are the processes that go on when we recharge a rechargeable battery.

Let us examine the set-up of more typical electrolysis cells. We place the material to be electrolyzed in a container, the electrolysis cell. The material to be electrolyzed will either be a molten salt or a solution containing a salt. Two electrodes are placed in the cell, and the electrodes are hooked up to a source of DC voltage. When a high enough voltage is used, chemical reactions start to happen at the electrodes.

The definition of cathode and anode remain the same in that the cathode is the electrode at which a reduction process occurs, and the anode is the electrode at which an oxidation process occurs. The signs assigned to the electrodes are different, however. In the case of an electrolytic cell, the cathode is negative and the anode is positive whereas in the voltaic cells we studied earlier, the anode was negative and the cathode was positive. Another difference between electrolytic and voltaic cells is that we needed two separate containers for the cathode and anode reaction in a voltaic cell. In an electrolytic cell, the two electrodes are placed in the same molten salt or solution.

Another difference between voltaic and electrolytic cells is that in a voltaic cell, the chemical reaction that occurred was the one that would spontaneously occur on its own. The reaction had a positive E°. We tapped into the electrical energy produced as electrons were transferred from the anode to the cathode. In the case of an electrolytic cell, the chemical reaction that occurs does so only when we provide electricity; the reaction is not spontaneous on its own. It has a negative E°. We overcome this unfavorable E° by using a voltage even greater to push the electrons in the opposite direction they would flow on their own.

Let's briefly describe the examples of electrolysis gone over in your textbook. In the first example, the authors consider the electrolysis of molten sodium chloride. First of all, we know that the reaction of sodium metal with chlorine gas to form sodium chloride is spontaneous. This reaction has a positive E°. The reverse reaction of sodium ions to form sodium metal and of chloride ions to form chlorine gas is not spontaneous. This reaction has a negative E°. By performing electrolysis on molten sodium chloride, we can get this nonspontaneous reaction to occur if we supply electricity with a high enough voltage to overcome the E° for the reverse reaction. The products of the electrolytic reaction will be sodium metal and chlorine gas.

Cathode:	$2Na^+ + 2e^- \rightarrow 2Na$ *(l)*
Anode:	$2Cl^- \rightarrow Cl_2$ *(g)* $+ 2e^-$
Overall	$2Na^+ + 2Cl^- \rightarrow 2Na$ *(l)* $+ Cl_2$ *(g)*

If, however, we electrolyze an aqueous solution containing a salt, the situation is more complicated. When we had the molten sodium chloride, the only species present were sodium ions and chloride ions. In the aqueous solution, we not only have the ions from the salt, but water is also present. Water can get involved in electrolysis. The half-reactions that can occur for water are

Reduction:	$2H_2O$ *(l)* $+ 2e^- \rightarrow H_2$ *(g)* $+ 2OH^-$ *(aq)*
Oxidation:	$2H_2O$ *(l)* $\rightarrow O_2$ *(g)* $+ 4H^+$ *(aq)* $+ 4e^-$

To illustrate how this affects the electrolysis of an aqueous solution of a salt, your textbook goes over the electrolysis of an aqueous solution of sodium iodide.

At the cathode, we have two possible reactions that could occur, one involving sodium ions and one involving water.

$2Na^+$ *(aq)* $+ 2e^- \rightarrow 2Na$ *(s)*	$E° = -2.714$ V
$2H_2O$ *(l)* $+ 2e^- \rightarrow H_2$ *(g)* $+ 2OH^-$ *(aq)*	$E° = -0.8277$ V

Which of these will occur? The sodium reaction is unfavorable by -2.714 V, whereas the water reaction is unfavorable only by -0.8277 V. The water reaction will require less of a voltage to get going, so it is the one that occurs.

At the anode, there are again two possible reactions that could occur, one involving iodide ions and one involving water.

$2I^-$ *(aq)* $\rightarrow I_2$ *(s)* $+ 2e^-$
$2H_2O$ *(l)* $\rightarrow O_2$ *(g)* $+ 4H^+$ *(aq)* $+ 4e^-$

Which of these will occur? In Appendix M, we find the following half-reactions and E°'s

I_2 *(s)* $+ 2e^- \rightarrow 2I^-$ *(aq)*	$E° = 0.535$ V
O_2 *(g)* $+ 4H^+$ *(aq)* $+ 4e^- \rightarrow 2H_2O$ *(l)*	$E° = 1.229$ V

These half-reactions are favorable as reductions. They will be unfavorable as oxidations. The iodide reaction will be unfavorable by 0.535 V and the water reaction will be unfavorable by

1.229 V. It will therefore be easier for the iodide reaction to occur. The oxidation that will occur will be the oxidation of iodide.

The net reaction that occurs is

Cathode	$2H_2O\ (l) + 2e^- \rightarrow H_2\ (g) + 2OH^-\ (aq)$
Anode	$2I^-\ (aq) \rightarrow I_2\ (s) + 2e^-$
Overall	$2H_2O\ (l) + 2I^-\ (aq) \rightarrow H_2\ (g) + 2OH^-\ (aq) + I_2\ (s)$

$E°$ for this reaction will be

$$E^o_{cell} = E^o_{cathode} - E^o_{anode} = -0.8277\ V - 0.535\ V = -1.37\ V$$

We therefore need to apply a voltage greater than 1.37 V to get this reaction to occur.

Example 20-11: Predict the reaction and $E°$ required for the electrolysis of an aqueous solution of cadmium fluoride.

In an aqueous solution of cadmium fluoride, there will be cadmium ions, fluoride ions, and water. The possible cathode reactions are one involving cadmium and one involving water:

$Cd^{2+}\ (aq) + 2e^- \rightarrow Cd\ (s)$ $E° = -0.403\ V$

$2H_2O\ (l) + 2e^- \rightarrow H_2\ (g) + 2OH^-\ (aq)$ $E° = -0.8277\ V$

It will be easier to overcome the unfavorable potential for the cadmium reaction than for the water reaction, so the cadmium reaction occurs.

The possible anode reactions are one involving fluoride and one involving water:

$2F^-\ (aq) \rightarrow F_2\ (s) + 2e^-$

$2H_2O\ (l) \rightarrow O_2\ (g) + 4H^+\ (aq) + 4e^-$

Which of these will occur? In Appendix M, we find the following half-reactions and $E°$'s

$F_2\ (s) + 2e^- \rightarrow 2F^-\ (aq)$ $E° = 2.87\ V$

$O_2\ (g) + 4H^+\ (aq) + 4e^- \rightarrow 2H_2O\ (l)$ $E° = 1.229\ V$

In this case, we can see that it will be easier to reverse the water half-reaction than to reverse the fluoride half-reaction, so the water half-reaction will occur.

The overall reaction that will occur is

Cathode:	$2[Cd^{2+}\ (aq) + 2e^- \rightarrow Cd\ (s)]$
Anode:	$2H_2O\ (l) \rightarrow O_2\ (g) + 4H^+\ (aq) + 4e^-$
Overall	$2Cd^{2+}\ (aq) + 2H_2O\ (l) \rightarrow 2Cd\ (s) + O_2\ (g) + 4H^+\ (aq)$

The $E°$ for this reaction is

$$E^o_{cell} = E^o_{cathode} - E^o_{anode} = -0.403\ V - 1.229\ V = -1.632\ V$$

We must therefore apply a voltage greater than $-1.632\ V$.

Try Study Questions 39 and 43 in Chapter 20 of your textbook now!

b) Relate the amount of a substance oxidized or reduced to the amount of current and the time the current flows (Section 20.8).

The amount of chemical reaction that occurs during electrolysis is dependent upon how many electrons are transferred. This can be calculated using the current and the Faraday constant. The current is the amount of charge that flows per unit of time. The usual unit for current is the ampere (A), which is equal to 1 coulomb/second. The current can be used to relate time and charge.

The Faraday constant is the charge carried by 1 mole of electrons; its value is 9.648309×10^4 coulombs/mole of electrons. Often, its value is rounded off to 96,500 C/mole e^-. The Faraday relates charge and electrons. The chemical process can be used to relate the number of electrons to an amount of a chemical species.

<u>Example 20-12:</u> A current of 2.20 A is passed through a solution containing Pb^{2+} for 2.00 hours, with lead metal being deposited at the cathode. What mass of lead is deposited?

We are given the current. 2.20 A = 2.20 C/s. The time that this was applied was 2.00 hours. If we convert the hours to seconds, we can figure out the charge in coulombs.

$$2.00 \text{ hours} \times \frac{60 \text{ minutes}}{1 \text{ hour}} \times \frac{60 \text{ seconds}}{1 \text{ minute}} = 7.20 \times 10^3 \text{ seconds}$$

$$7.20 \times 10^3 \text{ s} \times \frac{2.20 \text{ C}}{\text{s}} = 1.58 \times 10^4 \text{ C}$$

We also know the value of the Faraday. This can be used to figure out the number of moles of electrons transferred.

$$1.58 \times 10^4 \text{ C} \times \frac{1 \text{ mole } e^-}{96,500 \text{ C}} = 0.164 \text{ moles } e^-$$

We can then use the chemical equation for the reaction to figure out how this relates to the number of moles of lead. The chemical reaction is

$$Pb^{2+} \text{ (aq)} + 2e^- \rightarrow Pb \text{ (s)}$$

This tells us that two moles of electrons are transferred per mole of lead.

$$0.164 \text{ moles } e^- \times \frac{1 \text{ mole Pb}}{2 \text{ moles } e^-} = 0.0821 \text{ moles Pb}$$

Finally, the number of moles of lead can be converted to the number of grams of lead using the molar mass of lead.

$$0.0821 \text{ moles Pb} \times \frac{207.2 \text{ g Pb}}{1 \text{ mole Pb}} = 17.0 \text{ g Pb}$$

Try Study Questions 45 and 47 in Chapter 20 of your textbook now!

Other Notes

1. Inert Electrodes

There are many materials that do not make good electrodes. These include liquids, gases, solutions, nonmetals, and ionic materials. This does not keep us from wanting to perform and actually performing electrochemistry using these materials. What we do is to use a solid electrode made out of a material that conducts electricity but that will not participate in an oxidation or reduction reaction in the cell. This is called an inert electrode. We then bring the other material into contact with this inert electrode. Common materials used for inert electrodes are graphite, platinum, and gold.

2. Electrochemical Cell Notation

Chemists often use a shorthand notation to designate electrochemical cells. By convention, the anode is written on the left. The species involved in the half reactions are listed in order of how they occur in the half-reactions. A single vertical line is used to indicate a phase boundary, and a double vertical line is used to indicate the salt bridge.

<u>Example 20-13:</u> Use electrochemical notation to represent the following cells:

   a. An electrochemical cell is set up using the following half-cells: zinc with a
   solution of 0.1 M $ZnSO_4$ and copper with a solution of 0.1 M $CuSO_4$.

First, we must determine the anode and cathode reactions. In this case, Cu^{2+} is above Zn^{2+} in the table of standard reduction potentials, so the copper half-reaction will be the cathode reaction, and the zinc half-reaction will be the anode reaction:

  Cu^{2+} *(aq)* $+ 2e^-$ →Cu *(s)*
  Zn *(s)* → Zn^{2+} *(aq)* $+ 2e^-$

Because they are metals, the copper and zinc can serve as the electrodes. For the shorthand notation, we start with the anode (Zn). The anode reaction produces zinc ions. There is a phase change in going from zinc metal to the aqueous zinc ions. We then come to the salt bridge. The cathode reaction is that copper ions go to copper metal. There is another phase change in going from the aqueous copper ions to the copper metal.

  Zn *(s)*$|Zn^{2+}$ *(aq*, 0.1 M)$\|Cu^{2+}$ *(aq*, 0.1 M)$|$Cu*(s)*

b. An electrochemical cell is set up using the standard hydrogen electrode and a copper electrode in a solution of 0.1 M $CuSO_4$.

Cu^{2+} is above H^+ on the table of standard reduction potentials, so the copper half-reaction will be the cathode reaction.

  Cu^{2+} *(aq)* $+ 2e^-$ →Cu *(s)*
  H_2 *(g)* → $2H^+$ *(aq)* $+2e^-$

In this case, the copper can serve as one of the electrodes, but the other electrode will be the platinum electrode of the standard hydrogen electrode. Because the hydrogen reaction is the anode reaction, we will start the shorthand using the Pt electrode. The H_2 goes to H^+. There are phase changes between the Pt metal and the hydrogen gas and also between the hydrogen gas and the aqueous H^+ ions. There is then the salt bridge. The cathode side has copper ions forming copper metal. There is a phase change between the aqueous ions and the metal.

  Pt$|H_2$ *(g*, P = 1 atm)$|H^+$ *(aq*, 1.0 M)$\|Cu^{2+}$ *(aq*, 0.1 M)$|$Cu*(s)*

CHAPTER 21: The Chemistry of the Main Group Elements

Chapter Overview

In this chapter, you will learn more about the chemistry of the main group elements and their compounds. The main group elements are the elements in the tall columns of the periodic table. The s and p sublevels of the energy level corresponding to the row number on the periodic table are being filled in these elements. Main group elements usually react in order to obtain the same electron configuration as a noble gas. Compounds of main group metals with nonmetals are usually ionic. Compounds of one nonmetal with another nonmetal are usually molecular. Most of the chapter is devoted to describing the main group elements. Among the topics discussed are their properties, methods of preparation, some of their most important compounds, and uses of both the elements and their compounds. Much of the information of importance in this chapter must simply be memorized. The Study Questions in your textbook utilize various aspects of chemistry and the skills you have learned throughout the course. For example, a problem might deal with stoichiometry, thermochemistry, electrochemistry, or any other areas of chemistry we have studied.

Key Terms

In this chapter, you will need to learn and be able to use the following terms:

Aqua regia: a 1:3 solution of nitric acid and hydrochloric acid.

Bayer process: the process of separating aluminum oxide from iron and silicon oxides based on the amphoteric nature of aluminum oxide.

Brine: an aqueous solution (often a saturated solution) of sodium chloride.

Chlor-alkali industry: the industry based on the process of electrolyzing an aqueous solution of sodium chloride to form three important chemicals: chlorine gas, sodium hydroxide, and hydrogen.

Disproportionation: a reaction in which an element or compound is simultaneously oxidized and reduced.

Downs cell: a specially designed apparatus used in the electrolysis of molten sodium chloride to form sodium metal and chlorine gas.

Frasch process: a process to obtain sulfur from underground deposits. Superheated water and then air are forced into the deposit. This melts the sulfur and forces it to the surface.

Hall-Heroult process: the electrolysis of molten aluminum oxide in cryolite to produce aluminum.

Hard water: water containing significant quantities of ions, particularly calcium and magnesium.

Raschig process: the reaction of ammonia with sodium hypochlorite to form hydrazine.

Silica: SiO_2.

Silicate: a mineral based on tetrahedral SiO_4 units.

Silicone: a family of polymers containing silicon.

Slaking: the reaction of lime with water to produce calcium hydroxide.

Soda-lime process: a process to produce sodium hydroxide by reacting together sodium carbonate, calcium oxide, and water.

Three-center bond: a two electron bond spread out over three atoms such as that present in the BHB bridges in diborane, B_2H_6.

Water gas: the products of the reaction of coal with water, it consists of hydrogen and carbon monoxide.

Chapter Goals

By the end of this chapter you should be able to:

- **Relate the formulas and properties of compounds to the periodic table.**

    a) Predict several chemical reactions of the Group A elements (Section 21.2).

    Main group metals tend to react with nonmetals to form ionic compounds. For example, the metal sodium reacts with the nonmetal chlorine to form the ionic compound sodium chloride:
    $$2Na\ (s) + Cl_2\ (g) \rightarrow 2NaCl\ (s)$$

    The metal calcium reacts with the nonmetal bromine to form ionic calcium bromide:
    $$Ca\ (s) + Br_2\ (l) \rightarrow CaBr_2\ (s)$$

    In binary ionic compounds, the metals are cations usually with the electron configuration of the previous noble gas. The nonmetals are anions usually with the electron configuration of the next noble gas. We will consider ionic compounds in more detail in a later objective.

    The Group 1A and 2A metals react with the halogens to form ionic compounds. The two equations above are examples of this. Group 2A metals react with oxygen to yield the oxides of these metals. An example of this type of reaction is
    $$2Ca\ (s) + O_2\ (g) \rightarrow 2CaO\ (s)$$
    The reactions of the Group 1A elements with oxygen are more complicated and will be covered later.

    Nonmetals and metalloids tend to react with nonmetals to form molecular compounds. Often, there is more than one possible oxidation state for a given metalloid or nonmetal. The highest oxidation state possible is the same as the group number (for example, +7 for the halogens). Because several different oxidation states are possible, it is difficult sometimes to predict what the product of the reaction of a nonmetal or a metalloid with a nonmetal will be unless we have information about the amounts of reactants present and knowledge about similar compounds. For example, the following reactions of carbon with oxygen are possible:
    $$2C\ (s) + O_2\ (g) \rightarrow 2CO\ (g)$$
    $$C\ (s) + O_2\ (g) \rightarrow CO_2\ (g)$$

Example 21-1: Predict the products of the following reactions:

   a. potassium metal reacts with bromine liquid
Potassium is a Group 1A metal. Bromine is a halogen. These two elements react together to form a metal halide. The charge of the potassium will be 1+ because it is in Group 1A. The charge of the bromide ion will be 1– because it is in Group 7A. The balanced equation is
$$2K\ (s) + Br_2\ (l) \rightarrow 2KBr\ (s)$$

> b. magnesium metal reacts with oxygen gas
> Magnesium, a Group 2A metal, will react with oxygen to form a metal oxide. The charge of the magnesium ion will be 2+ and that of the oxide ion will be 2–. The balanced equation is
> $$2Mg \ (s) + O_2 \ (g) \rightarrow 2MgO \ (s)$$

Try Study Questions 5 and 13 in Chapter 21 of your textbook now!

b) Predict similarities and differences among the elements in a given group, based on the periodic properties (Section 21.2).

Elements in a given group almost always have the same number and arrangement of valence electrons. This plays a role in the compounds that these elements form. Within a group, we tend to find that the formulas of their compounds are similar. For example, the formula of calcium oxide is CaO and that of barium oxide is BaO.

Metallic character increases from the top of a group to the bottom.

Try Study Question 3 in Chapter 21 of your textbook now!

c) Know which reactions produce ionic compounds, and predict formulas for common ions and common ionic compounds based on electron configurations (Section 21.2).

Reactions between a metal and a nonmetal tend to form ionic compounds.

Main group ions having the same electron configuration as a noble gas are very common. The noble gas helium has an electron configuration of $1s^2$. The other noble gases all have an ns^2np^6 electron configuration. Metals tend to lose valence electrons to achieve a noble gas configuration whereas nonmetals tend to gain electrons to achieve a noble gas configuration.

Metals tend to form cations. Group 1A metals form ions with a 1+ charge. All of the compounds of Group 1A elements are ionic. Group 2A metals form ions with a 2+ charge. All of the compounds of Group 2A elements are ionic except for some involving beryllium. Many compounds of Group 3A elements contain 3+ ions. As one goes further down Group 3A, 1+ ions are also formed corresponding to losing only the lone p electron. The metals of Groups 4A and 5A, when they form ions, tend to form ones with +2 and +3 charges, respectively, corresponding to losing the outermost p electrons. (Higher oxidation states of +4 for Group 4A and +5 for Group 5A are possible in covalent compounds.)

Nonmetals, when they form ions, tend to form anions. Group 7A elements tend to form ions with a 1– charge. Group 6A elements tend to form ions with 2– charges. When Group 5A's nitrogen forms an ion, this ion has a charge of 3–.

Some common formulas for ionic compounds formed by various metal ions (M) and various nonmetal ions (X) are summarized in the following table.

	Hydrogen	X = Group 6A	X = Group 7A
M = Group 1A	MH	M_2X	MX
M = Group 2A	MH_2	MX	MX_2
M = Group 3A	MH_3	M_2X_3	MX_3

d) Recognize when a formula is incorrectly written, based on general principles governing electron configurations (Section 21.2).

In identifying incorrectly written formulas, the two key things to look for are 1) an incorrect charge required for an ion and 2) the positive oxidation number required for an element in a molecular compound exceeding the value of the group number.

Example 21-2: Identify the reason why each formula below is incorrect.

a. BaCl

This is a combination of a metal with a nonmetal; it is an ionic compound. We know the charge of a barium ion is 2+ because barium is in Group 2A, and we know the charge of a chloride ion is 1− because it is in Group 7A. The correct formula for their combination is $BaCl_2$, not BaCl. BaCl would imply either that barium has a charge of 1+ or that chloride has a charge of 2−, neither of which is correct.

b. PCl_6

Phosphorus is in Group 5A of the periodic table. The maximum positive oxidation number it can have is +5. Each chlorine will have an oxidation number of 1−. In this incorrect formula, the phosphorus would need to have an oxidation number of +6. This is not possible.

• **Describe the chemistry of the main group or A-group elements, particularly H; Na and K; Mg and Ca; B and Al; Si, N and P; O and S; and F and Cl.**

a) Identify the most abundant elements, know how they are obtained, and list some of their common chemical and physical properties.

Hydrogen

Hydrogen is the most abundant element in the universe. It has three isotopes: protium, deuterium, and tritium. Because it is so abundant, we are usually referring to protium when we talk about hydrogen. Tritium is a very rare isotope of hydrogen that is radioactive.

Isotope Name	Nuclear Composition	Symbol	Atomic Mass
Hydrogen (protium)	1 proton, 0 neutrons	^{1}H	1.0
Deuterium	1 proton, 1 neutron	^{2}H	2.0
Tritium	1 proton, 2 neutrons	^{3}H	3.0

Preparation of Hydrogen

Hydrogen can be prepared on a large scale by a number of methods:

Water Gas Reaction:
$$C\ (s) + H_2O\ (g) \rightarrow H_2\ (g) + CO\ (g)$$

Catalytic Steam Re-formation:
$$CH_4\ (g) + H_2O\ (g) \rightarrow 3H_2\ (g) + CO\ (g)$$

Catalytic Steam Re-formation Followed by the Water Gas Shift Reaction:
$$CH_4\ (g) + H_2O\ (g) \rightarrow 3H_2\ (g) + CO\ (g)$$
$$H_2O\ (g) + CO\ (g) \rightarrow H_2\ (g) + CO_2\ (g)$$

Electrolysis of Water:
$$2H_2O \; (l) \rightarrow 2H_2 \; (g) + O_2 \; (g)$$

Hydrogen can be prepared by many methods in the laboratory:

Metal + Acid $\rightarrow$ metal salt + H_2

Metal + H_2O or a base $\rightarrow$ metal hydroxide + H_2

Metal hydride + $H_2O \rightarrow$ metal hydroxide + H_2

Try Study Question 17 in Chapter 21 of your textbook now!

Properties of Hydrogen

Under standard conditions, hydrogen is a colorless gas. It is the least dense gas known. Hydrogen combines chemically with virtually every other element except the noble gases.

Sodium and Potassium

Sodium and potassium are the sixth and seventh most abundant elements by mass in the earth's crust. They are not found in their uncombined states in nature.

Preparation of Sodium and Potassium

Elemental sodium is prepared by means of electrolysis of molten sodium salts. The current major method of production of sodium is by electrolyzing molten sodium chloride in a Downs cell. The reaction carried out is
$$2NaCl \; (l) \rightarrow 2Na \; (l) + Cl_2 \; (g)$$

Sodium chloride, calcium chloride, and barium chloride are mixed together in the Downs cell. Adding the other salts to the sodium chloride lowers the melting point of the NaCl by about 200°C. The cathode is either copper or iron, and the anode is graphite. The sodium metal that forms has a lower density than the mixture of $NaCl$, $CaCl_2$, and $BaCl_2$, so the sodium floats on the surface where it can be drawn off and later allowed to cool and solidify.

While potassium can also be prepared by means of electrolysis, the primary method used to produce it is the reaction of potassium chloride with sodium gas.
$$Na \; (g) + KCl \; (g) \rightarrow K \; (g) + NaCl \; (l)$$

Properties of Sodium and Potassium

Sodium and potassium are silvery metals that are soft and easily cut with a knife. Their densities are a little less than water's. They melt below 100°C. They are both very reactive metals. They are never found in their elemental form in nature.

The alkali metals react with water, generating a basic solution and hydrogen gas in the process. For example, potassium reacts with water in the following reaction:
$$2K \; (s) + 2H_2O \; (l) \rightarrow 2NaOH \; (aq) + H_2 \; (g)$$

They also react directly with the halogens to produce ionic compounds.

Sodium and potassium react with gaseous oxygen, but the primary products are not the metal oxides. When sodium reacts with oxygen, sodium peroxide (Na_2O_2) is the primary product.

$$2Na\ (s) + O_2\ (g) \rightarrow Na_2O_2\ (s)$$

When potassium reacts with oxygen, potassium superoxide (KO_2) is the principal product.

$$K\ (s) + O_2\ (g) \rightarrow KO_2\ (s)$$

Because they react with air and water, sodium and potassium must be stored in a way that avoids contact with these. They are usually stored in a liquid such as mineral oil or kerosene.

Try Study Question 21 in Chapter 21 of your textbook now!

Magnesium and Calcium

Calcium and magnesium are the fifth and seventh most abundant elements on earth. They are very reactive and are not found uncombined in nature.

Preparation of Magnesium

Magnesium is prepared from seawater. The Mg^{2+} ion in seawater is precipitated by treating the seawater with calcium hydroxide. This allows for the isolation of the magnesium from the rest of what is dissolved in the water. The net ionic equation for this process is

$$Mg^{2+}\ (aq) + 2OH^-\ (aq) \rightarrow Mg(OH)_2\ (s)$$

The purified $Mg(OH)_2$ is then treated with hydrochloric acid to form magnesium chloride.

$$Mg(OH)_2\ (s) + 2HCl\ (aq) \rightarrow MgCl_2\ (aq) + 2H_2O\ (l)$$

The water is then evaporated and the resulting solid $MgCl_2$ is melted and then electrolyzed.

$$MgCl_2\ (l) \rightarrow Mg\ (s) + Cl_2\ (g)$$

Properties of Magnesium and Calcium

Both magnesium and calcium are fairly high-melting silvery metals. They are oxidized by a wide range of oxidizing agents to form ionic compounds that contain the M^{2+} ion.

They react directly with the halogens to form metal halides. For example, magnesium reacts with chlorine to form ionic magnesium chloride.

$$Mg\ (s) + Cl_2\ (g) \rightarrow MgCl_2\ (s)$$

They react directly with oxygen to form metal oxides and with sulfur to form metal sulfides:

$$2Mg\ (s) + O_2\ (g) \rightarrow 2MgO\ (s)$$
$$2Mg\ (s) + S\ (s) \rightarrow MgS\ (s)$$

They react with acids to produce hydrogen gas. For example, calcium reacts with hydrochloric acid as follows:

$$Ca\ (s) + HCl\ (aq) \rightarrow CaCl_2\ (aq) + H_2\ (g)$$

They also react with water. The result of this reaction is hydrogen gas and a basic solution.

$$Ca\ (s) + H_2O\ (l) \rightarrow Ca(OH)_2\ (aq) + H_2\ (g)$$

Try Study Question 25 in Chapter 21 of your textbook now!

Boron and Aluminum

Aluminum is the third most abundant element in the earth's crust. The other elements in Group 3A are relatively rare. Boron is a metalloid, and the rest are metals.

Preparation of Boron and Aluminum

Boron can be obtained by reacting B_2O_3 with Mg metal.
$$B_2O_3 \ (s) + 3Mg \ (s) \rightarrow 2B \ (s) + 3MgO \ (s)$$

The principal ore of aluminum is bauxite ($Al_2O_3 \cdot nH_2O$). The common impurities in bauxite are iron and silicon oxides. To separate the aluminum oxide from these, the bauxite is first treated with a hot concentrated solution of sodium hydroxide. Both the aluminum oxide and the silicon oxide react with this, leaving the iron oxide behind as a solid.
$$Al_2O_3 \ (s) + 2NaOH \ (aq) + 3H_2O \ (l) \rightarrow 2Na[Al(OH)_4] \ (aq)$$
$$SiO_2 \ (s) + 2NaOH \ (aq) + 2 H_2O \ (l) \rightarrow Na_2[Si(OH)_6] \ (aq)$$

The aluminum and silicon must still be separated. CO_2, which forms H_2CO_3, is added. The aluminum reacts with this to precipitate Al_2O_3, while the silicon remains in solution.
$$H_2CO_3 \ (aq) + 2Na[Al(OH)_4] \ (aq) \rightarrow Na_2CO_3 \ (aq) + Al_2O_3 \ (s) + 5H_2O \ (l)$$

Once the aluminum oxide has been purified, it can be used to obtain aluminum metal. The Al_2O_3 is mixed with cryolite (Na_3AlF_6) to reduce its melting point to 980°C. The molten Al_2O_3 in cryolite is electrolyzed to produce aluminum metal. This electrolytic process of obtaining aluminum from molten Al_2O_3 in cryolite is called the Hall-Heroult process.

Properties of Boron and Aluminum

Boron is very hard and resistant to heat and is a semiconductor. Aluminum is a relatively low-melting, soft metal with a high electrical conductivity. Even though its E° indicates that it should be easily oxidized, aluminum is resistant to corrosion. The reason is that the top layer of aluminum does react with oxygen to form aluminum oxide (Al_2O_3). This oxide forms a tough skin on the surface that protects the metal underneath from exposure to the air.

Silicon

Silicon is second in abundance in the earth's crust. It is a metalloid.

Preparation of Silicon

Silicon can be made in large quantities by heating pure silica sand with purified coke to approximately 3000°C in an electric furnace.
$$SiO_2 \ (s) + 2C \ (s) \rightarrow Si \ (l) + 2CO \ (g)$$

The silicon is purified even further. It is reacted with Cl_2 to form silicon tetrachloride.
$$Si \ (s) + 2Cl_2 \ (g) \rightarrow SiCl_4 \ (l)$$

The $SiCl_4$ is purified by distillation. The very pure $SiCl_4$ is then reduced back to silicon.
$$SiCl_4 \ (g) + 2Mg \ (s) \rightarrow 2MgCl_2 \ (s) + Si \ (s)$$

The magnesium chloride is dissolved in water and washed away. The silicon is then melted and cast into bars. It is then purified even further by a process called zone refining.

Nitrogen and Phosphorus

Nitrogen is found as N_2 in the atmosphere. It is the most abundant gas in the atmosphere, constituting 78% of the air by volume (and number of moles). Phosphorus is not nearly so abundant. In nature, it is found in solid compounds in the form of phosphate (PO_4^{3-}).

Preparation of Nitrogen and Phosphorus

Elemental nitrogen occurs naturally and can be obtained from the air.

Phosphorus is produced by the reduction of phosphate minerals in an electric furnace. The following equation shows the reduction of calcium phosphate.

$$2Ca_3(PO_4)_2 \ (s) + 10 \ C \ (s) + 6SiO_2 \ (s) \rightarrow P_4 \ (g) + 6CaSiO_3 \ (s) + 10CO \ (g)$$

Properties of Nitrogen and Phosphorus

Nitrogen is a colorless gas. Its does not react readily with other materials due to the large dissociation energy of its triple bond. Ammonia can be produced on a large scale by the reaction of H_2 with N_2 under high pressures and temperatures in the presence of a catalyst.

$$N_2 \ (g) + 3H_2 \ (g) <=> 2NH_3 \ (g)$$

Nitrogen also reacts with some metals to give metal nitrides.

$$3Mg \ (s) + N_2 \ (g) \rightarrow Mg_3N_2 \ (s)$$

The two major allotropes of phosphorus are white phosphorus and red phosphorus. Both are based on P_4 units. There are individual P_4 molecules in white phosphorus. In red phosphorus, the P_4 units are joined into chains.

Oxygen and Sulfur

Oxygen is the most abundant element in the earth's crust and is the second most abundant element in the atmosphere. Sulfur is fifteenth in abundance in the earth's crust.

Preparation of Oxygen and Sulfur

Commercially, oxygen (O_2) is obtained by the fractional distillation of air. In the lab, it can be produced by the electrolysis of water

$$2H_2O \ (l) \rightarrow 2H_2 \ (g) + O_2 \ (g)$$

and by the catalytic decomposition of metal chlorates

$$2KClO_3 \ (s) \rightarrow 2KCl \ (s) + 3O_2 \ (g)$$

The allotrope of oxygen called ozone, O_3, can be prepared by passing O_2 through an electric discharge or by irradiating O_2 with ultraviolet light.

Deposits of elemental sulfur occur in nature. Sulfur is obtained from these by the Frasch process. Superheated water and then air are forced into the deposit. The sulfur melts and is forced to the surface in the liquid state. It resolidifies when it cools down.

Properties of Oxygen and Sulfur

Oxygen (O_2) is a colorless gas at room temperature and pressure. In the liquid state, it is a pale blue liquid. It is paramagnetic.

Ozone (O_3) is a blue, diamagnetic gas with a very strong odor. Because it absorbs ultraviolet radiation in a particular region of the spectrum, this gas is important because its presence in the upper atmosphere protects life on earth from these damaging wavelengths. On the other hand, down in the level of the atmosphere where we live, ozone is a nasty pollutant.

Sulfur has numerous allotropes. The most common is the orthorhombic form. This is a yellow solid at room temperature and consists of S_8 molecules with the sulfur atoms arranged in a crown-shaped ring.

Fluorine and Chlorine

Fluorine and chlorine are the two most abundant halogens in the earth's crust. Fluorine is more abundant in the earth's crust, but chlorine is much more abundant in seawater. The halogens are always found in their combined states in nature.

Preparation of Fluorine and Chlorine

Fluorine is obtained by electrolysis of a fluoride salt.

While chlorine can be prepared by reacting chloride ions with strong oxidizing agents, the electrolysis of aqueous salt solutions is the method used industrially for producing chlorine.

Try Study Questions 63 and 65 in Chapter 21 of your textbook now!

Properties of Fluorine and Chlorine

Fluorine is a colorless gas at room temperature and pressure. It is a very strong oxidizing agent. It is the most reactive of all of the elements, forming compounds with every element except He, Ne, and Ar.

At room temperature and pressure, chlorine is a greenish-yellow gas.

b) Be able to summarize briefly a series of facts about the most common compounds of main group elements (ionic or covalent bonding, color, solubility, simple reaction chemistry) (Sections 21.3-21.10).

Hydrogen

Ionic metal hydrides are formed in the reaction of H_2 with a Group 1A or 2A metal. In these compounds, the oxidation number of hydrogen is -1. Examples of metal hydrides are NaH and CaH_2.

Molecular compounds are generally formed by direct combination of hydrogen with nonmetals. In these compounds, the oxidation number of hydrogen is $+1$. Examples are H_2O and HF.

Hydrogen can also form interstitial hydrides. These form when some metals absorb hydrogen into the spaces between the metal atoms. At high temperatures, the hydrogen can be driven out again.

Try Study Question 15 in Chapter 21 of your textbook now!

Sodium and Potassium

In all of their compounds, the Group 1A metals contain the element as a 1+ ion. Most sodium and potassium compounds are water-soluble.

Some important sodium compounds are sodium chloride, sodium carbonate, sodium bicarbonate, and sodium nitrate.

Magnesium and Calcium

Many compounds of the Group 2A elements have low water solubility. Common calcium minerals are limestone ($CaCO_3$), gypsum ($CaSO_4 \cdot 2H_2O$), fluorite (CaF_2), and apatites ($Ca_5X(PO_4)_3$ where X = F, Cl, or OH). Our teeth contain hydroxyapatite, $Ca_5(OH)(PO_4)_3$. When we use fluoride toothpastes, we are trying to convert the hydroxyapatite to fluoroapatite $Ca_5F(PO_4)_3$, a compound that is less soluble under acidic conditions and thus less susceptible to tooth decay.

Common magnesium minerals are magnesite ($MgCO_3$), talc ($3MgO \cdot 4SiO_2 \cdot H_2O$) and asbestos ($3MgO \cdot 4SiO_2 \cdot 2H_2O$). The mineral dolomite ($MgCa(CO_3)_2$) contains both magnesium and calcium.

Hard water contains dissolved metal ions, chiefly Ca^{2+} and Mg^{2+}.

Boron and Aluminum

The elements in Group 3A tend to lose three electrons to form the +3 oxidation state. Some of the heavier elements also form a +1 oxidation state.

One of the most common boron compounds is borax, $Na_2B_4O_7 \cdot 10H_2O$. Borax can be treated with sulfuric acid to produce boric acid, $B(OH)_3$.
$$Na_2B_4O_7 \cdot 10H_2O \text{ (s)} + H_2SO_4 \text{ (aq)} \rightarrow 4B(OH)_3 \text{ (aq)} + Na_2SO_4 \text{ (aq)} + 5H_2O \text{ (l)}$$

Boric acid can be dehydrated to boric oxide when strongly heated.
$$2B(OH)_3 \text{ (s)} \rightarrow B_2O_3 \text{ (s)} + 3H_2O \text{ (l)}$$

Boron forms a series of compounds with hydrogen. The simplest of these is diborane, B_2H_6, a gas at atmospheric pressure and room temperature. The bonding in this compound is very interesting. A Lewis structure is shown below:

The first oddity is that the two middle hydrogens form two bonds, not just one. The second oddity is that this species contains only 12 valence electrons, not 16. The way that this is rationalized is to imagine both of the BHB bridges as containing only two electrons each instead of four (as would be expected for two bonds). The two electrons in each of the BHB bridges are shared over three atoms. Such a bond is called a three-center bond.

Diborane can be made from sodium borohydride ($NaBH_4$) as follows:
$$2NaBH_4 \text{ (s)} + I_2 \text{ (s)} \rightarrow B_2H_6 \text{ (g)} + 2NaI \text{ (s)} + H_2 \text{ (g)}$$

$NaBH_4$ is itself made by reacting sodium hydride with a borate.
$$4NaH\ (s) + B(OCH_3)_3\ (g) \rightarrow NaBH_4\ (s) + 3NaOCH_3\ (s)$$

The most common aluminum compound is aluminum oxide (Al_2O_3). Aluminum halides are all solids. Aluminum chloride exists as $AlCl_3$ at atmospheric pressure and room temperature. The molecular formula of aluminum bromide is Al_2Br_6 and consists of a dimer of two $AlBr_3$ units. Aluminum iodide likewise exists as Al_2I_6.

Try Study Questions 35,37, and 41 in Chapter 21 of your textbook now!

Silicon

The bonding in silicon compounds is largely covalent and involves sharing of four electron pairs with neighboring atoms. The simplest oxide of silicon is SiO_2, commonly called silica. Quartz is a pure crystalline form of silica. Quartz is a high melting network solid in which each silicon atom is bonded to four oxygen atoms in a tetrahedral arrangement. Each oxygen is linked to two silicon atoms.

Silica (SiO_2) is resistant to attack by all acids except HF. The reaction with HF is as follows:
$$SiO_2\ (s) + 4HF\ (l) \rightarrow SiF_4\ (g) + 2H_2O\ (l)$$

Silica also dissolves slowly in hot, molten NaOH or Na_2CO_3 to give sodium silicate (Na_4SiO_4).
$$SiO_2\ (s) + Na_2CO_3\ (l) \rightarrow Na_4SiO_4\ (s) + 2CO_2\ (g)$$

Sodium silicate is sometimes called water glass. If sodium silicate is treated with acid, a gelatinous precipitate of SiO_2 called silica gel is obtained. This is a highly porous material that can absorb water to the extent of up to 40% of its own weight.

Silicate minerals are built from tetrahedral SiO_4 units, but these tetrahedral SiO_4 units can hook together in different ways. Orthosilicates contain SiO_4^{4-} anions. Examples of orthosilicates are calcium orthosilicate and olivine. In pyroxenes, there are chains of SiO_4 tetrahedra. Some minerals consist of sheets of SiO_4 tetrahedra that are formed by linking together many silicate chains. Examples of these sheet structures are present in mica and in many clay minerals.

Many minerals (such as mica, some clays, asbestos, feldspars, and zeolites) are aluminosilicates in which either a sheet of SiO_4 tetrahedra is bonded to a sheet of AlO_6 octahedra or in which some Si atoms in the silicate are replaced by Al^{3+} ions.

An example of a silicone polymer is polydimethylsiloxane, $[-(CH_3)_2SiO-]_n$. Its production starts with the reaction of silicon with methyl chloride.
$$Si\ (s) + 2CH_3Cl\ (g) \rightarrow (CH_3)_2SiCl_2\ (l)$$

This product reacts with water to produce $(CH_3)_2Si(OH)_2$. The $(CH_3)_2Si(OH)_2$ molecules react with each other to produce the silicone polymer.
$$(CH_3)_2SiCl_2 + 2H_2O \rightarrow (CH_3)2Si(OH)_2 + 2HCl$$
$$n\ (CH_3)_2Si(OH)_2 \rightarrow [-(CH_3)_2SiO-]_n + nH_2O$$

Try Study Question 43 in Chapter 21 of your textbook now!

Nitrogen and Phosphorus

The most common oxidation numbers of Group 5A elements in compounds are +3 and +5. Nitrogen can have oxidation number of anything from –3 to +5.

Ammonia (NH_3) is a gas at room temperature with a very penetrating odor. Ammonia is a weak base.

Hydrazine (N_2H_4) is a colorless, fuming liquid with an ammonia-like odor. It too is a weak base. Hydrazine is a strong reducing agent. It can be produced by the reaction of ammonia with sodium hypochlorite in a process known as the Raschig process:
$$2NH_3\ (aq) + NaClO\ (aq) \rightarrow N_2H_4\ (aq) + NaCl\ (aq) + H_2O\ (l)$$

There are many oxides of nitrogen. You should study Table 21.6 in your textbook, which lists the formulas and names of the various nitrogen oxides, their Lewis structures, descriptions, and the oxidation number of nitrogen in each.

Dinitrogen monoxide (nitrous oxide) (N_2O) is a nontoxic, odorless, and tasteless gas that produces a euphoric effect when inhaled. The oxidation number of nitrogen in this compound is +1. It can be produced from the decomposition of ammonium nitrate at 250°C:
$$NH_4NO_3\ (s) \rightarrow N_2O\ (g) + 2H_2O\ (g)$$

Nitrogen monoxide (NO) contains nitrogen in the +2 oxidation state. It has an odd number of valence electrons, and no satisfactory Lewis structure may be drawn for it.

Nitrogen dioxide (NO_2) is a brown gas. It too has an odd number of valence electrons, but we can draw a Lewis structure for it. Nitric acid decomposes to form NO_2.
$$4HNO_3\ (aq) \rightarrow 4NO_2\ (g) + 2H_2O\ (l) + O_2\ (g)$$

NO_2 is also present in air pollution. Under the conditions of an automobile engine, nitrogen and oxygen react to form nitrogen oxides. Nitrogen dioxide is formed in a two-step process:
$$N_2\ (g) + O_2\ (g) \rightarrow 2NO\ (g)$$
$$2NO\ (g) + O_2\ (g) \rightarrow 2NO_2\ (g)$$

Dinitrogen tetraoxide (N_2O_4) is a dimer of two NO_2 molecules joined together. N_2O_4 is colorless. There is an equilibrium that exists between NO_2 and N_2O_4, with N_2O_4 being favored at lower temperatures and NO_2 being favored at higher temperatures.
$$2NO_2\ (g) <=> N_2O_4\ (g)$$

One way to make nitric acid (HNO_3) is to treat sodium nitrate with sulfuric acid:
$$2NaNO_3\ (s) + H_2SO_4\ (aq) \rightarrow 2HNO_3\ (aq) + Na_2SO_4\ (s)$$

The large-scale production of nitric acid is now done by the oxidation of ammonia in a process called the Ostwald process.

Concentrated nitric acid is a strong oxidizing agent. It oxidizes most metals. The noble metals (Au, Pt, Rh, and Ir) are not attacked by nitric acid. If aqua regia (1:3 nitric acid: hydrochloric acid) is added to these metals, even they will react with the nitric acid.

Phosphine (PH_3) is the phosphorus analog of ammonia. It is poisonous, highly reactive, and has a faint garlic-like odor. It is made by the alkaline hydrolysis of white phosphorus.
$$P_4\ (s) + 3KOH\ (aq) + 3H_2O\ (l) \rightarrow PH_3\ (g) + 3KH_2PO_4\ (aq)$$

There are at least six different binary compounds of phosphorus and oxygen. Your book discussed two of these: P_4O_6 and P_4O_{10}. In P_4O_6, there is an oxygen inserted between each P-P bond in the P_4 structure. P_4O_{10}, commonly called phosphorus pentaoxide, is the most common phosphorus oxide. In P_4O_{10} each P in P_4 is surrounded tetrahedrally by oxygens. P_4O_{10} is a white solid at room temperature.

Phosphorus also forms a series of phosphorus sulfides. The most stable is P_4S_3. In this case, there has been a sulfur atom inserted into three of the P-P bonds of the P_4 tetrahedron.

There are many different phosphorus oxoacids; we shall discuss three of them. In orthophosphoric acid (sometimes called phosphoric acid) (H_3PO_4), a phosphorus atom with a +5 oxidation state is surrounded by three –OH groups and one oxygen. In phosphorous acid (H_3PO_3), one of the –OH groups has been replaced by a H and the oxidation number of the P is +3. In hypophosphorous acid (H_3PO_2), another –OH group has been replaced by another hydrogen, and the oxidation number of the phosphorus is +1. In each of these, only hydrogens in the –OH groups are ionizable.

By far, the most important phosphorus oxoacid is orthophosphoric acid. The starting material for its manufacture is white phosphorus. White phosphorus is burned in oxygen to give P_4O_{10}. The P_4O_{10} is then added to water to form H_3PO_4.

$$P_4 \; (s) + O_2 \; (g) \rightarrow P_4O_{10} \; (s)$$
$$P_4O_{10} \; (s) + 6H_2O \; (l) \rightarrow 4H_3PO_4 \; (aq)$$

Try Study Questions 47, 49, and 55 in Chapter 21 of your textbook now!

Oxygen and Sulfur

Hydrogen sulfide, H_2S, has a bent molecular geometry. It is a gas under standard conditions. It has a terrible smell and is poisonous. It is the compound with the "rotten egg" smell that many people associate with sulfur.

In compounds with metals, sulfur often occurs as the sulfide ion (S^{2-}). With the exception of the Group 1A sulfides, all metal sulfides are insoluble in water.

Sulfur dioxide is a colorless, toxic gas with a sharp odor. You might have detected it as the sharp odor in the air after lighting a match. It readily dissolves in water. It is produced by the combustion of sulfur (and sulfur containing compounds).

Sulfur dioxide can be further oxidized to sulfur trioxide.

$$2SO_2 \; (g) + O_2 \; (g) \rightarrow 2SO_3 \; (g)$$

Sulfur trioxide reacts with water to form sulfuric acid, H_2SO_4.

$$SO_3 \; (g) + H_2O \; (l) \rightarrow H_2SO_4 \; (l)$$

Sulfuric acid is the chemical produced in the greatest quantity by the chemical industry.

Try Study Questions 57 and 59 in Chapter 21 of your textbook now!

Fluorine and Chlorine

All of the halogens form compounds with hydrogen with the formula HX. In aqueous solution, these are the hydrohalic acids. All of these acids are strong acids except for HF.

HF reacts with SiO_2 and thus can be used to etch glass. Because of this, HF should not be stored in glass bottles. The reactions involved in this process are

$$SiO_2 \ (s) + 4HF \ (g) \rightarrow SiF_4 \ (g) + 2H_2O \ (l)$$
$$SiF_4 \ (g) + 2HF \ (aq) \rightarrow H_2SiF_6 \ (s)$$

While HCl can be synthesized by reacting H_2 and Cl_2, the usual laboratory method involves reacting NaCl with H_2SO_4:

$$2NaCl \ (s) + H_2SO_4 \ (l) \rightarrow Na_2SO_4 \ (aq) + 2HCl \ (g)$$

HCl has a sharp, irritating odor. Both gaseous and aqueous HCl react with metals and metal oxides to produce metal chlorides.

$$Zn \ (s) + 2HCl \ (aq) \rightarrow ZnCl_2 \ (aq) + H_2 \ (g)$$
$$ZnO \ (s) + 2HCl \ (g) \rightarrow ZnCl_2 \ (s) + H_2O \ (g)$$

The following oxoacids of chlorine are known.

Acid	Name	Oxidation Number of Cl
HClO	Hypochlorous	+1
$HClO_2$	Chlorous	+3
$HClO_3$	Chloric	+5
$HClO_4$	Perchloric	+7

Hypochlorous acid is formed when chlorine dissolves in water:

$$Cl_2 \ (g) + 2H_2O \ (l) \rightarrow H_3O^+ \ (aq) + HClO \ (aq) + Cl^- \ (aq)$$

Perchlorates are powerful oxidizing agents. Great care should be used when handling any perchlorate salt.

c) Identify uses of common elements and compounds, and understand the chemistry that relates to their usage (Section 21.3-21.10).

Hydrogen

The greatest use of hydrogen is in the production of ammonia gas:

$$2N_2 \ (g) + 3H_2 \ (g) \rightarrow 2NH_3 \ (g)$$

Sodium and Potassium

A certain amount of NaCl is essential in the diet of humans and other animals.

The electrolysis of aqueous NaCl is used to generate three important materials: chlorine, sodium hydroxide, and hydrogen. This process is called the chlor-alkali process.

$$2NaCl \ (aq) + 2H_2O \ (l) \rightarrow Cl_2 \ (g) + 2NaOH \ (aq) + H_2 \ (g)$$

Sodium carbonate (Na_2CO_3) can be used as a cleaning agent and sometimes goes by the name washing soda. It can also be used to produce sodium hydroxide in a process called the "soda-lime process."

$$Na_2CO_3 \ (aq) + CaO \ (s) + H_2O \ (l) \rightarrow 2NaOH \ (aq) + CaCO_3 \ (s)$$

Sodium bicarbonate ($NaHCO_3$) is used in cooking under the name baking soda.

Sodium nitrate can be used to produce potassium nitrate, which is used in gunpowder.

$$NaNO_3 \ (aq) + KCl \ (aq) \rightarrow KNO_3 \ (aq) + NaCl \ (s)$$

This reaction is run in hot water, where the sodium chloride is less soluble then either $NaNO_3$, KCl, or KNO_3. It can thus be forced to precipitate before any of the other materials.

As just mentioned, potassium nitrate is used in gunpowder. It is the oxidizing agent that is used to ignite both the carbon and sulfur in the gunpowder. These reactions produce gases, whose expansion can propel a bullet from a gun.

$$2KNO_3 \; (s) + 4C \; (s) \rightarrow K_2CO_3 \; (s) + 3CO \; (g) + N_2 \; (g)$$
$$2KNO_3 \; (s) + 2S \; (s) \rightarrow K_2SO_4 \; (s) + SO_2 \; (g) + N_2 \; (g)$$

Potassium superoxide is used in self-contained breathing apparati because its reaction with the exhaled CO_2 produces oxygen:

$$4KO_2 \; (s) + 2CO_2 \; (g) \rightarrow 2K_2CO_3 \; (s) + 3O_2 \; (g)$$

Try Study Question 23 in Chapter 21 of your textbook now!

Magnesium and Calcium

Because of its low density, metallic magnesium is primarily used in lightweight alloys. For example, magnesium is used in producing aircraft, automotive parts, and lightweight tools.

Calcium compounds have a variety of uses. Fluorite (CaF_2) is used in the steel industry to remove some impurities and improve the separation of the metal from impurities. It is also used in the production of hydrofluoric acid:

$$CaF_2 \; (s) + H_2SO_4 \; (l) \rightarrow 2HF \; (g) + CaSO_4 \; (s)$$

Apatites, $Ca_5X(PO_4)_3$, are used to produce phosphoric acid:

$$Ca_5F(PO_4)_3 \; (s) + 5H_2SO_4 \; (aq) \rightarrow 5CaSO_4 \; (s) + 3H_3PO_4 \; (aq) + HF \; (g)$$

Limestone (primarily $CaCO_3$) is spread on fields to neutralize acidic compounds in the soil and to supply Ca^{2+} ions to the plants. Limestone is also used to prepare lime (CaO). This is done by heating the limestone so that the carbonate will decompose.

$$CaCO_3 \; (s) \rightarrow CaO \; (s) + CO_2 \; (g)$$

Lime (CaO) is one of the components of mortar. The other components are sand and water. The hardening of the mortar is caused by chemical reactions. The first reaction that occurs is that the lime reacts with the water to produce calcium hydroxide. This process is called slaking. The calcium hydroxide then reacts over time with carbon dioxide in the air to form calcium carbonate. The calcium carbonate binds together the sand grains. This is what causes the mortar to harden.

$$Ca(OH)_2 \; (s) + CO_2 \; (g) \rightarrow CaCO_3 \; (s) + H_2O \; (l)$$

Boron and Aluminum

Borax ($Na_2B_4O_7 \cdot 10H_2O$) can be used as a flux in metallurgy.

Boric acid, $B(OH)_3$, is used as an antiseptic.

Boric oxide (B_2O_3) is used in the manufacture of borosilicate glass. The presence of boric oxide gives the resulting glass a higher softening temperature, makes it more resistant to acids, and makes the glass expand less when heated.

Sodium borohydride ($NaBH_4$) is used as a reducing agent in many organic syntheses.

Aluminum has many uses in its metallic form when alloyed with other elements to give it more strength. You are probably familiar with aluminum from its use in aluminum foil and aluminum cans, but it has many uses beyond this.

Compounds of aluminum are also useful. Dehydrated aluminum oxide (Al_2O_3) is known as corundum and is used as an abrasive in sandpaper and in grinding wheels.

Some gems are impure aluminum oxide. The contaminant in rubies is a small amount of Cr^{3+}. The contaminants in blue sapphires are Fe^{2+} and Ti^{4+}.

Silicon

Silicon compounds are used in bricks, pottery, porcelain, lubricants, sealants, computer chips, and solar cells.

Quartz crystals are used to control the frequency of radio and television transmissions.

Sodium silicate is used in household and industrial detergents because it can help maintain the desired pH. It is also used in adhesives and binders, especially for gluing corrugated cardboard boxes.

Silica gel is used as a drying agent. Packets of silica gel are often included in boxes of certain items in order to keep the items dry.

Mica, a sheet aluminosilicate, is used in furnace windows, as insulation, and as the glitter in some metallic paints. Zeolites contain regularly shaped tunnels and cavities. They are used as drying agents, as catalysts, and as water softening agents.

Silicone polymers are used in lubricants, peel-off labels, lipstick, suntan lotion, car polish, building caulk, and Silly Putty.

Nitrogen and Phosphorus

Elemental nitrogen can be used as an inert atmosphere for packaged foods and wine and to pressurize electric cables and telephone wires. Liquid nitrogen is useful as a coolant.

Both proteins and nucleic acids contain nitrogen-containing compounds.

Ammonia's primary use is as a fertilizer.

Hydrazine is used in wastewater treatment for chemical plants because it can remove strong oxidizing agents from the water. It is also added to the water in the boilers of large electricity-generating plants because it will react with oxygen dissolved in the water thus reducing the corrosion of the pipes in the boiler.

Dinitrogen monoxide is also known as laughing gas. It is sometimes used in minor surgery, such as some dental work, as an anesthetic. It is also used as a propellant and whipping agent in cans of whipped cream.

Nitrogen monoxide is important in many biochemical processes.

There are many uses for nitric acid, but its primary use is in the reaction of nitric acid and ammonia to make the fertilizer ammonium nitrate (NH_4NO_3).

Phosphorus in the form of phosphate groups is present in nucleic acids and phospholipids.

In "strike anywhere" matches, both an oxidizing agent ($KClO_3$) and a reducing agent (P_4S_3) are combined in the head of the match. The advantage of the "strike anywhere" match is that friction with almost anything can cause the match to ignite because both the oxidizing agent and reducing agent are present together. This is also a serious safety problem with them because they can lead to undesired fires as well. Today, most matches are "safety" matches. In these, the oxidizing agent (again $KClO_3$) is in the head of the match. The strip on the matchbook contains red phosphorus (as well as Sb_2S_3, Fe_2O_3, and glue). Because the oxidizing agent and reducing agent are separated, they must be brought together by rubbing the two together to start a fire. Many undesired ignitions of the matches are thus avoided.

The primary use of phosphoric acid is in making fertilizers. In addition, it is an ingredient in some carbonated soft drinks. Yet another use is to impart corrosion resistance to metal objects.

Phosphate salts also have many uses. Sodium phosphate (Na_3PO_4) is used in scouring powders and paint strippers due to the basicity of the phosphate anion. Sodium monohydrogen phosphate (Na_2HPO_4) is used in pudding mixes, in quick-cooking breakfast cereals, and in a process to make pasteurized cheese.

$Ca(H_2PO_4)_2 \cdot H_2O$ is used in baking powder. The weak acids $Ca(H_2PO_4)_2 \cdot H_2O$ and $NaAl(SO_4)_2$ react with basic $NaHCO_3$ to produce carbon dioxide gas. In dough, the carbon dioxide causes the dough to rise.

$$Ca(H_2PO_4)_2 \cdot H_2O \ (s) + NaHCO_3 \ (aq) \rightarrow 2CO_2 \ (g) + 3H_2O \ (l) + Na_2HPO_4 \ (aq) + CaHPO_4 \ (aq)$$

$CaHPO_4$ is used in toothpaste as an abrasive and polishing agent.

Oxygen and Sulfur

The major use of sulfur is the production of sulfuric acid (H_2SO_4).

The major use of sulfuric acid is the manufacture of superphosphate fertilizer. Treating phosphate rock with sulfuric acid produces a mixture of soluble phosphates.

$$Ca_3(PO_4)_2 \ (s) + 3H_2SO_4 \ (l) \rightarrow 2H_3PO_4 \ (l) + 3CaSO_4 \ (s)$$

In addition to $CaSO_4$, the mixture will also contain $CaHPO_4$, $Ca(H_2PO_4)_2$, and H_3PO_4.

Sulfuric acid is also used in the production of TiO_2 (a white pigment), iron, steel, petroleum products, synthetic polymers, and paper.

Fluorine and Chlorine

HF is used in the production of refrigerants, herbicides, pharmaceuticals, high-octane gasoline, aluminum, plastics, electrical components, and fluorescent light bulbs. It can also be used to etch or frost glass.

The reason why HF is used in aluminum production is that it is used to produce cryolite, Na_3AlF_6, in which the aluminum oxide is dissolved prior to electrolysis.

Both HF and F_2 are used in the separation of uranium isotopes in a gas centrifuge. The uranium oxide is first reacted with hydrofluoric acid. The UF_4 formed in this reaction is then reacted with F_2 to form UF_6. UF_6 is a volatile compound. Molecules of this compound containing the different isotopes of uranium can be separated from each other.

$$UO_2 \ (s) + 4HF \ (aq) \rightarrow UF_4 \ (s) + 2H_2O \ (l)$$
$$UF_4 \ (s) + F_2 \ (g) \rightarrow UF_6 \ (s)$$

Sodium hypochlorite is the active ingredient in laundry liquid bleach. Hypochlorite is an oxidizing agent. The bleaching is caused by the oxidation of dyes or stains to colorless compounds. Calcium hypochlorite is the chlorine compound usually used in swimming pools. In aqueous solution, it (like other hypochlorite salts) is in equilibrium with Cl_2.

ClO_2 is used for bleaching paper pulp in the paper industry.

Potassium chlorate ($KClO_3$) is the preferred oxidizing agent in fireworks and is also a component of matches.

Ammonium perchlorate (NH_4ClO_4) is used as the oxidizer in the solid booster rockets of the Space Shuttle.

Other Notes

1. While we are used to recognizing that there are similarities between the elements that are located in a vertical column on the periodic table, there are also some similarities between some elements that are diagonally situated on the periodic table. This is called the diagonal relationship.

2. Bromine is a red-orange liquid at room temperature and pressure. It is volatile, and a red-orange vapor of bromine is apparent in a closed container of bromine.

3. Iodine is a lustrous purple-black solid at room temperature and pressure. It is a volatile solid and sublimes to give a violet vapor.

CHAPTER 22: The Chemistry of the Transition Elements

Chapter Overview

In this chapter, we take a closer look at the transition elements, those elements between the main group elements. All of the transition elements are metals, and all except for mercury are solids at room temperature. They have a metallic sheen and conduct electricity and heat. They react with various oxidizing agents to give ionic compounds. We consider in greatest detail those elements in which the d orbitals are being filled. When these elements undergo oxidation, the highest energy level s electrons are removed first before any lower energy level d electrons are removed. Although a variety of oxidation states are possible, the most common oxidation states encountered are +2 and +3. As one moves across a row of the transition elements, there is first a small decrease in atomic radius until a minimum is reached around the middle of the transition elements. After that, the radius slightly increases as one continues to move from left to right. As one moves down a group of transition elements, the radii tend to get bigger, but in moving from the fifth to the sixth period, there is very little change at all; this is due to what is called the lanthanide contraction.

Most transition metals are not found in nature in their pure uncombined forms. Metallurgy is the general name given to the process of obtaining metals from their ores. Pyrometallurgy involves high temperatures. Iron can be obtained through pyrometallurgy. Hydrometallurgy uses aqueous solutions in its processes. Copper can be obtained by means of hydrometallurgy.

In a coordination complex, a metal ion is linked to some number of molecules or ions by means of coordinate covalent bonds. The most common complexes involve one of the following coordination geometries: linear, tetrahedral, square planar, or octahedral. You will learn a set of rules for naming coordination compounds and for writing their formulas. Many coordination complexes have isomers. There are structural isomers (coordination and linkage isomers), geometric isomers (*cis-trans* and *fac-mer* isomers) and optical isomers.

The model most often used to explain the magnetic properties and colors of coordination complexes is ligand field theory. This theory proposes that ligands destabilize a metal ion's d orbitals, raising their energies. The ligands cause some d orbitals in the ion to have different energies than others. In octahedral complexes, three of the d orbitals are at a lower energy than the other two. Depending upon the size of the energy difference between the d orbitals (which is related to the ligands present), metal ions with 4-7 d electrons in an octahedral complex can be either high spin or low spin. In a tetrahedral complex, two of the d orbitals are lower in energy than the other three. In a square planar complex, the d orbitals have levels containing 2, 1, 1, and 1 d orbitals.

When we shine white light on an object, the object will often absorb some wavelengths. What we see are those wavelengths that are not absorbed. The excitation of an electron from a d orbital to a higher energy d orbital in a complex often corresponds to the absorption of a photon in the visible region. Coordination complexes are therefore often colored. The wavelength absorbed is related to the energy difference between the d orbitals. We can derive a list of ligands called the spectrochemical series that lists ligands in order of the extent that they cause orbital splitting.

Key Terms

In this chapter, you will need to learn and be able to use the following terms:

Actinides: the elements from actinium to rutherfordium. These elements comprise the bottom row of the elements shown at the bottom of the periodic table away from the main body of the table.

Bidentate: a term used to describe a ligand that coordinates to a metal ion by means of two atoms and therefore two coordinate covalent bonds.

Chelating agent: a polydentate ligand.

Cis-trans **isomers:** isomers involving either octahedral or square planar complexes in which the two groups of interest are directly across from each other (*trans-*) or in adjacent positions at 90° (*cis-*).

Coordination complex (complex ion): a metal ion that has Lewis bases joined to it by coordinate covalent bonds.

Coordination compound: a compound containing a coordination complex.

Coordination geometry: the arrangement of donor atoms of the ligands around the central metal ion in a coordination complex.

Coordination isomers: isomers in which a coordinated ligand and an uncoordinated counterion are exchanged.

Coordination number: the number of monodentate ligands that can attach to a particular metal ion in a coordination complex.

Corrosion: the deterioration of a metal by a product-favored oxidation.

d to d transition: the excitation of an electron from one d orbital in a metal ion to a d orbital that has a different energy due to the presence of a ligand. The energy differences involved in d to d transitions often correspond to photons in the visible region of the spectrum.

Fac-mer **isomers:** isomers involving an octahedral complex in which three identical ligands are either lying at the corners of a triangular face of an octahedron and having 90° bond angles between any two of them (*fac-*) or where the two of the ligands are directly across from each other and a line connecting all three follows a meridian around the sphere of the complex (*mer-*).

High spin: a term used to describe complexes in which electrons are added to higher energy d orbitals before pairing occurs in lower energy d orbitals.

Hydrometallurgy: recovery of metals from their ores by reactions in aqueous solution.

Lanthanide contraction: the decrease in size that results from the filling of the 4f orbitals. This is responsible for the d-block elements of the sixth period having approximately the same radii as their counterparts in the d-block elements of the fifth period.

Lanthanides: the elements from lanthanum to hafnium. These elements comprise the top row of the elements shown at the bottom of the periodic table away from the main body of the table.

Ligand field splitting: the difference in energy between d orbitals in a metal ion in a coordination complex; this difference in energy arises when a ligand binds to the metal ion.

Ligand field theory: a theory of bonding in complex ions that focuses on repulsion of electrons in the metal ion by the electron pairs in the ligands.

Ligand: a group joined to the metal ion in a coordination complex.

Linkage isomers: isomers in which a ligand has two structurally different spots where it can attach to the metal ion.

Low spin: a term used to describe complexes in which electrons are paired in lower energy d orbitals before adding electrons to higher energy d orbitals.

Metallurgy: the process of obtaining metals from their ores.

Monodentate: a term used to describe a ligand that coordinates to a metal ion by means of one atom and therefore one coordinate covalent bond.

Optical isomers: nonsuperimposable mirror image isomers.

Ore: a material from which we can profitably obtain a desired metal or mineral.

Pairing energy: the energy that must be overcome in order to pair up electrons in the same orbital.

Polydentate: a term used to describe a ligand that coordinates to a metal ion by more than one atom and therefore more than one coordinate covalent bond.

Pyrometallurgy: recovery of metals from their ores by high-temperature processes.

Spectrochemical series: an ordering of ligands by the magnitudes of the splitting energies they cause.

Chapter Goals

By the end of this chapter you should be able to:

- ## Identify and explain the chemical and physical properties of the transition elements.

  ### a) Identify the general classes of elements: transition elements, lanthanides, and actinides (Section 22.1).

  The transition elements are located in the short columns on the periodic table and also in the two rows at the bottom below the main body of the periodic table. They are the d-block elements and the f-block elements. The elements lanthanum through hafnium comprise the lanthanides. The elements actinium through rutherfordium comprise the actinides. This chapter focuses mainly on the d-block elements.

  ### b) Identify the more common transition metals from their symbols and positions in the periodic table, and recall some physical and chemical properties (Section 22.1).

  The valence electrons in the d-block elements are the ns and (n–1)d electrons, where n refers to the row number of the periodic table.

  Some of the most important transition metals are titanium (Ti, atomic number 22), chromium (Cr, atomic number 24), iron (Fe, atomic number 26), nickel (Ni, atomic number 28), copper (Cu, atomic number 29), zinc (Zn, atomic number 30), silver (Ag, atomic number 47), tungsten (W, atomic number 74), platinum (Pt, atomic number 78), gold (Au, atomic number 79), and mercury (Hg, atomic number 80).

  The transition elements are solids at room temperature (except for mercury), often with high melting and boiling points. They are shiny and conduct electricity and heat. They react with various oxidizing agents to give ionic compounds.

  As one moves from left to right across a row of the d-block elements, the radii vary over a fairly narrow range. There is a small decrease to a minimum around the middle of the row. They then gradually increase across the row to the end of the transition metals.

  As one moves down a group in the d-block elements of the fifth and sixth periods, the radii are almost identical rather than increasing as is the norm as one moves down a group on the periodic table. The reason for this has to do with the protons that were added for the lanthanide elements. Let us consider what happens as we move across the entire sixth period. The elements at the beginning are larger than their fifth period counterparts. We first fill the 6s orbital and then the 4f orbitals. In this region, the atomic radii go down as predicted from the increase in effective nuclear charge. By the end of the f-block, the atomic radius has decreased to be approximately the same size as the last element before the last period's d-block elements (remember that there is no corresponding 3f sublevel in the previous period). The elements in the d-block of both periods are thus approximately the same.

The densities of the transition metals also go through a pattern across a period. The densities first increase as we move across a period, corresponding to the decrease in atomic radius. The densities then decrease as the radii again become larger. The most dense of all the elements are found in the transition metals of period six. The most dense elements are osmium (Os, atomic number 76) and iridium (Ir, atomic number 77).

The melting points of the transition metals rise to a maximum around the middle of the period and then descend again. The metal with the highest melting point is tungsten (W, atomic number 74). The metal with the lowest melting point is mercury (Hg, atomic number 80). Mercury is the only metal that is a liquid at room temperature.

c) Understand the electrochemical nature of corrosion (Section 22.1).

All metals undergo oxidation. When a transition metal gets oxidized and loses electrons, the s electrons are removed first followed by d electrons. The most common oxidation states of the first series of transition metals (fourth period on the periodic table) are +2 and +3, but there are many different positive oxidation states that are possible. Higher oxidation states are more likely further down on the periodic table. Many of the ions formed are paramagnetic – they have unpaired electrons. Many of the ions have colors.

Example 22-1: What is the electron configuration of Mn^{2+}? Is this ion paramagnetic?

The electron configuration of Mn is $[Ar]3d^5 4s^2$.

To form Mn^{2+}, the Mn must lose two electrons. The electrons that are lost first are the s electrons. The electron configuration of Mn^{2+} is $[Ar]3d^5$.

Something that is paramagnetic must have unpaired electrons. To determine this, we must look at the orbital box notation:

$$[Ar] \quad \uparrow \ \uparrow \ \uparrow \ \uparrow \ \uparrow$$
$$3d$$

There are unpaired electrons, so this ion is paramagnetic.

Try Study Question 1 in Chapter 22 of your textbook now!

One of the most important oxidations of a transition element is the corrosion of iron. Corrosion, in general, is the deterioration of a metal by a product-favored oxidation. The corrosion of iron is an electrochemical process. In order to have an electrochemical process, we need to have an oxidation at the anode, a reduction at the cathode, a way to transfer electrons from the anode to the cathode, and a way for ions to move to retain electrical neutrality. In the case of the corrosion of iron, the oxidation half-reaction takes place at one location on the iron, and the reduction takes place at another location. The metal itself, through the mobility of electrons due to the metallic bonding in the metal, provides the electrical connection that the wire provided in the electrochemical cells we studied in

Chapter 20. The movement of ions is allowed by having an electrolyte (present in the environment) in contact with both the anode and the cathode regions of the metal.

The oxidation half-reaction takes place at one spot on the piece of iron. The anode reaction is the oxidation of iron metal:

$$Fe\ (s) \rightarrow Fe^{2+}\ (aq) + 2e^-$$

The reduction half-reaction takes place at another spot on the piece of iron. It turns out that the cathode reaction depends upon the conditions that are present. If little or no oxygen is present, one of the following two half-reactions occurs:

$$2H^+\ (aq) + 2e^- \rightarrow H_2\ (g)$$
$$2H_2O\ (l) + 2e^- \rightarrow H_2\ (g) + 2OH^-$$

In the presence of excess oxygen and water, then the following half-reaction occurs

$$O_2\ (g) + 2H_2O\ (l) \rightarrow 4OH^-\ (aq)$$

If either the second or the third cathodic reaction occurs, we have both Fe^{2+} and OH^- present. These combine to form insoluble $Fe(OH)_2$.

$$Fe^{2+}\ (aq) + OH^-\ (aq) \rightarrow Fe(OH)_2\ (s)$$

In the absence of oxygen, this is the end of the process. Insoluble $Fe(OH)_2$ forms on the surface of the iron and actually protects the rest of the iron from further corrosion. If some oxygen, but not a large excess, is present, then magnetite (Fe_3O_4) is formed. Magnetite can be thought of as a mixed oxide, in which half of the iron is in the +2 oxidation state and half is in the +3 oxidation state. Its formula is a combination of FeO and Fe_2O_3.

$$6Fe(OH)_2\ (s) + O_2\ (g) \rightarrow 2Fe_3O_4 \cdot H_2O\ (s) + 4H_2O\ (l)$$
$$Fe_3O_4 \cdot H_2O\ (s) \rightarrow H_2O\ (l) + Fe_3O_4\ (s)$$

In the presence of excess oxygen, oxidation continues, and hydrated iron(III) oxide is formed. This is the red-brown material we commonly call rust.

$$4Fe(OH)_2\ (s) + O_2\ (g) \rightarrow 2Fe_2O_3 \cdot H_2O\ (s) + 2H_2O\ (l)$$

d) Know the general features of pyrometallurgy and hydrometallurgy, and describe the metallurgy of iron and copper (Section 22.2).

Only a few metals are found in nature in their elemental state. Most are found in some combined state. The relatively few minerals from which elements can be obtained profitably are called ores. Metallurgy is the name given to the process of obtaining metals from their ores. We shall examine two major types of metallurgy: pyrometallurgy and hydrometallurgy. Pyrometallurgy involves high temperatures. Hydrometallurgy involves aqueous solutions.

Iron Production (Pyrometallurgy):

The production of iron from its ores is carried out by means of pyrometallurgy. The iron ore (usually Fe_2O_3), coke (primarily carbon), and limestone ($CaCO_3$) are combined in a blast furnace at a high temperature. Hot air is forced in at the bottom of the furnace. Both the carbon in the coke and the carbon monoxide formed from the reaction of the carbon with the oxygen serve to reduce the iron in the Fe_2O_3 to elemental iron.

$$Fe_2O_3\ (s) + 3C\ (s) \rightarrow 2Fe\ (l) + 3\ CO\ (g)$$
$$Fe_2O_3\ (s) + 3CO\ (g) \rightarrow 2Fe\ (l) + 3CO_2\ (g)$$

The molten iron collects at the bottom of the furnace and is drawn off through an opening in the side of the blast furnace. This impure iron is called cast iron or pig iron. Another major function of the blast furnace is to remove silicate minerals from the ore. These minerals react with CaO formed from heating the limestone. The products of this reaction are less dense than the iron and also dissolve some other nonmetal oxides. These products float on the liquid iron and can be easily removed. This material is called slag.

Pig iron is not pure enough to be of much use, so it is purified further. This is done in a basic oxygen furnace. Pure oxygen is blown into the molten pig iron. This oxidizes nonmetals such as phosphorus, sulfur, and carbon to their oxides which are either gases that escape or react with basic oxides that are added or are used to line the furnace. The iron containing product of this process is called carbon steel. Carbon steel can be reheated and cooled repeatedly to achieve the desired flexibility, hardness, strength, and malleability. This process of reheating and cooling is called tempering.

Other transition metals can be added to the mixture during the steel-making process to yield different alloys that have particular desirable characteristics, such as stainless steel.

Production of Copper (Hydrometallurgy)

Copper ores are usually sulfides of copper. These ores, however, usually contain a low percentage of the copper-containing compound. The first step in the process of obtaining copper is to perform a separation to yield an ore with a higher percentage of copper. This process is called enrichment. The particular process used for copper enrichment is a process known as flotation. The ore is finely powdered. Oil is then added as well as soapy water. The copper sulfide particles are carried to the top of this frothy mixture whereas many of the other materials in the ore sink. The frothy mixture at the top thus contains most of the copper in the ore. It is skimmed off and used in the rest of the process.

In one method, the compound containing copper is chalcopyrite ($CuFeS_2$). We must get the copper separated from the iron and we must reduce the copper to elemental copper. The chalcopyrite is treated with copper(II) chloride. This forms copper(I) chloride, which is not soluble and iron(II) chloride, which is soluble. The copper containing compound can thus be separated from the iron.

$$CuFeS_2 \ (s) + 3CuCl_2 \ (aq) \rightarrow 4CuCl \ (s) + FeCl_2 \ (aq) + 2S \ (s)$$

The insoluble material is then treated with aqueous sodium chloride. The chloride ion reacts with the copper(I) chloride to form the complex ion $[CuCl_2]^-$.

$$CuCl \ (s) + Cl^- \ (aq) \rightarrow [CuCl_2]^- \ (aq)$$

This reacts with itself to form our desired product (copper metal) and copper(II) chloride.

$$2[CuCl_2]^- \ (aq) \rightarrow Cu \ (s) + CuCl_2 \ (aq) + 2Cl^- \ (aq)$$

Your textbook also describes a method of obtaining even more copper from a copper mine by treating copper-mining wastes with acidified water and the bacterium *Thiobacillus ferrooxidans*. The bacterium breaks down the trace amounts of iron sulfides in the rock and converts iron(II) to iron(III). The iron(III) ion oxidizes the sulfide ion of any copper sulfide to sulfate. The Cu^{2+} ion that started out as part of the insoluble sulfide can then go into solution with the sulfate ions. The copper ions are now in solution, away from the rest of the rock. The solution is collected, and the copper ions can be reduced to copper metal. This is done by means of a single displacement reaction with iron metal.

$$Cu^{2+} \ (aq) + Fe \ (s) \rightarrow Cu \ (s) + Fe^{2+} \ (aq)$$

The copper produced by either of these methods is very pure, but not pure enough for many uses. It is further purified by electrolysis. Thin sheets of pure copper metal and of the impure copper are the electrodes. These are immersed in a solution containing $CuSO_4$ and H_2SO_4.

Anode (Impure copper): $Cu\ (s) \rightarrow Cu^{2+}\ (aq) + 2e^-$

Cathode (Pure copper) $Cu^{2+}\ (aq) + 2e^- \rightarrow Cu\ (s)$

Thus, the pure copper electrode gets even more pure copper added to it. The impure copper electrode gradually goes away as the copper in it goes into solution (and eventually comes out of solution as copper metal on the pure copper electrode).

- ## Understand the composition, structure, and bonding in coordination compounds.

a) Given the formula for a coordination complex, be able to identify the metal and its oxidation state, the ligands, the coordination number and coordination geometry, and the overall charge on the complex (Section 22.3).

A coordination complex, also called a complex ion, is a metal ion that has Lewis bases joined to it by coordinate covalent bonds. Compounds containing a coordination complex are called coordination compounds. The groups joined to the metal ion are called ligands. A ligand that coordinates to the metal by means of one coordinate covalent bond between the metal ion and one atom of the ligand is a monodentate ligand. A ligand that binds to the metal ion by means of two coordinate covalent bonds, one to each of two atoms of the ligand, is a bidentate ligand. In general, a ligand that can attach to the metal ion by means of more than one atom is called a polydentate ligand. Polydentate ligands are also called chelating agents.

In the formula of a coordination compound, the metal ion and its ligands are placed in brackets. In the formula for a coordination complex, the full charge of the complex ion is placed outside the brackets, for example, $[Fe(H_2O)_6]^{3+}$.

The coordination number of a metal ion is the number of monodentate ligands that can attach to it. For example, the coordination number of the iron ion in $[Fe(H_2O)_6]^{3+}$ is six. The charge of a complex ion is the sum of all of the charges of the metal ion and ligands.

Example 22-2: For the complex ion $[Ag(NH_3)_2]^+$, answer the following questions:

      a. What is the metal ion?
The metal ion is the silver ion.

      b. What are the ligands in this complex ion?
The ligands are the ammonia molecules.

      c. What is the oxidation state of the silver ion?
NH_3 is a neutral molecule, so its charge is zero.
      charge of complex = charge of metal ion + total charge of all ligands
      $+1 = x + 2(0)$
      $+1 = x$

The oxidation number of the silver ion is +1.

      d. What is the coordination number of the silver?
Ammonia is a monodentate ligand. There are two ammonia molecules surrounding the silver ion. The coordination number is therefore 2.

Example 22-3: What is the oxidation number of the cobalt ion in $[CoCl_4]^{2-}$?

The charge of the entire complex ion is –2. The oxidation number of each chloride is –1.

$$-2 = x + 4(-1)$$
$$-2 = x - 4$$
$$+2 = x$$

The oxidation number of the cobalt is +2.

Example 22-4: What is the oxidation number of the copper in the compound $Na_2[Cu(NO_2)_4]$?

The sum of the oxidation numbers in the compound must add up to zero. There are two sodium ions outside the brackets. The charge of each sodium ion is +1, giving a total of +2 for the two sodiums. The complex ion must have a charge of –2.

Within the complex ion, each nitrite has an oxidation number of –1.

$$-2 = x + 4(-1)$$
$$+2 = x$$

The oxidation number of the copper is +2.

Try Study Question 9 in Chapter 22 of your textbook now!

The coordination geometry of a coordination complex is defined by the arrangement of donor atoms of the ligands around the central metal ion. While metal ions can have coordination numbers ranging from 2 to 12, the most common are 2, 4, and 6. The following table summarizes the possible coordination geometries for these coordination numbers.

Coordination Number	Generic Formula	Geometry	Bond Angles
2	$[ML_2]^{n\pm}$	Linear	180°
4	$[ML_4]^{n\pm}$	Tetrahedral OR Square planar	109.5° *cis*-90°, *trans*-180°
6	$[ML_6]^{n\pm}$	Octahedral	*cis*-90°, *trans*-180°

The square planar geometry is most common when the metal ion has eight d electrons.

b) Provide systematic names for complexes from their formulas, and write formulas if the name is given (Section 22.3).

The following rules are used to name compounds containing complex ions.

1. In naming a coordination compound that is a salt, name the cation first and then the anion.

2. When giving the name of the complex ion or molecule, name the ligands first, in alphabetical order, followed by the name of the metal. (When determining alphabetical order, the prefix is not considered.)

3. Ligands and their names:
       a) If a ligand is an anion whose name ends in –ite or –ate, the final e is changed to o.
       b) If the ligand is an anion whose name ends in –ide, the ending is changed to o.

c) If the ligand is a neutral molecule, its common name is usually used with several important exceptions: water as a ligand is referred to as aqua, ammonia is called ammine, and CO is called carbonyl. (Notice that ammonia as a ligand is called ammine with two m's in the name. This is different from the organic functional group amine, which has only one m in its name.)
d) When there is more than one of a particular ligand with a simple name, the number of ligands is designated by the appropriate prefix: di, tri, tetra, penta, or hexa. If the ligand name is complicated, the prefix changes to bis, tris, tetrakis, pentakis, or hexakis, followed by the ligand name in parentheses.

4. If the coordination complex is an anion, the suffix –ate is added to the metal name.

5. Following the name of the metal, the oxidation number of the metal is given in Roman numerals.

Example 22-5: Name the following coordination compounds.

a. $[Co(NH_3)_4Cl_2]Cl$

In the complex ion, the ligands are ammonia, which will be called ammine, and chloride, which will be called chloro. We list them in alphabetical order; ammine will come before chloro. There are four ammonias, so the ammonia portion will be tetraammine. There are two chlorides, so the chloride portion will be dichloro. So far, we have tetraamminedichloro.

Next comes the metal's name: cobalt. So far, we have tetraamminedichlorocobalt. We then specify the oxidation state of the metal. Because the counterion to the complex ion is only one chloride with a charge of –1, the charge of the complex ion must be +1. The ammonias are neutral, and the chlorides are each –1.

$$+1 = x + 4(0) + 2(-1)$$
$$+1 = x - 2$$
$$+3 = x$$

Cobalt's oxidation number is +3. The complex ion's name is tetraamminedichlorocobalt(III). The full name of the compound is tetraamminedichlorocobalt(III) chloride.

b. $K_2[Zn(CN)_4]$

In the complex ion, the ligands are cyanide ions. These become cyano in the name of the complex ion. There are four attached, so this part of the name will be tetracyano. The metal ion is zinc. Thus, we have tetracyanozinc. In addition, this complex ion is an anion, so we must add –ate to the name. It is tetracyanozincate. Finally, we must determine the oxidation number of the zinc. The charge of the complex ion must be –2 to balance out the two +1's from the potassiums. The charge of cyanide is –1. We calculate the charge of the metal ion as

$$-2 = x + 4(-1)$$
$$x = +2$$

The complex ion is tetracyanozincate(II). The compound is potassium tetracyanozincate(II).

c. $[Cu(en)_2]Cl_2$

En stands for ethylene diamine. This is a more complicated name than the other ligands we have looked at. To make it clear that we have two of these, we designate its portion of the name as bis(ethylenediamine). The metal ion is copper. The charge of the full complex ion is +2 to balance out the –2 from the chlorides. Because en is neutral, the oxidation number of the copper must be +2. The name of the complex is bis(ethylenediamine)copper(II) chloride.

Try Study Question 15 in Chapter 22 of your textbook now!

Example 22-6: Write formulas for the following complex ions:

    a. tetraamminedichloroplatinum(IV)
The metal is platinum with an oxidation number of +4. Tetraammine means that there are four ammonias, and dichloro means that there are two chlorides. The oxidation number of the metal is +2, the charge of each ammonia is 0, and the charge of each chloride is –1. The total charge will be the oxidation number of the metal plus the charges of the ligands:
    $+4 + 4(0) + 2(–1) = +2$
This ion is $[Pt(NH_3)_4Cl_2]^{2+}$.

    b. diamminetetrachlorocobaltate(II)
The –ate ending tells us that this is an anion. The metal ion is cobalt in a +2 oxidation state. Diammine means that there are two ammonias and tetrachloro means that there are four chlorides. The total charge is the oxidation number of the ion plus the charges of the ligands:
    $+2 + 2(0) + 4(–1) = –2$
This ion is $[Co(NH_3)_2Cl_4]^{2–}$.

Try Study Question 13 in Chapter 22 of your textbook now!

c) Given the formula for a complex, be able to recognize whether isomers will exist, and draw their structures (Section 22.4).

We ran across isomerism when we studied organic chemistry in Chapter 11. Some coordination complexes also exhibit isomerism. Structural isomers have the same molecular formula but different bonding arrangements of atoms. There are two major types of structural isomers for coordination compounds. The first type is called coordination isomerism. In coordination isomers, a coordinated ligand and an uncoordinated counterion are exchanged. In order for a complex ion to exhibit coordination isomerism, a negatively charged ligand and the counterion to a positively charged coordination complex must be different. Two coordination isomers listed in your textbook are $[Co(NH_3)_5Br]SO_4$ and $[Co(NH_3)_5SO_4]Br$. In the first case, bromide is a ligand attached to the cobalt ion, and sulfate is the counterion. In the second case, the sulfate is attached to the cobalt ion, and the bromide is the counterion.

Linkage isomers are another type of structural isomer. In these isomers, the ligand has two structurally different spots where it can attach to the metal ion. The two most common ligands with which this type of isomerism occurs are the thiocyanate ion (SCN^-) and the nitrite ion (NO_2^-). In the case of the thiocyanate ion, some complexes are formed by the attachment of the S to the metal ion (called S-bonded thiocyanate), whereas others are formed by the attachment of the N to the metal ion (called N-bonded thiocyanate). In the case of the nitrite ion, some complexes are formed by the attachment of one of the oxygens to the metal ion (called nitrito complexes), whereas others are formed by the attachment of the nitrogen to the metal ion (called nitro complexes). An example of two linkage isomers is

S-bonded   N-bonded

Stereoisomers have the same atom-to-atom bonding sequence, but the atoms have a different arrangement in space. There are two types of stereoisomers: geometric isomers and optical isomers (non-superimposable mirror images).

There are two types of geometric isomers that arise in complexes. One type is *cis-trans* isomers. These are possible in octahedral and square planar complexes. In each case, we focus on two sites that are occupied by different ligands than the other sites. If these two sites are arranged such that they are directly across from each other with a bond angle of 180°, then the complex is the *trans-* isomer. If they are arranged such that the bond angle is between them is 90°, then the complex ion is the *cis-* isomer. An example of *cis-trans* isomerism is

cis-   *trans-*

The other type of geometric isomer is *fac-mer* isomers. These are possible for octahedral complexes. In this case, there are again at least two different ligands present, but in this case, there are three sites occupied by the same ligand and the other three sites are occupied by other ligands (which may be identical to each other but are different from the other ligand). The *mer-* isomer has two of the three identical sites at a bond angle of 180° to each other. The *fac-* isomer has bond angles of 90° between any two of the three isomers of interest. An example of *fac-mer* isomerism is

fac-   *mer-*

Optical isomers are also possible for some complexes. Optical isomers exist when two species have identical composition but are nonsuperimposable mirror images of each other. These occur in octahedral complexes when the metal ion coordinates to three bidentate ligands or when the metal ion coordinates to two bidentate ligands and two monodentate ligands in a *cis-* relationship. An example of two complexes that are optical isomers is

Example 22-7: Identify the types of isomerism that are possible in the following complexes and draw the possible isomers.

  a. $[Co(NH_3)_5(NO_2)]^+$

This metal ion in this complex has six ligands surrounding it, so this is an octahedral complex. Five of the ligands are the same, and one is different. Stereoisomers are not possible for this combination because the formula does not fall into any of the categories for stereoisomers. The nitrite ion is one of the ligands. This is one of the two major ligands involved in linkage isomerism. It can either link via the N or via one of the O's. The nitro isomer is shown on the left, and the nitrito isomer is shown on the right.

  b. $[Cr(C_2O_4)_2(H_2O)_2]^-$

Oxalate is a bidentate ligand, and water is a monodentate ligand. This is an octahedral complex with four sites occupied by the two oxalates and two of the sites occupied by the waters. This is a candidate for *cis-trans* isomerism. The *cis*-isomer is shown on the left, and the *trans*- isomer is shown on the right.

Looking at the *cis*- isomer on the left, we can see that this is also a case where we have two identical bidentate ligands and two *cis*- monodentate ligands. The compound shown will have an optical (mirror image) isomer.

  c. $[Cr(NH_3)_2(H_2O)_3OH]^+$

We have an octahedral complex because we have six monodentate ligands bonded to the metal ion. Three of them are identical (H_2O), so this is a case where we can have *fac-mer* isomerism. The *fac*- isomer is shown on the left, and the *mer*-isomer is shown on the right.

Try Study Questions 19 and 21 in Chapter 22 of your textbook now!

d) Describe the bonding in coordination complexes (Section 22.5).

One way to picture metal-ligand bonding in a coordination complex is to picture the formation of a coordinate covalent bond between the ligand and the metal ion. In this model, a lone pair of electrons originally belonging to the ligand is pictured as being shared by the ligand and the metal ion in the complex. While this model explains some aspects of coordination complexes, it is not able to explain the color and magnetic properties of many coordination complexes. Ligand field theory is another model that is used to explain bonding in coordination complexes.

The ligand field model focuses on repulsion and destabilization of d-electrons in the metal ion. You should review the shapes and orientations of the five d-orbitals. These are shown in Figure 22.31 in your textbook. The key thing to notice is that two of the orbitals have lobes along the coordinate axes ($d_{x^2-y^2}$ and d_{z^2}) whereas three of the orbitals have lobes in between the coordinate axes (d_{xy}, d_{yz}, and d_{xz}). In an isolated atom or ion of the metal, all five d orbitals have the same energies. When a metal ion bonds to a ligand, there will be repulsions between the d electrons on the metal ion and electron pairs of the ligand. This destabilizes (raises the energy of) the d orbitals. It also results in the d orbitals no longer having all the same energy.

In an octahedral complex, the ligands bind to sites along the x, y, and z axes. Those orbitals with lobes along the axes ($d_{x^2-y^2}$ and d_{z^2}) are destabilized more than those that have lobes between the axes (d_{xy}, d_{yz}, and d_{xz}). In an energy diagram, the d_{xy}, d_{yz}, and d_{xz} orbitals are lower in energy than the $d_{x^2-y^2}$ and d_{z^2} orbitals. The difference in energy separating them is designated Δ_O. This is shown in Figure 22.22 in your textbook.

In a tetrahedral complex, the ligands bind to sites in between the axes. In this case, the d orbitals with lobes in between the axes are destabilized more than those along the axes. The $d_{x^2-y^2}$ and d_{z^2} orbitals are lower in energy than the d_{xy}, d_{yz}, and d_{xz} orbitals. This splitting pattern is shown on the left side of Figure 22.23 in your textbook. This is the reverse order from that in the octahedral complexes. The difference in energy separating the different types of orbitals in a tetrahedral complex is Δ_t. This energy difference is smaller in tetrahedral complexes than in octahedral complexes.

In a square planar complex, the ligands are located along the x- and y-axes. A more complicated splitting pattern emerges. Those orbitals with lobes between the x and z axes and between the y- and z-axes are the lowest in energy. Next comes the d_{z^2} orbital whose lobes are along the z-axis. Next is the d_{xy} orbital because its lobes are between the axes where the ligands are. At the highest energy is the $d_{x^2-y^2}$ orbital whose lobes lie along the same axes where the ligands are. There are thus four distinct energies for the orbitals: one for

d_{xz} and d_{yz}, one for d_{z^2}, one for d_{xy}, and one for $d_{x^2-y^2}$. Most square planar complex ions involve metal ions with eight d electrons. Thus, only the $d_{x^2-y^2}$ orbital is unoccupied. The energy between the d_{xy} and the $d_{x^2-y^2}$ orbital is designated Δ_{sp}. The ligand field splitting for a square planar complex ion is shown in Figure 22.23 in your textbook.

e) Use ligand field theory to determine the number of unpaired electrons in a complex (Section 22.5).

We now wish to turn to figuring out how the electrons in the orbitals in a metal ion in a complex ion are arranged.

Let us first consider the case of an octahedral complex. The presence of the ligands splits the d orbitals in the metal ion into two subgroups. Three d orbitals (d_{xy}, d_{xz}, and d_{yz}) are lower in energy than the other two ($d_{x^2-y^2}$ and d_{z^2}). The energy separating these two is Δ_O. The size of this energy difference is very important in determining how the orbitals fill. The first three electrons in the d orbitals go in as expected. They occupy the three lower energy d orbitals singly. For the next four electrons (d^4-d^7), there are two possible arrangements. Which occurs depends on the size of Δ_O. If Δ_O is large, then pairing occurs in the lower d orbitals before the higher d orbitals get any electrons. If, however, Δ_O is small, then instead of pairing in the lower orbitals, the electrons go into the higher d orbitals, and all five d orbitals get occupied by one electron before any pairing occurs. This second case where the higher energy d orbitals get electrons before pairing occurs is called high spin because the ion has more unpaired electrons than does the first case where the electrons got paired in the lower d orbitals, which is called low spin. For octahedral complexes with 8, 9, and 10 electrons, there again is only one configuration possible; we do not have to worry about high spin and low spin for these. See Figure 22.25 in your textbook for drawings showing the possible electron configurations for the d orbitals in metal ion in octahedral complexes containing anywhere from 1 to 10 d electrons.

Example 22-8: Determine the number of unpaired electrons in the metal ions in the following coordination complexes.

a. $[Cr(H_2O)_6]^{2+}$ high spin

This is an octahedral complex because there are six monodentate ligands surrounding the chromium ion. The charge of the chromium ion in this complex is 2+ because the total charge of the complex is 2+, and the charge of the water molecules is neutral. A chromium atom has the following electron configuration: $[Ar]3d^54s^1$. To form the 2+ ion, we have removed two electrons. The 4s electron is removed first followed by one of the d electrons. The electron configuration of the ion is thus $[Ar]3d^4$. We must place four electrons into the d orbitals. Because this complex is high spin, three electrons will go into the lower d orbitals, and one will go into the higher d orbitals.

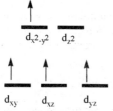

There are four unpaired electrons in this metal ion.

---

b. $[Co(H_2O)_6]^{3+}$ low spin

There are six monodentate ligands surrounding the cobalt ion; this is an octahedral complex. The charge of the cobalt ion is 3+ because the total charge of the complex is 3+, and that of the H_2O molecules is neutral. The electron configuration of a Co atom is $[Ar]3d^7 4s^2$. To form the 3+ ion, we remove three electrons: the two 4s electrons and one of the 3d electrons. The electron configuration of the ion is $[Ar]3d^6$. We are told that this complex is low spin. Once each of the lower 3d orbitals has one electron, we begin pairing the electrons in these orbitals before starting to fill the higher energy d orbitals.

$$\overline{} \quad \overline{}$$
$$d_{x^2-y^2} \quad d_{z^2}$$

$$\overline{\uparrow\downarrow} \quad \overline{\uparrow\downarrow} \quad \overline{\uparrow\downarrow}$$
$$d_{xy} \quad d_{xz} \quad d_{yz}$$

There are no unpaired electrons in this metal ion.

Δ_t between the d orbitals in the metal ions in tetrahedral complexes is small so they will be high spin. All five d orbitals will be occupied by one electron before pairing occurs.

Example 22-9:Determine the number of unpaired electrons in the following tetrahedral complex ion: $[Mn(CN)_4]^{2-}$.

The charge of the complex is 2–. The charge of each cyanide is 1–, so the charge of the nickel ion must be 2+. The electron configuration of a Mn atom is $[Ar]3d^5 4s^2$. For the 2+ ion, the 4s electrons have been lost, leaving an electron configuration of $[Ar]3d^5$. This is a tetrahedral complex. The order as to which d orbitals are higher energy and which are lower energy is the reverse of the order for octahedral complexes. In a tetrahedral complex, the ion will be high spin: all five d orbitals will be occupied by one electron before pairing occurs.

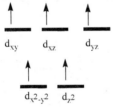

$$\overline{\uparrow} \quad \overline{\uparrow} \quad \overline{\uparrow}$$
$$d_{xy} \quad d_{xz} \quad d_{yz}$$

$$\overline{\uparrow} \quad \overline{\uparrow}$$
$$d_{x^2-y^2} \quad d_{z^2}$$

There are five unpaired electrons in this metal ion.

Try Study Questions 23 and 25 in Chapter 22 of your textbook now!

In square planar complexes, the issue of high spin and low spin does not usually come up. The most common square planar complexes have eight d electrons. The lower four orbitals are all occupied by two electrons, leaving the $d_{x^2-y^2}$ orbital as the only one unoccupied.

This table summarizes some of what we have been talking about:

Coordination Geometry	High or Low Spin
Octahedral	Low spin or high spin depending upon the ligand.
Tetrahedral	High spin.
Square planar	Low spin.

- **Show how bonding is used to explain the magnetism and spectra of coordination compounds.**

a) **Understand why substances are colored and be able to predict colors when you know the color of light absorbed (Section 22.6).**

It is possible to produce the various colors that we see using light of three colors: red, green and blue. These are called primary colors. We can mix light of these colors to produce light of other colors. Keep in mind that we are talking about mixing light, not paint. You may know that green paint may be produced by mixing yellow and blue paints. This is different from what we are talking about here. In mixing light, green is a primary color.

Color	Primary Colors Present
Red	Red
Green	Green
Blue	Blue
Yellow	Red and Green
Cyan	Green and Blue
Magenta	Red and Blue
White	Red, Blue, and Green

When we shine white light on an object, the object usually absorbs some colors and transmits or reflects others. The color we see is the color of the light that is not absorbed by the object.

Object Appears	Light Not Absorbed	Light Absorbed
Red	Red	Green and Blue
Green	Green	Red and Blue
Blue	Blue	Red and Green
Cyan	Green and Blue	Red
Yellow	Green and Red	Blue
Magenta	Red and Blue	Green
White	Red, Green, and Blue	None
Black	None	Red, Green, and Blue

Example 22-10: When leaves contain chlorophyll, they appear green. What does this imply about chlorophyll's absorption of light.

The colors we see are the colors not absorbed. Chlorophyll must not absorb green light to a significant extent. The colors that it must absorb are the other primary colors: red and blue.

Try Study Question 32 in Chapter 22 of your textbook now!
The answer to this problem is not in the back of your textbook. The color should be cyan.

b) **Understand the relationship between the ligand field splitting, magnetism, and color of metal complexes (Section 22.6).**

Ligands cause a splitting of the d orbitals into lower energy and higher energy d orbitals. Some ligands cause a greater difference in energy than others. In the following list, those to the left cause the least amount of splitting; the energy difference between the lower energy and the higher energy d orbitals is smaller. Those on the right cause large orbital splitting.

$$Cl^-, Br^-, I^- < C_2O_4^{2-} < H_2O < NH_3, en < phen < CN^-$$

This extent of splitting plays a role in the energy needed to excite an electron from a lower energy d orbital to a higher energy d orbital. If we compare two complexes containing the same metal ion but in which one contains ligands that cause large orbital splitting (for example, cyanide) and the other contains ligands that cause lower orbital splitting (for example, chloride), there is a greater energy difference between the d orbitals in the one with the ligands that cause large orbital splitting. It will therefore take more energy to excite an electron from the lower d orbitals to the higher d orbitals. A higher energy translates to a smaller wavelength (energy and wavelength are inversely proportional). This series of ligands is referred to as the spectrochemical series.

<u>Example 22-11</u>: Which of the following complexes would be expected to absorb light with a lower wavelength: $[Ni(H_2O)_6]^{2+}$ or $[Ni(NH_3)_6]^{2+}$?

Lower wavelength absorbed implies a higher frequency, which implies a higher energy. We are thus looking for the complex that has a higher Δ_O. Ammonia is further to the right in the spectrochemical series. It causes greater orbital splitting. The complex involving ammonia will absorb light of a lower wavelength.

These issues also play a role in determining whether an octahedral complex will be low spin or high spin. Those ligands on the right will cause the energy difference between the d orbitals to be greater. Thus, these ligands will be involved in more low spin complexes than will those on the left. The actual difference in energy, however, depends upon both the ligand strength and also the metal ion, thus we have not given you enough information to predict whether a specific complex will be high spin or low spin.

Sometimes, we can tell whether a particular complex is low spin or high spin by its magnetic properties. Recall that a paramagnetic material contains unpaired electrons whereas a diamagnetic material contains all paired electrons. In some cases, a high spin complex would be predicted to be paramagnetic whereas a low spin complex would be diamagnetic. Thus, whether the material turns out to be paramagnetic or diamagnetic in these cases tells us whether the material is high spin or low spin.

<u>Example 22-12</u>: $K_4[FeF_6]$ is paramagnetic. Is $[FeF_6]^{4-}$ a high spin or a low spin complex?

The charge of the iron must be 2+ because the full charge is 4– and each of the six fluoride ions has a charge of 1–. A neutral Fe atom has an electron configuration of $[Ar]3d^6 4s^2$. Fe^{2+} has the electron configuration $[Ar]3d^6$. On the left is the orbital filling diagram for the d orbitals in a high spin complex, and on the right is the diagram for a low spin complex.

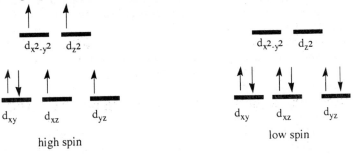

Because the species is paramagnetic, it must have unpaired electrons. This corresponds to the diagram on the left. The complex must be high spin.

Try Study Question 27 in Chapter 22 of your textbook now!

CHAPTER 23: Nuclear Chemistry

Chapter Overview

In this chapter, you will learn about nuclear reactions. In radioactive decay, an unstable nucleus emits radiation. The three most important types of radiation are α, β, and γ radiation. An α particle is a helium-4 nucleus. A β particle is an electron. γ radiation is a very high energy form of electromagnetic radiation. You will learn how to write equations for nuclear processes. The key to writing these equations is that the sum of the mass numbers on each side of the equation must be the same and the sum of the atomic numbers on each side of the equation must be the same. Often, a radioactive nucleus will decay to form another unstable nucleus that undergoes decay as well. This continues until finally a stable nucleus is formed. This is called a radioactive decay series.

Only certain nuclei are stable. At low atomic number, the stable nuclei tend to have approximately equal numbers of protons and neutrons. At higher atomic numbers, the number of neutrons is slightly greater than the number of protons. All elements beyond bismuth are unstable. Beta emission tends to occur for isotopes that have a high neutron to proton ratio. Positron emission or electron capture tends to occur for isotopes that have a low neutron to proton ratio.

The mass of a nucleus is less than the mass of the particles that went into making the nucleus. The reason for this is that some of the matter is converted to energy when the nucleus forms and is released. This energy is called the nuclear binding energy.

Nuclear decay obeys first-order kinetics. We can use the equations we learned back in Chapter 15 dealing with first-order kinetics and half-lives.

Many of the transuranium elements have been discovered by using neutron bombardment and also by using particle accelerators. Nuclear fission and nuclear fusion are two other types of nuclear reactions. In fission, a larger atom is split into two or more smaller nuclei. Nuclear fission reactions are examples of chain reactions. In fusion, two or more smaller nuclei are joined together to form a larger nucleus. Both nuclear fission and fusion release tremendous quantities of energy.

One measure of radioactivity is the number of decompositions in a given unit of time. The usual unit used is the curie, which corresponds to 3.7×10^{10} decompositions per second. The biological damage done by a type of radiation (measured in rem) is calculated by multiplying the energy of the radiation expressed in rads by a quality factor that depends on the type of radiation. α radiation has a quality factor of 20, whereas β and γ radiation each have a quality factor of 1.

You will also learn about various uses of radioactivity in science and medicine: radiocarbon dating, medical imaging, radiation therapy, radioactive tracers, isotope dilution, neutron activation analysis, and food irradiation.

Key Terms

In this chapter, you will need to learn and be able to use the following terms:

Activity: in radioactive decay, the number of disintegrations observed per unit time.

Alpha (α) radiation: a particle ejected from some nuclei in radioactive decay; an α particle has a mass number of 4 and an atomic number of 2. It is a helium nucleus.

Band of stability: the narrow range of neutron to proton ratios that lead to stable nuclei.

Beta (β) radiation: an electron that is ejected from some nuclei in radioactive decay. A β particle has a zero mass number and a charge of -1.

Chain reaction: a reaction in which one step of a reaction produces a product that can be used to cause further reaction to occur. Once a chain reaction has been initiated, it can continue on its own, given sufficient amounts of the reactants in sufficient concentration.

Daughter nucleus: the product nucleus in a nuclear reaction.

Gamma (γ) radiation: very high energy electromagnetic radiation emitted by some nuclei in radioactive decay. Gamma radiation consists of photons and thus has zero mass number and zero atomic number.

Half-life: the amount of time it takes for half of the atoms in a sample of a radioactive material to decay.

Mass defect: the difference in mass between a nucleus and the same number of separated protons and neutrons.

(n,γ) reaction: a nuclear reaction in which a neutron is captured by a nucleus following by the emission of a γ ray by the product nucleus.

Nuclear binding energy (E_b): the energy required to separate the nucleus of an atom into separated protons and neutrons.

Nuclear fission: the process by which a nucleus is split into smaller fragments.

Nuclear fusion: the process by which several small nuclei react to form a larger nucleus.

Nuclear reaction: a reaction in which the nucleus of an atom changes.

Nucleon: a proton or a neutron.

Parent nucleus: the reactant nucleus in a nuclear reaction.

Plasma: a state of matter consisting of unbound nuclei and electrons. This is the state of matter that is in nuclear fusion reactions.

Positron: the antimatter equivalent of an electron. A positron has a mass number of zero and a positive charge.

Radioactive decay series: a set of radioactive decays in which one radioactive isotope decays to form another radioactive isotope, which decays to form yet another radioactive isotope, and so forth until a stable nucleus is eventually formed.

Radiocarbon dating: a process to determine the age of the remains of a living thing by comparing the radioactivity (due to ^{14}C) per gram of carbon in the remains to the natural level of radioactivity due to ^{14}C per gram of carbon in a living thing.

Transuranium element: an element with an atomic number greater that that of uranium (92).

Chapter Goals

By the end of this chapter you should be able to:

- ## Identify radioactive elements and describe natural and artificial reactions.

   a) Identify α, β, and γ radiation, the three major types of radiation in natural radioactive decay (Section 23.1).

   Some atoms spontaneously emit a particle or energy. This is called radioactive decay. There are three major types of radioactive decay that you should know about: α, β, and γ. An alpha particle is the same as a helium-4 nucleus. It contains two protons and two neutrons. It is often symbolized as $_2^4 \text{He}$. Because it contains two protons but not two electrons, the charge of an alpha particle is +2.

   A beta particle is an electron that is ejected from the nucleus of the atom. You might wonder how an electron is present in the nucleus since we have always talked about electrons orbiting the nucleus. What happens in β decay is that a neutron in the nucleus breaks down to form both a proton and an electron. The electron that is formed, the β particle, is ejected from the nucleus. Because β particles are electrons, they are negatively charged.

   Gamma radiation is a very high energy form of electromagnetic radiation. Unlike alpha and beta particles, gamma radiation has no mass.

   The penetrating power of these types of radiation goes in the order $\alpha < \beta < \gamma$. Alpha radiation can be stopped by several sheets of paper. Beta radiation can penetrate several millimeters of living bone or tissue but no further; it can be stopped by aluminum that is at least 0.3 cm thick. Thick layers of lead or concrete are required to shield the body from γ radiation.

   b) Write balanced equations for nuclear reactions (Section 23.2).

   In a nuclear reaction, the nucleus of an atom changes. This often results in an isotope of one element being changed into an isotope of a different element. In other words, the identity of the atom changes. For example, a uranium atom might change into a thorium atom. Why is this so? The reason that the identity changes in a nuclear reaction is that we are changing the composition of the nucleus. This is different from ordinary chemical reactions because in chemical reactions, the nuclei remained the same and only the arrangement of the electrons and which atoms are connected to which other atoms changed.

   In writing balanced equations for nuclear reactions, one of the major rules we have always used with chemical reactions is not in effect. Previously, we had always indicated that if we had an atom of something on the left side, then we would need to have an atom of that same element on the right side. In nuclear reactions, the identity of the atoms can change.

   The mass numbers and the charges (indicated by the atomic numbers) must be balanced. The sum of the mass numbers on the left side of the equation must equal the sum of the mass numbers on the right side. Similarly, the sum of the charges on the left side must equal the sum of the charges on the right side.

Example 23-1: Complete the following nuclear equation.
$$^{230}_{90}\text{Th} \rightarrow {}^{4}_{2}\alpha + \underline{\hspace{1cm}}$$

We know that the mass numbers must be equal on both sides of the equation. The mass number on the left side is 230. The α particle on the right has a mass number of 4. The other particle must therefore have a mass number of $230 - 4 = 226$.

Similarly, the atomic numbers must be equal on both sides of the equation. The atomic number on the left side is 90. The α particle has an atomic number of 2. The other particle must have an atomic number of $90 - 2 = 88$. The element with atomic number 88 is radium.

The completed equation is
$$^{230}_{90}\text{Th} \rightarrow {}^{4}_{2}\alpha + {}^{226}_{88}\text{Ra}$$

Example 23-2: Bismuth-214 undergoes β decay. Write a nuclear equation for this process.

Bismuth has an atomic number of 83. The beginning nuclide is $^{214}_{83}\text{Bi}$.

A β particle is an electron. The mass number of an electron is zero. The charge is -1. We can represent it as $^{0}_{-1}\beta$. The nuclear equation thus far is
$$^{214}_{83}\text{Bi} \rightarrow {}^{0}_{-1}\beta + \underline{\hspace{1cm}}$$

Because the mass number of the β particle is zero, the mass number of the product must be the same as the original nuclide, 214. The sum of the atomic numbers on the right must be the same as that on the left, 83. The charge of the β particle is -1, so the other product must have an atomic number of 84 because $84 + -1 = 83$. The element with atomic number 84 is Po. The completed equation is
$$^{214}_{83}\text{Bi} \rightarrow {}^{0}_{-1}\beta + {}^{214}_{84}\text{Po}$$

Try Study Question 14 in Chapter 23 of your textbook now!

You should also learn how to write equations for the processes of positron emission and electron capture. A positron has the same mass as an electron but a positive charge instead of a negative charge. It can be symbolized by $^{0}_{+1}\beta$. When a positron is emitted by a nucleus, the mass number of the remaining nucleus is the same as it was before, but the atomic number goes down by 1.

Example 23-3: $^{13}_{7}\text{N}$ undergoes positron emission. What is the other product of this reaction?

First, let's write the equation with a blank for the product of interest.
$$^{13}_{7}\text{N} \rightarrow {}^{0}_{+1}\beta + \underline{\hspace{1cm}}$$

Because the mass number of the positron is zero, the other product will have the same mass number as the reactant, 13. The sum of the atomic numbers must be the same on each side of the equation, 7. The positron represents 1 of this, so the other product must represent the other 6. The element with an atomic number of 6 is carbon. The full equation is
$$^{13}_{7}\text{N} \rightarrow {}^{0}_{+1}\beta + {}^{13}_{6}\text{C}$$

In electron capture, a nucleus gains an electron. In the equation for this process, the electron appears on the left. Its mass number is 0, and its charge is –1. As in positron emission, the product nucleus has the same mass number as the original, but its atomic number is one less.

Example 23-4: Calcium-41 undergoes electron capture. Predict the product of this reaction.

Calcium-41 has a mass number of 41 and an atomic number of 20. It gains an electron.
$$^{41}_{20}\text{Ca} + ^{0}_{-1}\text{e} \rightarrow \underline{\quad}$$

Because the mass number of the electron is zero, the mass number of the product will be the same as the original nucleus, 41. The sum of the atomic numbers on the left is 20 –1 = 19. This will be the atomic number of the product. The element with this atomic number is K.
$$^{41}_{20}\text{Ca} + ^{0}_{-1}\text{e} \rightarrow ^{41}_{19}\text{K}$$

---

c) Predict whether a radioactive isotope will decay by α or β emission, or by positron emission or electron capture (Sections 23.2 and 23.3).

Only certain nuclei are stable. Whether a particular nucleus is stable or not has to do with the ratio of neutrons to protons. The narrow range of neutron to proton ratios that lead to stable nuclei is called the band of stability. At low atomic number (up to 20) the optimal ratio of neutrons to protons is approximately 1:1. Beyond calcium, the optimal ratio is greater than 1. Beyond bismuth (atomic number 83), there are no stable nuclei.

We can use these facts to predict which type of radioactive decay a nucleus will undergo. Such predictions will not always be correct, but they are a good starting point. If the atomic number of the nucleus is greater than 83, then the atomic number of the nucleus must come down. The best guess for one of these nuclei is alpha decay.

Nuclei with high neutron to proton ratios need to eliminate neutrons. The best guess for these is beta decay. For nuclei with atomic numbers around or less than that of Ca, a high ratio corresponds to greater than 1 neutron to 1 proton. The optimal ratio beyond this is greater than 1 neutron to 1 proton, but there is still a point above which β decay is predicted.

Nuclei that have a low neutron to proton ratio (less than 1 neutron to 1 proton) tend to decay by positron emission or by electron capture.

Example 23-5: Predict the type of decay that is probable for each of the following nuclei.

   a. ^{226}Ra
Radium has an atomic number of 88. This is above 83. α decay is most probable.

   b. ^{14}C
The atomic number of carbon is 6. The mass number of this isotope is 14. The number of neutrons is 14–6 = 8. The ratio of neutrons to protons is 8/6 = 1.3. This ratio is too high. The most probable type of decay is β decay.

   c. ^{7}Be
The atomic number of beryllium is 4. The mass number of this isotope is 7. The number of neutrons is 7 – 4 = 3. The ratio of neutrons to protons is 3/4 = 0.75. This is lower than the optimal 1:1 ratio. The most probable reaction is either positron emission or electron capture.

Try Study Question 20 in Chapter 23 of your textbook now!

d) Understand the nature and origin of γ radiation (Section 23.2).

Gamma radiation results from transitions between nuclear energy levels. Nuclear reactions often result in the formation of nuclei in which the nucleus is in an excited nuclear state. In relaxing to the ground state, the nucleus emits a photon of electromagnetic radiation. This is γ radiation.

- ## Calculate the binding energy and binding energy per nucleon for a particular isotope.

a) Understand how binding energy per nucleon is defined (Section 23.3.).

The nuclear binding energy, E_b, is the energy required to separate the nucleus of an atom into separated protons and neutrons. Nuclear binding energy is a positive quantity; it takes energy to separate the subatomic particles from each other. The nucleus is thus at a lower energy than the separated protons and neutrons.

The mass of a nucleus is always less than the sum of the masses of the separated protons and neutrons. This difference in mass, Δm, between isolated protons and neutrons and the combined nucleus is called the mass defect.

The mass defect arises because some of the mass of the separated protons and neutrons is converted into energy when the protons and neutrons are brought together to form the nucleus. This energy is released to the surroundings, leaving the nucleus with both a lower mass and a lower energy than the separated particles.

The nuclear binding energy corresponds to the energy released when the mass of the mass defect is converted to energy. This actual amount is given by Einstein's famous equation relating mass and energy:

$$E_b = (\Delta m)c^2$$

In comparing different nuclei, it is often useful to calculate the binding energy per mole of nucleons. Protons and neutrons are nucleons. A helium-4 nucleus contains 2 protons and 2 neutrons. It therefore contains 4 nucleons. The number of nucleons corresponds to the mass number of the nucleus.

<u>Example 23-6:</u> Calculate the nuclear binding energy per nucleon in forming 1 mole of Li-7. The molar mass of Li-7 is 7.016003 g/mole. (Your textbook gives the mass of a hydrogen-1 atom as 1.007825 g/mole and the mass of a neutron as 1.008665 g/mole.)

Lithium-7 contains 3 protons, 4 neutrons, and 3 electrons. First, we will calculate the mass defect. To do this, we will imagine one mole of lithium-7 atoms broken apart into 3 moles of hydrogen atoms (1 proton and 1 electron each) and 4 moles of neutrons.

$$^{7}_{3}\text{Li} \rightarrow 3\,^{1}_{1}\text{H} + 4\,^{1}_{0}\text{n}$$

Mass defect = mass of products − mass of reactants

$$= \left[3\left(1.007825\ \frac{\text{g}}{\text{mole}}\right) + 4\left(1.008665\ \frac{\text{g}}{\text{mole}}\right) \right] - 7.016003\ \frac{\text{g}}{\text{mole}}$$

$$= 0.042132\ \frac{\text{g}}{\text{mole}}$$

Because the final unit of energy will be in J (= kg m^2/s^2), we need to express this mass in kg/mole. This is 4.2132 x 10^{-5} kg/mole.

We can now calculate the nuclear binding energy using Einstein's equation:

$$E_b = (\Delta m)c^2$$

$$= \left(4.2132 \times 10^{-5} \frac{kg}{mole}\right)\left(2.99792 \times 10^8 \frac{m}{s}\right)^2$$

$$= 3.7866 \times 10^{12} \text{ J/mole} = 3.7866 \times 10^9 \text{ kJ/mole}$$

To calculate the binding energy per nucleon, we will divide the total nuclear binding energy by the sum of the protons and neutrons in the nucleus. In this case, this number is 7.

$$\text{Binding energy/nucleon} = \frac{3.7866 \times 10^9 \text{ kJ/mole}}{7}$$

$$= 5.4095 \times 10^8 \text{ kJ/mole nucleons}$$

Try Study Question 24 in Chapter 23 of your textbook now!

b) Recognize the significance of a graph of binding energy per nucleon versus mass number (Section 23.3).

The greater the binding energy per nucleon, the more stable the nucleus is (the lower in energy the nucleus is from the separated protons and neutrons). The binding energy per nucleon decreases in going from helium to lithium. It then increases to a maximum at a mass number of 56, corresponding to an isotope of iron. Following this, it decreases again.

• **Understand the rates of radioactive decay.**

a) Understand and use mathematical equations that characterize the radioactive decay process (Section 23.4).

Radioactive decay obeys first-order kinetics. The equations that characterize radioactive decay are therefore equations with which you are already familiar from our work in Chapter 15. If N is the number of atoms that remain at time t, N_0 is the number of atoms originally present at time zero, and k is the first-order rate constant, the following relationship holds.

$$\ln\left(\frac{N}{N_0}\right) = -kt$$

The number of disintegrations observed per unit time, measured using a device such as a Geiger-Müller counter, is called the activity (A) of the sample. The activity is directly proportional to the number of atoms of the radioactive material present. In our first-order kinetic equation, we therefore usually use the activity rather than the number of atoms.

$$\ln\left(\frac{A}{A_0}\right) = -kt$$

The half-life, $t_{1/2}$, of a radioactive isotope is the amount of time it takes for half of the atoms of that isotope to decay. The half-life for a first-order process is calculated with the equation

$$t_{1/2} = \frac{\ln 2}{k}$$

Example 23-7: ^{206}Tl emits β particles. The activity of a sample of ^{206}Tl initially has an activity of 8.0×10^4 disintegrations per second (dps). After 9.0 minutes, the sample has an activity of 1.8×10^4 dps.

a. What is the rate constant for this decay process?
Radioactive decay is a first-order process. The integrated rate law is

$$\ln\left(\frac{A}{A_0}\right) = -kt$$

We know that A is 1.8×10^4 dps, A_0 is 8.0×10^4 dps, and t is 9.0 min.

$$\ln\left(\frac{1.8 \times 10^4 \text{ dps}}{8.0 \times 10^4 \text{ dps}}\right) = -k(9.0 \text{ min})$$

$$-1.49 = -(9.0 \text{ min}) k$$

$$k = 0.17 \text{ min}^{-1}$$

The rate constant is 0.17 min^{-1}.

b. What is the half-life of ^{206}Tl?
The equation for a first-order half-life is

$$t_{1/2} = \frac{\ln 2}{k}$$

We know that k is 0.17 min^{-1}.

$$t_{1/2} = \frac{\ln 2}{k} = \frac{\ln 2}{0.17 \text{ min}^{-1}} = 4.2 \text{ min}$$

Example 23-8: Radon-222 has a half-life of 3.8 days. If the initial activity of a sample of radon-222 is 5.0×10^4 dps, what will be the activity of this sample after 2 days?

We are given the half-life, the initial activity, and the time. We are to determine the activity after this time. The equation involving activity and time is

$$\ln\left(\frac{A}{A_0}\right) = -kt$$

The key thing missing from this equation is the value of k. We can determine this from the half-life.

$$t_{1/2} = \frac{\ln 2}{k}$$

$$k = \frac{\ln 2}{t_{1/2}} = \frac{\ln 2}{3.8 \text{ days}} = 0.18 \text{ day}^{-1}$$

We can now use the first-order integrated rate law to solve for the activity after 2 days.

$$\ln\left(\frac{x}{5.0 \times 10^4 \text{ dps}}\right) = -(0.18 \text{ day}^{-1})(2.0 \text{ days})$$

$$\ln\left(\frac{x}{5.0 \times 10^4 \text{ dps}}\right) = -0.36$$

We need to solve for something in the natural logarithm. To undo the natural logarithm, we raise e to both sides of the equation.

$$e^{\ln\left(\frac{x}{5.0 \times 10^4 \text{ dps}}\right)} = e^{-0.36}$$

$$\frac{x}{5.0 \times 10^4 \text{ dps}} = e^{-0.36}$$

$$x = 3.5 \times 10^4 \text{ dps}$$

The activity remaining after 2.0 days is 3.5×10^4 dps.

Try Study Questions 30 and 36 in Chapter 23 of your textbook now!

b) Use the half-life to estimate the time required for an isotope to decay (Section 23.4).

With each half-life, half of the remaining atoms of a radioactive isotope decay. After one half-life, half of the original activity remains. After two half lives, half of that $(1/2)^2$, one quarter, of the original activity remains. After three half-lives, $(1/2)^3 = 1/8$ of the original activity remains. After n half-lives, $(1/2)^n$ of the original activity remains.

We can use the concept of half-life to estimate the time required for an initial activity to decay to particular level.

Example 23-9: The half-life of ^{206}Tl is 4.2 min. The initial activity of ^{206}Tl in a sample is 3.0 $\times 10^4$ dps. Estimate how long it will take for the activity to decrease to 1500 dps.

After 1 half-life, the activity will be $1/2(3/0 \times 104 \text{ dps}) = 1.5 \times 10^5$ dps.
After 2 half-lives, the activity will be $1/2(1.5 \times 104 \text{ dps}) = 7.5 \times 10^3$ dps.
After 3 half-lives, the activity will be $1/2(7.5 \times 103 \text{ dps}) = 3.75 \times 10^3$ dps.
After 4 half-lives, the activity will be $1/2(3.75 \times 103 \text{ dps}) = 1875$ dps.
After 5 half-lives, the activity will be $1/2(1875 \text{ dps}) = 938$ dps)

It will take between four and five half-lives for this amount of ^{206}Tl to decay to 1500 dps. This corresponds to $17 - 21$ minutes.

• Understand artificial nuclear reactions.

a) Describe procedures used to carry out nuclear reactions (Section 23.5).

Naturally occurring isotopes account for only a very small fraction of the radioactive isotopes that are known. The rest have been made by means of artificial nuclear reactions.

From 1919-1932, the primary method used to produce new isotopes was to bombard atoms with α particles. For example, oxygen-17 can be produced by this method:

$$_2^4\text{He} + _7^{14}\text{N} \rightarrow _8^{17}\text{O} + _1^1\text{H}$$

The principal drawback of this method is that in most cases α particles are scattered by target nuclei instead of colliding with them due to repulsions between the positively charged nuclei and the positively charged α particles.

One solution to this problem was to use higher energy particles that could overcome these repulsive forces. Instead of α particles, protons that had been accelerated by applying a high voltage were used. An example of a reaction involving high energy protons is

$$^{7}_{3}\text{Li} + ^{1}_{1}\text{p} \rightarrow 2^{4}_{2}\text{He}$$

Another major step forward was to use a neutral bombarding particle, a neutron. This avoids the electrostatic repulsion of a positive particle and a positive nucleus. Low energy neutrons are better to use than high energy neutrons because they tend to fuse with a target nucleus rather than breaking the nucleus apart. The product nucleus is in an excited state. It emits energy, in the form of γ radiation, to come to the ground state. A reaction in which a neutron is captured and γ radiation is emitted is called an (n,γ) reaction. An example of an (n,γ) reaction is

$$^{31}_{15}\text{P} + ^{1}_{0}\text{n} \rightarrow ^{32}_{15}\text{P} + \gamma$$

An (n,γ) reaction was involved in producing Np and Pu. ^{238}U absorbed a neutron to form ^{239}U. This isotope of uranium undergoes β decay to form ^{239}Np which itself undergoes β decay to form ^{239}Pu. Adding neutrons was also involved in producing ^{241}Am.

The majority of new isotopes today are produced using particle accelerators. Particle accelerators generate very high energy particles with which to bombard target nuclei. These high energy particles are produced by accelerating the particles to very high energies between charged plates. A number of different types of particles have been accelerated in particle accelerators: neutrons, helium nuclei, larger nuclei such as ^{11}B and ^{12}C, or even ions of calcium, chromium, cobalt or zinc. The high energy bombarding particle fuses with the nucleus of the target atom. This new nucleus will then break up into other nuclei. Based on the products that form when the nucleus breaks apart, it is possible to determine the composition of the nuclei that were formed.

Try Study Questions 48 and 52 in Chapter 23 of your textbook now!

b) Describe nuclear chain reactions, nuclear fission, and nuclear fusion (Sections 23.6 and 23.7).

In the process of nuclear fission, a nucleus is split into smaller fragments. When nuclear fission occurs, a tremendous amount of energy is released. Fission reactions involve chain reactions. For example, when ^{235}U is bombarded with a neutron, ^{236}U is formed. This ^{236}U then can split into smaller fragments such as ^{141}Ba and ^{92}Kr. In doing so, three neutrons are released:

$$^{235}_{92}\text{U} + ^{1}_{0}\text{n} \rightarrow ^{236}_{92}\text{U}$$
$$^{236}_{92}\text{U} \rightarrow ^{141}_{56}\text{Ba} + ^{92}_{36}\text{Kr} + 3^{1}_{0}\text{n}$$

The three neutrons produced can then collide with other ^{235}U atoms causing them to undergo fission, releasing more neutrons. The reaction thus continues and accelerates.

The three key parts of a chain reaction are the following three types of steps:

1) Initiation step. This is the initial reaction of a single atom to start the chain reaction. In the ^{235}U chain reaction, the initiation step is the one in which the first ^{235}U atom absorbs a neutron.

2) Propagation steps. These are the steps that continue the chain reaction. The ^{236}U formed breaks down into the fission products plus the three neutrons. These

neutrons then react with more ^{235}U, which forms ^{236}U, which breaks down, releasing more neutrons, etc.

3) Termination step. A termination step is a step that does not continue the chain. Not all of the neutrons produced will react with ^{235}U. Some may be captured by other materials that do not lead to more neutrons being produced. If this occurs enough, the reaction will not be able to continue.

Atomic bombs, such as those used in World War II, utilize nuclear fission. The chain reaction continues in an uncontrolled fashion causing a nuclear explosion.

Nuclear fission is also used in nuclear power plants. The key to a nuclear power plant being able to work is to allow the chain reaction to occur somewhat but to keep it from getting out of control. To do this, cadmium rods are inserted into the reactor. Cadmium absorbs neutrons, causing termination steps to occur. The rods are inserted to the point that the nuclear reaction occurs at a certain rate, but enough neutrons are absorbed to prevent the reaction from getting out of control.

In nuclear fusion, several small nuclei react to form a larger nucleus. An example of a fusion reaction is the following:

$$^{3}_{1}H + ^{2}_{1}H \rightarrow ^{4}_{2}He + ^{1}_{0}n$$

Nuclear fusion generates tremendous amounts of energy, even more than fission reactions. Fusion reactions provide the energy of stars (including our sun). Fusion reactions are used in nuclear bombs such as hydrogen bombs. We have not been able to use fusion reactions to meet our energy needs because there are several critical problems that have not been satisfactorily solved yet. First, very high temperatures must be used ($10^6 - 10^8$ K) to generate plasma, unbound nuclei and electrons. Secondly, this plasma must be contained. Also, there must be a way to tap into the energy released.

Example 23-10: Balance each of the following nuclear reactions and indicate whether the reaction indicates nuclear fission or fusion.

a. $^{235}_{92}U + ^{1}_{0}n \rightarrow ^{144}_{54}Xe + \underline{\hspace{1cm}} + 2^{1}_{0}n$

The sum of the atomic numbers on the left is 92. On the right, 54 of this is accounted for. This means that the remaining particle must have an atomic number of $92 - 54 = 38$. The element with an atomic number of 38 is Sr. The sum of the mass numbers on the left is $235 + 1 = 236$. On the right, we have already accounted for $144 + 2(1) = 146$. The remaining particle must have a mass number of $236 - 146 = 90$. The completed equation is

$$^{235}_{92}U + ^{1}_{0}n \rightarrow ^{144}_{54}Xe + ^{90}_{38}Sr + 2^{1}_{0}n$$

Because one nucleus has been split up into two nuclei, this process is nuclear fission.

b. $2\ ^{2}_{1}H \rightarrow \underline{\hspace{1cm}} + ^{1}_{0}n$

The sum of the atomic numbers on the left is 2. On the right, none of this is accounted for. The remaining particle must have an atomic number of 2. The element with an atomic number of 2 is He. The sum of the mass numbers on the left is 4. On the right, 1 is already accounted for. This leaves 3 for the remaining particle. The completed equation is

$$2\ ^{2}_{1}H \rightarrow ^{3}_{2}He + ^{1}_{0}n$$

Because nuclei combine to form a new nucleus, this is an example of nuclear fusion.

•   **Understand issues of health and safety with respect to radioactivity.**

Describe the units used to measure intensity and understand how they pertain to health issues (Section 23.8).

Activity is measured either in curies (Ci) or in becquerels (Bq). The unit used most in the United States is the curie.

   1 Bq = 1 decomposition per second (dps)

   1 Ci = 3.7 x 10^{10} dps = 3.7 x 10^{10} Bq

By itself, activity does not fully describe the radiation because it does not indicate the energy of the radiation or the damage that can be done to living tissue by the radiation. The amount of energy absorbed by living tissue is measured in rads or in grays (Gy).

   1 rad = 0.01 J of energy absorbed per kilogram of tissue

   1 Gy = 1 J of energy absorbed per kilogram of tissue = 100 rads

In considering the effect on living tissue, there is another factor that comes into play. This is the relative damage that a type of radiation will cause once it is *inside a living tissue*. This goes in the opposite direction as the penetrating ability. Recall that penetrating ability goes in the order that α radiation is least penetrating, β is next, and finally γ is the most penetrating of the three major types of radioactive decay products. Because α radiation is not very penetrating, it will not pass out of the tissue; it will thus deliver all of its energy right there and cause a large amount of damage. γ radiation, on the other hand, will pass right through the body. It does not deliver most of its energy in the body; most of it goes right on through to be delivered elsewhere in the universe. To indicate these different effects of radiation inside the body, a "quality factor" has been assigned to each of the types of radiation.

Type of Radiation	Quality Factor
α	20
Low energy protons and neutrons	5
β	1
γ	1

To determine the full effect of a type of radiation, one must take into account both the energy of the radiation and the quality factor. The biological damage expected for a type of radiation is calculated by multiplying together the energy and the quality factor. If the energy is expressed in rads, then the unit for the biological damage is expressed in rem. If the energy is expressed in grays, then the unit for the biological damage is expressed in sieverts (Sv).

Exposure to a small amount of radiation is unavoidable. Earth is constantly being bombarded with radioactive particles from outer space. There is also natural radioactivity associated with radioactive isotopes that occur naturally. The average dose of radioactivity a person in the United States receives is about 200 mrem per year = 0.200 rem per year. More than half of this comes from natural sources. About 90% of the remainder is due to medical procedures.

No detectable health effects are usually observed for a single radiation dose of less than 25 rem. On the other hand, half of the people exposed to a single radiation dose of 450 rem will die. It is still unclear what the overall health effects are for multiple smaller doses of radiation.

- **Be aware of some uses of radioactive isotopes in science and medicine (Section 23.9).**

Radiocarbon Dating

Radiocarbon dating is used to determine the ages of materials that are between about 100 and 40,000 years old. Carbon-14 is a β emitter with a half-life of 5730 years.

Carbon-14 is formed naturally in the upper atmosphere by nuclear reactions caused by cosmic radiation. The carbon-14 is then oxidized to radioactive $^{14}CO_2$. There is thus always a certain amount of radioactive $^{14}CO_2$ in the atmosphere. Plants, in the process of photosynthesis, incorporate CO_2 into their cells. As long as a plant (or an animal that is further up the food chain) is living, the radioactivity due to ^{14}C per gram of carbon in the living thing is the same as the radioactivity due to ^{14}C per gram of carbon in the CO_2 in the atmosphere. This natural level of radioactivity is about 14 disintegrations per minute per gram (14 dpm/g) of carbon. Once the living things dies, it stops taking in radioactive ^{14}C. The radioactive ^{14}C in the remains of the living thing decays without being replenished. The remains of the living thing will thus have a smaller fraction of ^{14}C than is present in the atmosphere. Based on how much is left, it is possible to determine the amount of time that has elapsed since the living thing died and the present time.

<u>Example 23-11:</u> A bone is determined to have an activity of 6.3 dpm/g of carbon. Approximately how long ago did the animal die?

We know that the half-life of ^{14}C is 5730 years. We can use this to determine the first-order rate constant.

$$t_{1/2} = \frac{\ln 2}{k}$$

$$k = \frac{\ln 2}{t_{1/2}} = \frac{\ln 2}{5730 \text{ years}} = 1.21 \times 10^{-4} \text{ year}^{-1}$$

We can now solve the integrated rate law for time. The current activity is 6.3 dpm/g. We will assume that the initial activity was 14 dpm/g.

$$\ln\left(\frac{A}{A_0}\right) = -kt$$

$$\ln\left(\frac{6.3}{14}\right) = -(1.21 \times 10^{-4} \text{ year}^{-1})\, t$$

$$-0.799 = -(1.21 \times 10^{-4} \text{ year}^{-1})\, t$$

$$t = 6.6 \times 10^3 \text{ years}$$

Try Study Question 38 in Chapter 23 of your textbook now!

Medical Imaging

In medical imaging, a radioactive material in administered to the patient. The particular isotope chosen depends upon the tissue whose image is desired. Different isotopes will be incorporated to different extents in different tissues. The radioactivity is then measured. A computer is used to analyze the pattern of radioactivity that is measured. From this, a meaningful image of the tissue can be constructed. In more than 85% of the diagnostic scans performed, the isotope used is Tc-99m.

Radiation Therapy

Gamma radiation is often used to treat cancers. The purpose of the radiation is to kill the cancer cells. Unfortunately, some healthy tissue is also damaged in the process, but the risk from this is lower than the risk of leaving the cancer alone.

Radioactive Tracers

A radioactive version of a material introduced into a system, such as a living thing or the environment. The scientist then determines where the radioactivity ends up. This allows the scientist to determine the fate of this material in the organism or the environment.

Isotope Dilution

In this technique, a known amount of a radioactive material is added to a sample of interest. After the radioactive material has been completely dispersed throughout the sample, a portion of the sample is removed, and the quantity of radioactivity in this portion is analyzed. From the amount of radioactivity in this portion, it is possible to determine the size of the sample of interest.

Neutron Activation Analysis

In this process, a sample is irradiated with neutrons. Most atoms in the sample will absorb a neutron to form a new isotope in which the nucleus is in an excited state. The nucleus returns to the ground state by releasing a γ ray. Each element will release γ rays of different energies, so it is possible to tell which elements are present in the sample by the energies of the γ rays produced. It is also possible to tell how much of each element is present by the number of γ rays produced at each particular energy. This method of analysis was used in analyzing the composition of moon rocks.

Food Irradiation

When food is irradiated with γ rays, the γ rays retard the growth of organisms such as bacteria, molds, and yeasts that can cause food spoilage. This food thus does not spoil for a much longer time than food that has not been irradiated.

When food is irradiated with even higher energy radiation in the 1 to 5 Mrad range, every living organism in the food is killed. Such food will not spoil indefinitely when sealed in plastic or aluminum foil packages.